DIGITAL SIGNAL PROCESSING IN TELECOMMUNICATIONS

Kishan Shenoi

Prentice Hall P T R
Upper Saddle River, New Jersey 07458

Library of Congress Cataloging-in-Publication Data

Shenoi, Kishan.
Digital signal processing in telecommunications / Kishan Shenoi.
p. cm.
Includes bibliographical references and index.
ISBN 0-13-096751-3 (cloth)
1. Signal processing--Digital techniques. I. Title.
TK5102.9.S53 1995
621.382′2--dc20 95-13586
CIP

Editorial/production supervision: *Patti Guerrieri*
Book and cover design, layout,
illustration, and typography: *Amita Shenoi*
Manufacturing buyer: *Alexis R. Heydt*
Acquisitions editor: *Karen Gettman*
Editorial assistant: *Barbara Alfieri*

Prentice-Hall Inc.
A Simon & Schuster Company
Upper Saddle River, NJ 07458

The publisher offers discounts on this book when ordered in bulk quantities. For more information, contact: Corporate Sales Department, Prentice Hall PTR, One Lake Street, Upper Saddle River, NJ 07458, Phone: 800-382-3419, Fax: 201-236-7141, e-mail: corpsales@prenhall.com

Printed in the United States of America
10 9 8 7 6 5 4 3 2 1

ISBN 0-13-096751-3

Prentice-Hall International (UK) Limited, *London*
Prentice-Hall of Australia Pty. Limited, *Sydney*
Prentice-Hall of Canada Inc., *Toronto*
Prentice-Hall Hispanoamericana, S.A., *Mexico*
Prentice-Hall of India Private Limited, *New Delhi*
Prentice-Hall of Japan, Inc., *Tokyo*
Simon & Schuster Asia Pte. Ltd., *Singapore*
Editora Prentice-Hall do Brasil, Ltda., *Rio de Janeiro*

CONTENTS

PREFACE

This book is intended to be of use to two broad groups of readers. The first group comprises those who have good familiarity with Digital Signal Processing (**DSP**) and/or Telecommunications. The second group consists of those who need at least a casual or, perhaps, detailed, knowledge of the applications of **DSP** in telecommunications. A typical member of the second group is someone who is involved in a design project utilizing **DSP** but whose area of expertise is a different branch of electrical engineering, such as software, hardware, systems engineering, or application-specific integrated circuit (**ASIC**) design. Included in the second group are students who wish to get a feel for how **DSP** is applied.

The book is comprised of nine chapters. The intent of Chapter 1 is to provide the spectrum of readership a preamble that puts the material in the subsequent chapters in perspective. Chapters 2 and 3 are targeted toward the second group and cover the fundamental concepts of communication theory and digital signal processing. Readers in the first group could skip directly to Chapter 4 or skim Chapters 2 and 3 to get a feel for the notation used. Chapters 4 through 9 are reasonably self contained and draw on material from the first three chapters. Those well versed in telecommunications would find the material useful in understanding the concepts of **DSP**; experts in **DSP** would find a description of some telecommunications concepts in a familiar jargon. Except for Chapters 1 and 5, one section of each chapter consists of selected exercises. For Chapters 2 and 3 these exercises are chosen to enhance the understanding of the material in a mathematical sense. Applying pencil and paper remains the best way to develop a proficiency in dealing with the mathematical, and sometimes abstract, notions introduced. The exercises in the later chapters assume the availability of some form of computing power, either a PC or a workstation, or some other form of desktop computing. These exercises are better described as suggestions for computer programs to simulate the structures and execute the algorithms described in the text.

Chapter 1 discusses some of the unique characteristics and thought processes associated with the telephony channel. In particular, the implications of the access portion of a telephone channel, the coding requirements for conversion between analog and digital formats, and the need for echo control in circuits that have substantial transmission delay. From the viewpoint of transmission, the subscriber's signal is affected first by the the cable plant that is used to phyically connect the station set to the network. This connection, called the subscriber loop, materially impacts the signal, especially when the subscriber is geographically distant from the central office, a distance that could be in excess of three miles. At the central office the signal experiences bandlimiting; the telephone network principally supports channels that have a (nominal) cutoff frequency of about 4 kHz. Furthermore, the signal is converted from analog to digital format using a nonlinear encoding process. The subscriber loop is full duplex, or "two-wire," with the cable pair supporting signals in both directions. The Network is "four-wire," assigning separate (possibly logically separate) paths for signals in the two directions. This split is achieved by a *hybrid* and the non ideal nature of the hybrid gives rise to the phenomenon of echo. The focus of Chapter 1 is an explanation of the principal characteristics and impairments of the subscriber loop, signal processing in the "line circuit," signal processing in the trunking network, and the need for echo control. Since telecommunications has its own jargon with several acronyms, which often times have lost their origin, an appendix is provided where several commonly used acronyms are expanded and a short description provided in some cases.

The fundamental concepts of communication theory and signal processing are presented in Chapter 2. In particular, the essentials of signal theory, transforms, and linear time-invariant systems are discussed. The principles of modulation, with emphasis on amplitude modulation, as well as the basic signal processing associated with data transmission, is treated in a unified, though simple, fashion. Chapter 3 extends these concepts to discrete-time and digital signal processing. The cornerstone of **DSP**, the sampling theorem, is discussed in detail and the notions of Fourier transforms and frequency response of discrete-time filters are developed as extensions of the concepts introduced in the derivation of the sampling theorem. Chapter 3 also covers the essentials of the **Z**-transform, and **FIR** and **IIR** filters. Finite-wordlength effects in A/D and D/A conversion, as well as in the implementation of digital filters, are treated in a generic fashion as additive noise.

Analog-to-digital and digital-to-analog conversion are implied whenever we use digital techniques to process real-world, i.e., analog, information-bearing signals. In telecommunications the main information signal is speech or speech-like. For such signals the conversion process can be tailored to achieve a desired behaviour. Chapter 4 discusses the principles of quantization, a fundamental component of an A/D converter. Quantization can be "uniform," as in traditional converters, or "companded," the term used to describe the nonuniform quantization charateristics used in A-law and μ-law converters. Quantization also plays a part in the digital implementation of discrete-time filters. The impact of "finite wordlength" in **IIR** digital filters is discussed in Chapter 4 and certain rules of thumb are presented to mitigate the deleterious effects of wordlength reduction.

Chapter 5 deals with the principles of efficient coding of speech signals, arguably the most predominant of information bearing (analog) signals. The fundamentals of differential encoding, adaptive quantizers, Adaptive Differential Pulse Code Modulation (**ADPCM**), linear predictive coding (**LPC**) techniques, and Digital Speech Interpolation (**DSI**) are introduced. The **ADPCM** algorithm described follows the world-wide standard agreed upon and additional performance information, not commonly available in the open literature, is provided. The applicability of **LPC** for speech compression is presented via an example—the **EIA-IS-54** standard for digital cellular telephony, which includes the description of the encoding of the speech signal. In the same vein, the application of **DSI** to speech compression is described via a discussion of the key facets of an available product, the **TC421** from DSC Communications Corp.

Chapter 6 covers techniques for echo control in detail. The approach taken is to describe the manner in which echo control is accomplished by the use of echo suppressors and echo cancelers. The latter is the method of choice in all new deployment. Echo cancelers are basically adaptive digital filters and one section of Chapter 6 is devoted to the underlying theory of adaptive filters used in this application.

The Discrete Fourier Transform (**DFT**), first introduced in Chapter 3, is treated in greater detail in Chapter 7 along with its companion, the Discrete Cosine Transform (**DCT**). From the viewpoint of telecommunications, the principal usage of the **DFT** and the **DCT** is in applications calling for a collection of bandpass filters. Two such applications are described. The popularity of the **DFT**, in a general sense, stems from the availability of algorithms, generically referred to as Fast Fourier Transforms (**FFT**s), which drastically reduce the computational burden associated with a **DFT**. The study of **FFT** algorithms is quite mature and the bibliography provides numerous articles and books that the interested reader can refer to for details. Bandpass filtering is usually accompanied by a change in sampling rate, that is interpolation, whereby the rate is increased, or decimation, whereby the rate is decreases. Chapter 7 provides an introduction to interpolation and decimation and discusses the use of "polyphase schematics," which are block diagram representations of signal processing accompanied by sampling rate changes.

Chapter 8 explains the principles of Delta Sigma Modulation, especially as applied to analog-to-digital and digital-to-analog conversion. The principal contribution of Chapter 8 is an explanation of the manner in which conversion wordlength can be traded-off with sampling frequency. Such converters, even when used to convert speech signals, operate at a high sampling rate, of the order of 1 MHz, and hence providing digital filters that operate at this high rate can be an expensive proposition unless the filters are simple. The use of the rectangular and triangular windows, simple filters that can be implemented very cost-effectively, is covered in detail and a guideline as to the performance of Delta Sigma Modulation in conjuction with such filters is quantified.

Digital filters are completely described by their transfer functions, that is, the coefficients of the polynomials that define the poles and zeros of the digital filter. The notion of designing a digital filter is thus equivalent to obtaining either the poles and zeros or the coefficients in such a manner that we obtain a stable filter whose frequency response approximates a desired shape. Chapter 9 describes one particular design method, especially useful for recursive filter design, and that is not readily available in text books, in detail.

One of the best ways to learn about **DSP** is to experiment with it. There are several software packages available that would automate many of the operations described in this book and provide wonderful, graphical, output so as to "visualize" the operation. Simulation of digital filters is quite straightforward since the description of the operations is in a form that is well suited for translation into a computer program. Further, computing power is sufficiently inexpensive such that desktop machines can run most, if not all, the programs available. To the reader who has such computing power available, I strongly recommend the actual coding of algorithms since the process of so doing does add insight into the computational and control complexity of an actual implementation.

I have been fortunate in my career to have had the opportunity to apply **DSP** techniques in a variety of applications. I have been doubly blessed in having many associates whose field of expertise was not **DSP**, but who displayed a fierce desire to understand the principles of **DSP** as applied to the projects we were working on jointly. By demanding that I explain, to their satisfaction, not just what **DSP** principle was being applied but why it was appropriate, they planted the seed of this book. Since 1985 1 have been involved in teaching "Communication Theory and Applications," "Digital Signal Processing I," and "Digital Signal Processing II," under the auspices of The University of California, Berkeley, Extension Program. The lecture notes and general course content I generated in that effort form the basis for the Chapters 2 and 3. Chapters 4 through 9 are related very closely to the research and development efforts I have been personally involved with, starting as a graduate student at Stanford in the early 1970s, and subsequently at the ITT-Advanced Technology Center, Granger Associates/DSC Communications Corporation, and continuing at present at Telecom Solutions Inc.

Over the past 20 years of involvement with **DSP** I have had the good fortune to be associated with some very bright people. I am indebted to my advisers at Stanford, the late Prof. A. M. Peterson and Dr. M.J. ("Sim") Narasimha, with whom I still have a close professional, and personal association; several colleagues at ITT, most notably Dr. B.P. Agrawal and Dr. Hyokang Chang, as well as Sarma Jayanthi and Doug Sutherland; Dr. Moon Song, Dr. Raphael Montalvo, Paul Yang, Pat Hanagan, Conne Skidanenko, and Helena Ho at Granger/DSC; and several others, too numerous to list. Each one has contributed something to my own understanding of the subject and thus has made a contribution to this book.

More than any professional colleague, the one person who has made a significant impact is my wife, Amita, who dedicated a great deal of her time to the production of this book for publishing. While maintaining a household, raising our pre-schooler, and meeting the needs of her clients, she found the time and energy for the design and desktop-publishing effort associated with the production of this book, and its cover. Although my name appears as the author, the book is as much her creation as it is mine.

KISHAN SHENOI
Saratoga, California

AN INTRODUCTION TO TELECOMMUNICATIONS

1

1.1 INTRODUCTION

The purpose of this chapter is to provide a general, brief, description of the Public Switched Telephone Network (**PSTN**), principally from the viewpoint of Transmission. By considering the path taken by an information signal from its source to its destination, we get a flavor of the modifications the signal undergoes – modifications that are usually associated with the degradation of information content in the signal. Transmission equipment is designed to minimize the deleterious impact of signal alteration, alterations in format that are mandatory to suit the transmission medium, or alterations in content that, in some sense, improve the quality if the signal. A general description of the **PSTN** will be useful in pointing out those areas where Digital Signal Processing (**DSP**) can be used.

In Section 1.2 we propose a model for the communications process. The model expresses the mind-set, or philosophy, underlying the study of information transmission. At a basic level, we consider information to be encapsulated in electrical signals, and impairments to information transfer as modifications of these signals. The modifications are described in terms of a *deterministic* alteration of signal form embodied by a **Linear Time Invariant** (**LTI**) transformation, and a probabilistic, or statistical, alteration modeled as the addition of **noise**. The communication medium, or path followed by the information, is called the ***channel***. The **LTI** aspect of a channel is usually that of a bandpass filter and thus represents the bandwidth of the channel or, rather, the bandwidth-limiting aspect of a channel. A channel could be as simple as a pair of wires or as complex as a telephone circuit completed between two telephones separated by thousands of miles. In fact, if one follows the path of a signal from its source to its destination, one could visualize the path as a concatenation of several smaller connections, each represented by an **LTI** signal modification and the addition of noise.

The telephone is arguably the most ubiquitous of communication equipment. Available in a variety of form factors and colors, it is easily recognizable. Telephone sets, or "station sets" as they are commonly referred to, are usually analog in nature and are powered directly by the provider of telephone service over the pair of wires connecting the telephone on the subscriber's premise to equipment operated by the telephone company. This connection is the means by which the subscriber requests service and is alerted to incoming calls. Further, information, typified by speech or speech-like signals, are carried in a full-duplex mode over the connection. This connection is called "access" (to the telephone network). The cable pair that provides this connection is called the subscriber loop. In Section 1.3 we provide an introduction to the notion of access and the implications of "two-wire" access. Two-wire terminology is employed when signals in the two directions of transmission are carried simultaneously over the cable pair and could thus interact; four-wire is the term used when the signals in the two directions are carried in separate cable pairs, either physically or logically (we could separate the two directions of transmission by using different frequency bands and thus share the same cable pair, for example).

1.2 A SIMPLE MODEL FOR THE COMMUNICATION PROCESS

The underlying concepts of the theory of electronic communications can be explained using a very simple model of the communication process. The primary purpose of introducing this model is to set the framework for the mathematical aspects of the subject. The model, though simple, will serve as a reference point for the study of basic signal theory and provide a rationale for studying the mathematical operations and certain important properties of the signals involved. The same model is also useful for studying general signal processing and for analyzing the distortion experienced by a signal whenever it is modified.

A communication system has one principal task. Information from a *source* must be transmitted to a *destination*. From the view of electronic communications, the information generated by the source will be considered as an electrical signal, either a voltage or current, which changes with time. The actual information to be transmitted may be of a nonelectrical nature, such as voice, which is in principle an acoustic signal. In such situations we assume the presence of a transducer that will convert the actual information into electrical form. Similarly, at the destination, a transducer will convert the received electrical signal into the appropriate form.

The information signal is delivered from the source to the destination employing an appropriate medium. This medium is referred to as the ***channel***. A channel may be as simple as a pair of wires or as complex as a complete telephone network. The essence of a channel is that it provides a restriction on the signals that can be transmitted through it and futhermore may alter the signal as it passes through. The restriction on the transmitted signal imposed by the channel is equivalent to saying that the channel is *matched* in some sense to the signals it can transport. For example, a channel associated with the use of electromagnetic wave propagation, i.e., radio, can support signals consisting of reasonably high frequencies. Speech signals, in electrical form, are comprised of reasonably low frequency components. Thus speech signals are not matched to radio channels. Typical telephone channels are geared toward supporting normal analog speech signals whereas one may wish to transport digital data signals over the telephone network. The process of altering the information signal to match the characteristics of the channel is referred to as ***modulation***. The recovery of the information-bearing signal is called ***demodulation***.

With this preamble, we can introduce the basic model for the communication process in Figure 1.1.

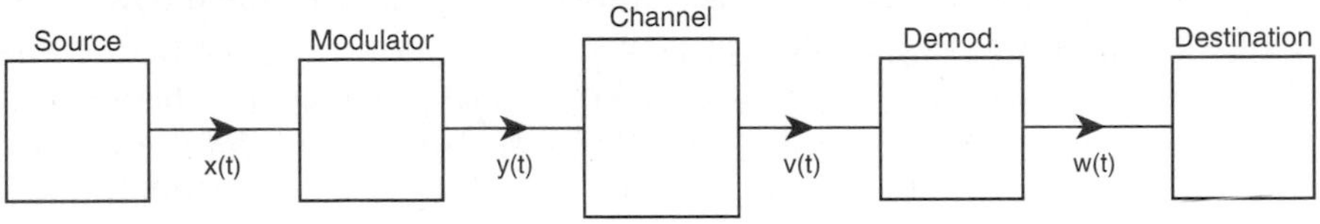

Fig. 1.1
Basic model for the communication process

The source generates an information-bearing signal labeled x(t). Now x(t) may not be matched to the channel, necessitating a transformation of the signal by the modulator into another signal labeled y(t). In principle this transformation is unique and invertible in the sense that x(t) can be recovered from y(t). The output of the modulation step, y(t), forms the input to the channel. One would like the output of the channel to be y(t), indicating that the channel is an ideal transmission medium. However, the effect of the channel is to distort the signal and the output of the channel, v(t), is thus not the same as y(t). The demodulation step converts v(t) into w(t) using the logical inverse of the modulation process. Since v(t) and y(t) differ, it is clear that w(t) and x(t) would differ as well. If the channel were an ideal medium then y(t) would be equal to v(t) and, assuming the demodulation process was indeed the logical inverse of the modulation step, w(t) would be the same as x(t). Note that there are several transformations the information undergoes in the passage from source to destination. Each transformation has the potential for introducing distortion. The mathematical theory of communication provides us with tools required to analyze these deformations and, in most cases, allows us to quantify them. A quantitative analysis is useful in gauging the effectiveness of the communication system and in describing the limitations imposed on the communication of information between the source and destination.

Quantitatively speaking, the overall distortion introduced by the communication system depicted by the model can be expressed as a signal $\eta(t)$, where

$$w(t) = x(t) + \eta(t) \tag{1.2.1}$$

In this definition of distortion we have, for the time being, ignored the fact that there would be some finite time delay related to propagation and any processing involved.

One measure of the efficacy of the communication system can be defined in terms of the strength of the error relative to the strength of the information-bearing signal. The most commonly used definition of strength is the power of the signal. Thus if $P_{w,}$ $P_{x,}$ and P_η denote the powers of w(t), x(t), and $\eta(t)$ respectively, then the *signal-to-distortion ratio* (**SDR**) can be expressed in two ways as either

$$\text{SDR} = \frac{P_w}{P_\eta} \quad \text{or} \quad \text{SDR} = \frac{P_x}{P_\eta} \tag{1.2.2}$$

Clearly, a large value for **SDR** indicates that there is little distortion between the input signal x(t) and the signal delivered to the destination, w(t). Furthermore, if the **SDR** is large then either of the ratios in Eq. (1.2.2) is applicable.

The difference as defined in Eq. (1.2.1) does not establish the nature of the distortion. Any difference between x(t) and w(t) is treated as distortion. There may, however, be differences between the signals caused by transformations that are, at least in principle, reversible. A better description of the overall transmission is

$$w(t) = \mathbf{H}\{x(t)\} + n(t) \tag{1.2.3}$$

In Eq. (1.2.3) the transformation **H{ }** indicates that the signal x(t) is modified in a calculable, and therefore probably controllable, manner. n(t) is an additive component that cannot be accounted for and is called ***noise***. If one assumes that the transformation does not modify the signal excessively, then n(t) represents the degradation encountered by the signal between source and destination. The efficacy of transmission can be quantified by the ***signal-to-noise ratio*** (**SNR**) as

$$\text{SNR} = \frac{P_w}{P_n} \quad \text{or} \quad \text{SNR} = \frac{P_x}{P_n} \tag{1.2.4}$$

where P_n is the power of the noise n(t). The noise signal is by nature unpredictable in value and is described in statistical terms; the power of the noise is equivalent to the variance of the probability distribution function used to describe the noise behavior. While either of the ratios in Eq. (1.2.4) could be used as the definition of **SNR**, the choice is usually that which is more easily measured or calculated. Again, in most cases a large **SNR** is desired and in this situation the two ratios do not differ substantially.

Fig. 1.2 depicts a suitable model for the channel. The channel is considered as the combination of a predictable, and thus quantifiable, transformation **H{ }** along with an unpredictable distortion that is modeled as an additive component, n(t). Further, for the purposes of this book, the transformation **H** will be considered as a ***Linear-Time-Invariant*** (**LTI**) operation. Clearly, the model for the channel shown is just as applicable to several subsystems, such as amplifiers and filters, in an overall communication system.

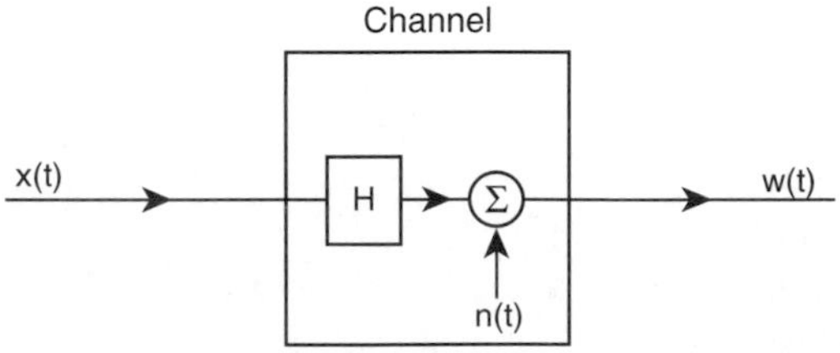

Figure 1.2
Channel model – combination of linear filter and additive noise

In the case where the information being transmitted is digital in nature, the signal-to-noise ratio may not be an appropriate measure. A better gauge of the quality of transmission would be the probability of error. In Chapter 2 we show how the probability of error can be estimated from the ratio of signal-to-noise powers and how the impact of the **LTI** transformation can be analyzed. Fig. 1.2 depicts more a mind-set than an actual mathematical entity. Whenever there is a modification of a signal, we consider the effect as a combination of a "deterministic" impact and an additive "random" contribution, or noise, where the latter is a catchall for all the effects that we cannot model, or do not understand, completely.

From a mathematical viewpoint, communication theory is the study of the overall process embodied in Fig. 1.1 and has three broad components. The first addresses the need to provide a framework for expressing and quantifying signals. Second is the need to define the predictable transformations that signals may encounter along with the means to analyze them. Third is the need to express how well a communication system is operating or can be expected to operate. These three items form the core of communication theory. There are several excellent references available on this vast subject and some of these are listed in the bibliography at the end of the chapter.

In dealing with the notion of signal processing, a model akin to Fig. 1.2 can be proposed to quantify the error introduced, or, equivalently, the deviation from the ideal situation. For example, suppose we wish to process a signal in a manner that can be described mathematically by $\mathbf{G}_\mathbf{I}\{x\}$. In practice we may not be able to achieve the system $\mathbf{G}_\mathbf{I}$ but will actually implement a system $\mathbf{G}_\mathbf{A}$, where the subscripts **I** and **A** are meant to convey the notion of "ideal" and "actual." The two will usually be different. For a variety of reasons such as stability, causality, realizability, component accuracy, nonideal components, finite wordlength (in digital filters), and so on, $\mathbf{G}_\mathbf{A}$ will be at best an approximation to $\mathbf{G}_\mathbf{I}$. In addition, there would be the effects of "interference" such as thermal noise, finite-precision arithmetic (in digital filters), and so on. Fig. 1.2 will be representative of the distortion if we interpret **H** (which we assume is linear) as the *difference* between $\mathbf{G}_\mathbf{I}$ and $\mathbf{G}_\mathbf{A}$ and the additive component constitutes the noise or nonsystematic error component.

1.3 SUBSCRIBER ACCESS TO THE TELEPHONE NETWORK

The general form of a telephone circuit is depicted on the next page in Fig. 1.3. A subscriber is connected to the network by a cable pair called the ***subscriber loop*** or ***subscriber line***. The equipment in the customer's premise is called a *station set* and is usually a telephone set or another device that "mimics" a telephone set such as a modem or facsimile machine. This cable pair terminates in the telephone company premises (called the **C**entral **O**ffice, or **CO**) in a device called the ***line circuit***, which is usually dedicated to that particular loop. Signals between line circuits are transported across the **PSTN** using transmission methods (also called "trunks") and switching systems (called "switches"). The **PSTN**, and the switching and transmission systems it is comprised of, is able to provide an "on-demand" semipermanent connection between line circuits and thus between end users. While the implications of the subscriber loop are covered in this section, some of the implications of the trunking, or transmission, fabric are covered in Section 1.4.

The term ***access*** is used to describe the means whereby subscribers' signals enter and leave the network and, for common telephony, involves the subscriber loop and the line circuit. Since signals originating as well as signals terminating at the station set are carried over the same pair of wires, this transmission is called ***two-wire*** (or 2-wire). Modern transmission systems, such as fiber optic cable systems, employ separate transmission paths for the *send* and *receive* information signals and are thus called ***four-wire*** (or 4-wire) schemes. The circuitry in the line circuit whose function is to do this "2-wire-to-4-wire" conversion is called *the hybrid.* In some cases the connection to the subscriber's premise is done in a 4-wire manner and called 4-wire access.

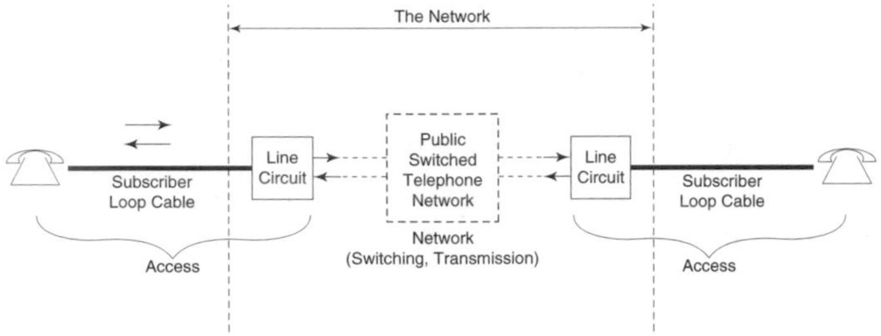

Fig. 1.3
Components affecting transmission of subscriber signals over the **PSTN**

The telephone set is an analog device and the information signals carried over the subscriber loop are analog in nature. For switching and transmission purposes it is better to have the information in the form of digital signals. The line circuit provides this conversion capability and thus includes an A/D converter, D/A converter, the associated anti-aliasing and replicate-reduction filters, and amplifiers to provide the appropriate signal levels.

In the rest of Section 1.3 we provide an overview of the access into the network. In Sections 1.3.1 and 1.3.2 we present some of the effects of the subscriber loop itself. Then in Section 1.3.3 we describe the specifications of the frequency response expected of the filters in the line circuit and in Section 1.3.4 discuss the levels of noise associated with conversions between analog and digital formats. Finally, in Section 1.3.5 some thoughts on the modern subscriber loop, and the trends in providing access to the Network in digital form, and at high data rates, are discussed. One of the consequences of the subscriber loop being analog and the network being digital is that we need to convert between two forms of transmission, 2-wire-to-4-wire, which gives rise to echoes. It will be clear that the very nature of the loop, and the potential variation from loop to loop, will guarantee that echoes will be present in some, if not most, telephone connections. The methods used to control echo are the subject of Chapter 6, which deals with the considerations inherent in the design of network echo cancelers.

The intent of this section is to provide the reader with a flavor of the functionality, in terms of signal processing, that is, or must be, provided at the interface point between the subscriber loop and the local switch where the loop terminates. For a more detailed description of the loop, the equipment used, and practices followed in engineering the loop plant, the reader is referred to Reeve [1.5].

1.3.1 The Line Circuit

Traditionally the line circuit is associated with the **BORSCHT** functions. **BORSCHT** is an acronym for the different functions that need to be implemented in the line circuit.

B is for **battery**, meaning that the line circuit provides a source of power to the telephone set, via the subscriber loop, nominally –48 volts with respect to earth-ground.

O is for the protection function. Since the subscriber loop connects the line circuit to the customer premise over external terrain, it is exposed to such hazards as lightning or *power cross*, where it may come into contact (accidentally, of course) with power lines. Appropriate **over-voltage** detection and protection are required to shield the (delicate) electronic circuits from these hazardous voltages.

R, is for **ringing**, referring to the method by which a telephone is alerted to an incoming call. The line circuit applies *ringing voltage*, typically 100 Volts rms at a frequency of about 100 Hz, to excite the ringer in the telephone set.

S stands for **supervision**. Idle telephone sets, characterized by having the handset on the cradle, appear as an open circuit at the end of the loop. In the early part of this century the handset was kept on a hook when not in use, hence the term ***on-hook*** for station sets that are idle. When the handset is taken from its cradle, the loop is closed and since the line circuit is applying battery voltage, a DC current will flow in the loop. This alerts the line circuit that the set has gone ***off-hook*** and is ready for service. Monitoring the loop current to ascertain the on-hook or off-hook status of subscriber equipment is termed supervision.

C represents the **codec** function. The A/D converter along with the anti-aliasing filter constitute the *coder* and the D/A converter along with its replicate-rejection filter are the *decoder*. The combination is called the *codec*.

H is the **hybrid** function that involves the 2-wire-to-4-wire conversion.

T is for the **testing** function. Using relays or equivalent methods, the connection between the loop and the line circuit can be (temporarily) broken and suitable test gear connected to each part separately. Thus the condition of the loop and the line circuit can be tested independently. These test-access relays are usually considered as part of the line circuit.

From the viewpoint of information-bearing signals the line circuit can be represented as shown in Fig. 1.4. While the **BORSCHT** functions are invaluable from a telephony system perspective, they are not all involved in actual signal processing of the information-bearing signals. Only the hybrid and codec are.

We can visualize the operation of the line circuit as follows. The digital signal from the far end comes via the network to the line circuit, where it is converted to analog. The hybrid attempts to transfer the entire *Receive* signal into the subscriber loop toward the subscriber. The near-end speech signal, s(t) originates at the local subscriber and enters the line circuit where the hybrid steers this signal toward the A/D converter, where it is converted into a digital format as the *Send* signal. Some of the Receive signal leaks into the Send path and will be perceived by the far-end subscriber as echo. Ideally all the Receive signal power would be coupled into the subscriber loop and there would be no echo. The efficacy of this coupling is governed by the *Balance Impedance*, Z_B.

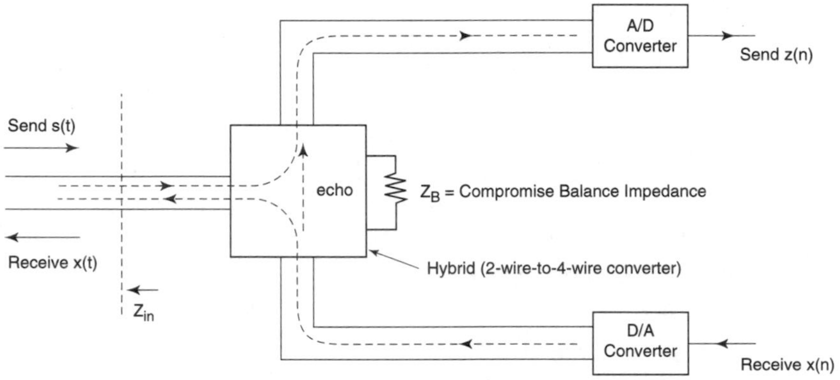

Fig. 1.4
Function of the *hybrid*, indicating presence of an echo path

The hybrid is usually implemented using transformers but the action can be described by the circuit of Fig. 1.5. The output of the D/A is launched into the loop through an impedance Z_s ("source impedance"). Some portion is reflected back according to the ratio of Z_s to the input impedance of the subscriber loop. An *estimate* of this can be generated by driving the same voltage though an impedance Z_s into an impedance Z_B. The signal from the subscriber is sensed by the amplifier and the estimate of reflection subtracted. For such an arrangement we would have Z_s Z_B and these would be as close to Z_{in} as we could possibly achieve. We can compute the effective transfer function, $H_e(f)$, as

$$H_e(f) = \frac{1}{2}\,\frac{Z_{in}(f) - Z_B(f)}{Z_{in}(f) + Z_B(f)} \qquad (1.3.1)$$

The quantity in parentheses is called the "***return loss*** of Z_{in} versus Z_B" and is a measure of how different the two impedances are. Thus the echo is determined by how well the input impedance to the subscriber loop can be measured or estimated and thus *matched*.

In practice subscriber loops consist of varying lengths of cable, the length of cable being equal to the distance between the subscriber's premise and the serving central office. This distance could be a few hundred feet, if the telephone was "next door," to several miles. Modern loop plant engineering methods employed in North America can ensure that loop lengths will be less than 12 thousand feet (12 kft) though older loops, of length much longer than 18 kft are still in use. The telephone set at the subscriber's end is another variable. The impedance presented as the load would depend on current, manufacturer of the telephone set, the number of extension lines, and so on. In short it is not possible to define **one** balance impedance that would match **all** loops. What is done is to define a few ***compromise balance networks*** such that for any given loop we can choose one that achieves a "reasonable" match.

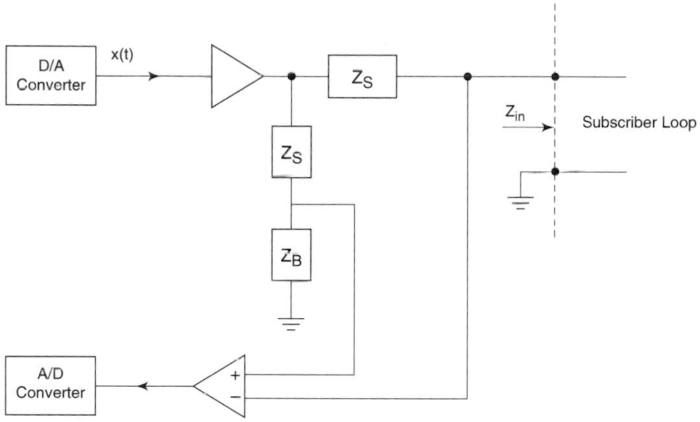

Fig. 1.5
Circuit model for computing echo transfer characteristics

1.3.2 Subscriber Loop Cable

The cable between the line circuit and the telephone introduces its own effect on signal transmission, modeled as a (nontrivial) frequency response. We can quantify this effect using the theory of transmission lines. A short segment of a transmission line can be modeled as a combination of series resistance, series inductance, shunt conductance, and shunt capacitance, as shown in Fig. 1.6.

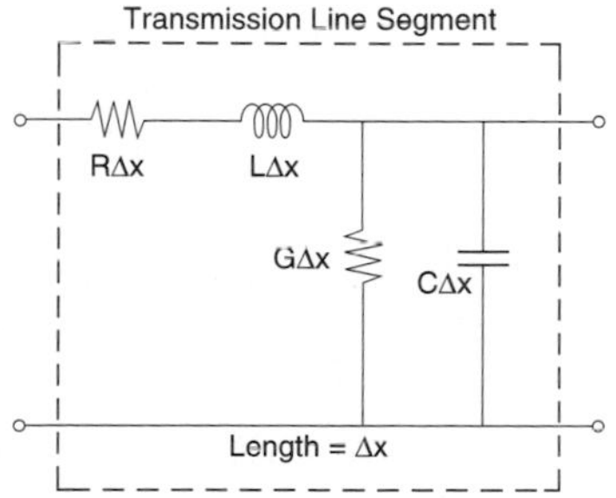

Fig. 1.6
Lumped circuit element model for a short segment of transmission line of length Δx

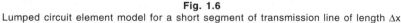

In this manner the cable is described by the parameters R, for resistance per unit length; L, for inductance per unit length; G for conductance per unit length; and C, for capacitance per unit length. In North America the cables used for providing new subscriber telephone service are predominantly 26 and 24-AWG with some 22-AWG and 19-AWG cable remaining from earlier construction. The diameter of the conductor decreases with increasing AWG and thus 26-AWG cable has the least diameter and thus the most series resistance (and least cost!). The other parameters, L, G, and C are affected by the construction of the cable, the material used to insulate the conductors from each other, and so on. These have been well documented and are available in several standards and requirements documents, for example (American National Standards Institute) **ANSI T1.601** [4.4.15] or, for specific brand-name cable, from the manufacturers' data sheets. These parameters allow us to compute the ***propagation constant***, γ, and the ***characteristic impedance***, Z_0, as

$$\gamma = \sqrt{(R + j\,2\pi f L)(G + j\,2\pi f C)} \qquad (1.3.2a)$$

$$Z_0 = \sqrt{\frac{(R + j\,2\pi f L)}{(G + j\,2\pi f C)}} \qquad (1.3.2b)$$

The transmission characteristics of the cable can be expressed in terms of these (complex) constants and the length of the cable. Suppose our source impedance was Z_s and the termination (load) impedance was Z_t. Then the frequency response between the source (driver) and the load can be computed as follows. Fig. 1.7 depicts the situation of a source, E_s, driving the load through the source impedance and the transmission line and we wish to determine $H_c(f)$, where

$$H_c(f) = \frac{E_t(f)}{E_s(f)} \qquad (1.3.3)$$

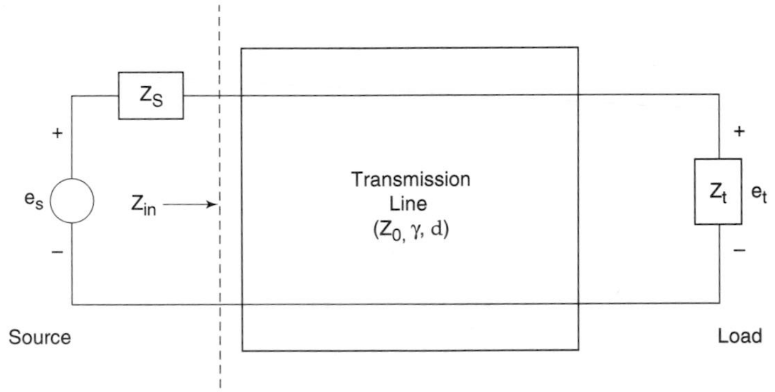

Fig. 1.7
Circuit model for computing frequency responses of a segment of cable of length d

The expressions for the transfer function are simplified by introducing the ***reflection coefficients***, ρ_s and ρ_t, which represent the return loss of Z_s and Z_t, respectively, versus the characteristic impedance Z_0, and which are defined as

$$\rho_s = \frac{Z_s - Z_0}{Z_s + Z_0} ; \rho_t = \frac{Z_t - Z_0}{Z_t + Z_0} \quad (1.3.4)$$

where the dependence on frequency is not explicitly indicated.

Application of transmission line theory will yield the following expressions for the frequency response, $H_c(f)$, and the input impedance looking at the load through the transmission line, Z_{in}:

$$H_c(f) = \frac{Z_t}{Z_t + Z_s} \frac{1 - \rho_s\rho_t}{e^{\gamma d} - \rho_s\rho_t e^{\gamma d}} \quad (1.3.5a)$$

$$Z_{in} = Z_0 \frac{1 + \rho_t e^{-2\gamma d}}{1 - \rho_t e^{-2\gamma d}} \quad (1.3.5b)$$

where the quantity d represents the length of the cable. The optimal choices for the source and termination impedances would be Z_0. Unfortunately, Z_0 cannot be achieved using lumped elements (capacitors, inductors, and resistors) much less in an economical fashion. Thus a compromise value is used. This value corresponds to an impedance of 900 ohms (resistive).

If the cable parameters are known, then Eqs. (1.3.5a) and (1.3.5b) provide the frequency response and input impedance of the cable, and Eq. (1.3.1) determines the echo transfer function. For 26-gauge **PIC** (Plastic Insulated Cable) the parameters of capacitance, resistance, inductance, and conductance (C, R, L, and G) are 0.083 μF, 87 ohms, 0.985 mH, and 0.168 μmho (per mile), respectively (at 70 degrees F and at a frequency of 1 kHz), and these parameters are approximately constant over the range of frequencies considered. **ANSI T1.601** [4.4.15] provides the values of these parameters as a function of frequency, at various temperatures, various insulation types, and various cable diameters (AWG).

The frequency response of the subscriber loop for three lengths, 10 feet, 9 kft, and 18 kft, is shown in Fig. 1.8 and the return loss in Fig. 1.9. The unit kft, or "kilofeet" may sound strange but is the most commonly used unit used in the context of subscriber loops.

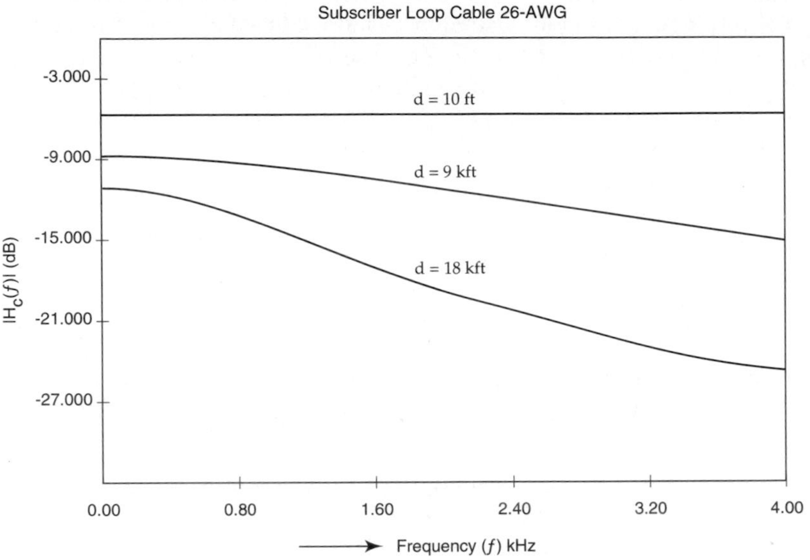

Fig. 1.8
Frequency response for three lengths of 26-AWG cable

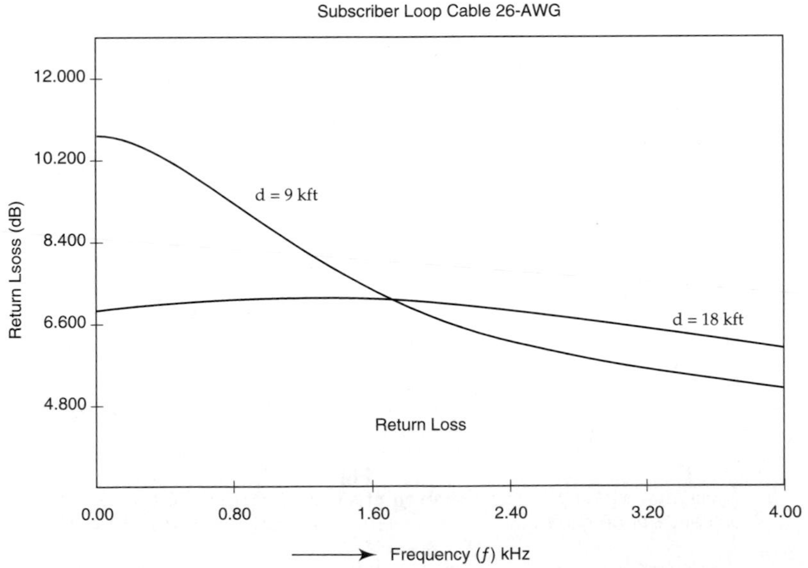

Fig. 1.9
Return loss as a function of frequency for 9-kft and 18 kft sections of 26-AWG cable

It is evident is that the cable introduces a lowpass characteristic. The variation in gain, versus frequency, over the 4-kHz range is as much as 6 dB for the 9-kft cable and 14 dB for the 18-kft cable. In addition to this gain variation is a "flat loss" in the cable. Of this, 6 dB arises because of the voltage division ratio of 2 between the source and termination impedances, even for a loop of zero length. This is usually not considered a loss and can be factored out. The flat loss over and above the nominal 6 dB is somewhat more easily dealt with using techniques of **AGC** (automatic gain control) than the nonflat frequency response. In fact this **AGC** function is addressed in the telephone set itself by using loop-current dependent resistors (the longer the loop the smaller will be the loop current since the battery voltage is nominally fixed at 48 volts). The variation in frequency response is more difficult. Special techniques known as ***conditioning*** a loop are employed whereby equalizers (which have a frequency response that is high-pass in nature to compensate the lowpass nature of the cable) and additional gain are employed. In practice these equalizers also address the phase response of the cable. The function of the equalizer is to make the frequency response as *constant* as possible.

The overall frequency shaping encountered by a signal traversing through the loop and line circuit is depicted by the blocks in Fig. 1.10.

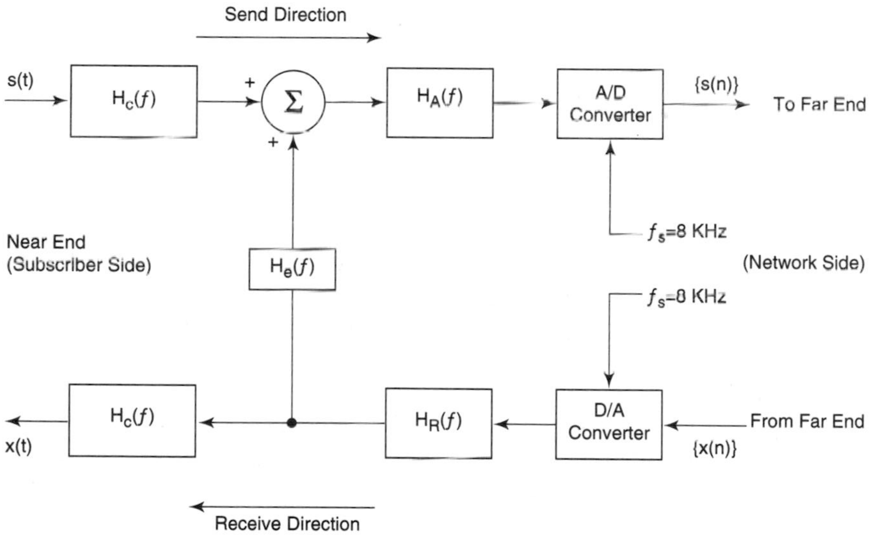

Fig. 1.10
Block diagram showing spectral shaping of signals by the cable and line circuit filters and the presence of an echo path. The converters necessitate limiting the bandwidth to 4 kHz

Subscriber loops that are longer than 18 kft are specially treated by adding **load coils** at uniform intervals. These load coils are in series with the cable and serve to equalize, to some extent, the amplitude response of the subscriber loop. Two loading schemes are used in the United States. The **H88** scheme, which originated in the

Bell System, uses 88-mH coils spaced 6000 feet apart. The **D66** scheme favored by the Rural Electrical Administration (**REA**) employs 66-mH coils at a spacing of 4500 feet. Such loops are called *loaded loops*. Modern network practice is to keep subscriber loops shorter than 12 kft of 24-AWG cable, or 9 kft of 26-AWG cable, and loading is not necessary in such cases. One consequence of loading is that whereas the amplitude response is "equalized" over the 0- through 4-kHz range, the net phase response is far from linear and cable response at frequencies higher than 4 kHz is well nigh zero. The interested reader is referred to Reeve [1.5], where a comprehensive treatment of the trials and tribulations of the subscriber loop, and how they have been addressed by the phone company, is provided.

1.3.3 Frequency Response of Filters in the Line Circuit

The impact of the cable is shown in Fig. 1.10 as $H_c(f)$ in both directions. This is true only if the load and source impedances are the same, a condition that cannot be guaranteed in practice. The load impedance is that of the telephone set, which is nominally 900 ohms but varies with loop current as well as from set to set, manufacturer to manufacturer, and so on. Also shown in Fig. 1.10 are two filters $H_A(f)$ and $H_R(f)$, which are associated with the A/D and D/A conversion processes. $H_A(f)$ serves as the anti-aliasing filter and $H_R(f)$ represents the replicate-reduction filter. The variation in subscriber loops, station sets, and so on, is quite significant and thus $H_c(f)$ is not a well-defined entity. The filters involved in the A/D and D/A conversion process, on the other hand, are quite controllable and are indeed specified in quite a rigid manner.

Both conversion filters are lowpass filters with a nominal cutoff frequency of 4 kHz. Considering that the signal in any direction will encounter an anti-aliasing filter and a replicate rejection filter (in two different line circuits), the manner in which these filters are specified is as a tandem. The passband characteristic of the tandem is ideally a gain of 0 dB and a constant group delay. The allowed deviations, as specified in the **CCITT** Recommendation **G.712** [4.1.4], are shown in Figs. 1.11 and 1.12 for the magnitude and group delay response, respectively.

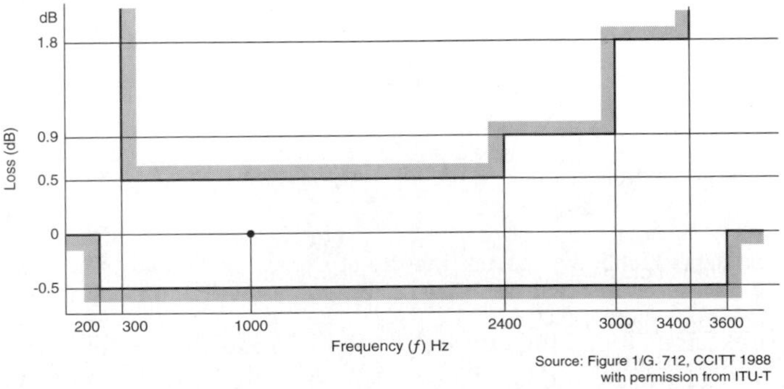

Fig. 1.11
Magnitude response limits of the A/D and D/A conversion filters

The stopband attenuation of each separately, and thus the tandem as well, would ideally be infinite (dB). The specification is provided as a curve that depicts the minimum attenuation versus frequency of the tandem and is shown in Fig. 1.13. The transition band is considered to be between 3400 Hz and 4600 Hz (4000 Hz ± 600 Hz) and the rolloff in this region is specified as a (better than) sinusoidal rolloff.

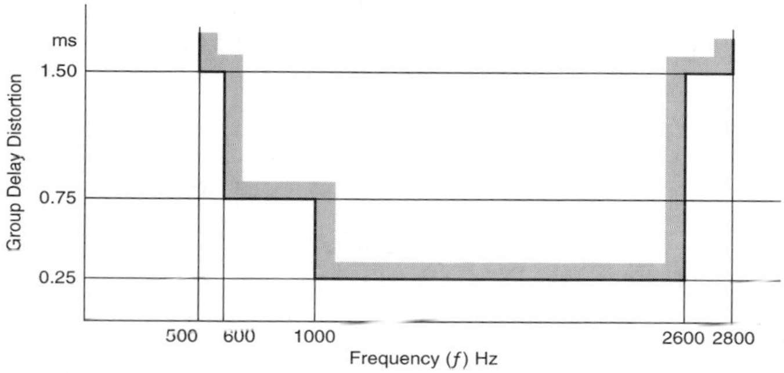

Source: Figure 2/G. 712, CCITT 1988
with permission from ITU-T

Fig. 1.12
Group delay distortion limits of the A/D and D/A conversion filters in the line circuit

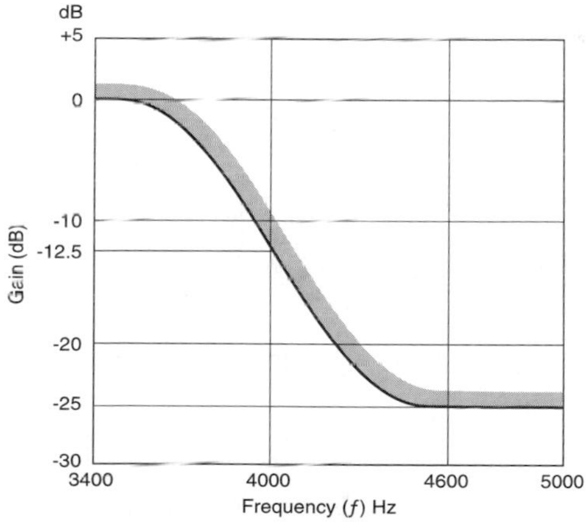

Source: Figure 4/G. 712, CCITT 1988
with permission from ITU-T

Fig. 1.13
Transition band and stopband requirements of the A/D and D/A conversion filters

How can the specification of the tandem be manipulated to obtain a specification for the individual filters? The approach taken in a channel bank, as specified in BELL PUB 43801 [4.6], is shown in Fig. 1.14, which shows the attenuation requirements of the *Send* (A/D) and *Receive* (D/A) filters. The attenuation is not allocated symmetrically, the *Send* direction requiring greater stopband attenuation. The reason for this is that aliasing, once it occurs, cannot be removed. Spectral replicates, while possibly problematic, are not an *in-band* distortion.

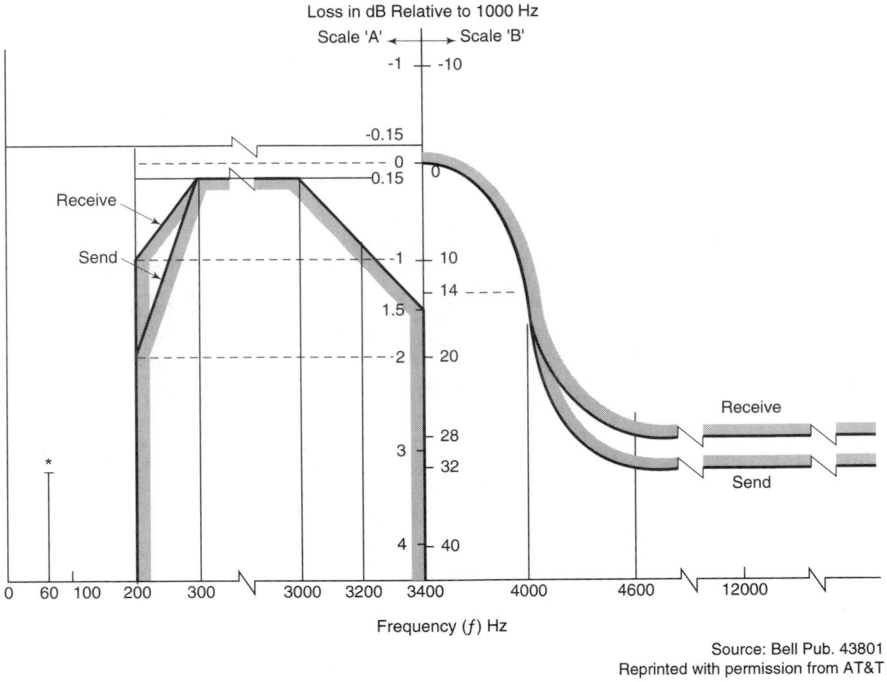

Fig. 1.14
Frequency response specifications for the *Send* and *Receive* filters

1.3.4 Quantization Effects in the Line Circuit

The requirements imposed on the conversion between analog and digital formats are not expressed directly in terms of the number of bits that must be used in the A/D and D/A converters, but in the form of a noise "mask." This mask is equivalent to a lower bound on the signal-to-noise ratio (**SNR**) that must be achieved in a tandem conversion, from analog to digital and back to analog. It is generally accepted that the level of noise that a human listener will tolerate is not a fixed quantity but is related to the signal level—at higher signal levels the human auditory system will tolerate a larger level of noise. In other words, the mask must reflect the ideal situation of a constant signal-to-noise ratio and indeed it does.

Assuming that the network were transparent from the viewpoint of transmission, with the integrity of the digital signal (bit-stream) maintained from A/D converter to D/A converter, then a signal traversing the network, as depicted in Fig. 1.3, would

sinewave signal we need to increase the power by a quantum amount before the amplitude reaches the next higher quantized output level. For random signals, whose sampled values follow a probabilistic law (probability density function), this discretization of levels is not so pronounced.

A second distinction between the two cases is evident from the maximum signal level at which the mid-level **SNR** must be achieved. The sinewave mask extends to an input level of 0 dBm0 whereas for a noise signal the mask extends to –3dBm0 and then drops to 0 dBm0. The reason for this stems, again, from the difference between a deterministic signal and a random signal. For a sinewave, the peak signal value, or amplitude, has a fixed, definite relation to the root mean square (rms) value (square root of the power). This ratio is 1.414, equivalent to 3 dB. The crashpoint of the A/D converter, the largest input level that can be encoded without clipping, is +3 dBm0 (actually 3.17 dBm0 and 3.14 dBm0 for the μ-law and A-law codecs, respectively). Therefore, a 0 dBm0 signal would have about –3 dB of "headroom." In the case of a random signal, large signal amplitudes can be observed (albeit with small probability) even when the signal level (say the rms value) is moderate. That is, the peak-to-rms ratio in the case of the random signal is greater than that of a sinewave (the notion of peak-to-rms value, and some of the implications, are discussed in Chapter 2). Thus if the (noise) signal power is much in excess of –3 dBm0, we will encounter, with not insignificant probability, sample values in excess of the crashpoint. As will be seen in Chapter 4, the effect of clipping is to have a marked degradation of **SNR** and hence the mask rolls off starting at this signal level.

1.3.5 Concluding Remarks on Subscriber Loops

In this section we described some of the issues of the loop plant and access. The emphasis was on what is termed Plain Old Telephone Service, or **POTS**. This emphasis implicitly assumed that the subscriber premise equipment is a station set, that is, a device that would mimic a conventional telephone. With the advent of digital switches and digital trunk transmission, especially fiber optic transmission, the network is geared up to provide services that are digital in nature and which could support very high data rates. In this scenario the subscriber loop is the "weak link" in the transmission chain. However, the investment in the existing embedded loop plant is quite enormous, billions and billions of dollars. It is thus more than likely that attempts to prolong the life cycle of the embedded base will continue well into the twenty-first century.

There are several techniques, both standardized and proposed, to better utilize the existing loop plant, or most of it anyway. The intent of these techniques is to carry information in digital form rather than the conventional analog form and thus the acronym **DSL**, for Digital Subscriber Loop, has been coined. One implementation of a **DSL** is Basic Rate **ISDN**, often called "2B+D," which uses the two-wire loop to transmit full duplex data at a rate of 160 kbps and which has been standardized to significant degree. The U.S. standard for the Basic Rate **DSL** is described in the **ANSI** Standard **T1.601**-1992 [4.4.16]. There are proposals to transmit even higher data rates on subscriber loops, rates as high as 6 Mbps. An excellent treatment of the problems associated with transmitting high speed digital data on existing loops is presented by Leichleider [1.1] and in several papers in the Special Issue of the *IEEE Journal on Selected Areas in Communications*, devoted to High-Speed Digital Subscriber Lines [1.4].

More recently, there has been much ado about the "Information Superhighway," which requires the provision, to each end user, of large bandwidth channels. These channels, much in excess of 6 Mbps, which is the accepted limit of the capabilities of the existing twisted-pair copper-cable loop plant, would require the investment in a new infrastructure based on coaxial cable or fiber optic cable extending to the customer premise. Two issues of the *IEEE Communications Magazine*, [1.2] and [1.3], have delved into the implications of such high speed access and the methodology to provide it.

1.4 THE TELEPHONY CHANNEL—TRUNK TRANSMISSION

The first electrical communication network was geared toward data. A widespread telegraph network providing a "real-time" adjunct to the "non-real-time" postal network was functional when Alexander Graham Bell invented the telephone in 1876. But then the ability to transmit speech signals over long distances using electrical signals revolutionized communications. Today, the telecommunications infrastructure is such that it is possible to provide real-time speech communications between any two people in the world with access to a telephone. This connectivity is provided by the Public Switched Telephone Network (**PSTN**). The **PSTN** can be viewed as an intricate system of switching centers interconnected by trunks—as depicted in Fig. 1.17.

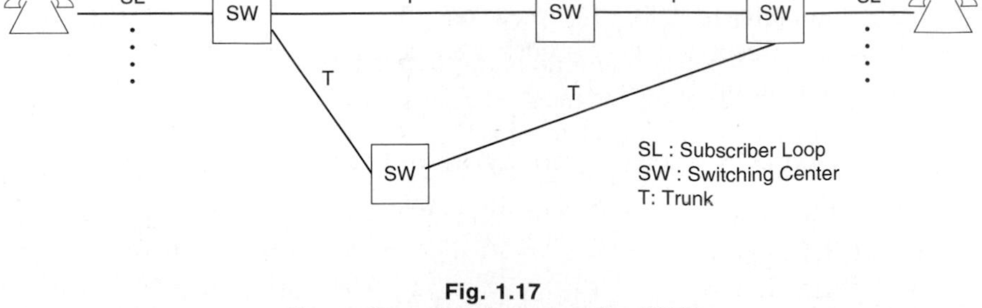

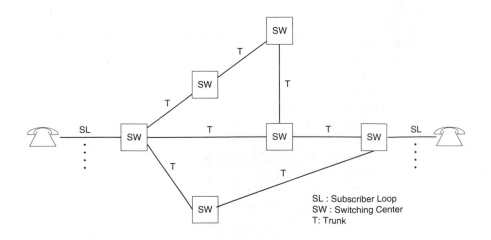

Fig. 1.17
Connecting subscribers over the **PSTN**

1.4.1 The Notion of a Trunk

Only the switching centers on the periphery connect to individual subscribers. A large number just provide (switched) paths between other switching centers. These interswitch connections are called ***trunks***.

The medium of an interswitch connection, or trunk, can be (electrical) cable, wireless (i.e., radio) or, increasingly, fiber optic cable. The transmission capacity of the trunking fabric is usually much more than required for a single speech channel. Thus several channels are multiplexed to better utilize the transport medium. Traditional long haul transmission, prior to about 1986, was predominantly analog in nature. Using the techniques of single side band (**SSB**) modulation, voice channels were assigned 4-kHz slots (8-kHz if both positive and negative frequencies are considered) in frequency division multiplex (**FDM**) assemblies. The **FDM** "unit" in North America is the ***Group*** signal, comprising 12 channels stacked in a frequency band between 60 kHz and 108 kHz. In European networks the **FDM** unit is the ***Supergroup***, which is comprised of 60 4-kHz channels stacked in the frequency bands between 312 kHz and 552 kHz. These units can be further stacked in frequency to create ***Mastergroups***, ***Jumbogroups***, and so on, in what is called the frequency division multiplex hierarchy.

The trend however, is toward digital transmission. Whereas the analog transmission system is eminently suitable for (analog) voice signals, emerging services involving the transport of digital data necessitate the usage of digital trunks. The promise of an **ISDN** (Integrated Services Digital Network) providing ubiquitous digital services is based on the assumption that the trunking network will, eventually, be purely digital in nature. In North America the digital hierarchy is based on the unit of the **DS0,** which corresponds to a 64-kbps data rate. 24 **DS0**s can be (time division) multiplexed, along with some overhead to compose a **DS1** signal at 1.544 Mbps; 4 **DS1** signals can be multiplexed, with some overhead to form a **DS2** signal at 6.312 Mbps; and a **DS3** signal at 44.736 Mbps is assembled from 7 **DS2** signals. Emerging standards targeted towards the use of fiber optic transmission define multiplexing schemes involving transmission data rates of N*51.84 Mbps. The hierarchy used in Europe and most of the rest of the world is slightly different. While the basic channel unit is still considered to be 64 kbps, the multiplexing scheme follows a different progression. 30 **DS0**s, along with overhead, are (time division) multiplexed into a 2.048-Mbps assembly, often referred to as an **E1** signal. Subsequent members of the hierarchy involve multiplexing of these **E1** signals into signals of increasing data rate. Descriptions of these multiplexing schemes are provided in **CCITT** Recommendation **G.704** (see [4.1.2]).

1.4.2 The DS1 and E1 Time Division Multiplex Assemblies

The multiplexing formats for combining 24 **DS0**s into a **DS1** aggregate and 30 **DS0**s into an **E1** aggregate are worthy of mention since the multiplexing schemes affect the information signal in different ways. The multiplexing formats for obtaining higher rate assemblies, such as the **DS2** from the **DS1** and so on, try to provide bit transparency and thus do not affect the **DS0** in any way (other than if there are errors in transmission).

In **DS1** the serial bit stream, nominally 1.544 Mbps, is constructed from its constituent **DS0**s in the following manner. The information is organized in frames of 193 bits over a 125-μsec interval for a net data rate of 1.544 Mbps. These 193 bits are comprise 192 information bits and one framing, or "F" bit. Recognizing that a **DS0** represents data at 64 kbps, and is usually created as an octet every 125 μsec (8-kHz sampling rate), the 192 bits per **DS1** frame can be associated with 24 separate **DS0**s or channels, as depicted in Fig. 1.18.

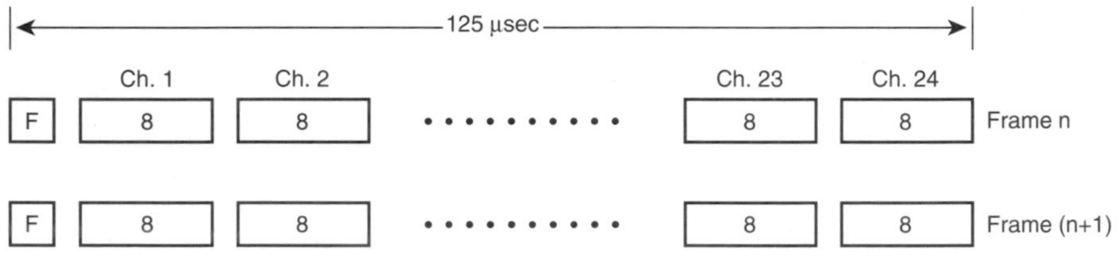

Fig. 1.18
The **DS1** format: 193 bits every 125 μsec; bit rate = 1.544 Mbps

The F bit follows a repetitive "framing" pattern that allows the receiver, having synchronized itself to this framing pattern, to identify which bits in the serial data stream are information bits. By convention, the first 8 bits after the F bit are associated with channel 1 (Ch. 1), the next 8 bits with channel 2, and so on, with channel 24 associated with the last 8 bits of the frame. Note that the notion of a channel can be expanded to include information streams of bit rates of the form N*64 kbps by associating n octets per frame with the channel. In traditional voice-oriented telephony we use 24 channels as the norm, with each octet representing a voice sample, corresponding to an underlying sampling rate of 8 kHz. This octet is a μ-law encoded word (see Chapter 4) and, again by convention, the most significant bit of the word is transmitted (received) first.

In the North American **DS1** standard there are two framing standards currently employed. Older equipment uses the so-called **SF** (for Super Frame) format. In the **SF** format all F bits are used for framing purposes and follow a pattern that has a period of 12 frames. Consequently we could number frames modulo 12. The **ESF** (for Extended Super Frame) format, which is the standard for all modern **DS1** equipment, employs a pattern that repeats every 24 frames, permitting the numbering of frames modulo 24. However, in **ESF** only every fourth F bit is used to specify the synchronization pattern. The remaining F bits are used for other purposes. Every other F bit (corresponding to a bit rate of 4 kbps) is used for providing a data communications channel between the transmitting and receiving equipment. The remaining F bits are used to incorporate an error-detection scheme by computing the **CRC** (for Cyclic Redundancy Code) check sum over the information bits of the prior

24 frames. The receiver can do likewise and a difference in the check-sum indicates that there was a transmission error (at least one). Modern equipment, although designed for **ESF**, must be able to operate in an **SF** mode when it connects to older equipment.

The ability of both the **SF** and **ESF** formats to number frames, modulo-12 and modulo-24, respectively, has an additional application. For channels that are so provisioned ("provisioning" is the term used when the nature or characteristics of a channel are programmed), the least significant bit of the octet is "robbed" every sixth frame. This bit position is used to convey signaling information associated with the channel and, in some sense, determines the state of the telephone call being carried in the said channel. Signaling information is of various types, but in its most basic usage, identifies a channel as being active ("off-hook") or idle ("on-hook"). As a result of bit-robbing, the least significant bit of the octet in a signaling frame (sixth frame, i.e., frames 6 and 12 in **SF** or frames 6,12,18, and 24 in **ESF**) can be corrupted. If a channel experiences several stages of multiplex-demultiplex-remultiplex, such as when it traverses a switching center, the least significant bit of the octet in all frames could become corrupted. Whereas a single multiplex operation would corrupt every sixth sample, there is no guarantee that if and when the channel is remultiplexed that this (corrupted) sample would fall in the sixth frame of the new multiplex assembly, implying that a different, good, sample could be corrupted by bit-robbing. This degradation of the least significant bit would occur even if the signal remained digital all through the network but, fortunately, in remaining digital the distortion is limited to the least significant bit. Robbed-bit signaling is the principal reason that in North America the notion of a digital channel is 56 kbps rather than 64 kbps, taking into account that the least significant bit might be altered as the signal traverses the network. Clearly, if the signal was converted to analog at any intermediate point as it traversed the network we would experience an accumulation of distortion and this accumulated distortion could be equivalent to the corruption of more than just the least significant bit.

Trunk transmission schemes that carry signaling in the same assembly as the information channel are referred to as **CAS** (for Channel Associated Signaling). Robbed-bit signaling is an example of **CAS**. In the modern network the trend, at least for interoffice trunks, is to route signaling information over a separate network altogether. Considering that switches are, in essence, large computer systems, capable of communicating with each other, it is but natural that a communications convention has been developed, called Signaling System Seven (**SS7**). **SS7** is a protocol for the communication between switches, used for the purposes of setting up and tearing down calls, transferring billing information, identitying the calling and called parties, and so on. Such a signaling scheme is generically called **CCS** (for Common Channel Signaling). Of importance to us, in the context of impact on the information channel, is that **CCS** obviates the need for robbed-bit signaling—all eight bits of the information octet can be transmitted intact, allowing the transmission of 64 kbps as opposed to 56 kbps.

The **E1** format for multiplexing **DS0**s involves taking 30 "information" octets and 2 "overhead" octets every 125 μsec to form a signal with a net bit rate of 2.048 Mbps. This is depicted in Fig. 1.19. Each of the octets is called a channel or "time slot" and are numbered from 0 through 31. Time slot 0, or TS0, and time slot 16, i.e., TS16, are the "overhead" octets. TS1 through TS15 and TS17 through TS31 constitute the 30 information **DS0**s. TS0 contains a specific pattern, the "frame alignment signal," in alternate frames. By locking on to this pattern the receiver can identify which bits of the incoming bit stream comprise TS0 through TS31 and can thus extract the information for each channel.

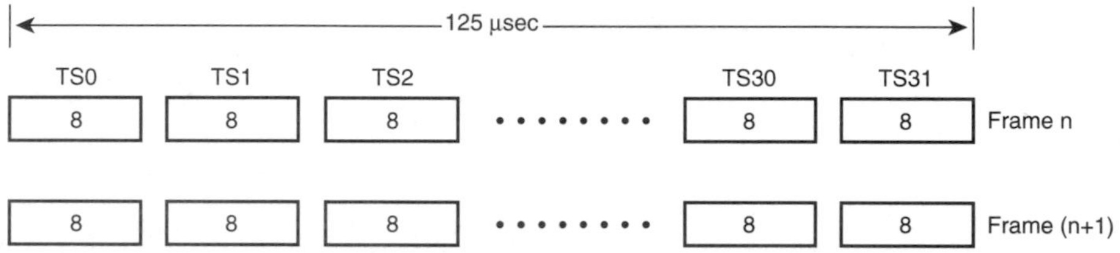

Fig. 1.19
E1 format: 32 octets every 125 μsec; bit rate = 2.048 Mbps

Time slot 16 serves a dual purpose. One is to insert a pattern that can be used to number frames modulo-16, a technique referred to a "multiframe alignment," between 0 and 15. In frame 0 TS16 will have the pattern 0000xyxx for the octet and it is the 4 zeros that identify that frame as frame 0. In each of the other frames, 1 through 15, time slot 16 carries signaling bits for 2 of the 30 channels and provides for the ability to specify 4 signaling bits per channel. Thus TS16 provides multiframe alignment and a conduit for the per-channel signaling information.

In stark contrast with **DS1**, which requires bit-robbing, the **E1** format allows the provision of **CAS** without disrupting any of the information bits. Thus in an all-digital environment, even with **CAS**, **E1** allows each channel to carry a full 64 kbps. A secondary advantage of the **E1** format is that it is octet oriented, compared with **DS1**, which is a mixture of bits (framing) and octets (information), which makes hardware design somewhat simpler since digital hardware tends to be organized in quanta of 8 bits and, further, the division by 193, required for keeping track of the frame boundary, is messier than division by a power of 2.

In a **CCS** environment, the need for carrying signaling information disappears and thus so does the need for multiframe alignment. Thus in such cases an **E1** could carry 31 channels (**DS0**s) of information.

1.4.3 Model of the Network as a Communications Channel

In a digital environment, the notion of a "bearer" channel is a signal of 64 kbps, i.e., a **DS0**. This channel could transport a speech or analog signal sampled at 8 kHz with each sample encoded using 8 bits (according to the μ-law in North America; the A-law characteristic is used in Europe and most of the rest of the world) or a digital signal of 64 kbps. Thus, from the viewpoint of a speech (or analog) signal, the simplest representation of the digital telecommunication network can be depicted in Fig. 1.20.

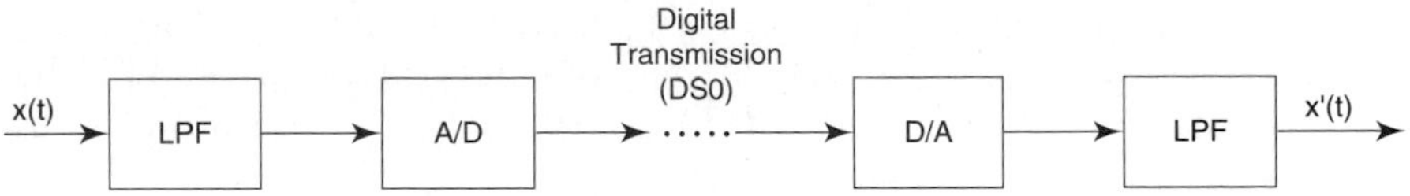

Fig. 1.20
Abstract view of transmission path over the **PSTN**

The signal processing chain comprises an anti-aliasing filter of nominal bandwidth 4 kHz followed by an A/D conversion (which introduces quantization noise). At the receiving end the digital signal is converted back to analog and lowpass filtered to remove the inherent spectral replicates. Until the trunking network is entirely digital in nature, there may be situations where, to traverse an analog transmission span, there may be additional D/A–A/D conversions involved. From a communication-theoretic point of view, the telephone channel can thus be viewed as a bandlimited channel with additive noise. Ignoring the nonlinear nature of the companded encoding scheme, the telephony transmission channel can be modeled as shown in Fig. 1.21.

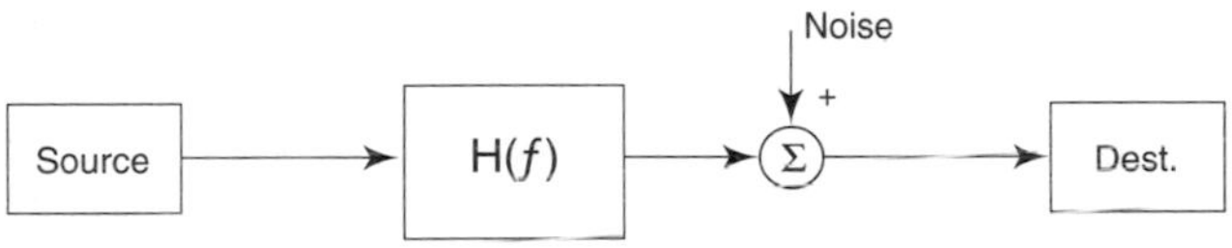

Fig. 1.21
Simplified model of transmission path

The channel is power limited to about 3 dBm0 (for a sinusoidal signal), bandlimited to about 4 kHz (actually a better model is a bandpass filter with passband between 300 Hz and 3.4 kHz) and corrupted by additive noise with a power level of approximately 30 dBrnc0 (the nomenclature dBrnc0 will be explained in Chapter 2). This additive noise is somewhat signal dependent but assuming that the noise and signal are independent is a reasonable approximation. The speech quality associated with this channel is referred to as ***toll quality***.

The ubiquity of the telephony channel has led to the development of devices that utilize the telephony channel for the transmission of digital data. Such devices, "modems," transmit digital data over the **PSTN** at data rates ranging from 300 bps to 19.2 kbps. Facsimile ("FAX") machines routinely transmit at 9.6 kbps. In fact, over certain trunk routes, at certain times of the day, the volume of data traffic may equal, or even exceed that of conventional voice traffic. This phenomenon is necessary to understand since it puts a restriction on the type and extent of signal processing that can be employed in the trunking system of the **PSTN**.

Encoding speech at 64 kbps may seem a waste, as will be discussed later, in Chapter 5. Speech could be encoded at much lower bit rates than 64 kbps and yet retain a **subjective** quality that would be indistinguishable from that corresponding to 64-kbps encoding. From an **objective** standpoint the use of lower bit rates would introduce distortion that would degrade the quality of the telephony channel and affect the performance of modems. This puts a certain constraint on the administration of a trunk group. Clearly, encoding at a lower bit rate would increase the overall traffic carrying capacity of, say a **DS1** (1.544 Mbps aggregate), but would place a restriction on the *type* of traffic the trunk group could carry. It does seem ironic that the encoding rule of a channel be mandated as 64 kbps to support a data rate of 19.2 kbps.

Consequently, voice compression schemes have not made major inroads into the trunking network of the general **PSTN**. In applications where the traffic is primarily voice traffic, compression devices are used extensively. A particular case is in the cellular telephone network. The conventional cellular telephone system involves *cell-sites* that maintain radio contact with the (mobile) telephone set. A terrestrial cable, usually providing transport at 1.544 Mbps, that is, the **DS1** rate, connects the cell-site with a central switching system. The use of voice compression on these **DS1** links is widespread, the reasons being mainly those of economics and availability. **DS1** links between the cell-site and the Mobile Telephone Switching Office (**MTSO**) (also referred to as the Mobile Switching Center, or **MSC**) could be expensive and difficult to obtain. With cellular traffic growing in leaps and bounds, the use of compression to increase the number of channels between cell and **MTSO** is natural from the viewpoints of economy and growth. With all the proposals for digital cellular telephony, wherein the link between the mobile (hand set) and the base station (cell-site) is limited to approximately 16 kbps, the use of low bit rate coding of the speech signal, to approximately 8 kbps, is a necessity rather than just an economic advantage, since some bandwidth is allocated for error correction.

1.5 OPPORTUNITIES FOR DIGITAL SIGNAL PROCESSING

The term "digital signal processing" may have a different meaning for different people. For example, a binary bit stream, such as a **DS1**, can be considered a "digital signal" and the various manipulations, or "signal processing," performed at the bit level by digital hardware may be construed as "digital signal processing." This is not the viewpoint taken in this book. Implicit in our definition of digital signal processing (**DSP**) is the notion of an information-bearing signal that has an analog counterpart. What are manipulated are samples of this implicitly analog signal. Further, these samples are quantized, that is represented using finite precision, with each word representative of the value of the sample of an (implicitly) analog signal. These manipulations, or filters, are arithmetic in nature—additions and multiplications—performed on these samples. We will include in our definition of **DSP** the processing associated with sampling, conversion between analog and digital domains, and changes in wordlength.

The first application of **DSP** in telephony is exemplified in what are generically called "channel banks." Channel banks are, in essence, A/D and D/A converters with the associated filtering for anti-aliasing and replicate-rejection. By converting several analog voice channels to digital format, time division multiplexing is used to achieve "pair-gain." For example, 24 voice channels in a 2-wire (normal subscriber loop) mode require 24 separate cable pairs. In the digital multiplexed mode, assuming a net digital data rate of 1.544 Mbps (**DS1**), the 24 channels can be transported in a 4-wire mode with one cable pair for each direction of transmission. The first channel banks, designed in the 1960s, used analog circuitry for the anti-aliasing and replicate-rejection filters. Whereas a simple analog filter can be constructed out of a few lumped circuit elements (resistors, capacitors, etc.) and a few active elements (transistors), the digital equivalent function requires several circuit boards of digital logic, if implemented in the then-available technology. Thus the channel banks of the 1960s, or even the early 1970s, used analog filters and *shared* a single A/D and D/A over several channels.

In the 1970s digital switching was being deployed on a wide scale in tandem switches, that is, switches that handled only trunks, and was being considered for end office applications as well. In end offices, the switch would have subscriber loops terminating on the switch and this meant that A/D and D/A conversion would have to be performed on a per-subscriber-line basis, rather than just on the trunk side (in channel banks). Such an application had the promise of a large market in terms of the number of devices needed for A/D and D/A conversion. This large perceived demand spurred intense activity in the development of "codec filters." Significant progress was made in the technology of integrated circuits including circuit techniques suitable for integration and fabrication techniques capable of putting both analog and digital circuitry in the same package.

Two distinct approaches found favor in the design of codec filters. The semi-conductor industry chose to implement the A/D and D/A converters in a conventional manner, with the conversion being done at the sampling rate of 8 kHz. The circuit techniques that were relevant included the design of good sample-and-hold circuits, comparators, and networks for providing the "breakpoint" voltages. The method of succesive-approximation, where the breakpoints are achieved using a D/A converter in a feedback loop, was the method of choice for the A/D converter. A new technique was developed for implementing the pre-A/D anti-aliasing and post D/A replicate-rejection filters referred to as "switched capacitor filters," whose theory is based on discrete-time signal processing. Early entries in the market were two chip solutions that had the filters in one chip and the converters in another, but it was in very short order that single chip "combo" solutions were introduced. The success of this methodology, which required considerable expertise in fundamental electronic design, is evidenced by the availability of combo chips that are quite inexpensive. In fact the availability of such low cost converters has led to the use of these devices in equipment, other than the telephony line circuits that they were originally intended for, such as modems, multiplexers, digital telephone sets, and other customer premise equipment. The notion of inexpensive codec filters, suitable for use in customer premise station sets, is one of the driving forces in the thrust toward an all-digital network.

The other path followed in the drive toward a single-chip codec filter was more closely associated with digital signal processing. By doing the actual conversion between analog and digital at a very high sampling rate, the in-band quantization noise performance could be achieved by converters of very low precision, 1 bit, to be specific. This method of conversion, called delta-sigma-modulation (which is described in Chapter 8), uses feedback to shape the quantization noise spectrum so that the majority of the noise power is out-of-band. In the A/D direction the 1-bit converter is followed by a chain of digital filters that removes the high frequency components, principally the out-of-band quantization noise, permitting a decimation in sampling rate to the requisite 8 kHz. In the D/A direction the 8 kHz sampling rate is increased by interpolation, using a chain of digital filters to perform the replicate rejection function, to the same high sampling rate (as the A/D) where a 1-bit D/A converter suffices. This requires a reduction in wordlength (to 1 bit) which is achieved using the notion of a Digital Delta-Sigma-Modulator. This "all-digital" path was favored by telecommunications equipment manufacturers that were designing their own codec filters.

One key advantage of the all-digital approach is the availability of the signals in both directions, in a form suitable for applying digital filters, namely in a uniform encoding format. This in turn means that the designer has the ability to include a modicum of echo canceling in the codec filter itself. A significant portion of the echo generated in the hybrid can be removed by using a short finite impulse response (**FIR**) between the D/A and A/D directions (echo cancelers are discussed in Chapter 6). At that time it was not cost-effective to make the filter adaptive. Nevertheless, locating the cancellation filter as close to the source of the echo as possible is *the* best possible choice and in the codec is the closest one gets to the hybrid. As an alternative to adaptive means for determining the filter coefficients, an approach that uses one choice of a predetermined collection of coefficient sets was deemed appropriate. In fact, a study conducted at the ITT Advanced Technology Center showed that, with a fourth-order **FIR** filter, just six sets of coefficients suffice to provide a median echo return loss (**ERL**) 6 dB in excess of that required in the line circuit. The study, which was (and still is) considered proprietary and thus never published, showed how these coefficients can be determined from simple impedance measurements at the time of installation or during routine maintenance.

The decade of the 1970s was an exciting one for **DSP** for other reasons as well as the advent of all-digital codec filters. The telecommunications industry was indeed developing means for processing and transmitting signals in a digital format. However, the information signals, speech for the most part, were considered inherently analog in nature, considering that the ingress and egress from the network was via the (analog) subscriber loop. At the same time the need for communicating purely digital bit-streams, typified by computer–computer communications, was growing by leaps and bounds. Unfortunately, from the viewpoint of the end user, the network was analog in nature, again considering the mode of access available was the (analog) subscriber loop. This led to the development and deployment of modems, devices that could convert (modulate) digital bit-streams into "voice-like" signals that could be transported across the "analog" network as though it were a speech signal (and hence the term "voice-band modem"). At the receiving end a companion device would convert the signal (demodulate) back to the form of a bit-stream. Digital signal processing techniques showed promise in the implementation of modems and it is safe to say that modern modems could not be implemented by means other than **DSP**. It is somewhat ironic that **DSP** was employed from the network viewpoint to process "analog" signals and convert them to and from a digital format while, from an end-user viewpoint, **DSP** was utilized to process an inherently digital bit-stream and facilitate the conversion to and from an analog format. The theory and practical aspects of modems are not addressed here; the interested reader is referred to Bingham [3.1].

The 1970s and early 1980s saw the advent of **LSI** (Large Scale Integration) integrated circuits, such as multipliers and adders, that could operate at a high (for that era) clock rate. These could be used to implement digital filters in cases where prior art called for analog filters and in fact usher in entirely new applications with **DSP** being the enabling technology. These components were still quite expensive and their size and power dissipation were such as to render their use impractical unless they could be shared over several channels. Consequently, **DSP** found applications in equipment handling trunks, where the cost of the equipment could be amortized over several channels and where the information traffic volume was high. When the cost of such processing was computed on a per-channel basis, and when the number of calls handled over the life expectancy of the equipment was factored in, such "expensive" devices were indeed very cost-effective. Two such applications are considered in this book. The transmultiplexer, described in Chapter 7, is an example where **DSP** techniques were used to replace "fixed" analog filters. The echo canceler, covered in Chapter 6, is an example where **DSP** is indeed the enabling technology —in order to be effective, the filter in an echo canceler needs to be adaptive, something in which analog filters are lacking.

The decade of the 1980s marked the blossoming of **DSP** as a fundamental technology, an important subject in its own right as opposed to an arcane branch of electrical engineering. There were two distinct drivers for this emergence. Telecommunications manufacturers were increasingly making use of Application Specific Integrated Circuits (**ASIC**s) for purposes of cost reduction. These **ASIC**s were used for a variety of reasons, not restricted to **DSP**. In the case of **DSP** though, because of the large nonrecurring expenses associated with each iteration of the **ASIC** design/fabrication cycle, considerable upfront analysis and simulation was required before the algorithm was cast in stone (or sand !) as it were. In parallel, the semiconductor industry was actively developing general-purpose platforms tailored for implementing **DSP** algorithms. The advantage of such devices was that algorithms could now be implemented as firmware. While a processor-based implementation would be, in general, more expensive than an **ASIC** implementation, the ability to correct "bugs" in a timely and inexpensive manner made **DSP** processors quite attractive as a design option. Further, a **DSP** processor platform provided greater freedom in the type of algorithms that could be implemented. **ASIC**s are well suited for "straight-line" algorithms but are not appropriate for implementing algorithms that called for "branches."

The decade of the 1990s and beyond will, no doubt, see an increasing utilization of **DSP** technology, especially the use of **DSP** processors, because of the freedom available in developing algorithms. The applications to which these designs are addressed would not be restricted to telecommunications. One of the objectives of this book is to provide some examples of how **DSP** has been used, albeit in one field, so that future designers may get a flavor of how the fundamental concepts of **DSP** can be applied.

1.6 REFERENCES AND BIBLIOGRAPHY

For convenience the references and bibliography have been grouped in four categories. The first consists of articles and books that contain material related to the subscriber loop. The second category comprises material that deals with telephony and general telecommunications. Several excellent references are available on the general subject of digital communications and a small subset is provided in the third category. The fourth group of references are standards documents.

1.6.1 Subscriber Loops

[1.1] Lechleider, J. W., "Line codes for digital subscriber lines," *IEEE Comm. Magazine*, Vol. 27, No. 9, Sep. 1989.

[1.2] Special Issue on The Twenty First Century Subscriber Loop, *IEEE Comm. Magazine*, Vol. 29, No. 3, Mar. 1991.

[1.3] Special Issue on Fiber optic Subscriber Loops, *IEEE Comm. Magazine*, Vol. 32, No. 2, Feb. 1994.

[1.4] Special Issue on High-Speed Digital Subscriber Lines, *IEEE Journal on Selected Areas in Comm.*, Vol. 9, No. 6, Aug. 1991.

[1.5] Reeve, W. D., *Subscriber Loop Signaling and Transmission Handbook—Analog*, IEEE Press, New York, 1992.

[1.6] T1E1.4 Working Group on Digital Subscriber Lines, *A Technical Report on High-Bit-Rate Digital Subscriber Lines (HDSL)*, Document T1E1.4/92-002R3, Report No. 28, Committee T1—Telecommunications, Feb. 1994.

1.6.2 Telephony and General Telecommunications

[2.1] Bellamy, J., *Digital Telephony*, John Wiley and Sons, New York, 1982.

[2.2] Briley, B. B., *Introduction to Telephone Switching*, Addison-Wesley Publishing Co., New York, 1983.

[2.3] Chorafas, D. N., *Telephony: Today and Tomorrow*, Prentice-Hall, Inc., Englewood Cliffs, NJ, 1984.

[2.4] McDonald, J. C., (Ed.), *Fundamentals of Digital Switching*, Plenum Press, New York, 1983.

[2.5] AT&T Bell Laboratories, *Transmission Systems for Communications*, Fifth Edition, Bell Telephone Laboratories, Inc., 1982.

[2.6] Bellcore and Bell Operating Companies, *Telecommunications Transmission Engineering*, Third Edition, BELLCORE, New Jersey, 1990 (a set of three volumes).

[2.7] Van Valkenberg, M. E., (Editor-in-Chief), *Reference Data for Engineers: Radio, Electronics, Computer, and Communications*, Eighth Edition, SAMS, Prentice Hall Computer Publishing, Carmel, IN, 1993.

1.6.3 Digital Communications

[3.1] Bingham, J. A. C., *The Theory and Practice of Modem Design*, Wiley Interscience, New York, 1988.

[3.2] Feher, K., Ed., *Advanced Digital Communications. Systems and Signal Processing Techniques*, Prentice-Hall, Inc., Englewood Cliffs, NJ, 1987.

[3.3] Gitlin, R. D., Hayes, J. F., and Weinstein, S. B., *Data Communications Principles*, Plenum Press, New York, 1992.

[3.4] Lee, E. A., and Messerschmitt, D. G., *Digital Communications*, Kluwer Academic Publishers, Higham, MA, 1988.

[3.5] Sklar, B., *Digital Communications*, Prentice-Hall, Inc., Englewood Cliffs, NJ, 1988.

1.6.4 Telecommunications Standards

The **CCITT** (The International Telegraph and Telephone Consultative Committee) operated under the aegis of the International Telecommunication Union (**ITU**) and generated "Recommendations," published every four years. These standards are followed quite closely by network providers and equipment manufacturers world-wide. The "Blue Book," published in 1988, was the last of the series. Since then the **CCITT** has been reorganized and the relevant standards setting body for Telecommunications is the **ITU-T**. Each individual country, or network provider, may in turn have its own set of standards. These are usually a variation of the **CCITT/ ITU-T** Recommendations as applied to the country's own unique situation. **ITU-T** publications can be purchased in their totality from the ITU Sales Service, Place des Nations, CH-1211 Geneva 20, Switzerland (Tel: 41-22-730-6141; Fax: 41-22-730-5194).

In North America the principal standards setting body is the American National Standards Institute (**ANSI**), which has accredited Committee **T1**. Within Committee **T1** there are several subcommittees, referred to as **T1x1** (where **x** is an appropriate letter, e.g., **T1A1**, **T1E1**, **T1X1**, etc.). Each subcommittee concentrates on particular aspects of Telecommunications. In the United States there are other organizations, such as the **EIA**, which generate requirements for their own field of endeavor. Care is taken to collaborate between these various organizations to ensure that the standards generated are not in conflict.

Prior to the creation of Committee **T1**, most standards for the North American Telecommunication Network were "de facto" standards published by the Bell System (AT&T). The origin of most of the standards could be traced to research and development efforts at Bell Laboratories. Following divestiture, the corporate entity AT&T retained Bell Laboratories, the long distance network, and the manufacturing arm, AT&T Technologies (formerly Western Electric Co.). The local exchange networks were segregated into seven Regional Bell Operating Companies. To support these **RBOC**s, a counterpart of Bell Laboratories was created, now called **BELLCORE**, to generate the appropriate standards, among other research and development, for the local exchange networks of the seven **RBOC**s.

An exhaustive list of all the standards is not provided here. Rather, we have concentrated on those standards that are pertinent to the material in the book. Not all the standards listed below are directly referenced in the body of the text.

[4.1] General Aspects of digital transmission systems; terminal equipments, **CCITT** Series G Recommendations.

[4.1.1] Recommendation **G.702**: Digital hierarchy bit rates.

[4.1.2] Recommendation **G.704**: Functional characteristics of interfaces associated with network nodes.

[4.1.3] Recommendation **G.711**: Pulse code modulation (**PCM**) of voice frequencies.

[4.1.4] Recommendation **G.712**: Performance characteristics of **PCM** channels between 4-wire interfaces at voice frequencies.

[4.1.5] Recommendation **G.721**: 32 kbit/s adaptive differential pulse code modulation (**ADPCM**).

[4.1.6] Recommendation **G.726**: 40, 32, 24, 16 kbit/s adaptive differential pulse code modulation (**ADPCM**).

[4.1.7] Recommendation **G.728**: Coding of speech at 16 kbit/s using low delay-code excited linear prediction (**LD-CELP**).

[4.1.8] Recommendation **G.164**: Echo Suppressors.

[4.1.9] Recommendation **G.165**: Echo Cancelers.

[4.2] Telephone Transmission Quality, **CCITT** Series P Recommendations.

[4.2.1] Recommendation **P.56**: Objective measurement of active speech level.

[4.3] Specifications of Measuring Equipment, **CCITT** Series O Recommendations.

[4.3.1] Recommendation **O.41**: Specification of a psophometer for use on telephone-type circuits.

[4.3.2] Recommendation **O.131**: Specification for a quantization distortion measuring apparatus using a pseudo-random noise stimulus.

[4.3.3] Recommendation **O.132**: Specification for a quantization distortion measuring equipment using a sinusoidal test signal.

[4.3.4] Recommendation **O.133**: Specification for equipment to measure the performance of PCM encoders and decoders.

[4.4] Standards Published by the American National Standards Institute.

[4.4.1] Digital Hierarchy—Electrical Interfaces, **ANSI T1.101**—1987.

[4.4.2] Digital Hierarchy—Format Specifications, **ANSI T1.107**—1988.

[4.4.3] Supplement (to [4.4.2]) **ANSI T1.107a**—1990.

[4.4.4] Digital Processing of Voice-Band Signals—32 Kbit/s **ADPCM** Line Format Standard, **ANSI T1.302**—1989.

[4.4.5] Digital Processing of Voice-Band Signals—Algorithm for 24, 32, and 40 Kbit/s **ADPCM**, **ANSI T1.303**—1989

[4.4.6] Compatibility Characteristics of 14/11 Bit Coders/Decoders—15 kHz Audio Signal, **ANSI T1.305**—1990.

[4.4.7] Digital Processing of Audio Signals—Algorithm and Line Format for Transmission of 7 kHz Audio Signals at 64/56 kbit/s, **ANSI T1.306**—1990.

[4.4.8] **DCME**—Interface Functional and Performance Specification, **ANSI T1.309**—1991.

[4.4.9] Digital Processing of Voice-Band Signals—Algorithms for 5-, 4-, 3-, and 2-bit/sample Embedded **ADPCM**, **ANSI T1.310**—1991.

[4.4.10] Interface Between Carriers and Customer Installations—Analog Voicegrade Switched Access Using Loop Reverse Battery Signaling, **ANSI T1.405**—1989.

[4.4.11] Interface Between Carriers and Customer Installations—Analog Voicegrade Special Access Lines Using Customer Installation Provided Loop-Start Supervision, **ANSI T1.407**—1990.

[4.4.12] Network Performance Standards—32 kbit/s **ADPCM** Tandem Encoding Limits, **ANSI T1.501**—1988.

[4.4.13] Advanced Digital Program Audio Services—Analog Interface and Performance Specifications, **ANSI T1.505**—1989.

[4.4.14] Network Performance Parameters for Circuit Switched Digital Services - Definitions and Measurements, **ANSI T1.507** - 1990.

[4.4.15] Integrated Services Digital Network (ISDN)—Basic Access Interface for Use on Metallic Loops for Application on the Network Side of the NT (Layer 1 Specification), **ANSI T1.601**—1992.

[4.5] Transmission Parameters Affecting Voiceband Data Transmission—Measuring Techniques. **BELL PUB. 41009**, *Bell System Data Communications Technical Reference*, May 1975.

[4.6] Digital Channel Bank. Requirements and Objectives. **BELL PUB. 43801**, *Bell System Technical Reference*, **AT&T**, Nov. 1982.

[4.7] Accunet T1.5 Service, Description and Interface Specification,

AT&T Technical Reference, **TR 62411**, **AT&T**, Dec. 1988 (updated, Dec. 1990).

[4.8] Asynchronous Digital Multiplexes. Requirements and Objectives, *Technical Reference* **TR-TSY-000009**, **BELLCORE**, 1986.

[4.9] Notes on the **BOC** Intra-Lata Networks, *Technical Reference* **TR-NPL-000275**, **BELLCORE**, 1986.

[4.10] Voice-Grade Switched Access Service—Transmission Parameter Limits and Interface Combinations, *Technical Reference* **TR-NWT-000334**, **BELLCORE**, 1990.

[4.11] Region Digital Switched Network Transmission Plan. *Science and Technology Series* **ST-NPL-000060**, **BELLCORE**, 1988.

THE MATHEMATICAL FRAMEWORK FOR COMMUNICATION THEORY AND SIGNAL PROCESSING

2

2.1 INTRODUCTION

The purpose of this chapter is to review the mathematical fundamentals that are involved in communication theory and signal processing. At an introductory level the mathematical concepts for the two disciplines are the same, the differences arising more from emphasis than principle. In presenting the material a certain inclination will be observed, a leaning that reflects the fact that all the applications discussed are related to the transmission of information.

Section 2.2 provides an introduction to Signal Theory, the mathematics involved with describing signals, manipulating signals, and expressing measures of signal strength. Manipulation of signals is discussed in Section 2.3. We will assume that all signal processing, unless specified otherwise, will be of the linear, time invariant form and we will use the terms "filter" and "linear time invariant system" interchangeably. In this section we introduce the notion of **convolution,** which is fundamental to the study of signal processing and communication theory.

Signals can be expressed as functions of time, called the ***time domain representation***. A signal could also be represented in other forms, that is, as functions of an independent variable other than time. This is achieved by manipulating the time function in a specific manner into the ***transform domain representation***. For these alternate forms of representation to be useful, there needs to be a one-to-one correspondence with the time domain and consequently we introduce the notion of a ***forward transform***, which converts the time function into a function of the alternate independent variable and the ***inverse transform*** that reverses the process. In Section 2.4 we introduce the notions of transforms and, especially, the **Fourier Transform** and **Fourier Series**. The independent variable in the ***Fourier domain*** has the dimensions of *cycles per second* (the reciprocal of time) and is called ***frequency***. A firm grasp of the concepts of transforms in general, and frequency in particular, is vital to the understanding of more advanced concepts in signal processing and communications. In this section we also discuss the notions of lowpass, highpass, bandpass, and bandstop filters. Related to convolution is the notion of modulation, which is also treated in Section 2.4.

All information-bearing signals are, by their very nature, *unpredictable* or random. By this we mean that the waveforms associated with these signals cannot be described in an *a priori* fashion by equations or graphs. Such signals are called, variously, ***random signals***, ***random processes,*** or ***stochastic processes.*** Study of stochastic processes must necessarily be done in a statistical sense whereby the actual waveform is not known *a priori*, but certain averages or expected values are used to predict what would happen to such a signal as it traversed a communication medium or channel. The foundation for the study of random signals involves **Probability Theory**, which is an entire discipline in and of itself. We provide a brief introduction to this vast subject in Section 2.5, which covers probability, random variables, and stochastic processes to the extent required in this book.

Throughout communication theory there is an underlying precept that the strength of a signal is an important attribute and the efficacy of communication Systems can be expressed in terms of the ratio of the strength of the desired signal to the strength of the unwanted signal (noise). The most common notion of strength is ***power***. In Section 2.6 we cover some of the issues related to the measurement, or estimation, of power. In telecommunications there is the notion of weighting the power measurement to reflect the sensitivity of the human auditory process to different frequencies. Two standards, called ***C-message*** and ***Psophometric*** weighting, have become prevalent and are introduced in that section.

The emergence of digital computers as the most profound, and powerful, influence on modern day civilization has been influenced to a significant degree by the ability to communicate between machines. This form of communication, called **Data Communications**, has developed in parallel with telecommunications. The emergence of digital electronics and Very Large Scale Integrated (VLSI) circuits has also influenced telecommunications, with the predominant method of communication shifting from analog to digital. In Section 2.7 we touch upon some of the considerations involved in transmitting data, that is digital signals, over channels of limited bandwidth and the impact of bandwidth and noise on the capacity of the channel to support digital communications. The intent of this section is not to provide a comprehensive discussion of data communications since there are numerous excellent books available on the subject. There is however, a marked analogy between some of the mathematics associated with data communication and the notions of sampling and quantization. The discussion in Section 2.7 concentrates on these aspects.

If the key points of the chapter are to be summarized, these would be the notions of the delta function, convolution, Fourier transforms and series, and frequency. It will be apparent that these concepts are not limited to "analog signals" or to any single branch of communication theory or signal processing. In Chapter 3 these same concepts are extended to the fields of discrete-time and digital signal processing. Chapters 4 through 9 then expand on the principles developed in Chapters 2 and 3 in a variety of applications.

2.2 BASIC SIGNAL THEORY

Signal theory is that branch of communication theory that involves the study of the representation of signals, the mathematical operations performed on signals, properties of signals, and so on. In this subsection we provide some of the elements of signal theory that will will be used later. This introductory segment also serves the purpose of establishing the notation used in the rest of the book. In particular, we introduce the notion of a signal; a description of some elementary functions; the definitions of peak value, energy, and power to quantify the notion of strength; units used to express the strength of a signal; the concept of reference levels and the unit dBm0 for signal power; some basic operations performed on and certain useful definitions and properties of signals. For a more comprehensive mathematical treatment of signal theory, the reader is referred to Franks [1.2] or Papoulis [1.7].

2.2.1 The Notion of a Signal

From an intuitive standpoint, a signal is an entity that conveys information. Mathematically, a signal is a function of an independent variable, usually time, that depicts the information content by its instantaneous value at any time instant. The term signal is quite generic and can also be used in reference to other time functions of importance. In the context of this book, an information signal refers to a current or a voltage.

The notation employed in this book comprises lowercase letter, sometimes with subscripts, to provide the identity, that is the name, of a signal. For example, x(t) would denote a signal called "x," with the relationship to time as the independent variable duly noted. Signals encountered in practical situations are usually real-valued though, from a theoretical standpoint, it is convenient to consider complex-valued functions as well. A complex-valued signal, as the name implies, requires the notion of a complex number to express the instantaneous value of a signal. If x(t) is a complex-valued signal then it can be expressed as

$$x(t) = a(t) + j\,b(t) \tag{2.2.1}$$

where a(t) and b(t) are real-valued and referred to as the real part and imaginary part of x(t), respectively. We write this as

$$\begin{aligned} a(t) &= \mathrm{Re}\{\, x(t) \,\} \\ b(t) &= \mathrm{Im}\{\, x(t) \,\} \end{aligned} \tag{2.2.2}$$

Nominally, all real-valued signals satisfy

$$- \quad < x(t) < + \tag{2.2.3}$$

In any practical application there is a finite limit to the voltage (or current) excursions of a signal, a limit imposed by the electronic circuits involved. Mathematically, however, it is advantageous to ignore these amplitude restrictions while performing an analysis. Rather, these limits must be factored into the process of interpreting the results.

There are several ways of describing signals. Signals encountered in communications are classified as being ***deterministic*** or ***random.*** In the former category fall the signals such as those used for test purposes and signals used for mathematically describing certain phenomena. In the latter category fall all the information-bearing signals such as speech, digital data, video, and so on. By nature, deterministic signals can be described in a manner so as to specify, directly or indirectly, the value of the signal at any prescribed instant of time. Random signals, on the other hand, can only be described in terms of certain properties that describe not so much the signal itself but the manner or process by which the signal is generated.

The common methods for defining deterministic signals are:

a) The signal, as a function of time, can be described by a formula. This direct, or explicit, specification of a signal allows one to compute the value of the signal at any given instant of time. For example,

$$x(t) = A\exp(j2\ ft + \phi) \tag{2.2.4}$$

For some signals, usually piece-wise constant in nature, the explicit specification of the time dependence can take the form, for example,

$$\begin{aligned} x(t) &= 0;\ t < -T \\ x(t) &= 1;\ -T\ \ t\ \ +T \\ x(t) &= 0;\ t > +T \end{aligned} \tag{2.2.5}$$

b) The signal can be described graphically. For example, consider the signal depicted in Fig. 2.1. Note that even though the functional value is not shown for all time, it is intuitively clear what the value is for time instants not shown. Similarly, for the signal, x(t), illustrated in Fig. 2.2, it is again quite obvious what the functional value is at time instants not shown. Generally speaking, graphical methods for specifying signals are useful only if we can discern a pattern that allows us to extend the graph to arbitrary values of the time axis. This notion of extending a graph is possible either if the signal is periodic or if the signal is of finite duration.

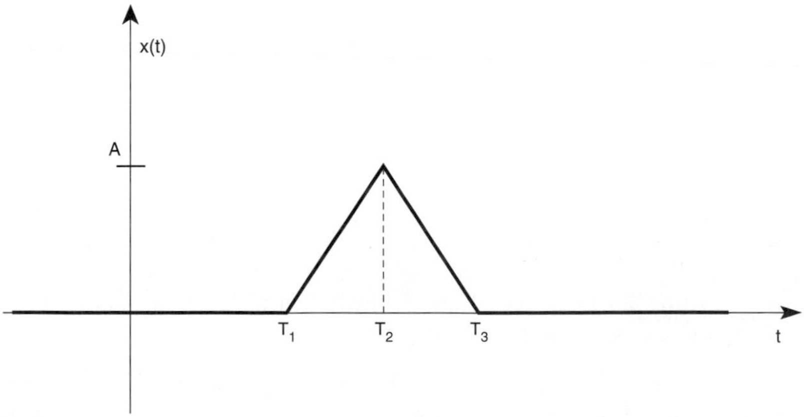

Fig. 2.1
Graphical representation of a finite-duration signal as a function of time (t)

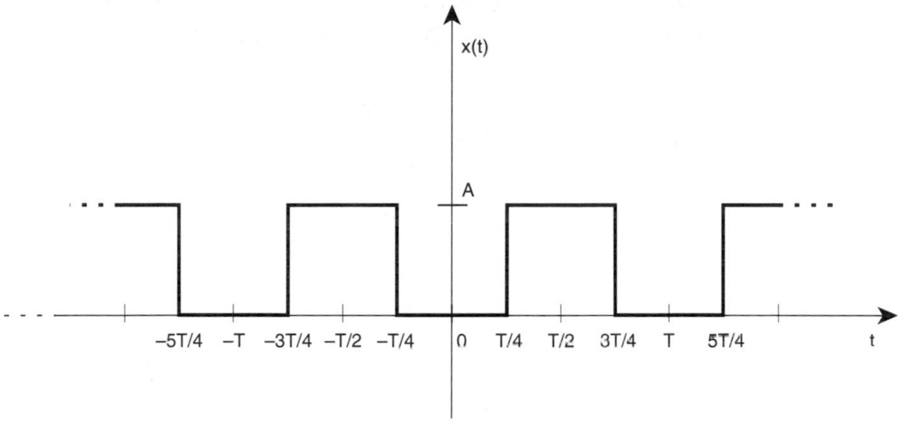

Fig. 2.2
Graphical representation of a periodic signal as a function of time (t)

A signal is said to be of finite duration if there is a finite interval T such that outside this interval the signal value is identically zero. The duration of the signal will then be the least value of T for which this zero-property holds. Clearly, the signal x(t) depicted in Fig. 2.1 is of finite duration and has a duration given by $(T_3 - T_1)$ time units.

A function x(t) is said to be periodic if for some T

$$x(t) = x(t + nT)\,;\ \forall\, n \text{ and } \forall\, t \qquad (2.2.6)$$

where n is an integer. In Eq. (2.2.6) the symbol ∀ is shorthand notation meaning "for all" or "for each and every." The period of the function is defined as the least value of T for which the relationship holds true. The signal x(t) depicted in Fig. 2.2 is clearly periodic with period T. Also, for example, the signal x(t) = cos(t) is periodic with period T=2π. The signal $x(t) = \cos(2\pi f_0 t)$ is periodic with period $T = 1/f_0$.

One of the key implications of periodicity is that knowledge of the values of the signal in the interval [−T/2,T/2] uniquely determines the value for any value of t. That is, the signal need be specified only over an interval of length equal to the period in order to specify the signal for all time. In a sense a periodic signal is a finite duration signal that repeats— the periodic signal is the sum of shifted versions of a finite duration signal for all possible shifts that correspond to an integer multiple of T, the (time) period. For example, if

$$x(t) = \cos(t);\ (T = 2\pi) \qquad (2.2.7)$$

then to evaluate x(t) at t=300 we can write

$$x(300) = x(\,48(2\pi) - 1.59) = x(\,-1.59) \qquad (2.2.8)$$

c) Signals can be described indirectly. For example, the signal y(t) is defined as the solution of a given differential equation in which the driving function, x(t), is specified using either an explicit function or a graph. That is,

$$y(t) + \sum_{k=1}^{N} b_k \frac{d^k y(t)}{dt^k} = \sum_{k=0}^{N} a_k \frac{d^k x(t)}{dt^k} \qquad (2.2.9)$$

Random signals, on the other hand, cannot be described explicitly. The approach taken here is to use a similar convention for naming random signals as for deterministic signals, such as x(t). To reinforce the notion that the signal is random in nature, the signal name is indicated in bold typeface or, if that would cause confusion, with an underscore. The description of the signal graphically could take the form shown in Fig. 2.3, where some care is taken to ensure that signal does look unpredictable. A graphical representation of a random signal does not attempt to specify a definite relationship between time and functional value. Means for specifying the behavior of a random signal utilize probabilistic techniques and are covered in Section 2.5.

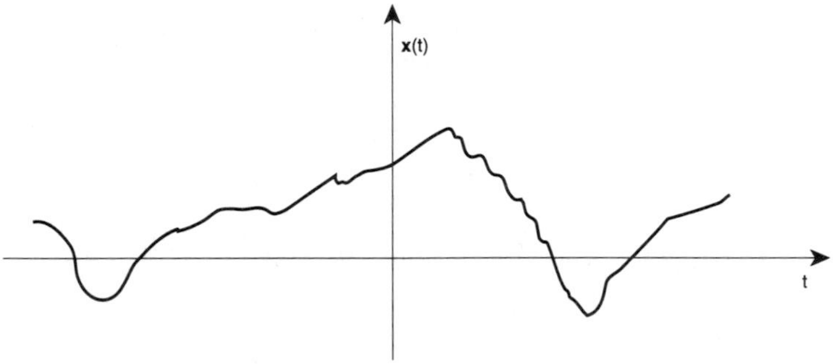

Fig. 2.3

2.2.2 Elementary Signals

There are some deterministic signals that are quite useful and which occur repeatedly in the study of communication theory and signal processing. These are discussed here.

a) The ***unit step*** function, u(t).

The unit step function, referred to by the name u(t), is specified graphically in Fig. 2.4 and mathematically as

$$u(t) = 1;\ t > 0$$
$$u(t) = 0;\ t < 0 \qquad (2.2.10)$$

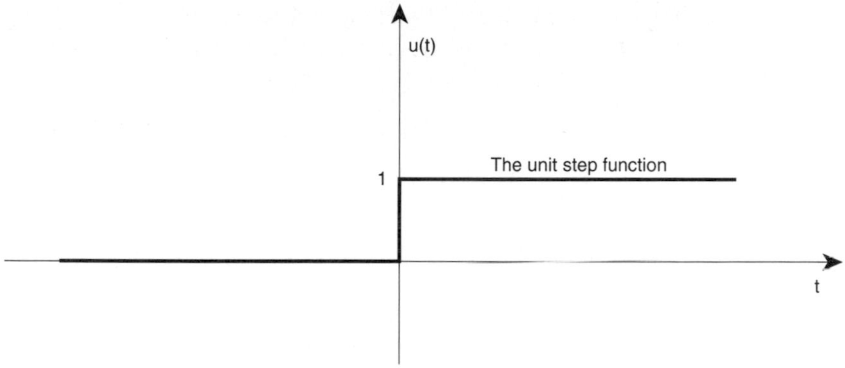

Fig. 2.4
The unit step function

The value of u(t) at t = 0 is intentionally vague. Depending on where the function is used, it is often convenient to specify the value at t = 0 as 1 or 0 or (1/2).

One of the principal uses of the step function is to delineate the start of a signal. For instance, suppose a signal x(t) is zero for $t < 0$. However, for $t > 0$ it matches a signal y(t) that can conveniently be specified for all values of t; then we can specify x(t) as the product of y(t) and the unit step function.

b) The ***unit delta*** function, $\delta(t)$.

There are several names assigned to this function other than "delta." These include "impulse function," "Dirac delta function," and "Sampling function." Strictly speaking it is not a physical entity but rather a concept and described by its defining properties rather than functional value. The mathematical definition appropriate for the delta function is indirect and goes as follows. If x(t) is a function that is continuous at t=0, then

$$\int_{-\varepsilon}^{+\varepsilon} x(t)\,\delta(t)\,dt = x(0) \tag{2.2.11}$$

where $e > 0$. One way to visualize a delta function is as the limiting case of a pulse. Consider the pulse $d_T(t)$ shown in Fig. 2.5. It is a narrow pulse of width T and amplitude such that the area under the curve is unity. That is,

$$\int_{-\infty}^{+\infty} d_T(t)\,dt = 1 \tag{2.2.12}$$

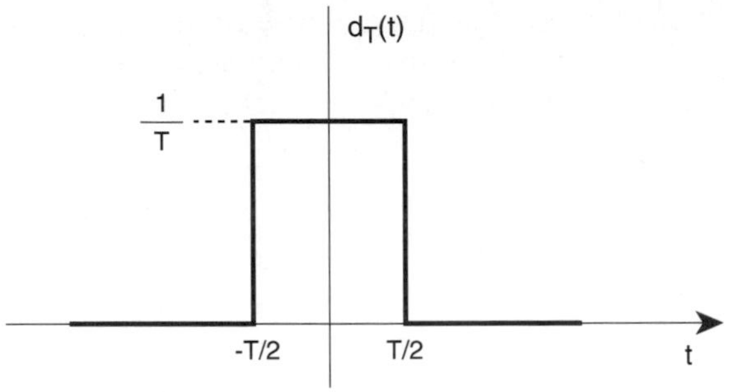

Fig. 2.5
Narrow pulse with unit area

The unit delta function $\delta(t)$ can be obtained as the limiting case of $d_T(t)$ as T tends to 0. That is, $\delta(t)$ is zero everywhere except at $t = 0$ but the area under the curve is unity. At $t = 0$, the functional value is infinite. Graphically, the delta function is represented by an arrow at the instant where it occurs. The numeral below the arrowhead indicates the area under the curve; unity for the unit delta function. This is depicted in Fig. 2.6.

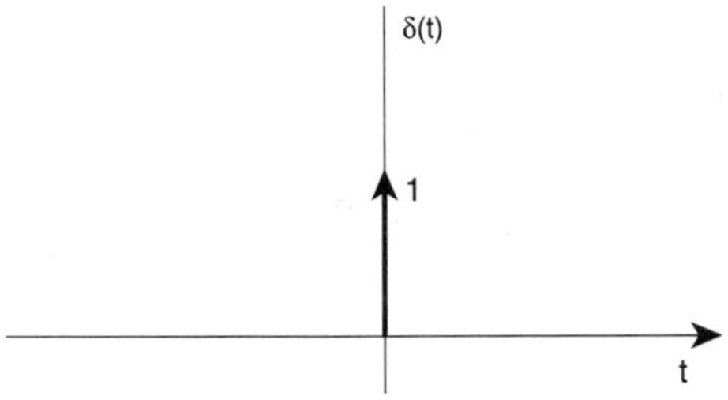

Fig. 2.6
Representation of a delta function

The definition of the delta function as the limiting case of a pulse can be related to the formal definition of Eq. (2.2.11) in the following way. If x(t) is continuous at $t = 0$ then, provided T is small enough,

$$x(t) \approx x(0) \text{ for } -\frac{T}{2} < t < +\frac{T}{2} \tag{2.2.13}$$

Consequently,

$$\int_{-(T/2)}^{+(T/2)} x(t)\, d_T(t)\, dt \approx x(0) \tag{2.2.14}$$

with equality in the limit as T tends to 0.

The need for introducing the concept of a delta function stems from a basic property of integration. The integral operator is fundamental to describing the action of physical systems such as filters. The delta function is the means for providing a "time-instant marker" that will not automatically integrate to zero.

c) The ***sinusoidal*** function.

The sinusoidal function occurs quite frequently and has several forms. The complex form of the sinusoidal function is given by

$$x(t) \;=\; A\,e^{j\Phi(t)} \tag{2.2.15}$$

where the entity A is refered to as the ***amplitude*** and the quantity F(t) as the total phase function. The phase F(t) can be written as

$$\Phi(t) \;=\; 2\,ft \;+\; \phi \tag{2.2.16}$$

where f is the ***frequency*** of the sinusoid and f is the ***initial phase*** (i.e., the phase at the origin, $t=0$). Other terms used to describe the sinusoidal function include "tone," "sinewave," and "complex exponential." When the frequency and amplitude are constant, i.e., independent of time, the term "pure tone" is often used.

In communications and signal processing, the sinusoidal signal plays an important role for at least two reasons. First, it is easy to generate in a reliable, repeatable manner. This makes it eminently suitable as a test signal. Second, when the sinusoidal signal passes through a filter, or channel, the frequency is not altered; if the input to the system is a sinusoid then the output is a sinusoid of the same frequency, albeit with a different amplitude and/or initial phase.

Applying the identity (the Euler formula)

$$e^{j\theta} \;=\; \cos(\theta) \;+\; j\,\sin(\theta) \tag{2.2.17}$$

the complex exponential can be expressed as

$$A\,e^{j(2\pi ft+\phi)} \;=\; A\cos(2\pi ft+\phi) \;+\; j\,A\sin(2\pi ft+\phi) \tag{2.2.18}$$

That is, both the real and imaginary parts of the complex sinusoid are in turn sinusoidal in nature. Real valued tones, which are usually expressed via the sine or cosine functions, can be expressed as either the real-part or imaginary-part of the complex exponential. The waveform of a real-valued sinewave takes the familiar shape shown in Fig. 2.7. The curve shown is for the "sine" function. The curve for the "cosine" function looks pretty much the same except that the origin must be shifted to the position indicated by the dashed line in Fig. 2.7. The cosine function is just a delayed version of the sine function with the delay corresponding to $(\pi/2)$ radians. Without loss of generality, the amplitude A is considered a real-valued, positive, number since the Euler formula can be used to take complex amplitudes expressed in polar notation and merge the "complex" nature of the amplitude with the phase angle.

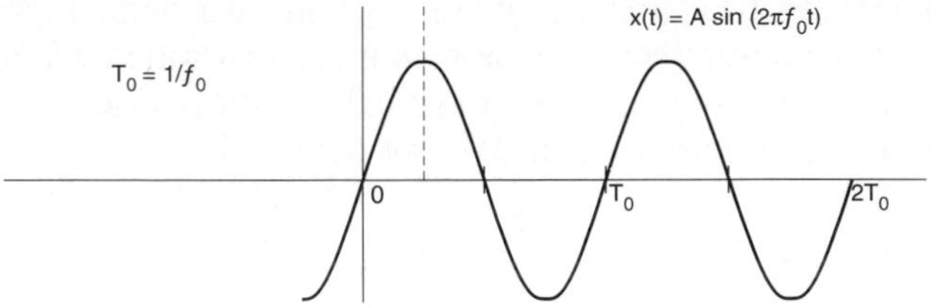

Fig. 2.7
Representation of the sinusoidal function. Dashed line indicates origin for the cosine function

Frequency, as used in the above definitions, has the units of *Hertz*, abbreviated as "Hz" or, in keeping with the decimal system, "kHz," "MHz," "GHz," etc. (for "kiloHertz," "MegaHertz," "GigaHertz," respectively). The choice of unit, Hz, kHz, etc., is for notational convenience. Some books utilize the radian frequency ω ("omega") that is related to frequency (Hz) by the relation

$$\omega = 2\pi f \qquad (2.2.19)$$

The units of ω are rad/s (radians/second) or Krad/s, Mrad/s, etc.

d) The ***exponential*** function.

The exponential function is defined by (with $a > 0$)

$$x(t) = A\,e^{-at}\,u(t) \qquad (2.2.20)$$

and is depicted in normalized form ($A = 1$ and $a = 1$) in Fig. 2.8. The reciprocal of the quantity a has the units of time and is called the "time constant" and is usually designated by the Greek τ (= 1/a). The time constant is the interval over which the amplitude of the signal drops by a factor of e. The exponential signal is associated with the transient response of stable systems and depicts the manner in which transients "die away." Strictly speaking the exponential signal is not of finite duration but it is clear that after about 6 time constants the value of the signal is well nigh zero.

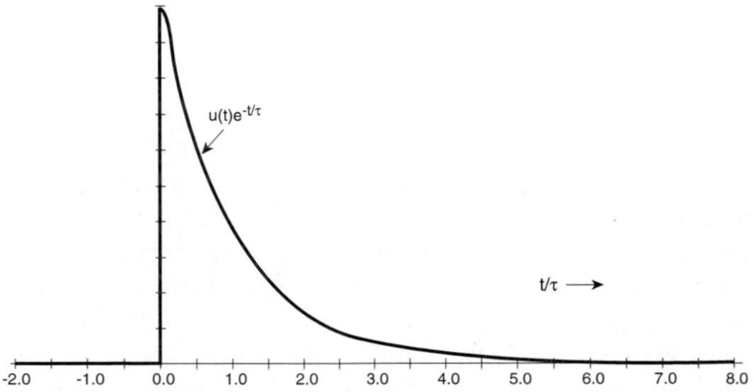

Fig. 2.8
Representation of the exponential function (of time) and the magnitude of its Fourier transform

e) The ***rect*** and ***sinc*** functions.

The rect and sinc functions are depicted in Fig. 2.9(a) and Fig. 2.9(b). The rect function is the prototypical finite duration signal and is often used for just that reason—to demarcate a finite duration. As will be shown in Section 2.4, the two functions form a Fourier transform pair, one a function of frequency and the other a function of time. The rect and sinc functions can be written as

$$\begin{aligned} \text{rect}(t;\, T) &= u(t + \frac{T}{2}) - u(t - \frac{T}{2}) \\ \text{sinc}(f\,T) &= \frac{\sin(\pi f\,T)}{(\pi f\,T)} \end{aligned} \qquad (2.2.21)$$

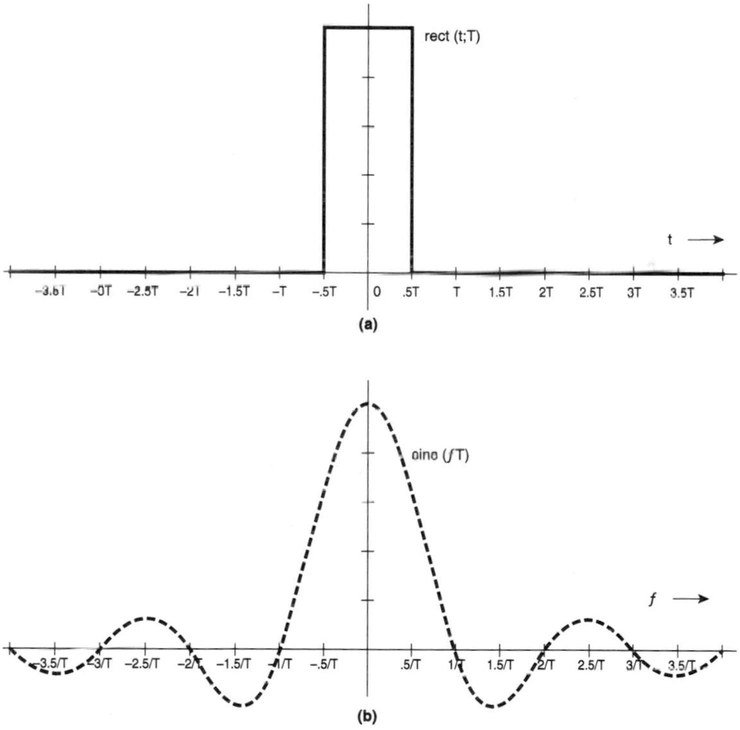

Fig. 2.9 (a)
The rect and sinc function shown as functions of a normalized independent variable.
For convenience the rect function has been shown as a function of time and the sinc function a function of frequency

2.2.3 The Notion of Strength of a Signal

Intuitively, a signal is strong if the voltage (or current) values that the signal takes on are large. There are a variety of ways to describe, quantitatively, the strength of a signal. Some of the common descriptors of strength are discussed in this section.

2.2.3.1 Maximum Value, Energy, and Power

a) Maximum Value M_x.

The maximum, or peak, value M_x of a signal x(t) is defined as

$$M_x = \max\{|x(t)|\} \tag{2.2.22}$$

where the maximum value is taken over the time interval of interest, usually (− ,+). This definition of strength is especially important in circumstances where electronic circuitry can saturate or clip. An amplifier, for example, can provide outputs of limited amplitude. The maximum value of the output signal being related to, among other factors, the power supply voltage. An analog-to-digital converter has an inherent maximum allowable input signal voltage related to the reference voltage used.

For deterministic signals the maximum value is often very easy to obtain by inspection. For example, consider the following elementary signals:

$$\text{Complex Sinusoid:} \quad x(t) = A e^{j(2\pi f t+\phi)} \Rightarrow M_x = |A| \tag{2.2.23}$$

$$\text{Step function:} \quad x(t) = A\,u(t) \Rightarrow M_x = |A| \tag{2.2.24}$$

Sometimes, even for seemingly simple signals, the peak value is not immediately obvious. Consider the case of the sum of two tones given by

$$x(t) = A_1 e^{j(2\pi f_1 t+\phi_1)} + A_2 e^{j(2\pi f_2 t+\phi_2)} \tag{2.2.25}$$

The maximum value will depend on the relationship between the two frequencies ϕ_1 and ϕ_2 as well as the relative phase difference. However, an upper bound for M_x can be found easily as

$$|x(t)| \le M_x \le |A_1| + |A_2| \tag{2.2.26}$$

and in most applications this upper bound ($|A_1|+|A_2|$) is as useful as the actual value of M_x.

In the case of random signals such as speech, the maximum value is not quite so easy to establish. Random signals, can exhibit very large values of instantaneous amplitude, albeit with very low probability. The logical approach to defining the peak value of a random signal is necessarily in terms of probability. If we write

$$M_{\mathbf{x}}(\beta) = A \quad \text{such that} \quad \text{Prob}\{|\mathbf{x}(t)| > A\} < \beta \tag{2.2.27}$$

the maximum value of a random signal **x**(t), parameterized by β, is that value A such that the probability of a signal (magnitude) excursion beyond A is less than β. If β is chosen as a small value, say 0.001, then

$$\text{Prob}\{\,|x(t)| > M_x(0.001)\} < 0.001 \tag{2.2.28}$$

implying that the probability of occurrence of signal values greater than M_x (0.001) is small (assuming that 0.001 is considered small). In practice, the information-bearing random signal may pass through an amplifier, or equivalent, which would go into saturation and perform an amplitude limiting function if the signal value exceeded some limit, say V. In this case, if $M_x(\beta) > V$, then the probability of experiencing saturation is less than β.

b) Energy E_x.

While appropriate for determining whether a system could pass the signal undistorted, i.e., without clipping, the maximum value does not provide a complete description of the strength of a signal. The definition of energy takes into account both the notion of signal duration and amplitude. The energy, E_x, of a signal x(t) is defined as

$$E_x = \int_{-\infty}^{+\infty} |x(t)|^2\, dt \tag{2.2.29}$$

Finite energy signals, i.e., signals for which $E_x <$, cannot be of infinite duration. More correctly, a finite energy signal must asymptotically approach zero as |t| tends to infinity. Conversely, all physical signals of finite duration have finite energy. Consider, however, the exponential signal x(t) defined by (with a > 0)

$$x(t) = A\, e^{-at} u(t) \tag{2.2.30}$$

The energy of x(t) can be evaluated as

$$E_x = \int_{0}^{+\infty} A^2\, e^{-2at} dt = \frac{A^2}{2a} \tag{2.2.31}$$

Clearly, assuming a > 0, x (t) is of finite energy even though, strictly speaking, it is not of finite duration. In most cases of finite energy signals that are not of finite duration, we shall see the asymptotic behavior resembles a decaying exponential and this can be considered "almost finite duration."

c) Power P_x.

Deterministic signals such as the sinusoidal signal or random signals cannot be classified as finite duration. Nor can they be classified as asymptotically zero for large t. Such signals have infinite energy as defined in Eq. (2.2.29). An alternate definition of strength is required. The power of a signal, defined as the average energy over a unit time interval, can be used to express the strength of signals that have infinite energy. Mathematically, the power P_x of a signal is defined as

$$P_x = \lim_{T \to \infty} \frac{1}{T} \int_{-(T/2)}^{+(T/2)} |x(t)|^2 \, dt \tag{2.2.32}$$

Rarely, if ever, is the expression in Eq. (2.2.32) used directly to compute the power of a signal. Other means, which yield the same numerical value as the definition, and that are based on some property of the signal are employed to compute the power. For example, consider the sinewave x(t) depicted in Fig. 2.7. If the time axis was split into equal intervals of duration T_0, the integral of $|x(t)|^2$ over each of the intervals would be the same. Consequently

$$P_x = \frac{1}{T_0} \int_0^{T_0} |x(t)|^2 \, dt \tag{2.2.33}$$

Therefore, for a real-valued sinewave of amplitude A,

$$x(t) = A \sin (2 \, ft + \phi)$$
$$P_x = 0.5 \, A^2 \tag{2.2.34}$$

If $\mathbf{x}(t)$ is a random signal with underlying mean value μ and variance σ^2, then the power P_x of x(t) is given by

$$P_x = \mu^2 + \sigma^2 \tag{2.2.35}$$

Random signals will be discussed in Section 2.5.

2.2.3.2 Units of Power

From basic electrical circuit theory one can recall that power is defined in terms of dissipation in a resistive impedance. In particular, the instantaneous power dissipation in the resistor R is given by

$$P(t) = \frac{v(t)^2}{R} = i(t)^2 R \quad \text{(watts)} \tag{2.2.36}$$

where v(t) is the voltage across the resistor and i(t) is the current flow through the resistor. Assuming that the values of voltage, current, and resistance are specified using volts, amperes, and ohms, respectively, the unit of power in Eq. (2.2.37) is watts.

More common than instantaneous power is the notion of average power, obtained using

$$P = \lim_{T\to\infty} \frac{1}{T}\int_{(T/2)}^{+(T/2)} P(t)\,dt \quad \text{(Watts)} \tag{2.2.37}$$

which is identical in form to Eq. (2.2.32). Note that the value of impedance, ie. R, affects the value of power directly. An equivalent method of establishing the dissipation in R is to define a "root-mean-square" (abbreviated rms) voltage or current from P_x. Depending on whether x(t) was a voltage or current,

$$V = \sqrt{P_x} \ \text{(volts) or } I = \sqrt{P_x} \ \text{(amps)} \tag{2.2.38}$$

from which the average dissipation in the resistor can be expressed as

$$P = \frac{V^2}{R} = I^2 R \quad \text{(watts)} \tag{2.2.39}$$

Both the current i(t) and voltage v(t) could be considered as the "same" signal from the viewpoint of information content since they can be obtained from each other via an application of Ohm's law provided the impedance is known. The power in watts, or the rms voltage (rms current) are thus equivalent methods for describing the strength of a signal provided the load impedance is specified. In signal theory we assume, unless otherwise specified, that the underlying resistance is R = 1 ohm, in which case the average power in Eq. (2.2.37) and the power P_x in Eq. (2.2.32) are the same.

An alternate method for describing the strength of a signal is the notion of "decibel" (abbreviated dB). The decibel unit is not a unit of power but, rather, the unit used to describe the ratio of two powers. If two signals, x(t) and y(t), have powers P_x and P_y (watts) respectively, then

$$D = 10 \log_{10}\left\{\frac{P_x}{P_y}\right\} \ \text{dB} \tag{2.2.40}$$

expresses the power of x(t) relative to y(t) in decibels. The unit dB is thus a representation of ***gain*** rather than power. The gain of amplifiers is usually provided in dB. If an amplifier provides a voltage amplification by a factor of g, then the amplifier gain can be expressed in dB via

$$G = 20 \log_{10}(|g|) \ \text{dB} \tag{2.2.41}$$

where the factor of 20 arises since power is a function of the square of the voltage.

The numerical value of gain, in dB, is the same whether gain refers to voltage or power **provided that the impedance levels are the same**.

Of special interest is when the signal y(t) has a power equal to 1 milliwatt. Application of Eq. (2.2.40) then provides a measure of the power of x(t) and the unit is referred to as "dBm," or "dB with respect to 1 milliwatt." Thus if x(t) has power P_x (watts), the power of x(t) in dBm is given by

$$D_x = 10 \log_{10}(10^3 P_x) \text{ dBm} \tag{2.2.42}$$

When we talk about signals, we normally imply voltage or current signals, whereas signal power is usually expressed in watts (or milliwatts) or dBm. The relationship between power and voltage, say rms voltage, requires the specification of resistance as well. Usually the resistance is implied or known. Otherwise the power of the signal is specified as "dBm into 900 ohms," for example, with the impedance level stated explicitly. Signal powers are typically of the order of milliwatts and dBm is an appropriate unit. The powers of noise, or interfering, signals are usually orders of magnitude smaller. For specifying noise powers another unit of power is defined, "dBrn," or "dB with respect to one picowatt." That is, if x(t) has power P_x (watts), the power of x(t), in dBrn, is given by

$$D_x = 10 \log_{10}(10^{12} P_x) \text{ dBrn} \tag{2.2.43}$$

2.2.3.3 Reference Levels and the Unit dBm0

Because of a variation in impedance level, or because of other considerations related to the electronic implementation of a function, the absolute power level (in watts) of a signal may vary as it traverses through a communication system. Now we would expect that the nominal power level of the signal be unchanged since the system is attempting to transmit the signal without any major alteration. However, the absolute signal level may vary. To separate the impact of implementation on the signal level, we can define a ***reference level*** at each section of the system. The reference level definition takes into account the specifics of the implementation; the power of the signal with respect to the reference can then be the same at each point in the transmission chain. The power of a signal with respect to the reference level is expressed using the unit "dBm0"; the reference level itself is referred to as the ***Transmission Level Power*** or "**TLP**."

The notions of dBm0 and **TLP** are best described via an example. Consider the situation depicted in Fig. 2.10. A signal (voltage) source with Thevenin equivalent resistance R_1 and Thevenin equivalent voltage x(t) drives a load impedance of R_1 (it is common for the source and load impedances to be equal). Suppose

$$\begin{aligned} x(t) &= A \sin(2\pi f_0 t) \\ A &= 2\sqrt{2} \text{ volts} \\ R_1 &= 1000\,\Omega \end{aligned} \tag{2.2.44}$$

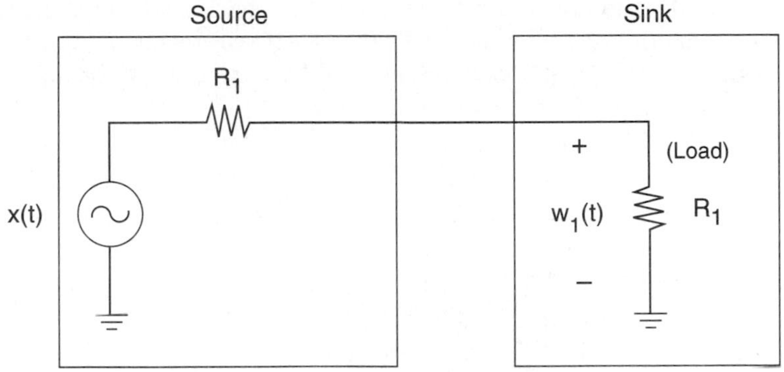

Fig. 2.10
Impedance matched delivery of power from source to sink

The voltage appearing across the load, $w_1(t)$, is given by

$$w_1(t) \;=\; \sqrt{2}\,\sin(2\pi f_0 t) \tag{2.2.45}$$

and the power dissipated in the load is 1 mW or 0 dBm. Now suppose that the load impedance is changed to R_2= 500 ohms. In order to have the source impedance appear to be R_2 ohms we need to introduce an amplifier as shown in Fig. 2.11 (alternatively a transformer could be used). The unity gain amplifier ensures that the voltage signal across the load is the same as before, $w_2(t) = w_1(t)$, but the power dissipated in the load is now 2 mW or 3 dBm (the approximation $\log_{10}(2) = 0.3$ is quite common). From the viewpoint of voltage, the amplifier has unity, or 0 dB, gain; however, from the viewpoint of power, the same amplifier has a gain of 2 , on a linear scale, or 3 dB. This difference is because of the differing impedance levels. Now, since the amplifier is introduced solely to accomodate a change in impedance level, from a system level viewpoint it is useful to consider the gain of the amplifier to be 0 dB (voltage or power).

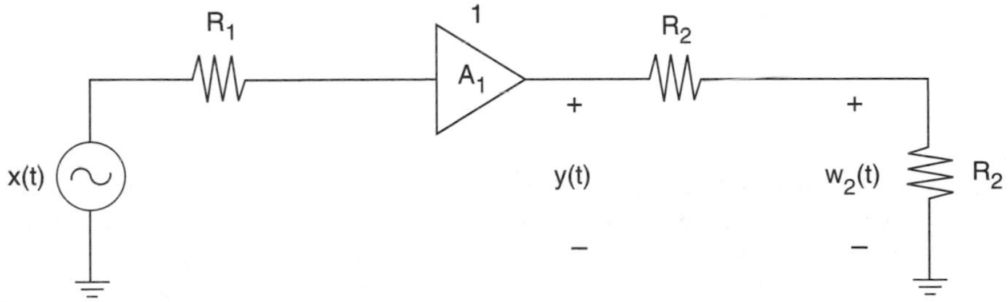

Fig. 2.11
Using an amplifier as a buffer to provide impedance matching

To achieve this end, suppose we define the reference level at the load in Fig. 2.10 as 0 dBm and the reference level at the load in Fig. 2.11 as 2 dBm. The nomenclature used would be 0 **TLP** and +2 **TLP** respectively. With this definition, the power levels at the load in both cases would be 0 dBm0 and, from a system level viewpoint, the gain of the amplifier is 0 dB.

Associated with the reference (power) level, one can define a reference voltage level that is equal to the rms voltage of a sinusoidal signal that would dissipate the reference power in the nominal impedance. With the particular choice of impedance levels in the above examples, both the 0 **TLP** and +2 **TLP** reference voltages would be 1 volt (rms). Such an artifact is necessary since there may be cases, such as analog-to-digital (A/D) converters and digital-to-analog (D/A) converters, where voltage has significance but power, in the absolute sense, does not.

A 0 dBm0 sinusoid is called a ***test-tone***. The frequency of the test-tone is nominally 1 kHz. In voice applications the sampling rate used for converting between analog to digital formats is 8 kHz. A 1-kHz signal thus gives rise to a digital sequence where there are 8 samples per cycle of the 1-kHz waveform. Further, these 8 samples are not all be arbitrary. By assuming a suitable sampling phase we can actually describe a 1 kHz signal completely by 2 values, say x_1 and x_2. The sequence of values comprising the 1 kHz sinewave is $(\ldots\ x_1, x_2, x_2, x_1, -x_1, -x_2, -x_2, -x_1, \ldots)$. Such a signal is very useful for a theoretical discussion. For purposes of testing, the short period (8 samples) makes this signal quite poor and thus other sinewave signals, with frequencies close to 1 kHz, are used. In North American practice the actual frequency used is either 1.01 kHz or 1.004 kHz; European practice is to use 810 Hz.

2.2.4 Mathematical Operations on Signals

Signal processing involves the manipulation of signals to provide either other signals or certain parameters of interest of the signal being manipulated. It is quite common to model (the majority of signal processing) as operations taking two signals, x(t) and y(t) and generating from them a third signal w(t). Of special importance are three mathematical operations: correlation, convolution, and modulation.

The operation of convolution is intimately connected with the notion of a linear time invariant system and is discussed in Section 2.3. To fully appreciate the implications of modulation, we have to first discuss the notion of frequency and Fourier transforms. Thus modulation will be covered in Section 2.4. In this section we cover some of the key points regarding correlation.

Correlation is an operation that is used to quantify how similar two signals are. The correlation between signals x(t) and y(t), also called the cross-correlation, is defined as

$$R_{xy}(\tau) = \int_{-\infty}^{+\infty} x(t)\, y^*(t+\tau)\, dt \tag{2.2.46}$$

and provides an average, i.e., an integral value, of how closely x(t) and an advanced (or delayed) version of y(t) match up to each other. The use of upper case R for correlation is conventional; the signals being correlated are indicated by subscripts. ***Autocorrelation*** is the correlation between a signal and an advanced (or delayed) version of itself, i.e.,

$$R_{xx}(\tau) = \int_{-\infty}^{+\infty} x(t)\, x^*(t+\tau)\, dt \tag{2.2.47}$$

The autocorrelation at a delay of zero, i.e., $\tau = 0$ is equivalent to the energy of the signal:

$$E_x = \int_{-\infty}^{+\infty} |x(t)|^2\, dt = R_{xx}(0) \tag{2.2.48}$$

For the case of real-valued signals the autocorrelation is an even function of τ and can be expressed as:

$$R_{xx}(\tau) = \int_{-\infty}^{+\infty} x(t)\, x(t+\tau)\, dt = \int_{-\infty}^{+\infty} x(t)\, x(t-\tau)\, dt = R_{xx}(-\tau) \tag{2.2.49}$$

To eliminate the effect of signal energy and to emphasize the similarity based on shape, the ***normalized correlation*** or ***correlation coefficient***, is defined as

$$r_{xy}(\tau) = \frac{R_{xy}(\tau)}{\sqrt{E_x E_y}} \quad \text{and} \quad r_{xx}(\tau) = \frac{R_{xx}(\tau)}{R_{xx}(0)} \tag{2.2.50}$$

Autocorrelation, plays an important role in communications because of the following property:

$$|r_{xx}(\tau)| \le r_{xx}(0) \quad \{r_{xx}(0) = 1\} \tag{2.2.51}$$

That is, the autocorrelation is maximum for zero delay or zero "lag." One example of the use of the property in Eq. (2.2.51) is depicted in Fig. 2.12. Suppose a known finite-duration signal, x(t), passes through an unknown delay and the output, say y(t), is observed. The delay can be estimated in the following manner. Suppose we were able to compute the cross-correlation between y(t) and x(t), $r_{xy}(\tau)$. Then, for some value of τ, $r_{xy}(\tau)$ achieves its maximum value and this value would be equal to the unknown delay.

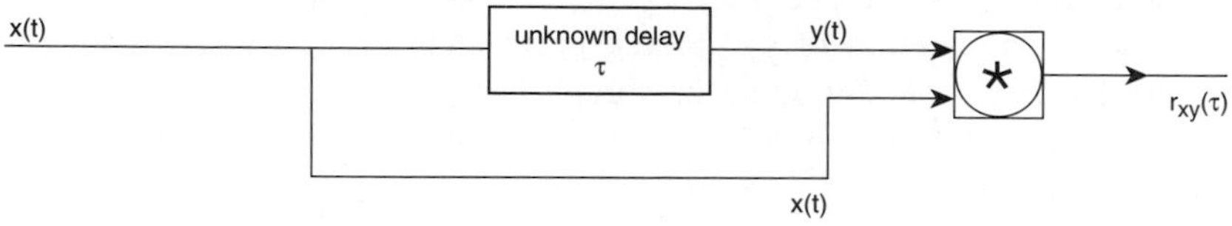

Fig. 2.12
Using autocorrelation to establish delay

Corresponding to Eq. (2.2.51) for autocorrelation, cross-correlation has the following property

$$R_{xy}(0) \leq \sqrt{R_{xx}(0)\, R_{yy}(0)} \tag{2.2.52}$$

Derivation of Eqs. (2.2.51) and (2.2.52) is quite simple and is based on the Schwartz inequality (see Taub and Schilling [1.9], for example).

That the definition of correlation and normalized correlation are appropriate for quantifying the similarity between two signals can be rationalized in the following manner. If x(t) and y(t) are finite energy signals, then we consider them to be "perfectly similar" if $y(t) = \alpha x(t)$ for some constant α. Now assume that $R_{xx}(0) = 1$, that is, x(t) is a unit-energy signal. We can estimate y(t) as a scaled version of x(t) and quantify the "estimation error" as

$$e(t) \;=\; y(t) - \alpha\, x(t) \tag{2.2.53}$$

The choice of α that minimizes the energy of the error can be derived as

$$\alpha \;=\; R_{xy}(0) \tag{2.2.54}$$

and the resulting (minimum) error is

$$E_{min} \;=\; R_{yy}(0)\,[\,1 - r_{xy}(0)\,] \tag{2.2.55}$$

Thus if $r_{xy}(0)$ is approximately 1, the (minimum) error is small, implying that the two signals are indeed similar. Thus the correlation (coefficient) is a good measure of the similarity of the two signals. Conversely, if we look at the correlation between the estimation error e(t) and x(t) we find that, with the optimal choice of α, that $R_{ex}(0) = 0$. That is, the (optimal) prediction error will be uncorrelated with x(t). When the correlation between two signals is zero, they are said to be *orthogonal* to each other. Thus the optimal prediction error is orthogonal to x(t).

To see how the property of Eq. (2.2.52) can be used, consider the example depicted in Fig. 2.13. One of two possible signals, $x_1(t)$ or $x_2(t)$ is transmitted. It is assumed that the two signals have the same energy. The receiver observes y(t) and must decide whether $x_1(t)$ or $x_2(t)$ was transmitted. This can be accomplished by computing the cross-correlation between the received signal y(t) and replicas of $x_1(t)$ and $x_2(t)$. The decision as to which was transmitted is based on which signal $x_1(t)$, or $x_2(t)$, has a higher correlation with y(t). This decision is very robust if $x_1(t)$ and $x_2(t)$ are orthogonal to each other. If the two signals are antipodal, i.e., $x_2(t) = -x_1(t)$, then only one correlation operation is required and the choice of $x_1(t)$ or $x_2(t)$ is based on the sign of the correlation Proof of Eq. (2.2.52) is based on the result

$$\left[\int_{-\infty}^{+\infty} |a(t)\, b(t)|\, dt\right]^2 \le \left[\int_{-\infty}^{+\infty} |a(t)|^2\, dt\right]\left[\int_{-\infty}^{+\infty} |b(t)|^2\, dt\right] \tag{2.2.56}$$

and the observation that

$$\int_{-\infty}^{+\infty} |x(t)|^2\, dt = R_{xx}(0) = \int_{-\infty}^{+\infty} |x(t+\tau)|^2\, dt \tag{2.2.57}$$

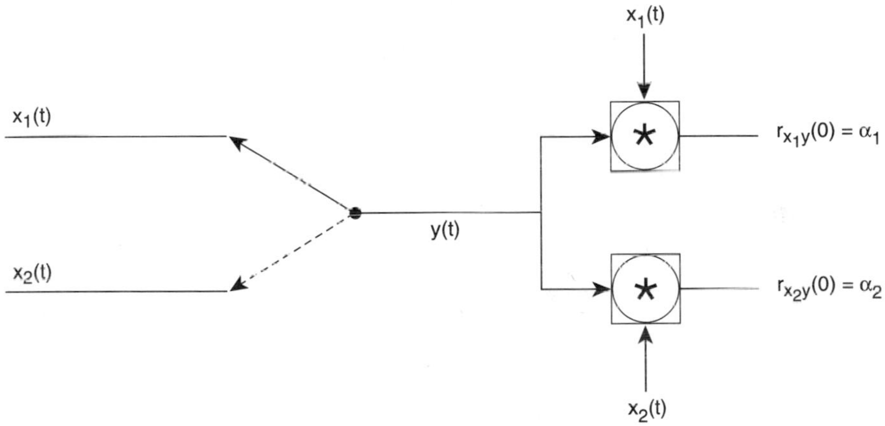

Fig. 2.13
Using correlation to decide which signal was transmitted

Defining autocorrelation by Eq. (2.2.47) is appropriate if $R_{xx}(0)$ is finite. Periodic signals do not fit this mold since such signals have either infinite energy or zero energy, where zero energy corresponds to a trivial case. The periodic signals of interest to us have finite power with one exception being the sampling function $\Theta(t)$ that is defined in the next section. For periodic signals with period T_0, an appropriate definition of autocorrelation is

$$R_{xx}(\tau) = \frac{1}{T_0}\int_0^{T_0} x(t)\, x^*(t+\tau)\, dt \tag{2.2.58}$$

For such signals $R_{xx}(0)$ corresponds to the power, P_x, rather than energy, but the autocorrelation defined by Eq. (2.2.58) has similar uses and properties as the autocorrelation of a finite energy signal.

2.3 LINEAR TIME-INVARIANT SYSTEMS

In most cases when a signal gets modified, for example in the process of transmission over a channel, we model the phenomenon as depicted in Fig. 2.14. The signal y(t), the modified signal, is visualized as the output of a system with x(t), the original signal, as input. This system will be assumed to be ***linear and time invariant*** (**LTI**). The nomenclature used is

$$y(t) = H\{x(t)\} \tag{2.3.1}$$

which indicates that the input has been altered by the linear system denoted by H.

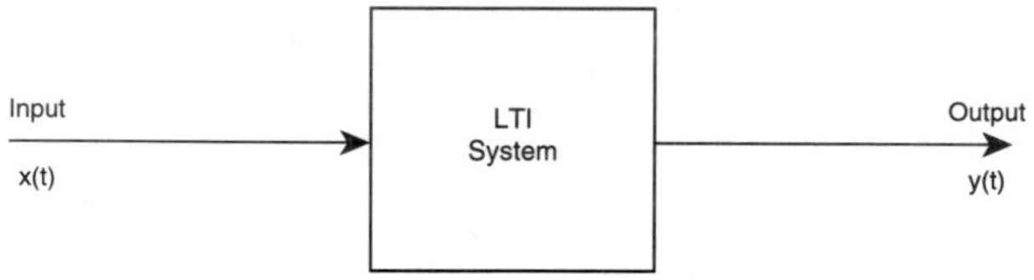

Fig. 2.14
Conceptual model for signal modification

2.3.1 Definition of Linearity and Time Invariance

By definition, a linear system satisfies a multiplicative and an additive property. Taken together, linearity implies that

$$\text{If } H\{x_i(t)\} = y_i(t) \text{ And If } x(t) = \underset{i}{\;}\alpha_i x_i(t)$$

$$\text{Then } H\{x(t)\} = y(t) = \underset{i}{\;}\alpha_i y_i(t) \tag{2.3.2}$$

where the $\{\alpha_i\}$ are constants.

The notion of time invariance is that the choice of time origin does not affect the input-output relationship of the system. That is, if the input is delayed (or advanced) by a certain amount then the output is delayed (or advanced) by the same amount. Mathematically,

$$\begin{aligned} \text{If} \quad & H\{x(t)\} = y(t) \\ \text{and if} \quad & w(t) = x(t - \tau) \\ \text{Then} \quad & v(t) = H\{w(t)\} = y(t - \tau) \end{aligned} \tag{2.3.3}$$

2.3.2 Fundamental Properties of LTI Systems

There are three fundamental consequences of linearity and time invariance that are stated below.

i) An **LTI** system can be described by a linear differential equation:

$$\sum_{i=0}^{N} \beta_i \frac{d^i y(t)}{dt^i} = \sum_{i=0}^{M} \alpha_i \frac{d^i x(t)}{dt^i} \tag{2.3.4}$$

where α_i and β_i are constants. Usually N is greater than M. The order of the system is the larger of N and M. Conversely, any system whose input-output relation can be described by a linear differential equation is linear and time invariant.

ii) If the input to an **LTI** is a sinusoidal function of underlying frequency f_0, then the steady state output will also be a sinusoid of the same frequency. Referring to the input and output as x(t) and y(t) respectively,

$$\begin{aligned} &\text{If} \quad x(t) = A \exp(j2\pi f_0 t) \\ &\text{Then} \quad y(t) = A\, H(f_0) \exp(j2\pi f_0 t) \end{aligned} \tag{2.3.5}$$

The **LTI** thus alters the amplitude of the input complex exponential by a factor $H(f)$ which is a complex-valued function of f, the frequency of the complex exponential.

This function is called the ***frequency response*** of the system. By allowing $H(f)$ to be complex valued, a simple, compact, description of the modification of (real) amplitude and phase angle is obtained. $H(f)$ can be written in the following polar notation as

$$H(f) = A(f) \exp(-j\beta(f)) \tag{2.3.6}$$

where $A(f)$ and $\beta(f)$ are real-valued functions of f. $A(f)$ and $\beta(f)$ are called the amplitude (or magnitude) and phase response, respectively. The "group delay" of the filter is another way of describing the phase response. The group delay is defined as the derivative, with respect to frequency, of the phase response, $\beta(f)$, scaled by a factor of 2π. The representation of frequency response in this fashion is appropriate when the signals involved are real-valued. For example,

$$\begin{aligned} &\text{If} \quad x(t) = C \cos(2\pi f t + \phi) \qquad \text{(real-valued)} \\ &\text{Then} \quad y(t) = C\, A(f) \cos(2\pi f t + \phi - \beta(f)) \end{aligned} \tag{2.3.7a}$$

If the phase response function is a linear function of frequency, $\beta(f) = 2\pi f\tau$, then Eq. (2.3.7a) can be written as

$$y(t) = C\, A(f) \cos[2\pi f(t-\tau) + \phi\,] \tag{2.3.7b}$$

The derivative of $\beta(f)$, scaled by a factor of 2π, is equal to the delay τ, and hence the name "group delay" for this derivative.

iii) The response of a system to a delta function is termed ***impulse response*** and given the name h(t). That is,

If $\quad x(t) = \delta(t) \quad$ (input = ***unit impulse***)

Then $\quad y(t) = h(t) \quad$ (output = ***impulse response***) $\quad$ (2.3.8)

Even though the delta function is not a physical signal, the impulse response, h(t), is representative of an observable response, usually considered a transient phenomenon. The utility of the impulse response lies in the fact that the response of a system to an input x(t) can be computed using the impulse response via the following ***convolution integral***

$$y(t) = \int_{-\infty}^{+\infty} x(s)\,h(t-s)\,dt = \int_{-\infty}^{+\infty} h(s)\,x(t-s)\,dt \tag{2.3.9}$$

where s is a dummy variable of integration. The form of the convolution integral in Eq.(2.3.9) is a direct consequence of linearity and time invariance. Since $h(\tau)$ is the response (output) at time τ as a consequence of an impulse at $\tau=0$, it can be viewed as the contribution to the output at time t from an impulse at time $(t-\tau)$, τ units of time in the past. That is, a unit impulse at $(t-\tau)$ will contribute $h(\tau)$ to the output at time t. Since a given (input) signal can be constructed as a linear combination (actually an integral) of delta functions as

$$x(t) = \int_{-\infty}^{+\infty} x(\tau)\,\delta(t-\tau)\,d\tau = \int_{-\infty}^{+\infty} \delta(\tau)\,x(t-\tau)\,d\tau \tag{2.3.10}$$

it follows that

$$y(t) = \int_{-\infty}^{+\infty} H\{\delta(\tau)\}\,x(t-\tau)\,d\tau = \int_{-\infty}^{+\infty} h(\tau)\,x(t-\tau)\,d\tau \tag{2.3.11}$$

Convolution is a very basic mathematical operation that is fundamental to the study of linear time invariant systems. The symbol we will use for convolution is (*), i.e.

$$w(t) = x(t)\ (*)\ y(t) \tag{2.3.12}$$

It is possible, in principle, to derive the underlying differential equation of an **LTI** from its frequency response and vice versa. Furthermore, as will be shown later in this chapter, the frequency response can be derived from the impulse response, and vice versa. That is, the frequency response, impulse response, and differential equation are equivalent descriptions of the system.

2.3.3 Computing the Response of an LTI

Most complicated (and possibly complex) signals can be broken down or decomposed into simpler *unit* signals. It is usually easier to compute the response of a system to each of these unit, or simple, signals. Consequently, if the response is known for each of the constituent simpler signals, the overall response can be determined by invoking the properties of linearity and time invariance. For example, consider the circuit in Fig. 2.15. The input-output relation is given by the differential equation

$$RC\frac{dy(t)}{dt} + y(t) = x(t) \tag{2.3.13}$$

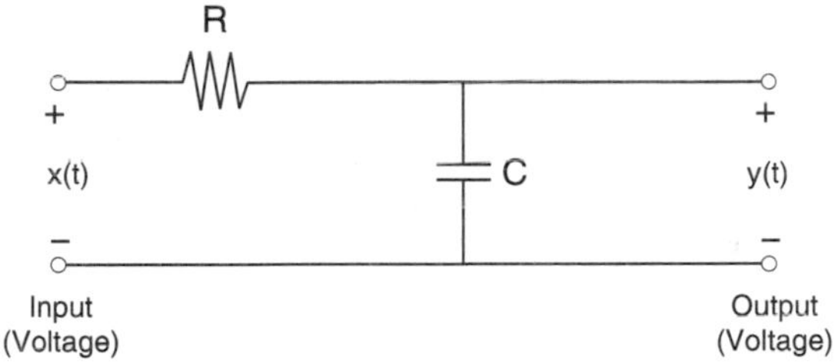

Fig. 2.15
A simple first-order linear time-invariant system

Solution of this differential equation is relatively straightforward when the input, x(t), is a unit step.

$$\begin{aligned} x(t) &= u(t) \\ y(t) &= (1 - \exp(-(t/RC))\, u(t) \end{aligned} \tag{2.3.14}$$

Now suppose that the input, x(t), was the pulse depicted in Fig. 2.16. Direct solution of the differential equation is cumbersome. However, x(t) can be decomposed into *simpler* components as

$$x(t) = A\, u(t-T) - A\, u(t-2T) \tag{2.3.15}$$

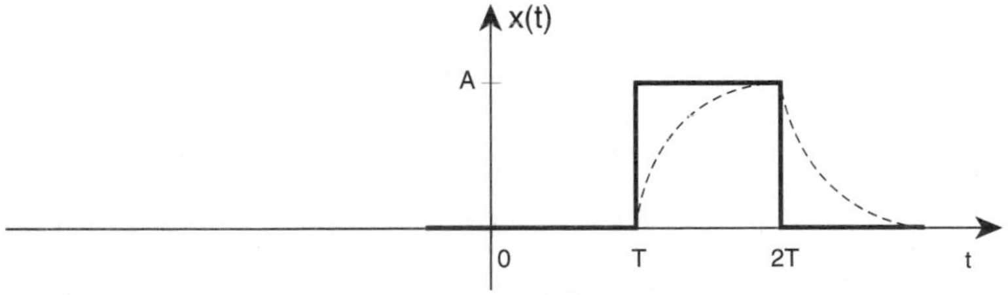

Fig. 2.16
Input and response of the first-order system of Fig. 2.15

Applying the notions of linearity and time invariance, the response y(t) can be derived from Eq. (2.3.14) as

$$y(t) = A[1-\exp(-(t-T)/RC)]u(t-T) - A[1-\exp(-(t-2T)/RC)]u(t-2T) \quad (2.3.16)$$

which is shown in Fig. 2.16 as a dotted line for the case where RC   T/3 .

A common representation of signals is via the Fourier Series, covered later in the chapter. The Fourier series representation of a periodic signal, x(t), with period T_0, is

$$x(t) = \sum_{k=-\infty}^{\infty} c_k e^{j2\pi k f_0 t} \quad \text{where } f_0 = \frac{1}{T_0} \quad (2.3.17)$$

The Fourier series expresses a signal as a linear combination of sinusoidal components.

The response of the system, with x(t) as input, is

$$y(t) = \sum_{k=-\infty}^{\infty} c_k H(kf_0) e^{j2\pi k f_0 t} \quad (2.3.18)$$

where H(f) is the frequency response of the system.

In some cases periodic signals are composed of simple waveforms replicated in time. For example, x(t) depicted in Fig. 2.17 can be written as

$$x(t) = \sum_k p(t-kT) \quad (2.3.19)$$

where p(t) is the pulse waveform shown in Fig. 2.18. The response of an **LTI** to the stimulus x(t) can be computed in many ways. One way is to compute the Fourier series representation and write x(t) in the form of Eq. (2.3.17) and compute the output using the frequency response as shown in Eq. (2.3.18). Alternatively, if the response of the system to the stimulus p(t) is known, say q(t), i.e.,

$$q(t) = H\{ p(t)\} \quad (2.3.20)$$

then the system output with x(t) as input will be given by

$$y(t) = \sum_k q(t-kT) \quad (2.3.21)$$

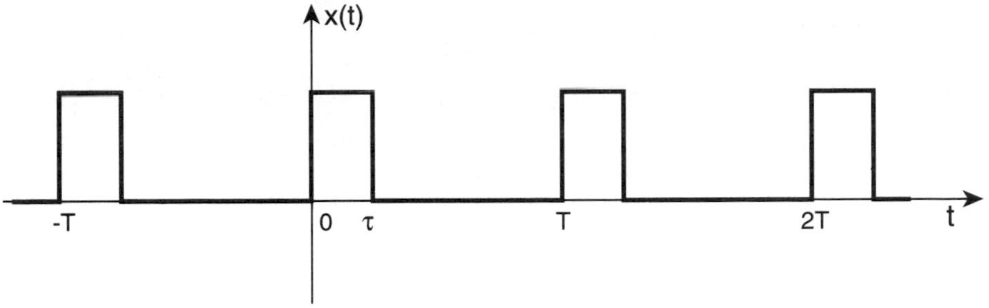

Fig. 2.17
A periodic signal with period T

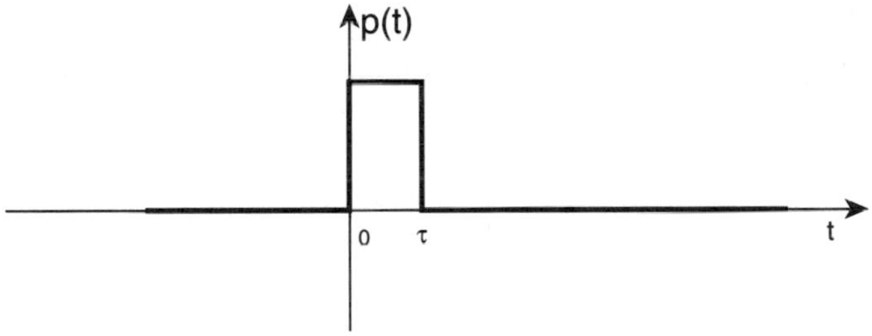

Fig. 2.18
Time-limited function that generates the periodic signal in Fig. 2.17

As a corollary of the example cited, it is possible to represent signals as the *output* of an **LTI** excited by a known input. For example, the signal x(t) in Fig. 2.17 can considered as the output of a linear system excited by a particular (periodic) input. In particular, an **LTI** with impulse response equal to p(t), with excitation Θ(t) depicted in Fig. 2.19, will have

$$\begin{aligned} x(t) &= H\{\Theta(t)\} = H\{\sum_k \delta(t-kT)\} \\ &= \sum_k H\{\delta(t-kT)\} = \sum_k p(t-kT) \end{aligned} \tag{2.3.22}$$

The representation of x(t) as the sum of shifted versions of a base signal p(t) is quite intuitive when p(t) is a finite duration signal of time width equal to (or less than) the period of x(t). What if p(t) is not time limited? Can an equivalent expression be derived for x(t) as in Eq. (2.3.22) where there is no *a priori* restriction on the base signal?

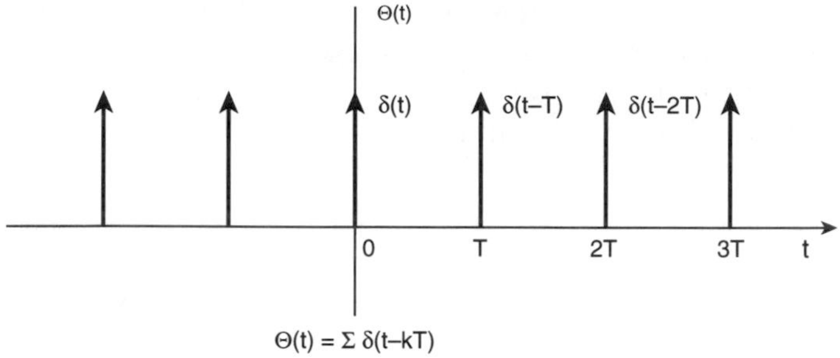

Fig. 2.19
The special function Θ(t)—a periodic train of unit delta functions

To answer this question consider a linear time-invariant system with the impulse response h(t) depicted in Fig. 2.20, which corresponds to the exponential decay of a first order filter. The impulse response h(t) is not finite duration but does die out exponentially. Now consider the function g(t) which is finite duration, of duration T, shown in Fig. 2.21 along with h(t). g(t) is created from h(t) by breaking up h(t) in time slots of length T units, shifting these by an integer multiple of T so as to be within the window [0,T], and summing these segments. Mathematically,

$$g(t) = \left(u(t) - u(t-T)\right) \sum_{k=-\infty}^{\infty} h(t-kT) \tag{2.3.23}$$

We can create a periodic signal, x(t), by taking g(t) and all possible delayed versions of g(t) where the delay is restricted to an integer multiple of T. The resultant x(t) is shown in Fig. 2.22 along with the original h(t). Now if the linear time invariant system is excited by the driving function Θ(t), the output is given by

$$y(t) = \sum_{k=-\infty}^{\infty} h(t-kT) \tag{2.3.24}$$

Examination of Eq. (2.3.23) leads us to the result

$$\sum_{k=-\infty}^{\infty} g(t-kT) = \sum_{k=-\infty}^{\infty} h(t-kT) \tag{2.3.25}$$

which means that the output of the **LTI**, whose impulse response is not quite time limited, excited by Θ(t) gives the same output signal as an **LTI** whose impulse response is time limited, specifically g(t). Thus, if we wish to represent a periodic signal as the output of an **LTI** excited by Θ(t), there is no unique choice of impulse response. If however, we insist that the impulse response be of finite duration, then there is a unique choice, namely g(t), the signal x(t) restricted to one time period (we could get other choices for g(t) if we did not fix the time origin). This "equivalence" of h(t) and g(t), where equivalence is in the sense that **LTIs** with h(t) and g(t) as impulse responses give the same output when excited by Θ(t), is similar to the situation that arises when we sample signals to create discrete-time signals (sequences). This similarity will become evident in our discussion of the sampling theorem in Chapter 3.

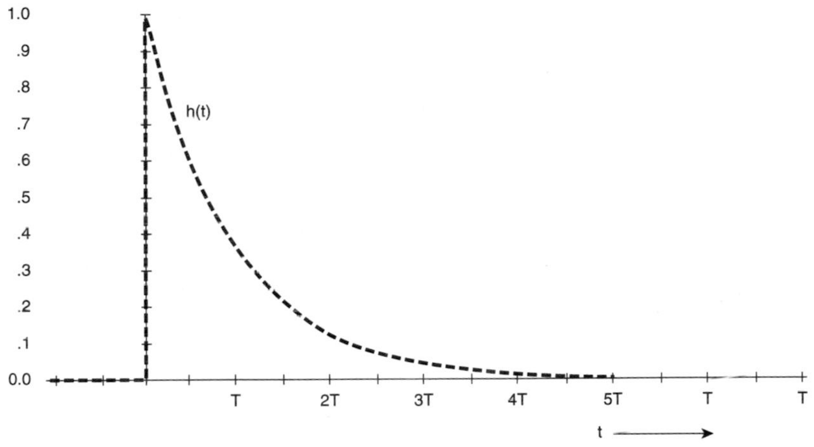

Fig. 2.20
Impulse response h(t) of a first-order filter

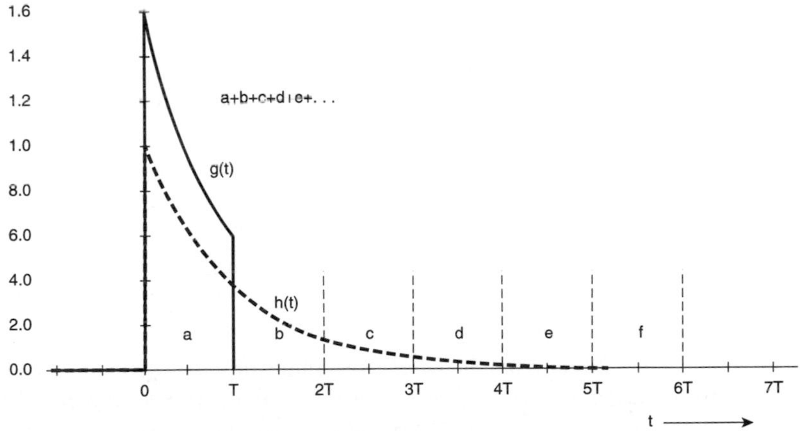

Fig. 2.21
g(t) is a finite duration signal constructed by summing h(t–kT) for all integer values of k

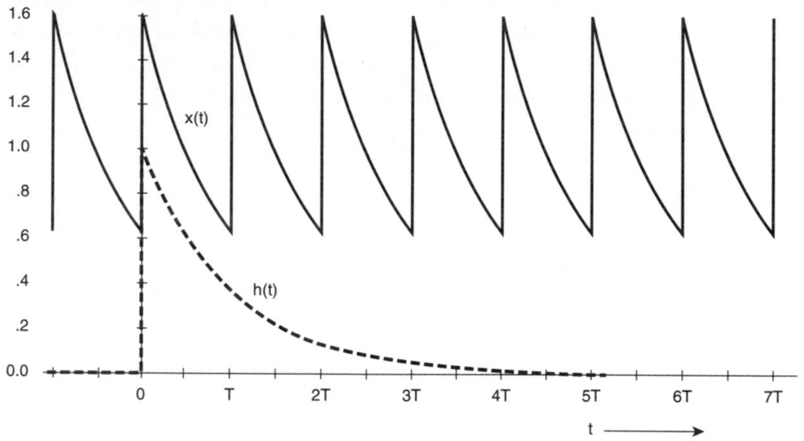

Fig. 2.22
Periodic signal x(t) created from g(t) of Fig. 2.21 by considering shifts of g(t), namely g(t–kT) for all integer k

Further implications of linearity and time invariance will become evident in the following sections and chapters. Such properties will be discussed there as appropriate. The study of linear time invariant systems is a subject in its own right and crosses many boundaries, from circuit theory to signal processing, from communication theory to control systems. The interested reader is referred to Kailath [1.3], for example, which provides an in-depth treatment of linear system theory.

2.4 FOURIER SERIES AND FOURIER TRANSFORMS—FREQUENCY DOMAIN REPRESENTATION OF SIGNALS AND SYSTEMS

In the previous section it was shown that representing a signal as the linear combination of elementary signals often helps in the computation of the response of an **LTI** if the response to each of the elementary signals is known or easy to establish. Some alternate representations are called transforms, or series. The basic idea underlying such alternate representations is to set up a one-to-one correspondence between the signal, or time function, and the transform by altering the independent variable or domain. A complex (i.e., complicated) operation in one domain usually translates to a simple operation in the other, or transform, domain. Transforms are an analytical tool that can be used to simplify the solution of problems arising in several branches of mathematics and electrical engineering.

Dealing with signals solely as time functions is reasonable if the time functions are simple, such as those of finite duration or those that are periodic. However, if we modify the signal we get a different waveshape and it is not easy to decide whether the alteration is damaging, benign, or even an improvement. Examples of benign modifications include the action of an (ideal) amplifier or a simple delay. Whereas a gain or a delay does not alter the signal shape in a drastic manner, removal of signal components might involve a drastic change in the waveform and it may not at all be obvious from the change in shape whether we indeed "improved" the signal. For example, we can improve a signal by removing interference, such as power-line hum. Furthermore, when we are dealing with information-bearing signals such as speech, which are random in nature, comparison of waveforms to ascertain whether the processing constituted an improvement or degradation is not very helpful. Developing a frequency domain representation of signals is extremely useful in *a priori* analysis of the impact of signal processing.

2.4.1 Fourier Series Representation of Periodic Signals

The Fourier series is an essential tool for handling periodic functions and provides a method for expressing a periodic function as the linear combination of sinusoidal functions. This is especially useful in the light of **LTI** systems since we know that the response of an **LTI** system to a sinusoid is a sinusoid of the same frequency.

2.4.1.1 Definition of the Fourier Series

If x(t) is periodic with period T_0, then x(t) can be expressed as a series

$$x(t) = \sum_{k=-\infty}^{\infty} c_k e^{j2\pi k f_0 t} \quad \text{where } f_0 = \frac{1}{T_0} \tag{2.4.1}$$

Eq. (2.4.1) describes an alternate representation of x(t) as the sequence $\{c_k\}$. That is, knowledge of $\{c_k\}$ is equivalent to knowledge of x(t); there is a one-to-one correspondence between the series and the time function. Further, the sequence $\{c_k\}$ can be computed from knowledge of x(t) over one time period. Strictly speaking, the equality in Eq. (2.4.1) does not necessarily hold true for all values of time, t. The correct descriptor is "equal almost everywhere." By this we mean that there may be some isolated instants of time, particularly where x(t) is discontinuous, where equality may not hold but the difference between the two sides of Eq. (2.4.1) is a signal of zero power. That is,

$$\int_0^{T_0} \left| x(t) - x'(t) \right|^2 dt = 0 \quad \text{where } x'(t) = \sum_{k=-\infty}^{\infty} c_k e^{j2\pi k f_0 t} \tag{2.4.2}$$

For all practical purposes, and throughout this book, we do not make a distinction between "equal" and "equal almost everywhere."

2.4.1.2 Calculating the Fourier Coefficients

The coefficients c_k can be obtained from x(t) using the following relation:

$$c_k = \frac{1}{T_0}\int_0^{T_0} x(t)\, e^{j2\pi k f_0 t}\, dt \tag{2.4.3}$$

The derivation of Eq. (2.4.3) is simple and is based on an important property of sinusoidal signals. Sinusoidal signals of differing frequencies have zero correlation if computed over a time interval that contains an integral number of periods of each of the two signals. Specifically, the time interval must be an integral multiple of the least common multiple (**LCM**) of the two associated time periods. Thus if we multiply both sides of Eq. (2.4.1) by $\exp(-j2m\pi f_0 t)$ and integrate over an interval T_0, the time period of x(t), which is an integer multiple of the time period of all sinusoids of frequency nf_0, we get

$$\int_0^{T_0} x(t)\, e^{-j2\pi m f_0 t}\, dt = \sum_{k=-\infty}^{\infty} c_k \int_0^{T_0} e^{j2\pi(k-m) f_0 t}\, dt \tag{2.4.4}$$

The zero-correlation property mentioned above can be expressed by the following relation

$$\int_0^{T_0} e^{j2\pi(k-m) f_0 t}\, dt = \begin{cases} 0 \text{ if } m \neq k \\ T_0 \text{ if } m = k \end{cases} \tag{2.4.5}$$

Eq. (2.4.3) is a direct consequence of Eqs. (2.4.4) and (2.4.5).

The complex form of the Fourier series in Eq. (2.4.1) is primarily for convenience. If x(t) is real-valued, then the Fourier series can be written as

$$x(t) = \frac{a_0}{2} + \sum_{k=1}^{\infty} \{a_k \cos(2\pi k f_0 t) + b_k \sin(2\pi k f_0 t)\} \tag{2.4.6}$$

Eq. (2.4.6) is the format used in books on Engineering Mathematics. The coefficients $\{a_k\}$ and $\{b_k\}$ are given by

$$a_k = \frac{2}{T_0}\int_0^{T_0} x(t)\cos(2\pi k f_0 t)\, dt \tag{2.4.7a}$$

$$b_k = \frac{2}{T_0}\int_0^{T_0} x(t)\sin(2\pi k f_0 t)\, dt \tag{2.4.7b}$$

The derivation of the relations in Eq. (2.4.7) is quite similiar to the derivation of Eq. (2.4.3).

If x(t) has any symmetry, either odd or even, then Eq. (2.4.6) can be simplified. In particular,

$$\begin{aligned} &\text{If } x(t) \text{ is even:} \quad b_k = 0 \\ &\text{If } x(t) \text{ is odd:} \quad a_k = 0 \end{aligned} \tag{2.4.8}$$

The Fourier series provides a description of a periodic signal in terms of sinusoids. Each such component will have a frequency of the form kf_0. The term with k=0, or zero frequency will be referred to as the DC component; the term with k=1 is called the fundamental and the frequency f_0 is referred to as the fundamental frequency; for other values of k, the components are called harmonics.

The complex form of the Fourier series is usually more compact. For real-valued signals the complex form exhibits the following relationhip:

$$c_k = c_{-k}^{*} \tag{2.4.9}$$

That is, the coefficients exhibit conjugate symmetry when viewed as functions of the index.

2.4.1.3 Parseval's Relation for Fourier Series

The zero-correlation property has another interesting and useful corollary. Namely, the power of the signal, x(t), is simply the sum of the powers of the components.

$$P_x = \frac{1}{T_0}\int_0^{T_0} x(t)\,x^*(t)\,dt = \sum_{k=-\infty}^{\infty} \left|c_k\right|^2 \tag{2.4.10}$$

Eq. (2.4.10), which relates power in the time and frequency domains, is one instance of Parseval's relation. Parseval's relation, in general, relates strength of a signal in the time and transform domains.

2.4.1.4 An Application of Fourier Series

Fourier series expansions are a useful tool to determine the ***quality*** of a sinusoidal function. For example, the output of an ideal oscillator may be the perfect sinewave shown in Fig. 2.23, which shows one period of the function

$$x(t) = A\sin(2\pi f_0 t) \tag{2.4.11}$$

The Fourier expansion for x(t) is quite simple

$$x(t) = \sum_{k=-\infty}^{\infty} c_k e^{j2\pi k f_0 t} \quad \text{where } f_0 = \frac{1}{T_0} \tag{2.4.12}$$

$$c_1 = j\frac{A}{2} \text{ and } c_{-1} = -j\frac{A}{2} \text{ and } c_k = 0 \text{ for } k \neq \pm 1$$

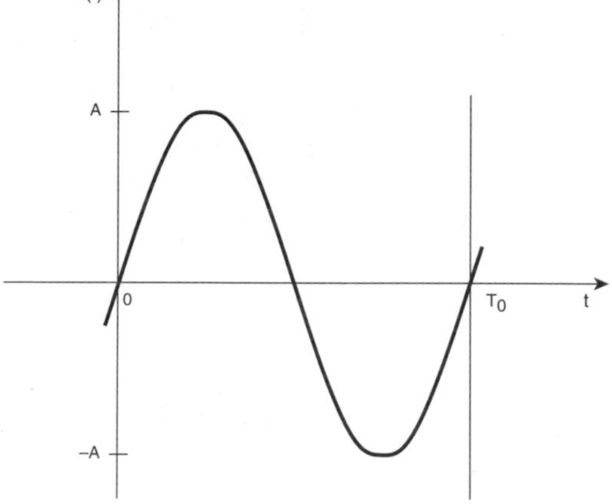

Fig. 2.23
One cycle of a perfect sinewave

The purity of the sinewave is quantified in Eq. (2.4.12), which indicates that there is no power in any of the harmonics. One observation from Eq. (2.4.12) is that since x(t) is real-valued, the coefficients c_k and c_{-k} are complex conjugates of each other, as expected.

Now suppose that the signal x(t) is modified by a device as depicted in Fig. 2.24. The input-output characteristic of G is shown in Fig. 2.25, with the corresponding output y(t) shown in Fig. 2.26, respectively. That is, the device G distorts the sinewave. The operation performed by G is dependent only on the instantaneous value of the input x(t); the output does not depend on past values of either x(t) or y(t). Such a device is said to be a ***memoryless nonlinearity***. One characteristic of memoryless nonlinearity is the preservation of the periodicity; the output will have the same period as the input. A second characteristic is that the signal distortion introduced is equivalent to the addition of harmonics. The Fourier series is a convenient way to express the ***impurity*** of y(t); equivalently, from the power of all the harmonics, one can quantify the distortion of a signal purporting to be a sinusoid.

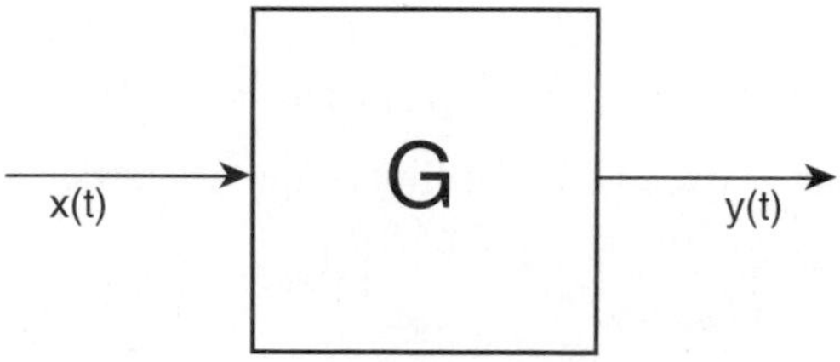

Fig. 2.24
Signal modification by a memoryless nonlinear device

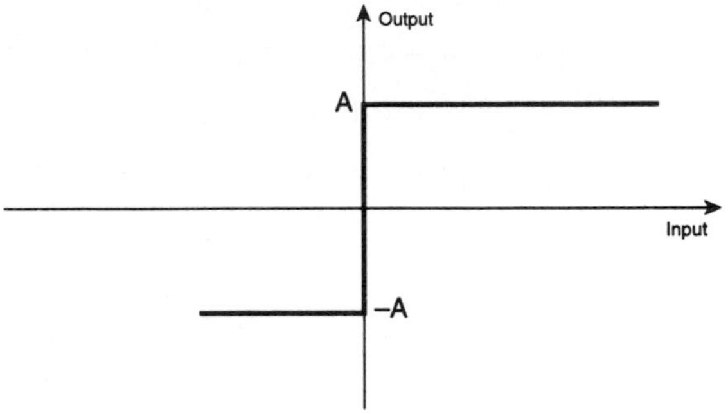

Fig. 2.25
Input-Output characteristic of an ideal comparator

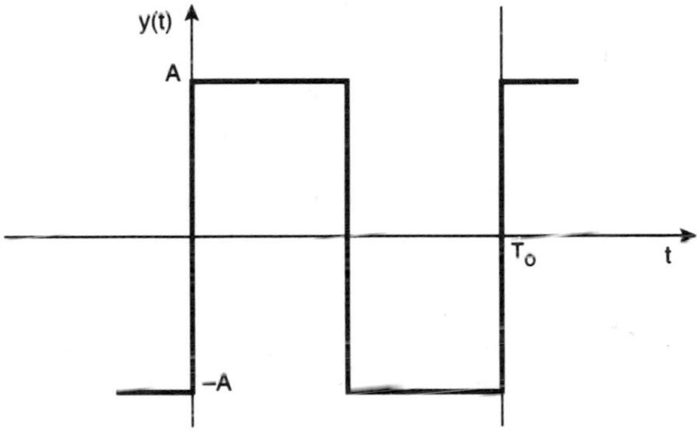

Fig. 2.26
Square wave at output of a comparator excited by the sinewave of Fig. 2.24

Example. The nonlinearity depicted in Fig. 2.25 is an ideal comparator where the output is equivalent to the sign of the input; the amplitude, +A or –A is arbitrary. In particular,

$$y(t) = \begin{cases} +A \text{ for } 0 \le t < (T_0/2) \\ -A \text{ for } (T_0/2) \le t < T_0 \end{cases} \tag{2.4.13}$$

The Fourier coefficients, $\{c_k\}$, can be computed as

$$c_k = \frac{A}{T_0}\int_0^{(T_0/2)} e^{-j2\pi k f_0 t}\,dt - \frac{A}{T_0}\int_{(T_0/2)}^{T_0} e^{-j2\pi k f_0 t}\,dt \tag{2.4.14}$$

which, after some algebra, yields

$$c_k = -c_{-k} \quad \text{(odd symmetry)}$$

$$c_k = 0 \quad \text{for even k} \tag{2.4.15}$$

$$c_k = -\frac{2A}{\pi k} e^{-j\frac{\pi k}{2}} \sin\left(\frac{\pi k}{2}\right) \quad \text{for odd k}$$

Combining terms for indices $\pm k$, one can show that the *sine* form of the Fourier series of the square wave y(t) is given by

$$y(t) = \sum_{\text{odd}\,k>0} \frac{4A}{\pi k} \sin(2\pi k f_0 t) \tag{2.4.16}$$

The power of the fundamental, i.e., the sinewave of frequency f_0, and the power of the harmonics are given by

$$\text{Fundamental}\,(k=1)\,\text{power} = \frac{8A^2}{\pi^2} \tag{2.4.17a}$$

$$\text{Harmonics}\,(k=3,5,\ldots)\,\text{power} = \sum_{n=1}^{\infty} \frac{8A^2}{\pi^2 (2n+1)^2} \tag{2.4.17b}$$

Observe that the squarewave has ***only odd*** harmonics. This is typical of nonlinearities that affect the signal symmetrically. In fact, it is easy to show that nonlinearities that are asymmetric introduce even harmonics. Equivalently, the presence of even harmonics implies that the signal has been acted upon by a nonlinear device whose input-output characteristic is not symmetric.

2.4.1.5 A Special Case of Fourier Series

The signal $\Theta(t)$ depicted in Fig. 2.19 is special. It can be used, as shown in Section 2.3, to represent periodic signals, that is pulse trains, as the output of an **LTI** system, with prescribed impulse response, when excited by $\Theta(t)$.

Applying the relation given in Eq. (2.4.3), the Fourier coefficients, assuming that a series expansion is indeed valid, are given by

$$c_k = \frac{1}{T_0} \int_{-(T_0/2)}^{+(T_0/2)} \delta(t)\, e^{-j2\pi k f_0 t}\, dt = \frac{1}{T_0} = f_0 \tag{2.4.18}$$

where the sampling property of the delta function has been invoked. Consequently, the function $\Theta(t)$ can be written as the sum of delta functions or the sum of complex exponentials (i.e., the Fourier series):

$$\Theta(t) = \sum_{n=-\infty}^{\infty} \delta(t - nT_0) = f_0 \sum_{k=-\infty}^{\infty} e^{j2\pi k f_0 t} \tag{2.4.19}$$

The second form does not converge to $\Theta(t)$ in the traditional sense because of the discontinuous behavior of $\Theta(t)$. We will, however, assume that both representations in Eq. (2.4.19) are valid descriptions of $\Theta(t)$.

2.4.2 The Fourier Transform

The Fourier series provides a description, in terms of sinusoids, of signals that are periodic in nature. For signals that are not periodic in nature, the Fourier transform is used to obtain a frequency domain representation.

2.4.2.1 The Fourier Transform Pair

Given a signal x(t), we define the Fourier transform, $X_F(f)$, by the relation (also called the forward transform)

$$X_F(f) = \int_{-\infty}^{+\infty} x(t)\, e^{-j2\pi f t}\, dt \tag{2.4.20}$$

By convention, the time function is represented by the lowercase letter and the Fourier transform by the corresponding uppercase letter. The subscript F associated with the Fourier transform is for convenience only. The intention is to distinguish the Fourier transform from any other transform, considering the convention that the uppercase notation may be used there as well. When it is obvious from the context which transform is being used, the subscript can be omitted.

The Fourier transform $X_F(f)$ is a complex-valued function of a real variable f. The dimension of f is equivalent to the reciprocal of time and thus we consider the variable f to represent ***frequency***. Strictly speaking, f is just a variable associated with the transform operation; assigning a physical attribute to f should be done carefully.

The inverse transform, whereby the time function, x(t), is computed from the transform, $X_F(f)$, is given in Eq. (2.4.21).

$$x(t) = \int_{-\infty}^{+\infty} X_F(f)\, e^{+j2\pi f t}\, df \tag{2.4.21}$$

Note the similarity between the forward and inverse transforms. This similarity is often used in proofs, derivations, and other mathematical manipulations involving the Fourier transform. The functions x(t) and $X_F(f)$, one a function of "t" (or "time") and the other a function of "f" (or "frequency") together constitute a Fourier transform pair. The Fourier transform is a useful tool for relating functions of independent variables that have reciprocal dimensionality.

A convenient shorthand notation for the inverse and forward transforms is

$$x(t) = \mathbf{F}^{-1}\{ X_F(f) \} \quad \text{and} \quad X_F(f) = \mathbf{F}\{ x(t) \} \tag{2.4.22}$$

2.4.2.2 Uniqueness of the Fourier Transform

The Fourier transform is unique in the following sense. If $X_F(f) = \boldsymbol{F}\{x(t)\}$ and $y(t) = \boldsymbol{F}^{-1}\{X_F(f)\}$, where the computations of the forward transform and inverse transform are done according to Eq. (2.4.20) and Eq. (2.4.21) respectively, then we expect x(t) and y(t) to be the same. They indeed are the same in the sense of being equal *almost everywhere* (a.e.) whereby the difference is a signal of zero energy. That is,

$$\int_{-\infty}^{+\infty} |x(t) - y(t)|^2 \, dt = 0 \tag{2.4.23}$$

The Fourier transform is usually defined only for signals of finite energy. That is, the transform of x(t) is defined only if x(t) satisfies

$$E_X = \int_{-\infty}^{+\infty} |x(t)|^2 \, dt < \infty \tag{2.4.24}$$

and it is in this context that the uniqueness of the Fourier transform can be defined in terms of equality *almost everywhere.* Signals that have infinite energy, but finite power, need to be treated with caution. The Fourier transform of such signals may contain delta functions. Because of the peculiar nature of delta function, equality a.e. in the sense of Eq. (2.4.23) is difficult to establish because the difference between two instantiations of infinity is not defined.

2.4.2.3 A Special Transform Pair

Consider the special case when the time signal is a delta function,

$$x(t) = \delta(t) \tag{2.4.25}$$

Applying the forward transform yields

$$X_F(f) = \int_{-\infty}^{+\infty} \delta(t) \, e^{-j2\pi f t} \, dt = 1 \tag{2.4.26}$$

This indicates that the delta function has frequency components at all frequencies. In fact, the narrower the time waveform, the greater is the range of frequencies where the signal has significant (frequency) content. It is a general rule that "time-limited" waveforms cannot be frequency-limited and vice versa. The delta function is an extreme example of a time-limited waveform (zero duration) and has a very "wide" transform. The terminology we use for "duration" in the frequency domain is "bandwidth." Thus a narrow time function will have a wide bandwidth. From Eq. (2.4.26), assuming the inverse transform is valid,

$$\int_{-\infty}^{+\infty} e^{+j2\pi ft}\, df = \delta(t) \qquad (2.4.27)$$

Considering the similarity between the forward and inverse transforms, Eq. (2.4.27) can be manipulated to show that if x(t) is a constant (DC), the Fourier transform is a delta function in the frequency domain. This is in keeping with the previous remark concerning the relationship between the time and frequency domains. A signal that is "limited" in one domain is necessarily not limited in the other. Normally a signal that is "wide" in one domain is "narrow" in the other. In fact, the Fourier transform of a constant (say unity) is given by

$$\int_{-\infty}^{+\infty} e^{-j2\pi f\tau}\, d\tau = \delta(f) \qquad (2.4.28)$$

Eq. (2.4.28) can be derived from Eq. (2.4.27) by recognizing that "f" and "t" are dummy variables and can be manipulated by the substitution "t"= "$-f$."

2.4.2.4 Elementary Properties of the Fourier transform

Several simple properties of the Fourier transform follow directly from the definition as an integral. A few other properties are described below. These properties often make seemingly complicated mathematical manipulations quite simple and are helpful in building intuition and understanding.

a) ***Complex conjugate.*** Suppose x(t) and $X_F(f)$ are a Fourier transform pair and let y(t) be the complex conjugate of x(t). Then

$$X_F^*(f) = \int_{-\infty}^{+\infty} x^*(\tau)\, e^{+j2\pi f\tau}\, d\tau = Y_F(-f) \qquad (2.4.29)$$

That is, the transform of the complex conjugate of a signal is the complex conjugate of the frequency-reversed version of the transform of the signal. As a corollary, the transform of a real-valued signal, in which case $x(t) = x^*(t)$, exhibits a conjugate symmetry as a function of frequency. That is, if x(t) is real-valued,

$$\overset{*}{X_F}(f) = X_F(-f) \Rightarrow$$
$$\mathrm{Re}\{X_F(f)\} = \mathrm{Re}\{X_F(-f)\} \text{ and } \mathrm{Im}\{X_F(f)\} = -\mathrm{Im}\{X_F(-f)\} \tag{2.4.30}$$

b) ***Odd and even symmetry***. If x(t) is real-valued, the Fourier transform $X_F(f)$ can be written as

$$X_F(f) = \int_{-\infty}^{+\infty} x(t)\cos(2\pi f t)\,dt - j\int_{-\infty}^{+\infty} x(t)\sin(2\pi f t)\,dt \tag{2.4.31}$$

If x(t) is an even function, then x(t)sin($2\pi f$t) is an odd function and hence the imaginary part of $X_F(f)$ is zero. If x(t) is an odd function, then so is x(t)cos($2\pi f$t) implying that the real part of $X_F(f)$ is zero. That is, for real-valued x(t),

$$\begin{aligned} x(t) \text{ even} &\Longleftrightarrow \mathrm{Im}\{X_F(f)\} = 0 \\ x(t) \text{ odd} &\Longleftrightarrow \mathrm{Re}\{X_F(f)\} = 0 \end{aligned} \tag{2.4.32}$$

As a corollary, if x(t) is complex-valued and exibits conjugate symmetry, then the transform will be real-valued and if x(t) is conjugate asymmetric, then the transform will be purely imaginary.

c) ***Delay.*** Given that $X_F(f)$ is the transform of x(t), what is the transform of a delayed version of x(t)? If

$$y(t) = x(t-\tau) \tag{2.4.33}$$

then $Y_F(f)$ is given by

$$Y_F(f) = \int_{-\infty}^{+\infty} x(t-\tau)\,e^{-j2\pi f t}\,dt = X_F(f)\,e^{-j2\pi f \tau} \tag{2.4.34}$$

The derivation of Eq. (2.4.34) implied the substitution t = (t – τ) which, together with the notion of the infinite limits, is a common technique employed with Fourier transforms.

Note that delaying a signal does not change the magnitude of the Fourier transform. That is, when the complex-valued nature of a transform is depicted in polar notation, the impact of delay is simply an alteration in the phase angle and further, this change in phase angle corresponds to a linear function of frequency.

The similarity between the forward and inverse transforms can be used to derive the following result.

$$\mathbf{F}^{-1}\{X_F(f-f_0)\} = x(t)\,e^{j2\pi f_0 t} \tag{2.4.35}$$

A "delay" in the frequency domain, or frequency shift, corresponds to multiplying the signal in the time domain by the complex exponential with frequency equal to the shift in the transform domain. A special case of Eq. (2.4.35) is when x(t) is DC, i.e., a constant. Combining Eqs. (2.4.35) and (2.4.28) yields the following important transform pair:

$$\mathbf{F}\{e^{j2\pi f_0 t}\} = \delta(f - f_0) \quad \text{and} \quad \mathbf{F}^{-1}\{\delta(f - f_0)\} = e^{j2\pi f_0 t} \qquad (2.4.36)$$

d) ***Time reversal.*** Denote the time reversed version of the signal x(t) by y(t). Then

$$Y_F(f) = \int_{-\infty}^{+\infty} x(-t)\, e^{-j2\pi(-f)(-t)} dt = X_F(-f) \qquad (2.4.37)$$

As a corollary, if x(t) is real valued,

$$\mathbf{F}\{x(-t)\} = X_F^*(f) \qquad (2.4.38)$$

e) ***Parseval's Relation***. Parseval's relation was introduced in the discussion on Fourier series as a means for relating the computation of the strength of a signal in either the time or transform domain. A similar relation holds good in the case of the Fourier transform. In particular,

$$E_x = \int_{-\infty}^{+\infty} x(t)\, x^*(t)\, dt = \int_{-\infty}^{+\infty} x(t) \int_{-\infty}^{+\infty} X_F^*(f)\, e^{-j2\pi f t} df\, dt$$

$$= \int_{-\infty}^{+\infty} X_F^*(f) \int_{-\infty}^{+\infty} x(t)\, e^{-j2\pi f t} dt\, df = \int_{-\infty}^{+\infty} X_F(f)\, X_F^*(f)\, df \qquad (2.4.39)$$

Key to the derivation of Parseval's relation is the interchange of the order of integration. This is quite valid provided $E_x <$. Since the energy of the signal can be computed as an integral in the frequency domain, the entity $|X_F(f)|^2$ is called the ***energy spectral density*** function. An alternate term that does not quite reflect Parseval's relation but does describe $|X_F(f)|^2$ literally is ***squared magnitude function***. $|X_F(f)|^2$ and $|x(t)|^2$ are both representative of the strength of the signal; the former indicates the concentration of energy in the frequency domain and the latter in the time domain.

f) ***Autocorrelation and the squared-magnitude function.*** In Section 2.2 the autocorrelation of a real-valued signal, x(t), was defined by

$$R_{xx}(\tau) = \int_{-\infty}^{+\infty} x(t)\, x(t + \tau)\, dt \qquad (2.4.40)$$

Since the variable τ is equivalent to *time* (actually time shift or delay), the Fourier transform of $R_{xx}(\tau)$ can be computed as

$$S_{xx}(f) = \int_{-\infty}^{+\infty} R_{xx}(\tau)\, e^{-j2\pi f \tau} d\tau \qquad (2.4.41)$$

Substituting Eq. (2.4.40) into Eq. (2.4.41) and performing some simple manipulations yields

$$S_{xx}(f) = |X_F(f)|^2 \qquad (2.4.42)$$

That is, the Fourier transform of the autocorrelation function is equal to the squared-magnitude function.

g) ***Convolution and the Fourier transform***. One of the principal application of transforms in general is to simplify certain mathematical operations. Consider the operation of convolution defined in Section 2.3, Eq. (2.3.9). If x(t) and y(t are convolved to produce w(t), then we can write

$$w(t) = x(t)\ (*)\ y(t) = \int_{-\infty}^{+\infty} x(\tau)\, y(t-\tau)\, d\tau \qquad (2.4.43)$$

Applying the definition of the forward Fourier transform to both sides of Eq. (2.4.43) yields

$$W(f) = \int_{-\infty}^{+\infty} w(t) e^{-j2\pi f t} dt = \int_{-\infty}^{+\infty} \int_{-\infty}^{+\infty} x(\tau)\, y(t-\tau)\, e^{-j2\pi f t} dt\, d\tau \qquad (2.4.44)$$

As before, when dealing with multiple integrals, we would expect some simplification arising from the change of the order of integration. From a mathematical standpoint this is valid only if the individual integrals exist (that is, are well-defined and finite). We usually assume that this is true and take the liberty of changing the order of integration at will though the reader is cautioned that changing the order, especially when there are delta functions involved, may not always be allowed. By interchanging the order and some algebra we get

$$W(f) = X(f)\, Y(f) \qquad (2.4.45)$$

The seemingly complicated operation of convolution in the time domain is reduced to the simpler operation of multiplication in the frequency domain. This is probably the most important feature of the Fourier transform.

h) ***Multiplication and the Fourier transform.*** The definitions of forward and inverse transform are very similar. Consequently we can expect that multiplication in the time domain corresponds to convolution in the frequency domain. This is indeed true. To verify this, let

$$w(t) = x(t)\, y(t) \qquad (2.4.46)$$

That is, w(t) is the product of x(t) and y(t). Taking the Fourier transform and performing some algebra yields

$$W(f) = X(f)\ (*)\ Y(f) \qquad (2.4.47)$$

2.4.2.5 Fourier Transforms of Finite Power Signals

Up to now the discussion on Fourier transforms has, implicitly, considered signals of finite energy. Signals that have infinite energy need to be treated with care. Deterministic signals of infinite energy and finite power will always be periodic or, at worst, almost periodic. The Fourier transform of periodic signals can be obtained using the Fourier series as an intermediate step. If x(t) is periodic with period T_0 then

$$x(t) = \sum_{k=-\infty}^{\infty} c_k e^{j2\pi k f_0 t} \quad \text{where } f_0 = \frac{1}{T_0} \tag{2.4.48}$$

and using Eq. (2.4.36) we get for the Fourier transform

$$X(f) = \sum_{k=-\infty}^{\infty} c_k \delta(f - k f_0) \tag{2.4.49}$$

Finite (nonzero) power signals have delta functions in the transform domain and consequently need to be handled with care. For example, the squared magnitude function would not make sense since the square of a delta function is not defined. The Fourier transform of periodic signals consists solely of impulses and thus is said to have a ***discrete spectrum*** or ***line spectrum***.

The need for specifying finite energy can best be explained using an example. Suppose x(t) is the unit step function, u(t). Then x(t) does not have finite energy and it is almost periodic. We thus expect to see delta functions in the Fourier transform $X_F(f)$. If we follow the definition of the Fourier transform, we run into the problem of evaluating $\exp(-j2\pi f t)$ as t —> , which is not well defined. The approach described below is one example of the care taken when values are infinite or not defined.

Consider the function y(t) defined by

$$y(t) = \begin{cases} e^{-at} & \text{for } t>0 \\ -e^{+at} & \text{for } t<0 \end{cases} \tag{2.4.50}$$

In Eq. (2.4.50) the value of "a" is positive and the value at $t = 0$ is 0. The Fourier transform of y(t) can be computed as

$$Y_F(f) = \int_0^{+\infty} e^{-at} e^{-j2\pi f t} dt - \int_{-\infty}^{0} e^{+at} e^{-j2\pi f t} dt = \frac{1}{a + j2\pi f} - \frac{1}{a - j2\pi f} \tag{2.4.51}$$

Now the unit step function can be written in terms of y(t) as

$$u(t) = \frac{1}{2} + \lim_{a \to 0} \left[\frac{y(t)}{2}\right] \tag{2.4.52}$$

Assuming that the Fourier operation and limit operation are interchangeable, taking the Fourier transform of both sides of Eq. (2.4.52) yields:

$$\mathbf{F}\{u(t)\} = \frac{1}{2}\delta(f) + \frac{1}{j2\pi f} \tag{2.4.53}$$

2.4.2.6 Examples of Transforms

Example 1. Consider the time function x(t) that corresponds to a rectangular pulse of duration T. This is the function rect(t;T) that was defined earlier in Eq. (2.2.21) and depicted in Fig. 2.9.

Computation of the Fourier transform yields

$$X(f) = AT\frac{\sin(\pi f T)}{\pi f T} = AT\,\mathrm{sinc}(fT) \tag{2.4.54}$$

This is a very important Fourier transform pair. Fig. 2.9 shows the behavior of the two functions using a normalized time/frequency scale. The time scale is normalized by T, the duration of the rectangular pulse, and the frequency scale by the reciprocal of T. Of special importance is that $X(f)$ passes through zero at frequency values that are nonzero integer multiples of (1/T).

The similarity of the forward and inverse Fourier transform allows us to compute the impulse response of an ideal lowpass filter by inspection. Specifically, the frequency response of an ideal lowpass filter with cutoff frequency f_c and the corresponding impulse response are given by

$$H(f) = \mathrm{rect}(f; 2f_c) \quad \text{and} \quad h(t) = 2f_c\mathrm{sinc}(2f_c t) \tag{2.4.55}$$

Example 2. The **LORAN** system used for navigational purposes involves the transmission by designated stations of a signal from which the receiver, by listening to the transmissions from several stations, can triangulate its position. The signal transmitted can be modeled as a series of pulses where each pulse can be written as

$$p(t) = A(t/T)^2\exp[-2(t/T)]\sin(2\pi f_0 t)u(t) \tag{2.4.56}$$

where u(t) is the unit step function. The **LORAN** pulse p(t) is depicted in Fig. 2.27. p(t) can be written as the product of an envelope signal x(t) and the sinusoidal carrier signal, where x(t) is given by

$$x(t) = A(t/T)^2\exp[-2(t/T)]\,u(t) \tag{2.4.57}$$

The Fourier transform, $X(f)$, of x(t) can be computed as

$$X(f) = \frac{AT}{4}\left[\frac{1}{1 + j\pi f T}\right]^3 \tag{2.4.58}$$

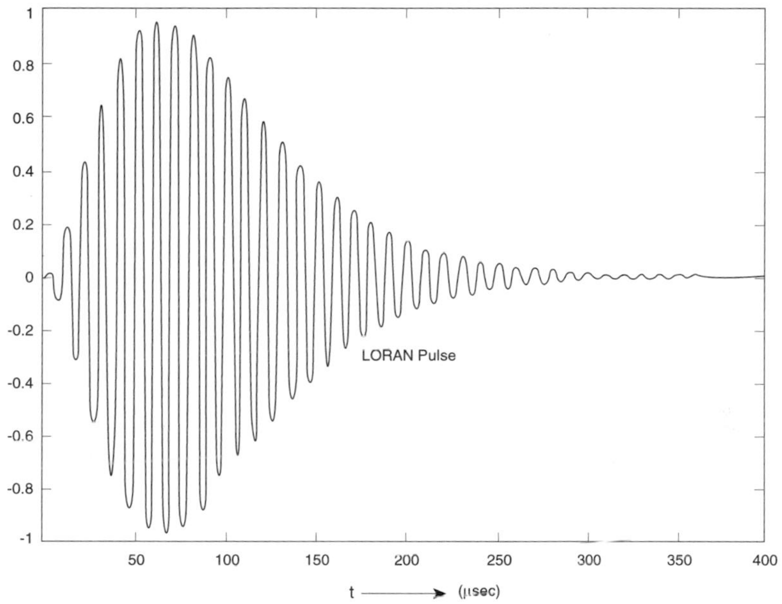

Fig. 2.27
Waveform of an isolated **LORAN** pulse

Since multiplication in the time domain corresponds to convolution in the frequency domain, the transform, $P(f)$, of the **LORAN** pulse p(t) can be computed as

$$P(f) = \frac{1}{2j}\left[X(f-f_0) - X(f+f_0)\right] \tag{2.4.59}$$

The carrier frequency of the **LORAN** transmission, f_0, is 100 kHz and the time-constant T that determines the envelope of the LORAN pulse is 65 μsec. The graph of $|P(f)|$ is shown in Fig. 2.28 for positive values of frequency. Examination of Eq. (2.4.58) and Eq. (2.4.59) would confirm that the frequency content of the **LORAN** pulse is cetered around f_0 (100) kHz and dies away, as the frequency excursion from the center increases, at a rate that is proportional to the cube of the frequency deveiation. Thus the pulse is reasonably narrowband; the spectrum, $|P(f)|$, is about 21 dB down at a frequency offset of 5 kHz from the center frequency. The spectrum shown in Fig. 2.28 is also indicative of the spectrum |X(f)|, which can be obtained by translating |P(f)| shown in the figure from centered at 100 kHz to centered at zero frequency.

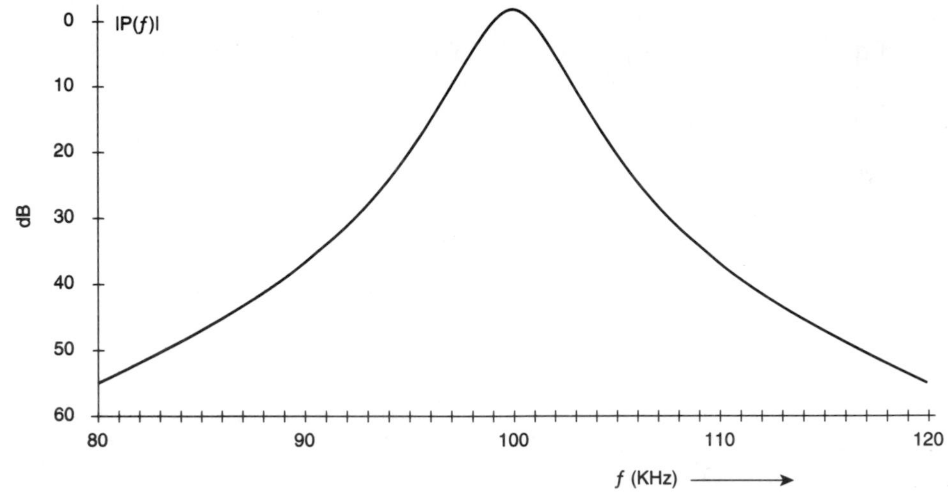

Fig. 2.28
Transform (magnitude) of the **LORAN** pulse for positive values of f

Example 3. In data transmission the shape of the pulse used to launch a data symbol is quite crucial and an attempt is made to control the frequency characteristic of the pulse. A common target frequency response is the so-called raised cosine pulse. The raised cosine pulse, G(f), can be described as follows. First, G(f) is an even function of frequency; second, it is almost equal to rect(f_0); third, the deviation from the ideal lowpass characteristic can be modeled as a cosine rolloff. For positive frequencies, G(f) is depicted in Fig. 2.29.

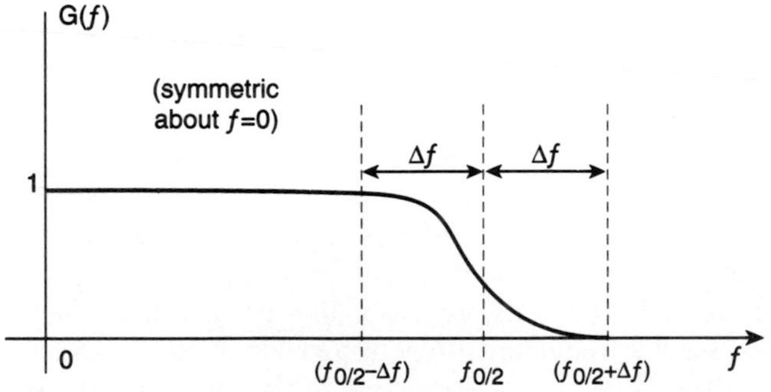

Fig. 2.29
Frequency domain raised cosine pulse

$$G(f) = \frac{1}{2}\{1+\cos[(\frac{\pi}{2\Delta f})(f-\frac{f_0}{2}+\Delta f)]\} \text{ for } \left|f-(f_0/2)\right| \le \Delta f \quad (2.4.60)$$

Eq. (2.4.60) describes the behavior of G(f) as it rolls off from unity at low frequencies to 0 at high frequencies (above f_0). The functional values of G(f) for other frequencies is obvious from Fig. 2.29. The parameter Δf is called the excess bandwidth since it represents how much beyond the edge of the ideal lowpass filter G(f) extends. When $\Delta f = (1/2)f_0$, the pulse is considered to have 100% excess bandwidth. This is the situation used most often and for 100% excess bandwidth the corresponding shape in the time domain can be obtained (after quite some mathematical manipulation) as

$$g(t) = \frac{\sin(2\pi f_0 t)}{2\pi}\{\frac{1}{t} + \frac{f_0}{1-2f_0 t} - \frac{f_0}{1+2f_0 t}\} \quad (2.4.61)$$

The raised cosine pulse (in the frequency domain) is a practical compromise—for efficient data transmission, we would really like to have a (pulse) signal that was both time-limited and frequency-limited, a condition that is mathematically not possible. One of the nice features of the impulse response, g(t), is that it goes through zero at ($1/f_0$) and further, after the first zero crossing at ($1/f_0$), there are zero crossings at multiples of $1/(2f_0)$. This behavior is useful and will be revisited when we discuss in data transmission systems in Section 2.7 and a graph of g(t) is provided there in Fig. 2.65.

Example 4. The Fourier transform can also be used to compute Fourier series in the following way. Suppose x(t) is periodic and we express x(t) as the (infinite) sum of shifted values of a base signal. As we have seen before, the choice of the base signal is not unique. In particular, let

$$x(t) = \sum_{k=-\infty}^{\infty} h(t-kT) = h(t)(*)\Theta(t) \quad (2.4.62)$$

where we shall assume that $T = 1/f_0$. Then

$$x(t) = h(t)(*)\Theta(t) \Rightarrow X(f) = H(f)\Xi(f) \quad (2.4.63)$$

and since, from Eq. (2.4.19) and Eq. (2.4.36)

$$\Theta(t) = f_0 \sum_{k=-\infty}^{\infty} e^{j2\pi k f_0 t} \Rightarrow \Xi(f) = f_0 \sum_{k=-\infty}^{\infty} \delta(f-kf_0) \quad (2.4.64)$$

it follows that

$$x(t) = f_0 \sum_{k=-\infty}^{\infty} H(kf_0)\, e^{j2\pi k f_0 t} \tag{2.4.65}$$

That is, the Fourier series of x(t) and be obtained by evaluating the Fourier transform at multiples of f_0, the fundamental frequency (including DC), of the base (aperiodic) signal that generates x(t). As a corollary, if g(t) and h(t) are two different base signals that can generate x(t) (either h(t), or g(t), or neither, could be of finite duration), then as per our earlier discussion in Section 2.3,

$$\sum_{k=-\infty}^{\infty} g(t-kT) = \sum_{k=-\infty}^{\infty} h(t-kT) \Rightarrow G(kf_0) = H(kf_0) \tag{2.4.66}$$

That is, all base signals for the periodic signal x(t) are such that their Fourier transforms are equal on the frequency grid corresponding to multiples of the fundamental frequency f_0. That is, kf_0, where k is any integer.

2.4.3 Filters

The traditional notion of a filter is that of an entity that removes unwanted signal components. Filtering action is most conveniently described in the frequency domain. The desired signal is restricted to a certain set of frequencies while the *unwanted* signal components resides in another set of frequencies. For filtering action to be possible, the two sets of frequencies must be nonintersecting. Modern approaches to signal processing use the term filter in a more general context. Any action on a signal that modifies the spectral content is called filtering. This includes the enhancement, or suppression of certain features of a signal and is usually achieved by the use of linear time-invariant systems. There are situations where the system may change with time in a particular manner; such systems are called adaptive filters. In this section we describe fixed filters. That is, linear time-invariant operations achieved by devices that do not change with time.

2.4.3.1 Ideal Filters

Consider the case depicted in Fig. 2.30. The signal x(t) contains two components, $x_D(t)$ and $x_U(t)$ representing the desired signal and unwanted signal respectively. The intended action of the filter is to remove the unwanted component and provide at its output the signal $y_D(t)$, which is supposed to be the same as the desired signal $x_D(t)$. In order for this to be achieved, it must be possible to discriminate between the two signal components. The separation between $x_D(t)$ and $x_U(t)$ can be expressed in the frequency domain as shown in Fig. 2.31, which assumes that $x_D(t)$ is "lowpass," with all its energy/power located in the frequency domain in the band $[-f_d, f_d]$. The interfering signal $x_U(t)$ is assumed to have all its energy/power in the frequency range outside $[-f_d, f_d]$.

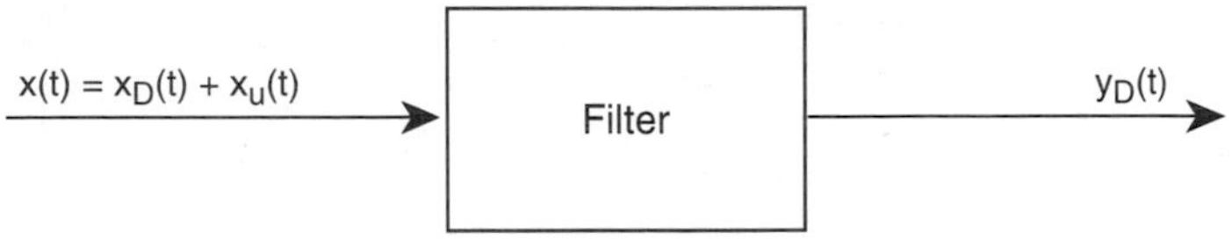

Fig. 2.30
Filtering a signal to remove unwanted components

The interfering signal can be removed if the filter has a frequency response as shown in Fig. 2.32, which shows the magnitude characteristic of an ***ideal lowpass filter*** with cutoff frequency f_d. An ideal lowpass filter has $A(f) = 1$ in the frequency range $[-f_d, f_d]$ and has $A(f) = 0$ for $|f| > f_d$. This still does not guarantee that the output, $y_D(t)$ is equal to $x_D(t)$. For this equality, we require that the phase response of the (ideal) filter be $\beta(f) = 0$. We normally extend the notion of equality to allow for a fixed, finite, delay. That is, $y_D(t) = x_D(t-\tau)$, for some τ, is considered as "$y_D(t)$ is equal to $x_D(t)$" for all intents and purposes. Consequently, we define the frequency response of an ideal lowpass filter by $H(f) = A(f)\exp(-j\beta(f))$ where

$$\begin{aligned} A(f) &= 1 \text{ for } |f| \leq f_d \text{ (passband)} \\ A(f) &= 0 \text{ for } |f| > f_d \text{ (stopband)} \\ \beta(f) &= 2\pi f\tau \text{ for } |f| \leq f_d \text{ (passband)} \end{aligned} \tag{2.4.67}$$

The phase response in the stopband, where $A(f)=0$, is of no consequence.

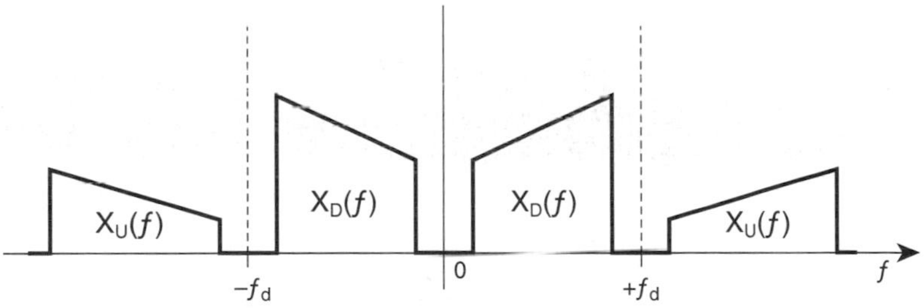

Fig. 2.31
Frequency domain representation of the desired and unwanted signal components as distinguished by their frequency content

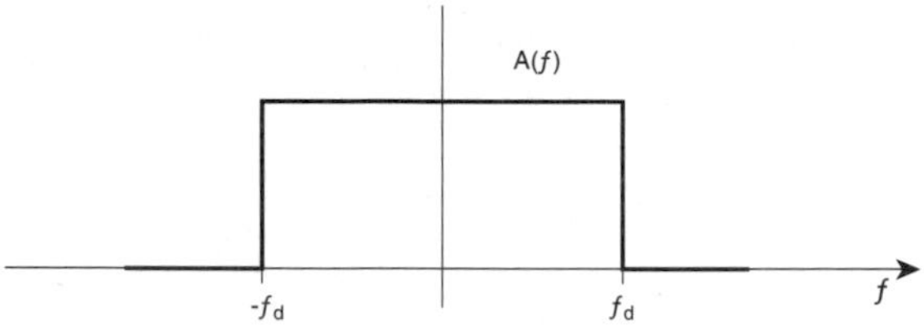

Fig. 2.32
Frequency response of an ideal lowpass filter with cutoff frequency f_d

Similarly we define ideal highpass, bandpass, and bandstop filters by frequency responses whose magnitude characteristics are shown in Fig. 2.33 (a, b, c). The choice of adjective describing the response is quite obvious. Each of the filters is characterized by a band of frequencies that it ***passes unchanged***, albeit with a delay, called the passband, and a range of frequencies that are rejected completely, called the stopband. The two-level shape of the magnitude frequency response has given these filters the name "brickwall." Ideal filters are a useful mathematical abstraction that help in analyzing and visualizing the processing of actual filters employed in signal processing. The extent of the passband is called the "bandwidth" of the filter. For defining bandwidth it is convenient at times to consider only positive frequencies—for real-valued impulse responses we obtain the negative frequency behavior from symmetry. The terminology used, for example, for the lowpass characteristic of Fig. 2.32, is "bandwidth = f_d (one-sided)" where, if it is obvious from context, the "one-sided" is dropped.

Achieving a brickwall characteristic is not feasible, but ideal filters are useful mathematical notions for conceptualizing the impact of filters on signals. Consequently the definition of "bandwidth" varies with application. Since the intent is to define the frequency range that is "passed," the common definition of bandwidth is via the frequency deviation at which the magnitude response drops x dB from the "center." A common choice for x is 3 dB (i.e., half-power). The passband would be the frequency range over which the response is within x dB from the center where the response is nominally 0 dB. Similarly, the stopband is defined as the region where the response is less than y dB from the center of the passband. y is usually a large number (in dB) and rather than use a negative sign the response is treated as an attenuation and the terminology is "y dB down." A wonderful treatment of filters is provided in Daniels [1.1] and Vlach [1.10].

Realizable analog filters do not exhibit the "flat" passband nor the perfect linear phase characterisic. The deviation of A(f) from unity (0 dB) in the passband is called "amplitude distortion" and the deviation from linear phase of the phase response $\beta(f)$ is called "phase distortion." An alternative to quantifying phase distortion is to consider the group delay, which is the derivative of the phase response (scaled by 2π to make the units that of "time"). If the phase is linear, the group delay will be constant; "group delay distortion" then is the deviation from flatness of the group delay response.

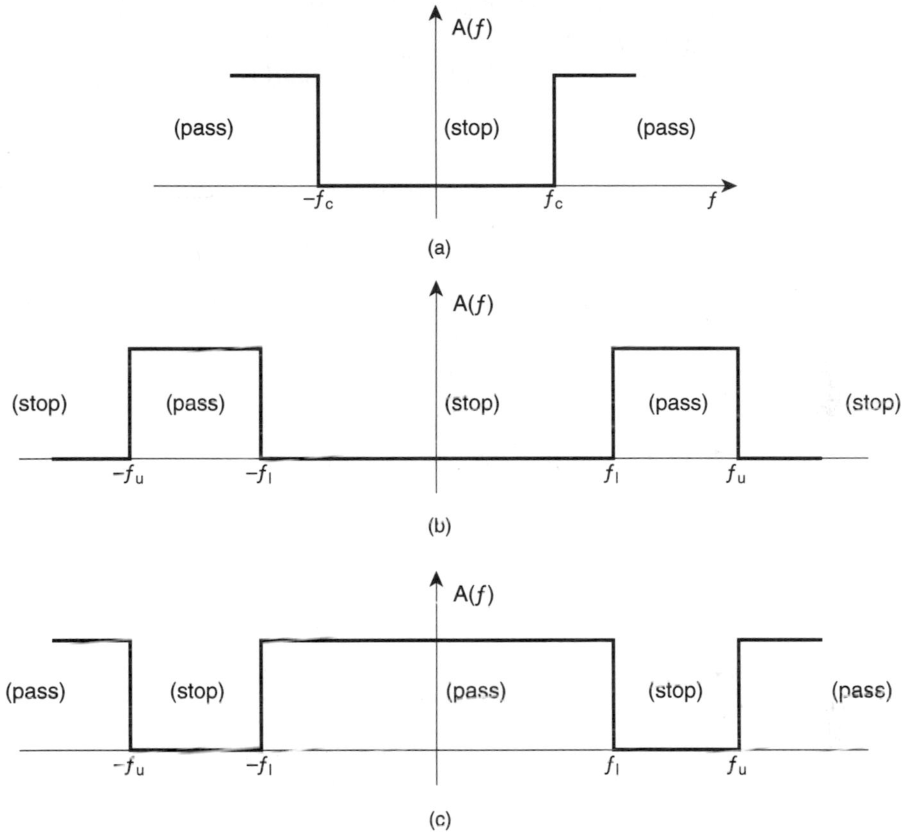

Fig. 2.33
Frequency (magnitude) response of (a) ideal highpass filter; (b) ideal bandpass filter; (c) ideal bandstop

2.4.3.2 Allpass Filters

Filters that provide a frequency response with $|H(f)| = A(f) = 1$ for all frequency are called allpass filters. The trivial case is

$$H(f) = 1 \tag{2.4.68}$$

corresponding to no filtering action whatsoever. $|H(f)| = 1$ is also similar to no filtering in the sense of removing frequency components, but comprises processing in that the phase response of a nontrivial allpass filter is not "zero" . The principal use of allpass filters is to "correct" the phase distortion introduced by other filters that were designed to meet a magnitude criterion.

A very special case of an allpass filter is the (ideal) Hilbert transformer, which has the frequency response

$$H(f) \;=\; -\,j\ \mathrm{sgn}(f) \tag{2.4.69}$$

That is,

$$A(f) \;=\; 1 \quad \text{and} \quad \beta(f) \;=\; (\pi/2)\ \mathrm{sgn}(f) \tag{2.4.70}$$

Since $H(f)$ exhibits conjugate symmetry, the corresponding impulse response, and hence filter, are real-valued. The form of Eq. (2.4.70) gives rise to the terminology "90-degree phase shifter."

The ideal Hilbert transformer is a useful analytical tool for the following reason. Suppose x(t) was a real-valued signal with Fourier transform $X(f)$. Now suppose that y(t) was the "Hilbert transform" of x(t), that is,

$$Y(f) \;=\; H(f)\ X(f) \tag{2.4.71}$$

Then the complex signal v(t) given by

$$v(t) \;=\; x(t) \;+\; j\ y(t) \tag{2.4.72}$$

is such that

$$V(f) \;=\; X(f)\ U(f) \tag{2.4.73}$$

When we have a (complex) signal with a "one-sided" Fourier transform, that is a Fourier transform that is identically zero for negative (or positive) frequency, the real and imaginary parts of the time signal constitute a Hilbert transform pair. The complex signal, such as v(t) in Eq. (2.4.72) is called the "analytic" signal associated with the real (or imaginary) part, x(t). As a point of interest, the signal $\exp\,(j2\pi f_0 t)$ [$= \cos(2\pi f_0 t) + j\ \sin(2\pi f_0 t)$] has a Fourier transform given by $\delta(f - f_0)$, which is one-sided. From this we can conclude that "cosine" and "sine" form a Hilbert transform pair.

2.4.4 Principles of Modulation

There are two basic precepts underlying the principles of modulation. First is that the information-bearing signal is to be transmitted over a channel that has limited bandwidth. Second is that the information signal is not matched to the channel but has a bandwidth less than that of the channel.

Consider an information signal x(t) of finite bandwidth. In the frequency domain the spectral components of x(t) can be depicted as shown in Fig. 2.34. The finite bandwidth assumption can be expressed in terms of the Fourier transform X(f) as

$$|X(f)| \;=0 \ \ \text{for}\ |f| > f_c \tag{2.4.74}$$

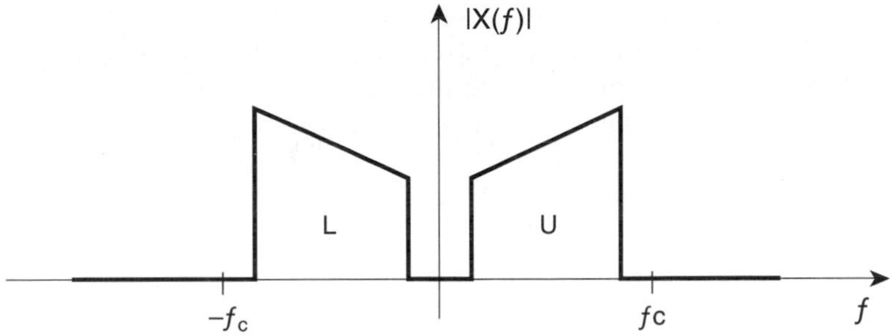

Fig. 2.34
Pictorial representation of a baseband signal of finite bandwidth

If x(t) is real-valued then the Fourier transform will exhibit conjugate symmetry and the positive and negative frequency components, denoted by "U" and "L" in Fig. 2.34, will not be independent. U and L are referred to as the *upper* and *lower* sidebands respectively. In principle, x(t) can be recovered from either upper or lower (or both together) sidebands though the associated signal processing may be quite complicated. In fact, the complex-valued signal whose Fourier transform is zero for negative frequencies and is equal to $X(f)$ for positive frequencies is the analytic signal associated with x(t).

The frequency response of a typical channel is depicted in Fig. 2.35. Most channels exhibit a bandpass characteristic with the notion of a *center frequency*, f_0, and a bandwidth of B (or 2B if one considers both positive and negative frequencies). Assuming that the associated impulse response is real-valued, the frequency response of the channel, $H(f)$, will display conjugate symmetry—the magnitude and phase responses will be even and odd functions of frequency, respectively.

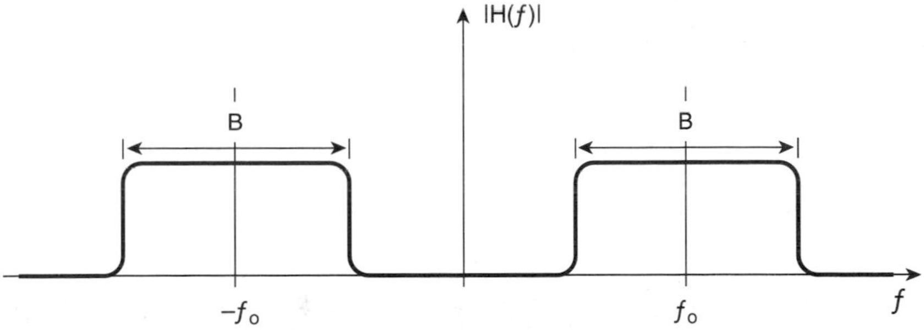

Fig. 2.35
Bandpass nature of a typical channel

Clearly, the channel is best suited for carrying, or transmitting a signal whose power is concentrated in a band B centered at f_0. One such signal is a sinusoid, say $w_0(t)$, given by

$$w_0(t) = A\cos(2\pi f_0 t + \theta) \tag{2.4.75}$$

where A is the amplitude, f_0 the frequency (in Hz), and θ the initial phase that is determined by the choice of time origin. The output of the channel, with $w_0(t)$ as input, will be of the same form as in Eq. (2.4.75) with no modifications other than, possibly, a linear gain (amplification) and/or shift in the reference phase angle. The signal $w_0(t)$ is called the ***carrier*** for reasons that will become obvious in the following discussion.

More generally, consider the time function w(t) given by

$$w(t) = A(t)\cos[\Phi(t)] \tag{2.4.76}$$

where A(t) is the amplitude (function) and $\Phi(t)$ is the phase (function). Further, we define the term ***instantaneous frequency*** as the derivative of the phase function

$$\psi(t) = \frac{1}{2\pi}\frac{\partial}{\partial t}\Phi(t) \tag{2.4.77}$$

Then if the amplitude function is (approximately) a constant and the phase function is (approximately) a linear function of time $(2\pi f_0 t + \theta)$ (or the instantaneous frequency is almost constant), then w(t) should be easily transmitted through the channel.

Modulating the carrier, $w_0(t)$, by the information-bearing signal involves altering the amplitude, phase, or instantaneous frequency by an amount that is proportional to the signal x(t) referred to, respectively, as **amplitude modulation**, **phase modulation**, or **frequency modulation**. In the ensuing discussion we will concentrate on amplitude modulation. Taub and Schilling [1.9], for example, provides a comprehensive treatment of all three forms of modulation.

2.4.4.1 Amplitude Modulation

As the name implies, amplitude modulation involves altering the amplitude function of the carrier $w_0(t)$ in a manner that can be traced to the information-bearing signal x(t).

The common form for defining amplitude modulation is expressed in Eq. (2.4.78)

$$w(t) = Ax(t)\cos(2\pi f_0 t) \tag{2.4.78}$$

where the the amplitude function A(t) is directly related to the information-bearing signal x(t).The Fourier transform of w(t), $W(f)$, is depicted graphically in Fig. 2.36. To see how this form arises, note that w(t) is the product of a sinusoidal signal and x(t). In the frequency domain this is equivalent to the convolution of the transforms of the sinusoid and $X(f)$. Consequently,

$$\begin{aligned} W(f) &= X(f)\ (*)\ (A/2)[\delta(f - f_0) + \delta(f + f_0)] \\ &= (A/2)[X(f - f_0) + X(f + f_0)] \end{aligned} \tag{2.4.79}$$

Note that the upper and lower sidebands of x(t) both appear in the modulated signal. The name given to this form of amplitude modulation is ***Double Side Band Suppressed Carrier***, abbreviated as **DSBSC**. The reason for this nomenclature is that there is no signal component in w(t) that is related to the carrier. There is another form of amplitude modulation that maintains a certain level of carrier component that is called just **DSB** amplitude modulation.

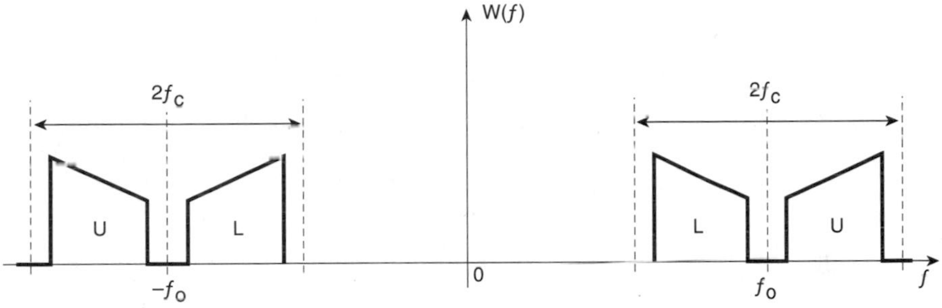

Fig. 2.36
Spectral respresentation of a double sideband amplitude modulated signal (suppressed carrier)

The process of demodulating the **DSBSC** signal to yield x(t) can be understood with reference to Fig. 2.37. A local replica of the carrier signal is used to multiply (i.e., modulate) the signal w(t). Suppose

$$w_L(t) = \cos(2\pi f_1 t + \phi) \tag{2.4.80}$$

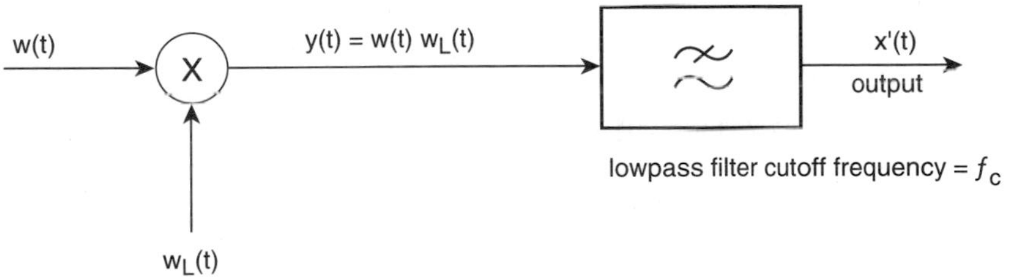

Fig. 2.37
Coherent demodulation of an amplitude modulated signal

Nominally the frequency f_1 would be equal to f_0. The demodulated signal, x'(t), would be different from x(t) because of the different frequency references at the modulator (transmitter) and demodulator (receiver). The initial phase term, ϕ, indicates the possibility of the local oscillator being out of phase with the transmitter oscillator. The product signal is lowpass filtered to obtain x(t). It is important to recognize the conditions under which x'(t) is the same as x(t). The product signal can be written as

$$y(t) = w(t)w_L(t) = 0.5A\, x(t)\, \{\cos[2\pi(f_0 - f_1)t - \phi] + \cos[2\pi(f_0 + f_1)t + \phi\,]\,\} \tag{2.4.81}$$

The action of the lowpass filter is to reject the spectral components centered around $\pm (f_0+f_1)$. Thus the demodulated output is

$$x'(t) \simeq 0.5A\, x(t) \cos[2\pi(f_0 - f_1)t - \phi] \tag{2.4.82}$$

The output of the lowpass filter will be equal to the information signal only if the difference in frequency (f_0-f_1) is zero. Also, any phase difference between the modulator oscillator and the local oscillator will reduce the amplitude of the recovered signal. This technique is called **phase-coherent** demodulation. Ensuring that the receiver and transmitter are in perfect synchronization of both frequency and phase is what makes the implementation of phase-coherent demodulation difficult and thus **DSBSC** is rarely, if ever, used for analog signals. For the transmission of data signals, additional information inherent in the data signal (such as the data rate, pulse shape, and hence the time interval between transitions) can be used to synchronize the receiver and transmitter.

Fig. 2.36 displays a certain redundancy. To see this, observe that **DSB** modulation carries both sidebands of $X(f)$ at both positive and negative frequencies and thus occupies a bandwidth of twice what should be required. Considering only positive frequencies, $X(f)$ has a bandwidth of f_c whereas $W(f)$ occupies a bandwidth of $2f_c$.

Consider what happens if the **DSBSC** signal w(t) is applied to the bandpass filter $H_1(f)$ shown in Fig. 2.38. Denote the output $y_1(t)$. The spectral content of $y_1(t)$ is depicted in Fig. 2.39. The filtering process is equivalent to picking one set of sidebands and hence the term "single sideband" (**SSB**). Since the upper sideband appears at positive frequencies (just as did the baseband version), this form of **SSB** is called "upright." Clearly, we could have chosen a bandpass filter characteristic that corresponded to picking the other sideband. The latter form of **SSB** is called "inverted" since the lower sideband appears at positive frequency (but at negative frequencies in the baseband version). The **SSB** signal, upright or inverted, does indeed contain the entire information content of x(t) represented by the upper and lower sidebands "U" and "L." Single sideband modulation allows us to multiplex several individual signals in the frequency domain in an efficient manner and is the basis for the group signal discussed in Chapter 7.

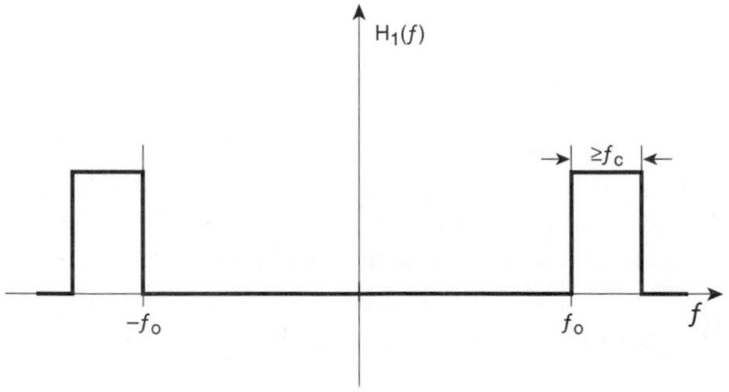

Fig. 2.38
Bandpass filter for creating an "upright" **SSB** signal from **DSBSC**

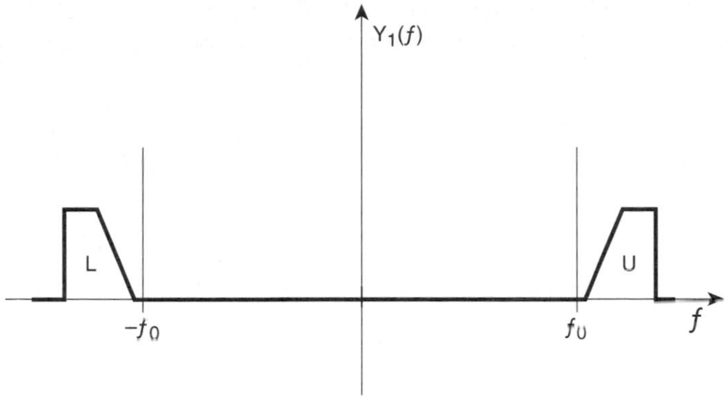

Fig. 2.39
Frequency domain representation of "upright" **SSB**

Amplitude modulation schemes are considered **linear** since the spectral content of x(t) is preserved. **AM** constitutes the process of shifting the spectral content of x(t) from baseband (around DC) to be centered around f_0, the carrier frequency (or close to f_0).

Double sideband modulation is apparently inefficient in the use of spectrum. Since **SSB** uses half the bandwidth, **DSB** can be considered as 50% efficient. One way to make **DSB** 100% efficient is to modulate two different signals into the same bandwidth! This technique is called Quadrature Amplitude Modulation (**QAM**). Suppose $x_1(t)$ and $x_2(t)$ were two separate signals, both bandlimited to f_c. A composite **AM** signal can be created using two carriers of the same frequency. The two carriers are distinguished by having a phase difference of 90 degrees and can be represented by $\sin(2\pi f_0 t)$ and $\cos(2\pi f_0 t)$. The composite signal will be

$$y(t) = x_1(t)\cos(2\pi f_0 t) + x_2(t)\sin(2\pi f_0 t) \tag{2.4.83}$$

It is easily verified that the bandwidth occupancy of this composite signal is no different that the **DSB** signal corresponding to a single carrier though there is an increase in power. At the receiver, coherent demodulation is employed by multiplying y(t) by each carrier separately:

$$\begin{aligned} y_1(t) &= 2y(t)\cos(2\pi f_0 t) \\ y_2(t) &= 2y(t)\sin(2\pi f_0 t) \end{aligned} \tag{2.4.84}$$

Simple algebra indicates that the signals $y_1(t)$ and $y_2(t)$ are

$$\begin{aligned} y_1(t) &= x_1(t) + \{\text{ terms with freq. } 2f_0 \} \\ y_2(t) &= x_2(t) + \{\text{ terms with freq. } 2f_0 \} \end{aligned} \tag{2.4.85}$$

and that the information signals $x_1(t)$ and $x_2(t)$ are obtained by (lowpass) filtering. Thus the bandwidth efficiency of **DSB** is improved using **QAM**. By convention the cosine carrier is called the ***in-phase*** component and the sine carrier the ***quadrature*** component.

As described, **QAM** has the same drawbacks as **DSBSC** in the sense that an accurate replica of the transmitter's carrier, both in-phase and quadrature, must be available at the receiver. Consequently **QAM** is used to transmit data signals, never analog signals such as speech (unless the analog signal has been converted into a digital form by an analog-to-digital conversion process).

2.5 PROBABILITY, RANDOM VARIABLES, AND STOCHASTIC PROCESSES

All information-bearing signals are ***random*** in nature. Deterministic signals, such as sinusoids, do not, by themselves, convey any information since the receiver knows exactly what to expect. Such signals are better suited for test and characterization purposes. Furthermore, all interference signals, such as noise, are also random in nature, the precise waveform unknown. Consequently, the study of information transmission, noise, signal bandwidth, power, interference, and so on, must necessarily be statistical in nature. Such signals are called ***random processes*** or ***stochastic processes*** and will be represented, from the viewpoint of terminology, as $\mathbf{x}(t)$. That is, they are treated as functions of a time variable though with the understanding that the value of the function at any instant of time cannot be specified as in the case of a deterministic signal.

In order to study random processes, it is necessary to provide a foundation. This foundation is based on probability theory and the theory of random variables. The subject matter is quite deep and a comprehensive treatment of these topics is beyond the scope of this book. The level of detail and mathematical rigor is intentionally kept simple though sufficient for the applications discussed in subsequent sections. There are several excellent references available, such as Papoulis [1.6]. The reader is advised that in reviewing the theory from any of the popular texts on the subject, the key concepts that are pertinent to the material in this and subsequent sections are the following: conditional probability and independence; Bayes theorem and total probability; probability density functions (**pdf**s) and cumulative distribution functions (**cdf**s); discrete and continuous random variables; expectation, mean value, and variance; multiple random variables, joint probability density functions, and correlation; the Central Limit Theorem; discrete and continuous random variables; autocorrelation and power spectral density; and the notions of stationarity and ergodicity.

The principal uses of the theory of probability, random variables and stochastic processes, at least for the applications covered in this book, are threefold. First, we use statistical methods to evaluate the strength of random signals, which may be information-bearing, or noise (interference). Second, the theory provides us a means for establishing the frequency content. The power spectral density (**PSD**) of a random process describes the spread of power over frequency. This allows us to establish, or estimate, the impact of filtering on a process. Thus the region in frequency where an information-bearing signal has (most of) its power would logically define the "passband" of a filter. The region in frequency where noise dominates would intuitively determine where the stopband of a filter should be placed. Third, in the analysis of data transmission, the figure of merit used is the probability of error, P_e. That is, we compute the likelihood that information transmission would be corrupted as a function of signal strength and noise power.

2.5.1 Random Variables, Probability Density Functions and Expectations

2.5.1.1 The Probability Density Function (pdf)

The calculation of power, probability of events, and other such parameters associated with a random variable **x**, are all closely tied to the probability density function (**pdf**). Viewing a random variable as an entity that can take on values on the real line ***R*** (between – and +), the **pdf** quantifies the likelihood that the value assumed by **x** will lie in a specified range. In particular, the probability of an event in ***R*** corresponding to the set (a,b] is given by

$$\Pr\left[\,(a\,,\,b]\,\right] = \int_a^b p_{\mathbf{x}}(x)\,dx \tag{2.5.1}$$

Two natural properties that the **pdf** must satisfy are

$$p_{\mathbf{x}}(x) \;\geq\; 0$$
$$\int_{-\infty}^{\infty} p_{\mathbf{x}}(x)\,dx \;=\; 1 \tag{2.5.2}$$

2.5.1.2 Expectation, Mean Value, and Variance

Rather than just probability, we often use certain ***statistics*** associated with random phenomena. These statistics are used for prediction and for establishing performance measures and are often more meaningful from a physical viewpoint than the **pdf**. As the term implies, the ***expectation*** operator extracts an average or *expected* value and is defined in the following manner. If $\mathbf{y} = g(\mathbf{x})$ is a function of a random variable then $E\{\mathbf{y}\}$, or equivalently $E\{g(\mathbf{x})\}$ can be obtained as

$$\mathbf{E}\{g(\mathbf{x})\} \;=\; \int_{-\infty}^{\infty} g(x)\,p_{\mathbf{x}}(x)\,dx \tag{2.5.3}$$

The **mean**, "**m**" of a random variable is defined as $E\{\mathbf{x}\}$ and is given by Eq. (2.5.3) with $g(\mathbf{x}) = \mathbf{x}$. The physical interpretation of the mean of a random variable is that of the DC or invariant component of the voltage or current that the random variable is supposed to model. The electronics used in implementing communications circuits usually employ transformers or coupling capacitors that block the DC component of signals. If we define a random variable $\mathbf{y} = \mathbf{x} - \mathbf{m}$, the the expected value of **y** is given by

$$\mathbf{E}\{\mathbf{y}\} = \mathbf{E}\{\mathbf{x} - \mathbf{m}\} = \mathbf{E}\{\mathbf{x}\} - \mathbf{m} = 0 \tag{2.5.4}$$

That is, **y** is a zero-mean random variable. Note that if $\mathbf{y} = \mathbf{x} + a$ (where a is any constant, not necessarily the mean value of **x**) then

$$p_{\mathbf{y}}(y) = p_{\mathbf{x}}(y - a) \tag{2.5.5}$$

which shows that the **pdf** of **y** is a shifted version of the **pdf** of **x** and retains the same shape and, as will be clear later, retains the same variance (another useful statistic) as **x**. The mean value of **y**, however, will be ($\mathbf{m}$ + a).

Since the random variables we examine are often associated with voltage or current, it is of interest to derive a statistic that expresses the *strength* of the voltage (or current). The expected value of the square of a random variable is equivalent to the notion of (average) power. Further, it is common to make a distinction between "AC" power and "DC" power; the DC power being representative of the constant, or invariant, component of the signal. We thus define a statistic called the ***variance*** of the random variable (or variance associated with the probability distribution) denoted by σ^2 as

$$\sigma^2 = \mathbf{E}\{[\mathbf{x} - \mathrm{E}\{\mathbf{x}\}]^2\} = \mathbf{E}\{\mathbf{x}^2\} - \mathbf{m}^2 \tag{2.5.6}$$

That is, the variance is the expected value of the square of the random variable after the mean (or DC value) has been removed. In keeping with the notion of root-mean-square (rms) associated with voltages and currents, we define ***standard deviation*** as σ, the squareroot of the variance.

From the definition of variance (standard deviation) in terms of expectation, which is a **linear** operator based on integration, it can be shown that if $\mathbf{y} = \alpha\mathbf{x}$ then

$$\sigma_{\mathbf{y}}^2 = \alpha^2 \sigma_{\mathbf{x}}^2 \tag{2.5.7}$$

One interpretation of variance is as a measure of the "width" of the **pdf** of the random variable about the mean value. Two different **pdf**s are depicted in Fig. 2.40. Both **pdf**s have the same mean value but it is clear that the random variable $\mathbf{x}_1$ whose **pdf** is depicted in Fig. 2.40(a) will have a smaller variance than the random variable $\mathbf{x}_2$ whose **pdf** is depicted in Fig. 2.40(b). A heuristic explanation for this is that large values of $(x-\mu)$ will be weighted more heavily by the latter **pdf** than the former.

The notion of *tail probability* is another way of looking at the width of a **pdf**. Tail probability relates to the likelihood of observing values of the random variable far removed from the mean. That is, $\Pr[\{|\mathbf{x}-\mathbf{m}| > \alpha\}]$. If, for large values of α, the probability is substantial, then the **pdf** is wide. In Fig. 2.40 the hatched area under the curve is representative of the tail probability. Clearly, for a given value of α , the **pdf** in Fig. 2.40(b) has the larger tail probability and is thus construed as being wider. Tail probability can be related to variance in a quantitative fashion, as shown next.

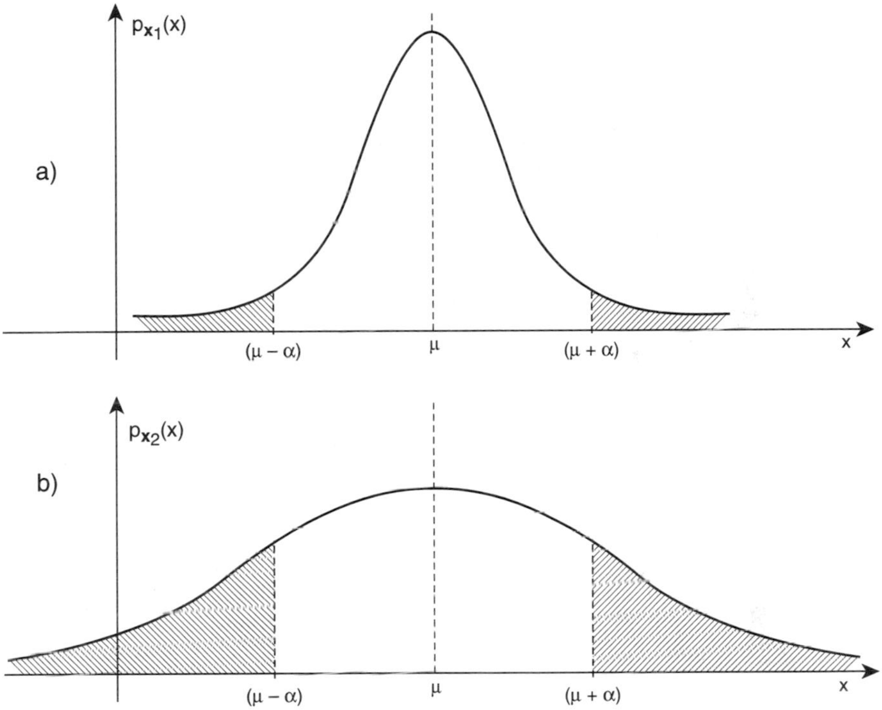

Fig. 2.40
Two **pdfs** that have the same mean but different variances

One of the more common inequalities relating tail probability and variance is the Chebyshev inequality (see Papoulis [1.6]) and is expressed as

$$\Pr\{ |\mathbf{x} - \mathbf{m}| \geq \alpha\sigma \} \leq \alpha^{-2} \tag{2.5.8}$$

Chebyshev's inequality is valid for all **pdf**s. Second, it is not necessarily a "tight" inequality. A common rule of thumb is the "4σ rule" whereby we assume that the probability of observing values of the random variable that are further removed from the mean than 4σ is approximately zero. If we have no information regarding the shape of the **pdf**, other than it is zero-mean, then the "peak" value of the random variable is taken as 4σ. Clearly, the shape of the **pdf** does play an important part in determining the peak value and the wider the **pdf** the greater is the likelihood of observing large values. In the context of data transmission, where the **pdf** represents the deleterious impact of interference, a wide **pdf** translates to a large probability of error. In the context of A/D conversion, a wide **pdf** translates to a greater probability of saturation, which occurs when the signal being converted is greater than a nominal maximum value.

Another interpretation of variance arises in estimating or guessing the value that a random variable will take on. Suppose the guess as to the value the random variable will take on is α. Then $(\mathbf{x} - \alpha)$ can be treated as the "error" and a (commonly used) measure of the strength of the error is the expected value of the square of the error. That is, denoting the error measure by E,

$$E = \mathbf{E}\{ (\mathbf{x} - \alpha)^2 \} \tag{2.5.9}$$

To see what the best, or optimal, value is for α, the value that minimizes E, we can write Eq. (2.5.9) as

$$\begin{aligned} E &= \mathbf{E}\{ [(\mathbf{x} - \mathbf{m})+(\mathbf{m} - \alpha)]^2] \} \\ &= \mathbf{E}\{(\mathbf{x} - \mathbf{m})^2\} + 2(\mathbf{m} - \alpha)\, \mathbf{E}\{(\mathbf{x} - \mathbf{m})\} + (\mathbf{m} - \alpha)^2 \\ &= \sigma_{\mathbf{x}}^2 + (\mathbf{m} - \alpha)^2 \end{aligned} \tag{2.5.10}$$

In Eq. (2.5.10) we have used the fact that $(\mathbf{x} - \mathbf{m})$ is a zero-mean random variable. Clearly, the best value for α is **m**, the mean value, and the corresponding error measure E is then equal to the variance of **x**. For this reason the mean or expected value is the ***minimum-mean-square*** estimate of **x**. Further, the minimum-mean-square-error is equal to the variance. The association can then be made between predictability and variance; a low value for variance being equivalent to high predictability. In fact, if the variance is zero (the case of a "trivial" random variable) then from Chebyshev's inequality $\Pr[\{|\mathbf{x} - \mathbf{m}| > \varepsilon\}] = 0$ for all $\varepsilon > 0$. The statements, loosely speaking, "the **pdf** is concentrated around the mean" , and "the variance is small," and "the phenomenon is predictable," and "the signal ac power is small," are all equivalent.

2.5.1.3 Three Important Random Variables

The three random variables that commonly arise in telecommunications are the Gaussian random variable (also called normal), the uniform random variable, and the exponential random variable. These are described below.

a) The Gaussian Random Variable

The Gaussian random variable is characterized by a **pdf** of the form

$$p_{\mathbf{x}}(x) = \frac{1}{\sigma\sqrt{2\pi}} \exp\left[-\frac{1}{2}\left(\frac{x-\mu}{\sigma}\right)^2\right] \tag{2.5.11}$$

where σ and μ are parameters, the physical interpretation of which will be clarified later. The Gaussian, also referred to as the **Normal**, random variable occurs very often in communication theory. Whenever we have a noise or noise-like signal, unless we have a better characterization, we assume that the behavior is Gaussian. Since it occurs so often, the shorthand notation used for the expression of the **pdf** in Eq. (2.5.11) is N(μ, σ). A graphical representation of the **pdf** is shown in Fig. 2.41.

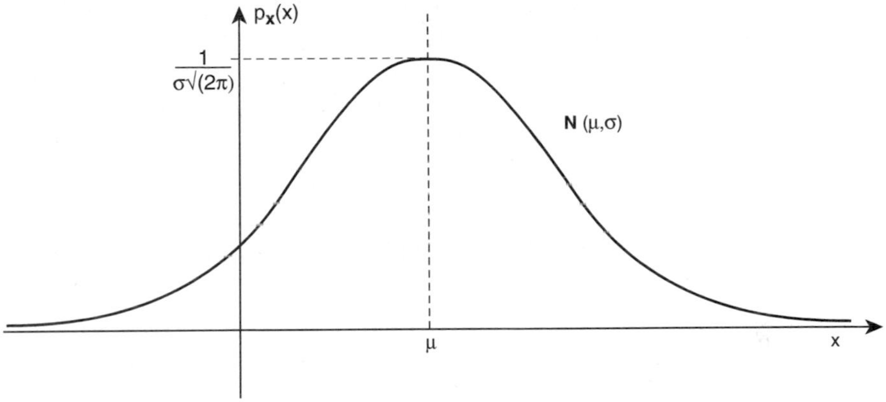

Fig. 2.41
Probability density function of a Gaussian random variable, N(μ, σ)

Calculating the probability of events using the Gaussian **pdf** is not usually possible in closed form. The applicability of the Gaussian density in a wide variety of applications has led to the tabulation of the numerical integration of the **pdf** and is available in most reference books. The function **erf**(x) is related to probability calculations associated with a N(0, 1) random variable, **x**, and is defined as

$$\mathbf{erf}(x) = \frac{2}{\sqrt{\pi}}\int_0^x \exp(-y^2)\,dy = \Pr\{|\mathbf{x}| \le x\sqrt{2}\} \tag{2.5.12}$$

The mean of the Gaussian random variable can be evaluated as

$$\mathbf{m} = \int_{-\infty}^{\infty} x\, p_{\mathbf{x}}(x)\,dx = \frac{1}{\sigma\sqrt{2\pi}}\int_{-\infty}^{\infty} x \exp\left[-\frac{1}{2}\left(\frac{x-\mu}{\sigma}\right)^2\right]dx \tag{2.5.13}$$

When dealing with the Gaussian **pdf**, a common substitution is

$$t = (x-\mu)/\sigma \tag{2.5.14}$$

This substitution essentially converts a N(μ, σ) random variable into a N(0, 1) random variable and is the basis of the applicability of the **erf**(x) tables for computing probability. The substitution transforms Eq. (2.5.13) into

$$\mathbf{m} = \frac{\mu}{\sqrt{2\pi}}\int_{-\infty}^{\infty} \exp\left[-\frac{1}{2}t^2\right]dt + \frac{\sigma}{\sqrt{2\pi}}\int_{-\infty}^{\infty} t \exp\left[-\frac{1}{2}t^2\right]dt \tag{2.5.15}$$

The two integrals in Eq. (2.5.15) can be evaluated by inspection. Note that the first involves the integral from $-$ to $+$ of a function that is a **pdf**, in fact the **pdf** corresponding to N(0, 1), and we know that this integral should be unity. The second integral involves the integral of an odd function over symmetrical limits and is thus zero. Hence

$$\mathbf{m} = \mu \tag{2.5.16}$$

Thus for a Gaussian random variable with **pdf** N(μ, σ), the mean is just the parameter μ. If one looks at the graph of the N(μ, σ) distribution, it would appear as the bell-shaped curve depicted in Fig. 2.41 with the "central" point of the curve at μ, the mean value.

The calculation of the variance of **x** is quite straightforward. By definition (and substituting m for E{**x**})

$$\mathrm{Var}\{\mathbf{x}\} = \frac{1}{\sigma\sqrt{2\pi}}\int_{-\infty}^{\infty}(x-\mu)^2 \exp\left[-\frac{1}{2}\left(\frac{x-\mu}{\sigma}\right)^2\right]dx \tag{2.5.17}$$

Using the same substitution as in Eq. (2.5.14) reduces Eq. (2.5.17) to

$$\mathrm{Var}\{\mathbf{x}\} = \frac{\sigma^2}{\sqrt{2\pi}}\int_{-\infty}^{\infty} t^2 \exp\left[-\frac{1}{2}t^2\right]dt = \sigma^2 \tag{2.5.18}$$

It should be noted that the two parameters μ and σ provided in the **pdf** N(μ, σ) are directly related to the mean and variance of the random variable. Further, if a random variable is known to be Gaussian, the mean and standard deviation (or variance) completely define the **pdf**.

The last step in the evaluation of the variance in Eq. (2.5.18) uses the following results related to the ***Gamma*** function. Since the integral of a **pdf** is unity, we can show that

$$\frac{1}{\sqrt{2\pi}}\int_{-\infty}^{\infty} \exp\left[-\frac{1}{2}t^2\right]dt = 1 \Rightarrow \int_0^{\infty} t^{-\frac{1}{2}}e^{-t}dt = \sqrt{\pi} \tag{2.5.19}$$

The ***Gamma*** function is defined as

$$\Gamma(s+1) = \int_0^{\infty} t^s e^{-t}dt \tag{2.5.20}$$

and can be manipulated easily from the relations

$$\Gamma(s+1) = s\,\Gamma(s) \text{ and } \Gamma(0) = 1 \text{ and } \Gamma(\tfrac{1}{2}) = \sqrt{\pi} \tag{2.5.21}$$

As a consequence, if n is an integer greater than 0,

$$\Gamma(n+1) = n! \tag{2.5.22}$$

Integrals combining powers of x and the Normal density can be evaluated using the Gamma function and symmetry as:

$$\frac{1}{\sqrt{2\pi}}\int_{-\infty}^{\infty} t^n \exp\left[-\frac{1}{2}t^2\right]dt = \frac{2}{\sqrt{2\pi}}\Gamma\left(\frac{n+1}{2}\right) \quad \text{for } n = 2, 4, \ldots$$

$$\frac{1}{\sqrt{2\pi}}\int_{-\infty}^{\infty} t^n \exp\left[-\frac{1}{2}t^2\right]dt = 0 \quad \text{for } n = 1, 3, 5, \ldots \tag{2.5.23}$$

b) The Uniform Random Variable

x is a uniform random variable if the **pdf** is of the form

$$p_{\mathbf{x}}(x) = \begin{cases} \alpha & \text{for } a \le x \le b \\ 0 & \text{otherwise} \end{cases} \tag{2.5.24}$$

For $p_{\mathbf{x}}(x)$ in Eq. (2.5.24) to be a valid **pdf**, the constraints are that $a < b$ and that

$$\alpha = \frac{1}{b-a} \tag{2.5.25}$$

The **uniform random variable** is completely characterized by an interval [a, b]. That is, the only values that the random variable can assume, with non zero-probability, lie in the range [a,b]. Fig. 2.42 depicts the **pdf** of a uniform random variable. The uniform **pdf** is the **pdf** of choice when one deals with the analysis of quantization and in other applications where it is known *a priori* that the values taken on by the random variable are bounded and assured to lie within a specific interval.

The mean value of **x** is given by

$$\mathbf{m} = \frac{1}{b-a}\int_a^b x\,dx = \frac{b+a}{2} \tag{2.5.26}$$

That is, the mean value is the midpoint of the interval.

To calculate the variance, we first compute $E\{\mathbf{x}^2\}$.

$$\mathbf{E}\{\mathbf{x}^2\} = \frac{1}{b-a}\int_a^b x^2\,dx = \frac{1}{3}(b^2 + ab + a^2) \tag{2.5.27}$$

The variance is then calculated as

$$\mathrm{Var}\{\mathbf{x}\} = \mathbf{E}\{\mathbf{x}^2\} - \mathbf{m}^2 = \frac{1}{12}(b-a)^2 \qquad (2.5.28)$$

where $(b-a)$ is the width of the interval over which the **pdf** is nonzero.

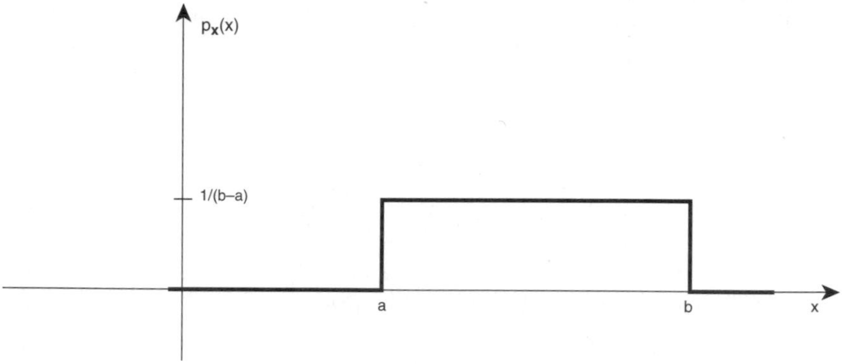

Fig. 2.42
Probability density function of a uniform random variable

c) The Exponential Random Variable

A random variable is considered to be of the exponential family if the **pdf** is of the form

$$p_\mathbf{x}(x) = \frac{\lambda}{2} e^{-\lambda|x|} \qquad (2.5.29)$$

Samples of speech signals can be characterized, approximately, by an exponential **pdf**. If **x** is a random variable of the **exponential** type, the mean value can be computed as

$$\mathbf{m} = \mathbf{E}\{\mathbf{x}\} = \frac{\lambda}{2}\int_{-\infty}^{+\infty} x\, e^{-\lambda|x|}\, dx = 0 \qquad (2.5.30)$$

Evaluation of the integral can be done by inspection. Since the integrand is an odd function and the limits of integration are symmetric, the integral must be zero. The evaluation of the expected value of $\mathbf{x}^2$ is outlined below.

$$\mathbf{E}\{\mathbf{x}^2\} = \frac{\lambda}{2}\int_{-\infty}^{+\infty} x^2 e^{-\lambda|x|}\, dx = \lambda\int_{-\infty}^{+\infty} x^2 e^{-\lambda|x|}\, dx = \frac{2}{\lambda^2} \qquad (2.5.31)$$

The form of the **pdf** in Eq. (2.5.29) guarantees a zero-mean. The parameter λ thus completely describes the density function.

The Gaussian, uniform, and exponential densities are shown together in Fig. 2.43 where it has been assumed that all three correspond to zero mean and unit variance. Clearly, the uniform random variable exhibits boundedness while the Gaussian and exponential variables exhibit "tails." Of these the exponential density seems to be more spread out. The consequence of this is that the exponential random variable will exhibit a higher peak value than the Gaussian case, even if the two have equal variance (power).

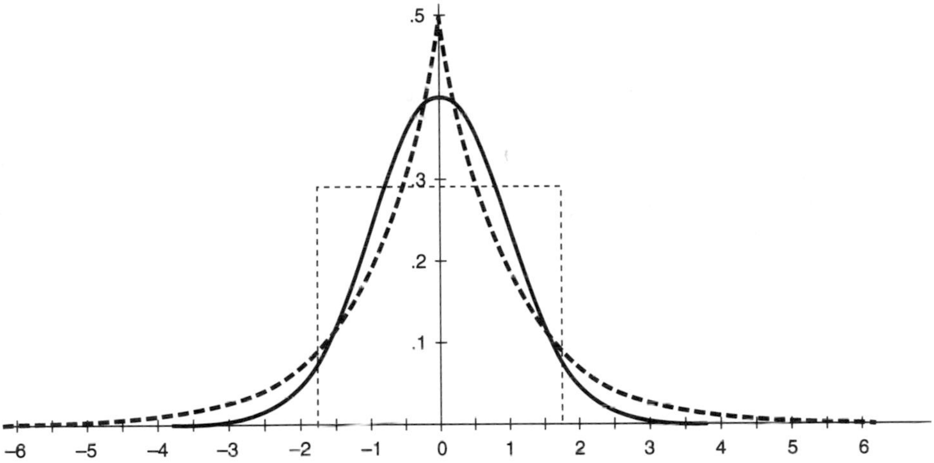

Fig. 2.43
Gaussian, uniform, and exponential **pdfs**, all normalized to zero mean and unit variance

2.5.1.4 Conditional Probability

In probability calculations, as in most problems, it is usually easier to split a complex problem into simpler, more manageable chunks. The notion of conditional probability is exceptionally useful for this as is evidenced by the following example.

In the study of the process of quantization taken up in a later chapter the situation depicted in Fig. 2.44 arises. The process of quantization entails the assignment of a single value, the *quantized* value, for all values of the input between two extremes. The I/O characteristic of the quantizer resembles a staircase with the function expressed mathematically as

$$\mathbf{x} \in (x_n, x_{n+1}) \Rightarrow \mathbf{y} = y_n \qquad (2.5.32)$$

The difference between the input and output is the quantization error. Clearly the quantization error behavior depends on the input range and specifically

$$\varepsilon = \mathbf{x} - y_n \quad \text{for} \quad x \in (x_n, x_{n+1}) \qquad (2.5.33)$$

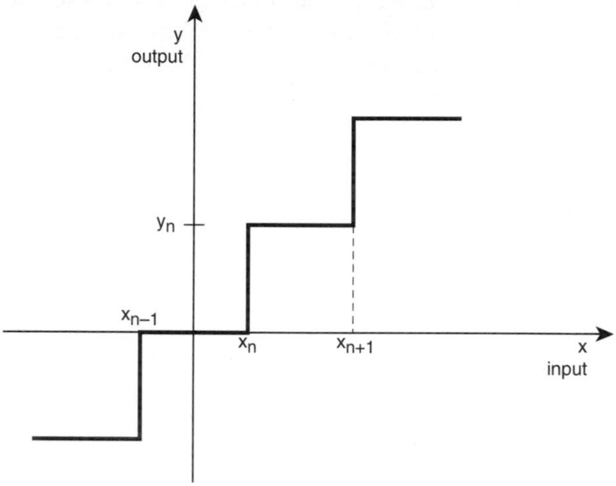

Fig. 2.44
Staircase function of a quantizer

To study the statistics of the quantization error ε it helps to define the conditional probability density of **x** given that **x** lies in the range $(x_n, x_{(n+1)})$. If we denote this event by $\mathbf{A}_n$ then the conditional density is given by

$$
\begin{aligned}
p_{\mathbf{x}\,|\,\mathbf{A}_n}(x) &= \frac{p_{\mathbf{x}}(x)}{\Pr\{\mathbf{A}_n\}} \quad \text{for} \quad x \in (x_n, x_{n+1}) \\
&= 0 \qquad\qquad\qquad \text{otherwise}
\end{aligned}
\tag{2.5.34}
$$

The conditional density satisfies all the properties of a regular probability density function and we could define (conditional) expectation, (conditional) mean value, and (conditional) variance in much the same way as for a regular **pdf**.

What is the *best* value for y_n? This is the same as choosing y_n to minimize the expected value of the square of $(\mathbf{x} - y_n)$ given that $\mathbf{x}\ \varepsilon\ (x_n, x_{n+1})$. Consequently it is not surprising to see that

$$
y_n = \mathbf{E}\{\mathbf{x}\,|\,\mathbf{A}_n\} = \int_{x_n}^{x_{n+1}} x \frac{p_{\mathbf{x}}(x)}{\Pr[A_n]} dx \tag{2.5.35}
$$

and the quantization error power, or variance, given that $\mathbf{A}_n$ has occured, is

$$
\mathbf{E}\{(\mathbf{x} - y_n)^2\,|\,\mathbf{A}_n\} = \mathbf{E}\{\mathbf{x}^2\,|\,\mathbf{A}_n\} - (\mathbf{E}\{\mathbf{x}\,|\,\mathbf{A}_n\})^2 \tag{2.5.36}
$$

The theorem on total probability provides the rationale for computing the overall (i.e., unconditional) variance, or power, of the quantization error as:

$$\sigma_\varepsilon^2 \;=\; \underset{n}{} \; \mathbf{E}\{(\mathbf{x}-y_n)^2 \,|\, \mathbf{A}_n\}\ \Pr\{\mathbf{A}_n\} \qquad (2.5.37)$$

The choice of y_n given in Eq. (2.5.35) guarantees that the average or mean value of the quantization error is zero.

2.5.1.5 Correlation of Random Variables

The correlation between two random variables $\mathbf{x}_1$ and $\mathbf{x}_2$ is defined as the expected value of the product of the two and denoted as R_{x1x2}. That is,

$$R_{\mathbf{x1x2}} \;=\; \mathbf{E}\{\mathbf{x}_1\mathbf{x}_2\} \;=\; \int_{-\infty}^{+\infty}\int_{-\infty}^{+\infty} x_1x_2\,\mathbf{p}(x_1,x_2)\,dx_1dx_2 \qquad (2.5.38)$$

where we assume that the joint **pdf** $\mathbf{p}(x_1,x_2)$ is well-defined.

Correlation is an entity that describes how closely $\mathbf{x}_1$ and $\mathbf{x}_2$ are related. In fact, if we try to predict $\mathbf{x}_2$ in terms of $\mathbf{x}_1$ by a linear function such as $\alpha\mathbf{x}_1$, then the optimal value for α is related to the correlation between the two random variables. Defining the optimal value of α as that which minimizes the power or expected value of the square of the prediction error, then since

$$\mathbf{E} \;=\mathbf{E}\{\,(\mathbf{x}_2-\;\alpha\mathbf{x}_1)^2\} = \mathbf{E}\{\mathbf{x}_2{}^2\} - 2\,\alpha\,\mathbf{E}\{\mathbf{x}_1\,\mathbf{x}_2\} \;+\; \alpha^2\,\mathbf{E}\{\mathbf{x}_1{}^2\} \qquad (2.5.39)$$

the optimal value for α is given by

$$\alpha \;=\; \mathbf{E}\{\mathbf{x}_1\,\mathbf{x}_2\}\,/\,\mathbf{E}\{\mathbf{x}_1{}^2\} \qquad (2.5.40)$$

The random variables $\mathbf{x}_1$ and $\mathbf{x}_2$ are said to be ***uncorrelated*** if

$$\mathbf{E}\{\mathbf{x}_1\,\mathbf{x}_2\} \;=\; \mathbf{E}\{\mathbf{x}_1\}\,\mathbf{E}\{\mathbf{x}_2\} \qquad (2.5.41)$$

The notion of being uncorrelated can be translated loosely into the statement that knowledge of the observed value of $\mathbf{x}_1$ does not assist us in predicting the optimal value of $\mathbf{x}_2$. If they are uncorrelated then the optimal estimate of $\mathbf{x}_2$ would be $\mathbf{E}\{\mathbf{x}_2\}$ and knowledge of $\mathbf{x}_1$ is moot. One observation regarding the optimal value of α given by Eq.(2.5.40) is that the prediction error $(\mathbf{x}_2 - \alpha\mathbf{x}_1)$ is uncorrelated with $\mathbf{x}_1$. That is,

$$\mathbf{E}\{(\mathbf{x}_2-\;\alpha\mathbf{x}_1)\mathbf{x}_1\,\} \;=\; \mathbf{E}\{\mathbf{x}_1\,\mathbf{x}_2\} -\; \alpha\,\mathbf{E}\{\mathbf{x}_1{}^2\} \;=0 \qquad (2.5.42)$$

The property that the optimal prediction error, assuming a squared-error criterion, is uncorrelated with $\mathbf{x}_1$ is referred to as the ***orthogonality principle.***

The property of being uncorrelated is weaker than independence. Independence of $\mathbf{x}_1$ and $\mathbf{x}_2$ requires that

$$\mathbf{p}(x_1,x_2) \;=\; \mathbf{p}(x_1)\,\mathbf{p}(x_2) \qquad (2.5.43)$$

It is easy to show that if $\mathbf{x}_1$ and $\mathbf{x}_2$ are independent then they are definitely uncorrelated. The converse is not always true. One case where uncorrelated random variables are indeed independent is when the two are jointly Gaussian random variables.

The ***correlation coefficient*** between two random variables is the correlation normalized by their standard deviation(s). To remove the effect of the DC, or mean values, of the two random variables, the correlation coefficient is defined as

$$r_{12} = \frac{\mathbf{E}\{(\mathbf{x}_1 - \mathbf{m}_1)(\mathbf{x}_2 - \mathbf{m}_2)\}}{\sigma_1\sigma_2} \tag{2.5.44}$$

Clearly, if the two random variables are uncorrelated then the correlation coefficient between the two will be zero.

2.5.2 Stochastic Processes

Information-bearing signals, as well as signals that interfere with the communications process, are inherently random in nature. Deterministic signals, though useful for the purposes of testing and characterizing systems, do not by themselves convey information in the conventional sense. The mathematical description of information-bearing signals requires the analytical tools developed for entities called ***stochastic processes.*** Other terminology used in referring to stochastic processes are ***random processes, random signals, stochastic signals,*** and ***random functions***, terms that are often used interchangeably.

A formal definition of stochastic processes is beyond the scope of this book. The interested student is referred to, for example, Papoulis [1.6], where an an in-depth treatment of the subject is provided.

2.5.2.1 Interpretation of a Random Process

One approach to interpreting a random process is as a collection of functions of time $\{\mathbf{x}(t; \xi); \xi \in S\}$ where S is a suitable index set for identifying the different functions. There could be an uncountably infinite number of such functions but we require that these functions have some common *statistical* properties. One could imagine that at time equal to minus infinity, one of these functions was chosen. Thus what we observe would be this chosen function. From this viewpoint a random process is observed as a particular signal or time function. At any instant of time t, the observed value, $\mathbf{x}(t)$, would depend on the (initial) random choice of time function and thus could be treated as a random variable. This approach permits us to visualize a random process as a time function. Complex valued random processes can be defined by assuming that the time functions are complex valued.

The second approach is to consider a random process as a collection of random variables $\{\mathbf{x}_t\}$ where the number of random variables in the collection could be uncountably infinite. The random variables are distinguished by an indexing variable "t." The choice of "t" is intentional. A time function $\mathbf{x}(t)$ could be constructed (observed) by choosing the functional value at time t to be the (random) value taken on by the random variable $\mathbf{x}_t$. This approach allows us to view a random process as a sequence of random variables. If the random process is complex valued, then each $\mathbf{x}_t$ would be complex valued or, equivalently, a pair of random variables representing the real and imaginary parts.

The two approaches are equivalent provided certain consistency conditions are satisfied. The specific conditions are not described explicitly in our treatment of random processes. They are however implied in various mathematical manipulations described. Suffice it to say that the conditions will always be satisfied (by assumption) for all random processes we will consider. We will also consider only real-valued random processes. The extension to complex-valued processes is straightforward but cumbersome in notation since the real and imaginary parts need to be considered jointly.

2.5.2.2 Statistics Associated with Stationary Stochastic Processes

The actual waveform observed in a single instantiation of a random process, is a one-time phenomenon. The statistical behavior, as encapsulated in certain statistics permit us predict how well, or how badly, a system will perform when either the information-bearing signal or the interference is random in nature. These statistics are defined in terms of averages, or expected values, of functions of the random process. The principal statistics of interest are the mean and autocorrelation.

We shall assume that the processes we deal with are stationary. The notion of stationarity is equivalent to that of time invariance. That is, we expect that the choice of time origin has no impact on the behavior of the random process; essentially, time is relative.

One form of time invariance that applies only to the first-order (mean value) and second-order (autocorrelation) order statistics and is usually sufficient for the applications we consider is ***wide sense stationarity*** (**WSS**). A random process is **WSS** if the mean value is constant and the autocorrelation ($R_{xx}(\tau)$) depends only on the difference between the time instants t_1 and t_2 under consideration. For a **WSS** process,

$$\mathbf{m} = \mathbf{E}\{\mathbf{x}(t)\} \quad \text{(independent of t)} \tag{2.5.48}$$

$$R_{xx}(\tau) = \mathbf{E}\{\mathbf{x}(t)\mathbf{x}(t+\tau)\} \quad \text{(independent of t)} \tag{2.5.49}$$

where $R_{xx}(\tau)$ is the autocorrelation function.

One consequence of stationarity is that the autocorrelation function is an even function of t. The quantity t has the same dimensions as time.

Note: Unless otherwise stated, we will assume that the mean value of the random process is zero.

In defining the expectations comprising the statistics of a process we have implicitly used the interpretation of a random process as a collection of random variables. Expectations are thus defined in terms of integrals associated with the **pdf**.

These averages are referred to as ***ensemble averages*** to distinguish them from ***time averages*** that may be computed from observed values of the process.

In practice we never have the opportunity to observe all possible instantiations of a random process. At best, the waveform *observed* is one of the uncountably infinite number of possible choices. The question arises as to whether we can establish the statistics of the process based on the observation of a single waveform. The term ***ergodic*** is applied to those stochastic processes that possess the property whereby the statistics of the process can be obtained by observation of a single (complete) waveform. For an ergodic process time averages are equal to ensemble averages. That is, if x(t) is any one waveform comprising the process then

$$\mathbf{E}\{\mathbf{x}(t)\} = \lim_{T\to\infty} \frac{1}{2T}\int_{-T}^{+T} x(t)\,dt \qquad (2.5.50)$$

and

$$R_{\mathbf{xx}}(\tau) = \mathbf{E}\{\mathbf{x}(t)\,\mathbf{x}(t+\tau)\} = \lim_{T\to\infty} \frac{1}{2T}\int_{-T}^{+T} x(t)\,x(t+\tau)\,dt \qquad (2.5.51)$$

All ergodic processes are stationary.

Of special interest is the value of the autocorrelation at "zero lag" or $\tau = 0$. $R_{\mathbf{xx}}(0)$ is the expected value of the square of the random variable $\mathbf{x}_t$ and can be interpreted as the power of the stochastic process. This is intuitively satisfying since if $\mathbf{x}(t)$ is stationary, $\mathbf{E}\{\mathbf{x}_t^2\}$ is constant (independent of time). The ergodic assumption allows us to relate the autocorrelation of a random process with the definition of autocorrelation used with deterministic finite-power signals.

Note: We will assume, unless otherwise stated, that all random processes we encounter are ergodic. With this assumption, we can treat a random process in much the same way as we treat a deterministic finite-power signal with the following caveats. First, deterministic finite-power signals are periodic (or almost so). Second, the exact functional values of a deterministic signal are known for all instants of time. Neither of these conditions is true for a random phenomenon. However, in terms of averages, time or ensemble, similar results apply.

2.5.2.3 Frequency Domain Descriptions of a Random Process

The assumption that a random process is ergodic implies that it cannot be of finite nonzero energy. To see this note that we require that behavior, when viewed as a random variable, be independent of time (stationarity). Hence we cannot guarantee that the time function be asymptotically zero at t goes to infinity. We can however, and we indeed do so, assume that the process has finite power. Thus, if x(t) is any one of the time function choices, then

$$E_x = \lim_{T\to\infty} \int_{-T}^{+T} |x(t)|^2\, dt \to \infty$$

$$P_x = \lim_{T\to\infty} \frac{1}{2T} \int_{-T}^{+T} |x(t)|^2\, dt = \sigma_x^2 \qquad (2.5.49)$$

where σ_x^2 is the power of the signal (i.e., process) (we have assumed a zero-mean process). One consequence of the energy being infinite is that the Fourier transform applied directly to the time waveform is not the preferred method for evaluating the frequency domain description of the process.

The need for describing a process in the frequency domain exists because most signal processing is accomplished by systems that are described in terms of frequency response. If we recognize that we are interested in the power (and mean value in some circumstances) of the resultant waveform at the output of the filter when excited by the random process, rather than the actual waveform itself, then an alternate view of spectral content is available. The power spectral density of a random process is defined as the Fourier transform of the autocorrelation function. Namely,

$$S_{xx}(f) = \mathbf{F}\{R_{xx}(\tau)\} = \int_{-\infty}^{+\infty} R_{xx}(\tau)\exp(-j2\pi f\tau)\, d\tau \qquad (2.5.50)$$

The autocorrelation function is better behaved than any of the individual time functions and is the ***same*** for each possible choice of time waveform comprising the process. This is one of the reasons for defining the spectral content via the transform of $R_{xx}(\tau)$. The second reason for this choice is evident from the following relation:

$$P_x = R_{xx}(0) = \int_{-\infty}^{+\infty} S_{xx}(f)\, df \qquad (2.5.51)$$

Since $R_{xx}(0)$ is the power of the process, the area under the curve of $S_{xx}(f)$ provides the value of P_x. Eq. (2.5.51) is the basis for the statement "$S_{xx}(f)df$ is the power contained in the range $(f, f+df)$" which is why we refer to $S_{xx}(f)$ as a power density function.

2.5.2.4 Filtering Stochastic Processes

Consider a linear time-invariant system described by the frequency response $H(f)$ as depicted in Fig. 2.45. The input is the random process **x**(t) and the resultant output is **y**(t). **y**(t) is a random process consisting of the ensemble of functions that are related to the ensemble of functions of **x**(t) by the convolution operation

$$y(t) = x(t)\,(*)\,h(t) \qquad (2.5.52)$$

where h(t) is the impulse response of the linear system. That is, for each realization of the input x(t) there will be a corresponding realization of y(t), which is related to x(t) by Eq. (2.5.52).

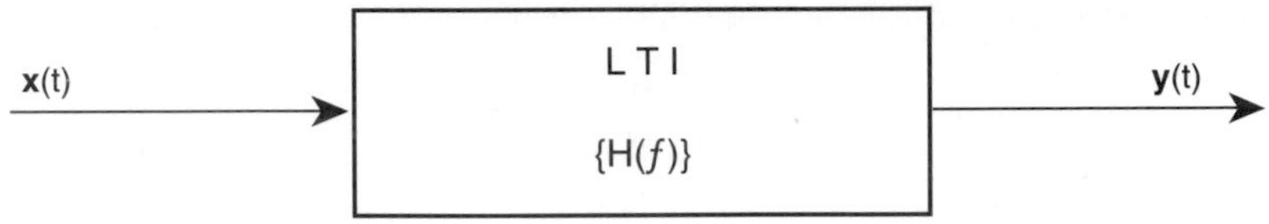

Fig. 2.45
Filtering a random process yields another random process

The autocorrelation of y(t) can be computed as

$$
\begin{aligned}
R_{yy}(\tau) &= \lim_{T\to\infty} \frac{1}{2T} \int_{-T}^{+T} y(t)\, y(t+\tau)\, dt \\
&= \lim_{T\to\infty} \frac{1}{2T} \int_{-T}^{+T} \int x(t-\eta)\, h(\eta)\, d\eta \int x(t+\tau-\zeta)\, h(\zeta)\, d\zeta\, dt \\
&= \int \int R_{xx}(\tau-\zeta+\eta)\, h(\eta)\, h(\zeta)\, d\eta\, d\zeta
\end{aligned}
\tag{2.5.53}
$$

In the frequency domain, this translates to

$$S_{yy}(f) = S_{xx}(f)\, |H(f)|^2 \tag{2.5.54}$$

which provides a description of the modification to the spectrum achieved by a filter. Of particular importance is that the phase response of the filter does not come into play when we are considering the power spectral density of the input/output process.

2.5.2.5 White Noise

White noise is a special random process that derives its name from the concept that white light consists of all the visible frequencies in equal proportion. White noise is the electrical equivalent in the sense that it has equal power components at all frequencies. The power spectral density of a white noise process is of the form

$$S_{xx}(f) = N_0/2 \tag{2.5.55}$$

That is, $S_{xx}(f)$ is a constant. The $N_0/2$ is conventional terminology.

White noise is a mathematical abstraction and not a physical phenomenon. To sce why, we first evaluate the autocorrelation function of white noise as

$$R_{xx}(\tau) = (N_0/2)\ \delta(\tau) \tag{2.5.56}$$

Clearly, the power of white noise, given by the autocorrelation at zero-lag, is infinite and thus not realizable. It is useful, however, to use the concept of white noise to model realizable processes. For example, if a process y(t) is known to have a power spectrum given by $S_{yy}(f)$, then it can be modeled as the output of a filter with frequency response (squared-magnitude) characteristic $|H(f)|^2 = [S_{yy}(f)]$, excited by white noise of unit power spectral density. That is,

$$S_{xx}(f) = 1$$
$$S_{yy}(f) = |H(f)|^2 S_{xx}(f) \qquad (2.5.57)$$

This is depicted in Fig. 2.46.

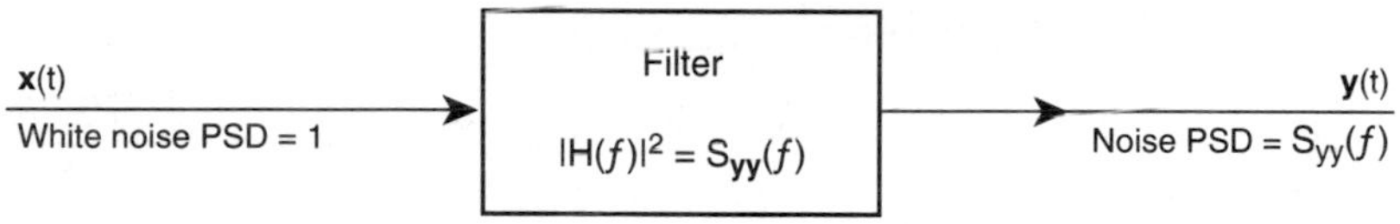

Fig. 2.46
Generating noise with power spectral density $S_{yy}(f) = |H(f)|^2$ by filtering white noise

2.5.3 Impact of Noise in an AM Transmission Scheme

In the discussion on amplitude modulation and demodulation in the previous section it was implicitly assumed that the channel did not modify the modulated signal between the transmitter and receiver. This would obviously not be true in general. To a first approximation, the channel simply attenuates the signal and adds noise. This is equivalent to saying that the bandpass nature of the channel corresponds to an ideal bandpass filter with (one-sided) bandwidth greater than $2f_c$ in the **DSB** case or f_c in the case of **SSB**. For simplicity, consider the **DSB** case using a coherent demodulation scheme depicted in Fig. 2.37 in which the filter is ideal. The filter limits the bandwidth of the signal as well as that of the additive noise. Thus the recovered signal can be expressed as

$$\xi(t) = G\,x(t) + \eta(t) \qquad (2.5.58)$$

where G is the gain (actually $G < 1$, corresponding to an attenuation) of the channel and $\eta(t)$ is the effective additive noise. Assuming that the additive noise prior to demodulation is white, at least over the bandwidth of the channel, $\eta(t)$ will have a flat spectrum over $(-f_c, f_c)$ and thus a total power of $N_0 f_c$ where $N_0/2$ is the spectral content in watts/Hz of the additive noise. The signal-to-noise ratio, **SNR**, a measure of how clean the demodulated signal is, can be written as

$$\text{SNR} = \frac{G\sigma_x^2}{N_0 f_c} \qquad (2.5.59)$$

where σ_x^2 is the power of the transmitted information signal x(t). Note that amplifying

$\xi(t)$ to make the power of the signal component of $\xi(t)$ equal to σ_x^2 does not alter the **SNR**. A common example of this fact is when we turn up the volume of a radio receiver when the received signal is weak, the background hiss (noise) increases proportionately. Actually, since all electronic devices, especially active amplifiers, add their own noise the amplification will actually degrade the **SNR** from that expressed by Eq. (2.5.59).

This discussion describes one of the fundamental properties of **AM**, namely that the additive noise degrades the signal-to-noise-ratio in a manner that cannot be alleviated by linear (i.e., amplification) processing and further that the degradation is generally proportional to the bandwidth utilized.

2.6 ESTIMATING THE POWER OF A SIGNAL

The strength of a signal is one of its key attributes. In fact, one of the basic premises in communication theory is that an optimal system is one in which the power of a signal is maximized relative to the power of all interference. If an analytical description of the signal were available then the power, or any other measure of strength, could be computed. In many cases the signal is only observed and some processing is required to estimate the strength. One example of such a circumstance is in measuring instruments where the technique for estimating power needs to be quite robust and applicable to a wide variety of signal types.

2.6.1 Different Approaches to Power Estimation

Consider the case where the signal, x(t), is a sinusoid

$$x(t) = A\cos(2\pi f_0 t + \theta) \tag{2.6.1}$$

For a sinusoidal signal the power is obtained analytically in terms of the energy over one time period, i.e.,

$$\sigma_x^2 = \frac{1}{T_0}\int_0^{T_0} x^2(t)\,dt = \frac{1}{T_0}\int_0^{T_0}[A\cos(2\pi f_0 t)]^2 dt = \frac{1}{T_0} \tag{2.6.2}$$

where we utilize the fact that the average value of a sinusoid, evaluated over an integral number of periods, is zero. Based on this computation we consider the scheme shown in Fig. 2.47.

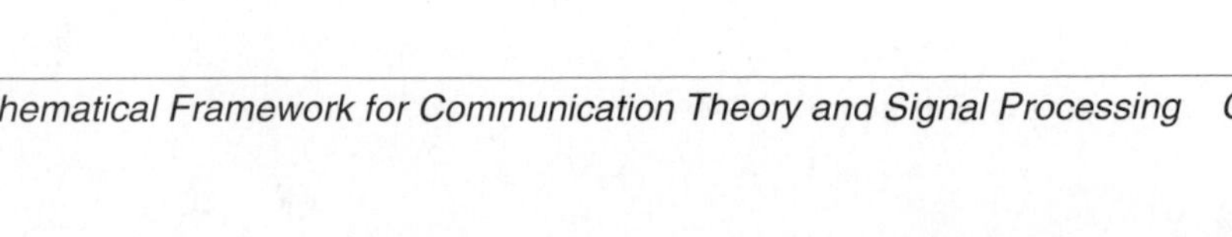

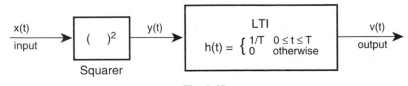

Fig. 2.47
Approximating the estimate of power of a signal by a finite average

The idea is to apply an **LTI** system to the square (or squared-magnitude in the case of complex-valued signals) of x(t). The **LTI** is equivalent to a moving average filter that averages, i.e., integrates, the input signal over an interval of T units. Thus if T is an integral multiple of the period of x(t) then the output v(t) will be a constant and equal to the power of x(t). This scheme provides the exact value of power for all periodic signals whose period is a submultiple of the averaging time T. If T is large, then the averaging time will encompass a large number of complete periods and one partial period. Consequently, v(t) will be approximately constant and will be a reasonably accurate estimate of the power. The scheme provides inexact values of power if the input has a fundamental frequency that is small, comparable to (1/T), since in this case the interval T encompasses a partial period and only a few complete periods of the input. If the input is an ergodic random process, the output of the filter, v(t), tends, in the limit, to the power of the process; the larger the value of T, the better the estimate.

If the signal is periodic, then the action of this scheme can be analyzed by first expressing the periodic input signal, x(t), as a Fourier Series

$$x(t) = \sum_{k=-\infty}^{\infty} c_k e^{j2\pi k f_0 t} \quad \text{where} \quad f_0 = \frac{1}{T_0} \tag{2.6.3}$$

The power of x(t), σ_x^2, is given by

$$\sigma_x^2 = \frac{1}{T_0}\int_0^{T_0} |x(t)|^2 \, dt = \sum_{k=-\infty}^{\infty} |c_k|^2 \tag{2.6.4}$$

where Parseval's relation is invoked to express the power of x(t) as the sum of the powers of its constituent complex exponential components.

The action of squaring x(t) is equivalent to applying a memoryless nonlinearity and therefore y(t) will be periodic and its Fourier series can be written as

$$y(t) = |x(t)|^2 = \sum_{k=-\infty}^{\infty} \beta_k e^{j2\pi k f_0 t} \tag{2.6.5}$$

where the Fourier coefficients $\{\beta_k\}$ of y(t) are related to $\{c_k\}$ of x(t). Of special interest is

$$\beta_0 = \sum_{k} |c_k|^2 \tag{2.6.6}$$

That is, the DC component of y(t) is exactly equal to the power of x(t). y(t) also has components at multiples of f_0. That is, β_k could be nonzero for indices k other than zero. The output of the **LTI** in Fig. 2.47 will be equal to the power of x(t) if, and only if, these components are removed. With the specific choice of impulse response in Eq. (2.6.3), the frequency (magnitude) response is given by

$$|H(f)| = |\text{sinc}(fT)| \tag{2.6.7}$$

which is depicted in Fig. 2.48 (not to scale).

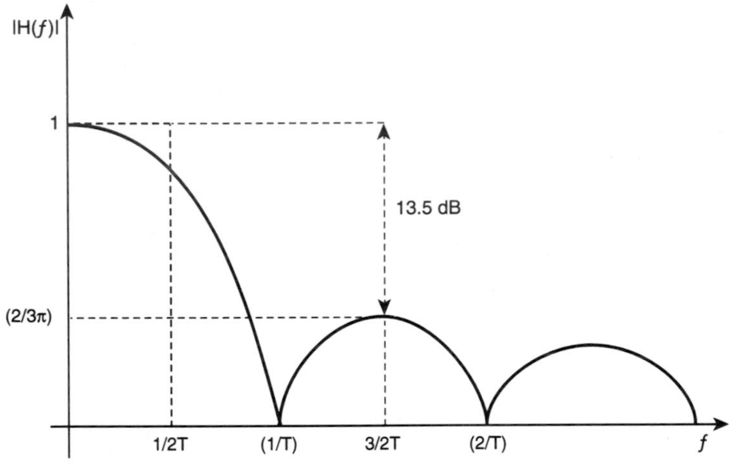

Fig. 2.48
Frequency response (magnitude) of moving average filter

The filter is a coarse approximation to an ideal lowpass filter with cutoff frequency (1/2T). The filter does provide nulls at frequencies that are multiples of (1/T). From Eq. (2.6.5) it is clear that y(t), apart from its DC component, has terms corresponding to sinusoids with frequencies that are multiples of $f_0 = 1/T_0$. Consequently, if T is an integer multiple of T_0, then the filter will annihilate these components and the output will be a constant, exactly equal to σ_x^2.

The attenuation provided by the filter follows a (sin(x)/x) pattern. Therefore, if $T_0 << T$ then the first harmonic component of y(t), which will be a component of frequency f_0, will be attenuated significantly. This is equivalent to saying that if the number of periods encompassed is (N+ε), where ε is a fraction, then the error introduced by the fractional period will be small if N is large. If the harmonics of y(t) were at frequencies other than multiples of (1/T) then these could "leak" into the output and corrupt the power estimate.

Nevertheless, the concept of **averaging** or **lowpass filtering** the square of a signal, x(t), provides a signal, v(t), that is approximately constant at a value representative of the power of x(t). The narrower the passband of the filter, the better the approximation; the higher the frequency corresponding to the lowest harmonic component of the squared signal, the better the approximation. It can never be overemphasized that the measurement will always be an approximation to the true (analytical) power value.

The general form of such power measurement schemes is shown in Fig. 2.49.

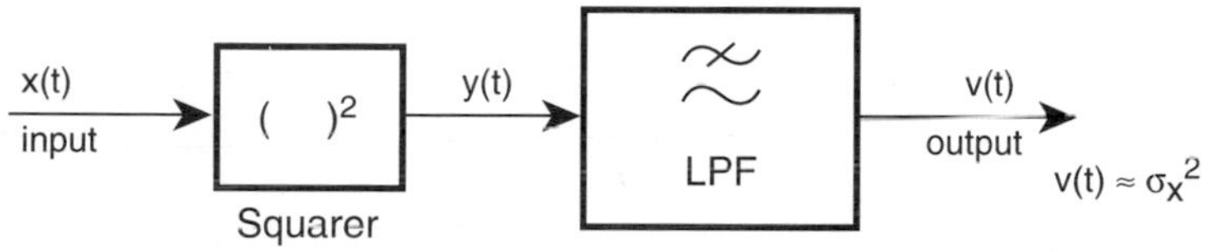

Fig. 2.49
Power estimation by squaring and lowpass filtering

In using such a scheme, the following points should be kept in mind.

i) The accuracy of such a scheme depends on the bandwidth of the filter. The narrower the bandwidth of the filter, f_c, the more accurate is the power estimate. A narrow bandwidth is equivalent to a long integration time. For example, the filter impulse response in Fig. 2.47 corresponds to an integration time of T and a bandwidth of, roughly, $f_c = (1/T)$.

ii) The accuracy also depends on the frequency content of the signal being measured. If the major components of power are at frequencies that are much greater than f_c then the power measurement will be more accurate. If the signal, x(t), has most of its power at low frequencies then the scheme would have to be modified. The applicability of the scheme does require some knowledge of the power spectral density of the signal.

iii) By taking a time average we are implicitly assuming that the signal being measured is ergodic.

2.6.2 Weighted Power Measurement

The estimation of power discussed earlier is referred to as ***flat*** measurement since all frequency components of the signal are treated equally. In telecommunications systems the measurement of noise power is usually ***weighted***. The idea of weighting is to treat the noise in a manner that mimics the human auditory system. That is, the signal components at different frequencies are weighted differently. This is achieved by applying a specific filter to the signal prior to making a power measurement as depicted in Fig. 2.50. Since we are considering power, the frequency response of the weighting filter is specified only in terms of magnitude response, because the effect of the filter on the power spectral density is expressed completely by $|H(f)|^2$.

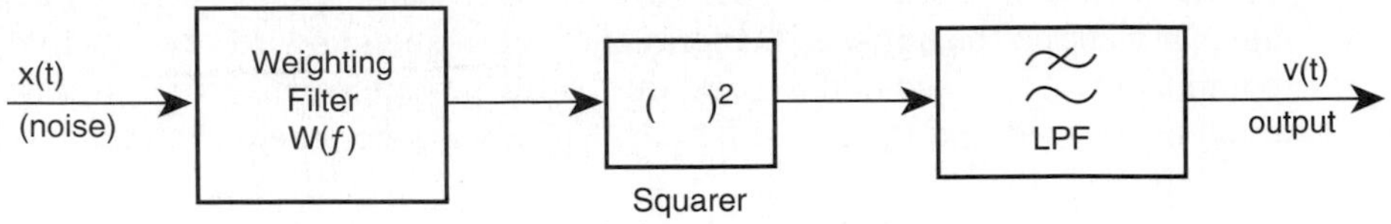

Fig. 2.50
Prefiltering to provide a frequency-weighted power estimate

There are two weighting schemes that are commonly used. In North America the weighting is referred to as ***C-message***; in Europe and most of the rest of the world the weighting is based on the ***psophometric*** or ***p-message*** characteristic. The two are quite similar and both attempt to weight the noise spectrum in a manner representative of the weighting applied by the human ear. The frequency response of the **C**- and **p**-message filters is shown in Fig. 2.51. Note that both filters "deweight" low frequencies, below 300 Hz, and high frequencies, above 3400 Hz, which is in keeping with the notion of a telephone channel being a bandpass filter with these frequencies as the extremes of the passband.

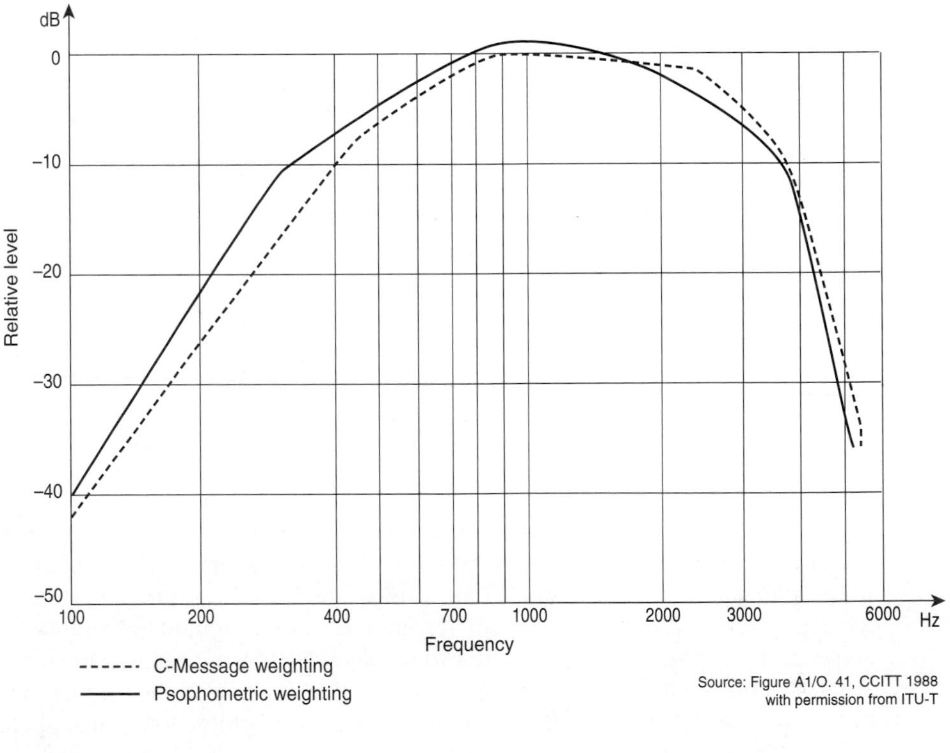

Fig. 2.51
Spectral characteristics of **C**-message and **p**-message weighting filters

Details on the C-message and psophometric weighting filter responses and the manner by which to calibrate equipment is described in **CCITT** Recommendation **O.41** [1.11]. Greater detail on the C-message filter and the requirements imposed on measurement devices to conform to C-message measurements can be found in Bell Pub. 41009 [1.12].

When a power level is specified following a weighted measurement, it is conventional to include a "C" or a "p" in the unit to indicate a C-message or psophometric weighted measurement. For example, it is common North American practice to express noise power in dB relative to 1 picowatt (1 pW is equivalent to –90 dBm or 90 dB smaller than 1 milliwatt). Noise power relative to 1 pW is given the unit **dBrn**, which stands for "dB relative to reference noise"; when C-message weighting is applied, the unit is modified to **dBrnC**. When powers are specified relative to a transmission level point, the units are **dBrn0** (unweighted) and **dBrnC0** for C-message weighted measurements and constitute decibels with respect to a reference noise level (i.e., 1 pW) at a 0 **TLP**. The unit dBrnC0 occurs very often and is verbalized as "debrinko." European practice is to express noise power in picowatts and when psophometric weighting is applied the units used are **pWp** or **pW0p** where the latter unit indicates that the measurement is at a 0 **TLP**.

The **CCITT** also specifies what is called a *3-kHz flat* weighting characteristic to mimic the expected frequency response of a voice channel typical to telephony. The weighting is indeed "flat" with 3-dB points at 300 Hz and 3.4 kHz and a response that is within ± 0.25 dB between 400 Hz and 2600 Hz when considered relative to a reference frequency of 1020 Hz. At frequencies above 3400 Hz and below 300 Hz, the rolloff is specified as greater than 24 dB/octave.

One difference between the **different** weightings is the reference frequency at which the filter response is 0 dB (normalized). The "center" frequencies are 1 kHz, 800 kHz, and 1020 Hz for the C-message, psophometric, and flat weighting filters, respectively. If the signal x(t) is (approximately) white noise over the frequency range [0,4] kHz (one-sided) then the relationships between the the three weightings are approximately

$$\begin{array}{lll} 0 \text{ dBm0 (flat)} & -2 \text{ dBm0C (C-mess.)} & -2.5 \text{ dBm0p (psoph.)} \\ 90 \text{ dBrn0 (flat)} & 88 \text{ dBrnC0 (C-mess.)} & -2.5 \text{ dBm0p (psoph.)} \end{array} \quad (2.6.8)$$

That is, 1 milliwatt at a 0 **TLP** of noise whose power spectral density is flat in the 300 to 3400 Hz band will be measured as 0 dBm with *flat* weighting; C-message weighting will reduce the value by about 2 dB and psophometric weighting by about 2.5 dB, respectively. This "equivalence" does account for the effect of having different reference frequencies for defining the weighting.

2.6.3 Short-Term Power Measurement

Squaring a signal and taking the lowpass filtered version as an estimate of the power of the signal provides the basis for the definition of ***short-term power***. If the impulse response of the lowpass filter is h(t), then the output v(t) is given by

$$v(t) = \int_{-\infty}^{t} y(\tau)h(t-\tau)\,d\tau = \int_{0}^{\infty} h(\tau)y(t-\tau)\,d\tau \tag{2.6.9}$$

where y(t) is the square of the input signal x(t). If x(t) is complex-valued, then y(t) is the magnitude-squared version of x(t). The output v(t) is thus a weighted average of y(t) over time. The temporal extent to which this weighting extends depends on the effective length of the impulse response or the memory of the lowpass filter. Strictly speaking this memory is infinite; for practical purposes however, it is useful to define this length in the following way. Let

$$E_h = \int_{0}^{\infty} |h(t)|^2\,dt \tag{2.6.10}$$

denote the energy of the impulse response, which is assumed to be real-valued and causal. Then, given any suitable percentage value P, say P=99%, there corresponds a length, L units of time, such that

$$\frac{P}{100} E_h \le \int_{0}^{L} |h(t)|^2\,dt \le E_h \tag{2.6.11}$$

The quantity L, parameterized by P, is considered the length of the response.

For all practical purposes then, the output v(t) is a weighted average of y(t) over the past L units of time. If x(t) is periodic or an ergodic process then v(t) will be, roughly, constant; if x(t) is finite energy then v(t) will peak when the time window of L units brackets the time interval where x(t) is nonzero; if x(t) has time varying characteristics, then v(t) would represent a measure of the strength of x(t) on a short-term basis as a pattern with respect to time.

This particular viewpoint is useful in defining a ***level*** for signals such as speech that tend to be bursty in nature. In a typical (speech) conversation there could be extended pauses of silence and brief pauses between words. Defining the power of such a signal as a long-term average, including periods of silence in the measurement, would not be reflective of the characteristics of the speech signal. To account for these periods of silence, the following approach is taken for measuring the level of a speech signal.

Consider that a speech signal, x(t), is applied to the power measurement device represented in Fig. 2.49 or 2.50. The output, v(t), will not be constant but will have a form depicted in Fig. 2.52.

One of the principal uses for a short-term power measurement, especially for speech signals, is to establish whether the signal is ***active*** or ***silent***. If some suitable threshold is chosen, the speech signal is considered active when the short-term power signal, v(t), is above the threshold and silent when v(t) is below the threshold. While this distinction is logical, it does not take into account one of the properties peculiar to speech signals. The nature of speech is such that, almost regardless of the threshold chosen, there is some speech content in a short interval beyond the instant that the power drops below the threshold. To account for this phenomenon, the speech level is computed as the average of v(t) over the intervals where the signal is clearly active (v(t) is greater than the threshold), as well as a hangover time beyond the instant when the short-term power dips below the chosen threshold. This hangover time is typically of the order of 60 msec. A secondary implication of including a hangover time is that momentary dips in signal power that occur between syllables are *bridged.* With reference to Fig. 2.52, the speech signal is considered active not only over the regions labeled T_A, but also for a short time T_H beyond T_A.

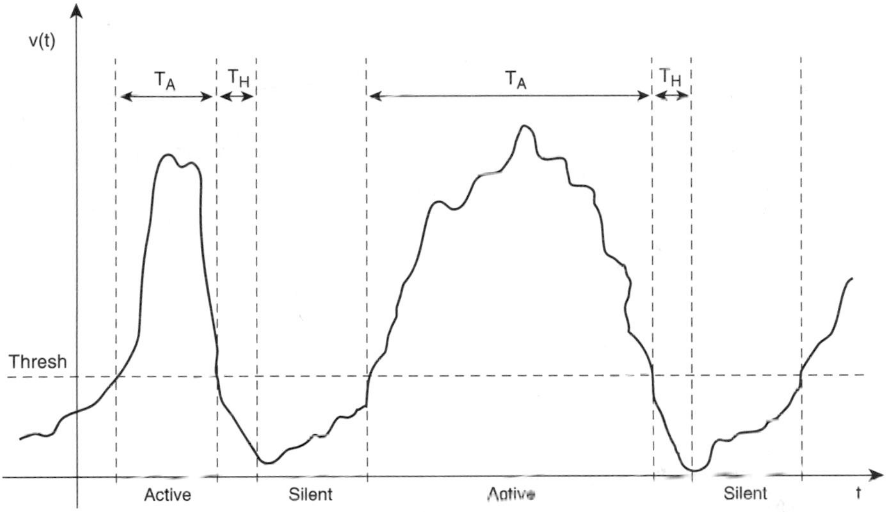

Fig. 2.52
Typical trends exhibited by short-term power measurements on a speech signal

The choice of threshold and method of averaging the short-term power over the active periods is somewhat arbitrary. The **CCITT**, in Recommendation **P.56** [1.13] provides a specification called the "objective measurement of active speech level." The principal reason for specifying the algorithm is that some degree of conformity can be achieved in different equipment designed by different engineers and built by different manufacturers.

One instance where determining when the speech signal is active or silent arises in speech compression devices that perform what is called "speech interpolation." Since these methods are usually performed using digital signal processing, the term **DSI** for Digital Speech Interpolation is used. The intention of **DSI** is to reserve transmission bandwidth for a speech signal only during those intervals when it is active. When the speech is silent, no transmission bandwidth is assigned. At the receiver periods of silence are filled in using some form of noise to mimic the effect of a channel that did not use **DSI**. Clearly, determining whether the speech is active or silent is crucial to the success of such a bandwidth compression scheme. If speech is erroneously classified as *active* then bandwidth, a valuable resource, is being used to transmit only background noise; if the speech is erroneously classified as silence then the receiver will be obliterating sections of speech and replacing it with noise, giving rise to a condition whereby the speech is *clipped* and sounds "choppy." In Chapter 5 we describe the operation of a speech compression unit that utilizes **DSI** and elaborate on the use of short-term power for determining the allocation of bandwidth.

2.7 PRINCIPLES OF DATA TRANSMISSION

The previous sections assumed that the information-bearing signal was analog, such as a speech signal. In this section we discuss some of the principles underlying the transmission of data, typified by binary signals that, at any instant of time, represent one of two values. These two values are associated with logic levels "**1**" and "**0**." Other nomenclature for these two levels include "**HIGH**" and "**LOW**," "**MARK**" and "**SPACE**," and so on. The principal distinction we make between analog and data signals is that for analog signals we attempt to reproduce the entire waveform at the receiver; for data signals it suffices that the level at any time epoch be reproducible. Data signals also incorporate the notion of "time-discretization" in the sense that the data value changes only at certain instants of time and, for all practical purposes, can be considered to be constant between these instants.

Typical binary signals are shown in Fig. 2.53 and Fig. 2.54. In Fig. 2.53 the two binary levels corresponding to **1** and **0** are taken, arbitrarily, to be +V and –V (volts) respectively. These data values change every T units of time (sec). The information content of the signal is the sequence of bits ..,**1**,**0**,**0**,**1**,**0**,.. In general, the information content of the signal can be characterized by the sequence of values $\{b(n) ; n = \ldots -1,0,1,2, \ldots\}$ together with the notion of ***bit time***, T. While Fig. 2.53 depicts a waveform that remains constant between transitions, this is not necessary. Knowledge of the sequence and the bit time allows us to reproduce any waveforms we may so choose, up to an uncertainity of the absolute location of the time origin, which is usually not of interest anyway. The only constraint on the waveform is that it be readily distinguishable what the value of the binary signal, either 1 or 0, is present in each bit time. Fig. 2.54 depicts two other waveforms that could be used to represent the same sequence of bits.

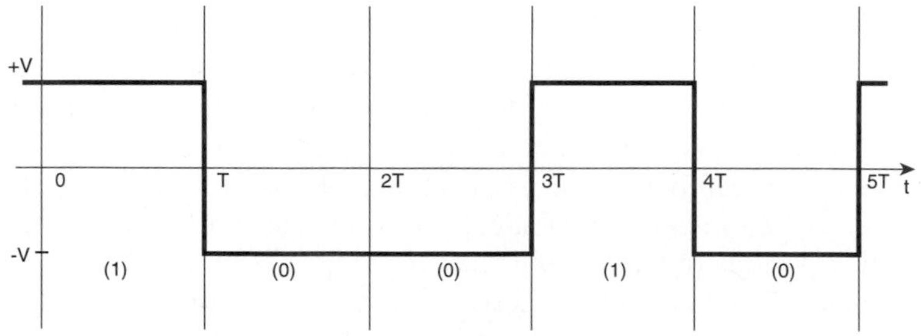

Fig. 2.53
Waveform of a data signal corresponding to the bit sequence ..., 1, 0, 0, 1, 0, ...

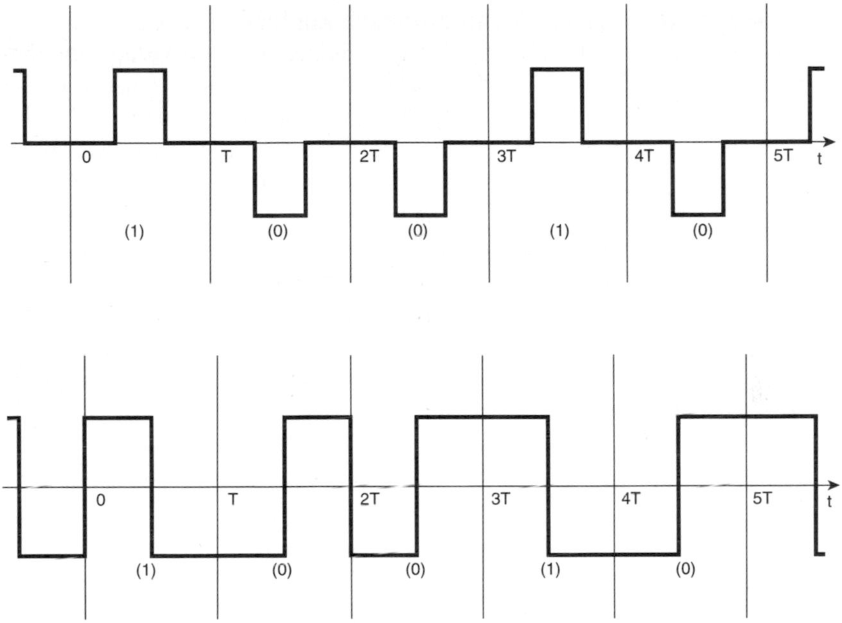

Fig. 2.54
Alternate waveforms for representing a data signal corresponding to the bit sequence ..., 1, 0, 0, 1, 0, ...

2.7.1 Elementary Data Transmission

A simple data transmission system is depicted in Fig. 2.55. The binary information is converted into a electrical signal x(t) depicted in Fig. 2.53 and a suitable line driver (amplifier) launches the signal over the channel. In this simple case the channel is nothing more complicated than a cable pair of some length. The receiver uses a buffer amplifier to account for loss of signal amplitude over the cable and samples the waveform every T sec. Based on the value of the sample the receiver decides whether a **1** or a **0** was transmitted in that bit time.

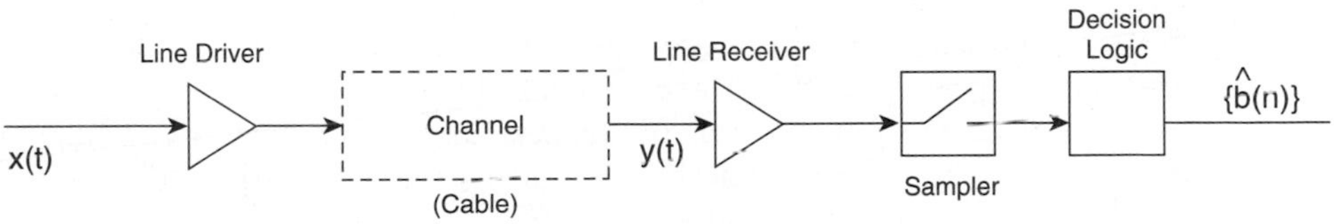

Fig. 2.55
A simple data transmission scheme

If the cable is short, the received waveform is substantially the same as the transmitted waveform. That is, if the bandwidth of the channel is very large, and linear phase, the transmitted signal will not be altered significantly in waveshape. Given this situation, one expects that the sampled values at the receiver correspond to the transmitted bit-stream. This is indeed true provided that the receiver and transmitter have the same time base. That is, both sides must have the same notion of bit time. This equality of time base, or ***synchronization***, is crucially important. If the transmitter and receiver have different notions of T, the bit time, then we conceivably have the situation where the transmitter has launched N bits in a time interval whereas the receiver is expecting (N–1) or (N+1) bits in the same (absolute) time frame. In a practical situation this effect of dropped bits, or stuffed bits, can be catastrophic, much more so than a bit received in error.

We shall assume that the transmitter and receiver are in synchronization not only from the viewpoint of bit time, but also phase, whereby the receiver also has knowledge of the (absolute) time instant where a bit starts and ends and thus can choose its sampling instant to correspond to, for example, the middle of the bit interval, mid-bit sampling, or the end of the interval. This is a major assumption. Practical data receivers often use complex processes to establish the notion of bit time and bit-start/bit-end.

When the cable is short the sampling phase, that is, the relative position within the bit interval where the sample is taken, is not crucial. If the cable length is increased, then the channel modifies the shape of the signal. That is, the received signal, y(t), in Fig. 2.55 does not appear similar to the transmitted signal x(t). As a first approximation, the effect of the cable can be modeled as that of a simple RC circuit considered in Section 2.3 and depicted in Fig. 2.15. The effective frequency response of the channel is

$$H(f) = \frac{Y(f)}{X(f)} = \frac{1}{1 + j2\pi f\tau} \qquad (2.7.1)$$

where $\tau = RC$ (seconds) is the time constant of the RC network. $H(f)$ can be viewed as a lowpass filter with (3–dB) cutoff frequency $f_c = (1/2\pi\tau)$ (Hz). The effect of the lowpass filter is to round out the edges of the waveform x(t). If τ is reasonably small compared to T, the bit time, then y(t) will take the shape shown in Fig. 2.56. The important message in the figure is that first there is a spillover of signal between bit times and second that the relative position within the bit time that the signal is sampled is no longer unimportant. Clearly, it would be advantageous to sample the value of y(t) towards the end of the bit time to minimize the impact of previously transmitted bits. Nevertheless, the sampled values will not be binary valued; there will be some component from prior bit times. This deleterious effect is called ***inter-symbol-interference*** (**ISI**) and is directly related to the fact that the channel has finite bandwidth. The narrower the bandwidth the more the **ISI**.

Because we know that the data is binary-valued, the detection procedure of determining what data was transmitted, a 1 or a 0, is achieved by decision logic that is called data slicing. A data slicer is a comparator that determines a suitable threshold level and decides that the transmitted bit was a **1** if the sample was above this threshold, and **0** otherwise. An intuitively satisfying value for the threshold is the midpoint between the positive and negative peak excursions. Observation of the waveform indicates that there is some impact of the instant at which the signal is sampled, i.e., the sampling phase. Generally speaking, if there is no **ISI** then the sampling phase is moot; if there is some **ISI** then it is better to sample the signal toward the end of the bit time to allow the impact of prior bits to "die away."

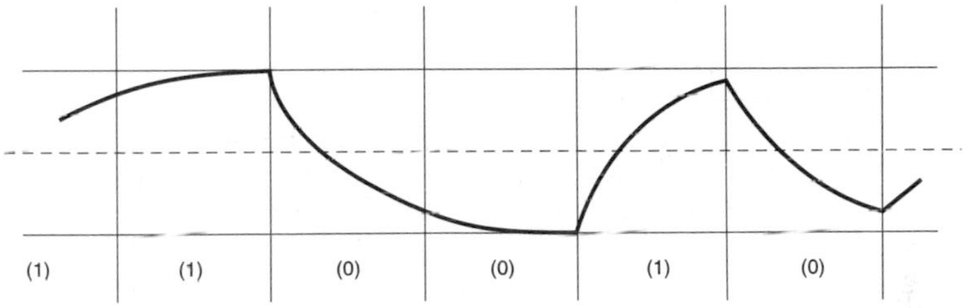

Fig. 2.56
Received waveform for the binary sequence ..., 1, 0, 0, 1, 0, ...

If one looks at a typical sampling instant, say t=nT, then the signal value observed at the sampling instant will depend not just on the transmitted waveform in bit interval n, but also on all prior intervals, albeit with decreasing impact as we go farther back in time. If we draw out the waveforms corresponding to every possible pattern we get a pattern of lines of the form shown in Fig. 2.57. Such a diagram is called an ***eye pattern***. The clear space between the lines at t=nT is called the ***eye opening***. If **ISI** is minimal then we have a wide eye opening; the opening decreases with increasing intersymbol interference. If the sampling is not precisely at the time instant t=nT then the receiver is not utilizing the widest opening of the eye. The eye opening for a transmitted data signal, i.e., y(t), can be observed on an oscilloscope that has some storage capability by transmitting a random bit pattern and synchronizing the sweep of the oscilloscope to correspond to the bit time. The implication of a wide eye is that the data slicer makes the correct decision more often than not; a narrow eye indicates that the transmission is susceptible to noise and/or imperfect sampling phase.

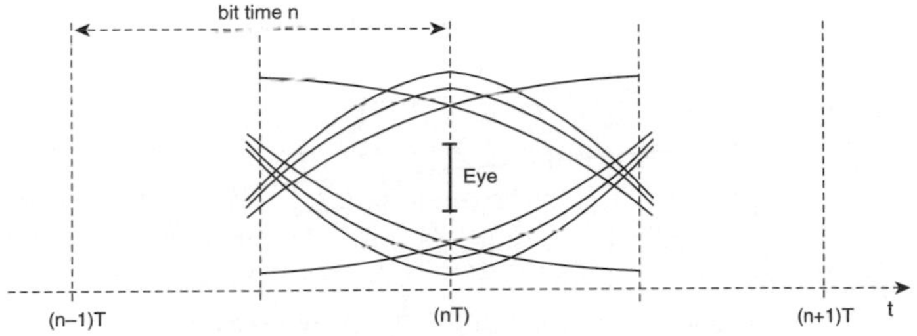

Fig. 2.57
Typical eye pattern indicating an open eye

As we increase the length of the cable, i.e., reduce the channel bandwidth, the waveform y(t) is distorted to a much greater extent and clearly the eye opening is smaller. When the channel bandwidth is reduced beyond a certain point the eye closes, indicating that **ISI** is so great that data cannot be transported over the channel with any degree of reliability. This conclusion is not restricted to the simple scheme discussed but is a general result that is applicable to all data transmission systems and is formulated more precisely in the following discussion of Nyquist's theorem.

2.7.2 Nyquist's Theorem Applied to Data Transmission

Nyquist's theorem provides an upper bound to the rate at which we can transmit information over a bandwidth limited channel. It is a statement as to the limit imposed by intersymbol interference. Details on the theorem can be found in Nyquist [1.5] or Korn [1.4] and in other references given in the bibliography at the end of the chapter.

In order to discuss Nyquist's result, we need to introduce the notion of a **symbol**. In the previous section we concentrated on the case where the information in any bit time was binary valued, either a **1** or a **0**. The notion of a symbol extends the possible values that the information can take to a multiplicity of levels, usually finite, and usually a power of 2. For example a ternary symbol takes on three values, say "1," "0," and "–1"; a quarternary symbol takes on four values, say "–3," "–1," "1," and "3." An N-level symbol is one that takes on values from an alphabet of N possibilities. By convention, the values used are represented by integers, positive and negative, and such that if "i" is a value then so is "–i" (symmetry). From a bit-stream we can create a symbol stream by taking n successive bits, in blocks, corresponding to an alphabet of size 2^n. The symbol rate will be (1/n) times the bit rate and the symbol-time will be n times the bit time. The information content would be n bits per symbol.

Nyquist's theorem states that the maximum symbol rate that a channel of bandwidth f_c Hz (one-sided) can support is $2f_c$ symbols per second. This extraordinary result is independent of the size of the symbol alphabet. Taken at face value, this result implies that we can transmit very high (infinite) bit rates over any channel that has nonzero bandwidth just by making the size of the alphabet large (enough). This is clearly not practical since making the alphabet size large would involve delay since several bits would have to be buffered (n bits for an alphabet size of 2^n), and secondly Nyquist's theorem does not include the impact of additive noise. The proper interpretation of Nyquist's result is that of an upper bound on the maximum symbol rate that can be transmitted without intersymbol interference.

We can justify this result in the following way. Suppose the size of the alphabet is N and the symbol rate is $f_0 = 1/T_0$. Then the transmitter launches a pulse every T_0 units of time where the pulse launched at time $t=nT_0$ is representative of the symbol to be transmitted at $t = nT_0$. In the most general case there are N different pulse shapes. For simplicity of analysis we shall assume that there is one underlying pulse shape, p(t), and the symbol information is carried in the amplitude of the pulse. That is, the pulse launched for symbol at $t=nT_0$ is written as

$$\text{For symbol at } t = nT_0 \text{ pulse is } b(n)\, p(t - nT_0) \tag{2.7.2}$$

implying that the symbol at $t = nT_0$ is characterized by an amplitude, b(n). The value taken on by b(n) will be one of N different voltages corresponding to the symbol alphabet. For the purposes of analysis, we model the pulse generation as the output of a filter whose impulse response is p(t) excited by an impulse at time $t = nT_0$ of strength b(n). That is

$$b(n)\, p(t - nT_0) \;=\; [b(n)\, \delta(t - nT_0)](*)\, p(t) \tag{2.7.3}$$

Consequently the overall output of the transmitter can be modeled as that of the filter excited by an impulse train. A block diagram of the transmitter is shown in Fig. 2.58. This method of encoding the identity of a symbol in terms of the amplitude of a pulse is called Pulse Amplitude Modulation (**PAM**).

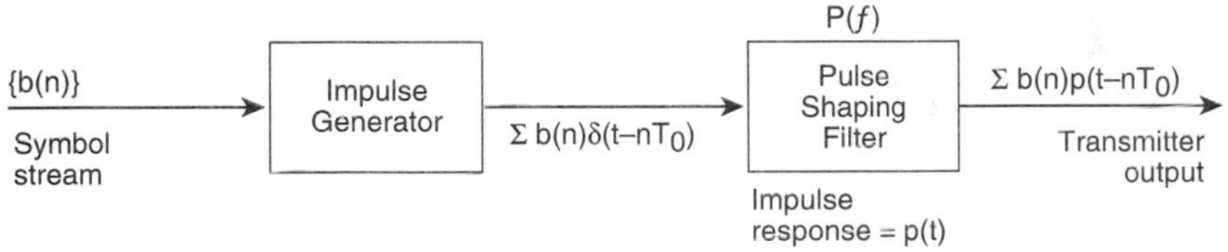

Fig. 2.58
Model for Pulse Amplitude Modulation (**PAM**)

With this representation, the overall scheme can be modeled as in Fig. 2.59. The transmitter generates a train of pulses *modulated* by the symbol stream {b(n)}; these pulses are modified by the channel whose frequency response is expressed by the function H(*f*); the front end of the receiver may introduce additional frequency shaping via a response Q(*f*); the subsequent waveform is sampled at the symbol rate; the decision logic determines, based on this sample value, which of the N possible alphabet was transmitted in that symbol interval.

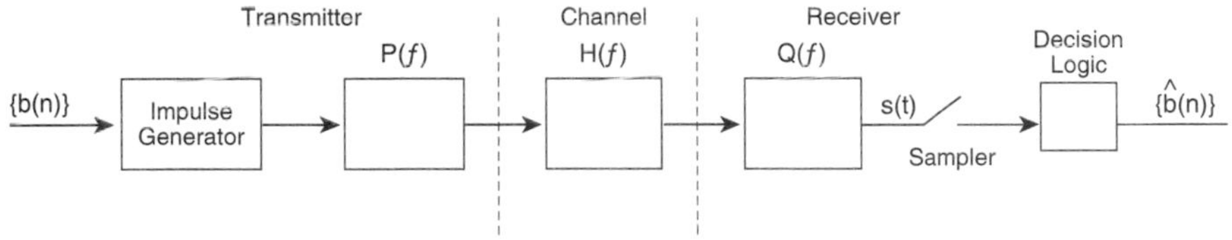

Fig. 2.59
Model for **PAM**-based symbol transmission and decoding

The decision logic takes the form shown in Fig. 2.60. In our discussion of quantization in Chapter 4 the input-output characteristic shown will be called a "mid-riser" since there is a step at the origin. The units of the abscissa are normalized to indicate that the gain of the channel and absolute voltage levels of the symbol waveforms have been accounted for. If there is no **ISI** or additive noise then the only sample values observed are ±1, ±3, ... and the decision is error-free. Errors are introduced if the contribution of the **ISI** and/or noise exceeds 1 normalized unit.

If we combine the frequency response of the three entities into a single filter, $G(f)$, the analysis model for the data transmission scheme is depicted in Fig. 2.61. For each symbol transmitted, the receiver experiences the waveform $g(t-nT)$. The sample at time $t=mT_0$ will consist of contributions from **all** transmitted symbols. If $g(t)$ is a *causal* impulse response then the **ISI** is a function of all the symbols transmitted **prior** to the current one. That is,

$$s(mT_0) = \sum_n b(n)\,g(mT_0 - nT_0) \tag{2.7.4}$$

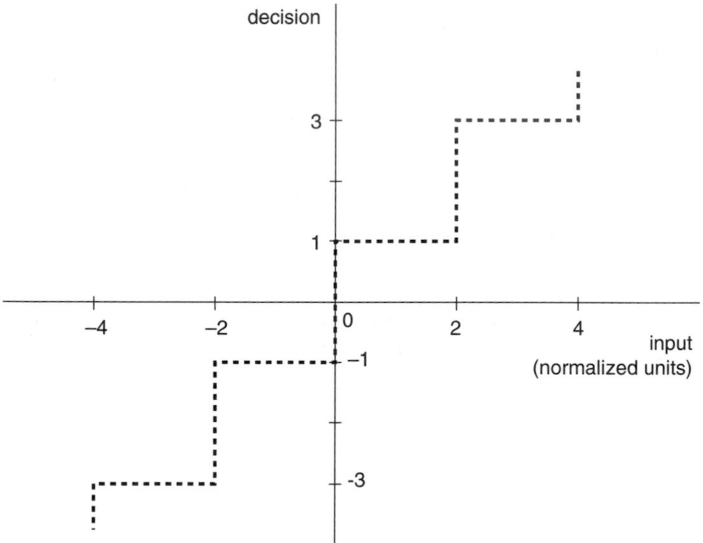

Fig. 2.60
Input-output characteristic of the decision logic (a mid-riser quantizer)

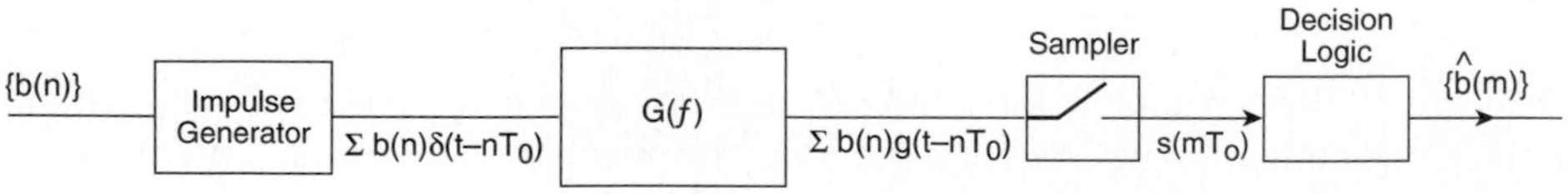

Fig. 2.61
Model for analyzing intersymbol interference in **PAM**

For the sample at time mT_0 to be representative of a single symbol only, say b(m), the condition on the overall response of the system can be expressed as

$$g(nT_0) = \begin{cases} \text{constant } (=1) & \text{for } n = 0 \\ 0 & \text{for all other } n \end{cases} \tag{2.7.5}$$

Some points regarding the conclusion expressed by Eq. (2.7.5) are in order. First is that we have assumed that the sampler is in perfect synchronization with the transmitter. This is not always true. Second, there is no delay between the sender and receiver. This is artificial—a fixed delay of $(kT_0 + t_0)$ can always be accounted for in the sampling process; the zero delay assumption is solely for notational convenience. Third, by assuming that the constant in Eq. (2.7.5) is 1, that there is a known fixed gain (or loss) that has been corrected for. In practice, data transmission is preceded by a training sequence whereby the transmitter and receiver achieve synchronization, determine the necessary gain, and establish the transmission delay. In simple schemes these parameters are assumed to be constant for the duration of the actual data transmission segment; in more complex schemes adaptive techniques are used to continually monitor and account for any changes and/or inaccuracies in the initial estimate.

An impulse response satisfying Eq. (2.7.5) is shown in Fig. 2.62. In particular, the value at the sampling instants nT_0 is zero except for n = 0. This implies that the symbol launched at time mT_0 contributes to the sampled voltage only at the sampling instant $t = mT_0$. Equivalently, at the sampling instant mT_0 the voltage value depends only on the symbol at mT_0 and the contribution from all other symbols will be zero. Eq. (2.7.5) is thus the condition of zero-**ISI**.

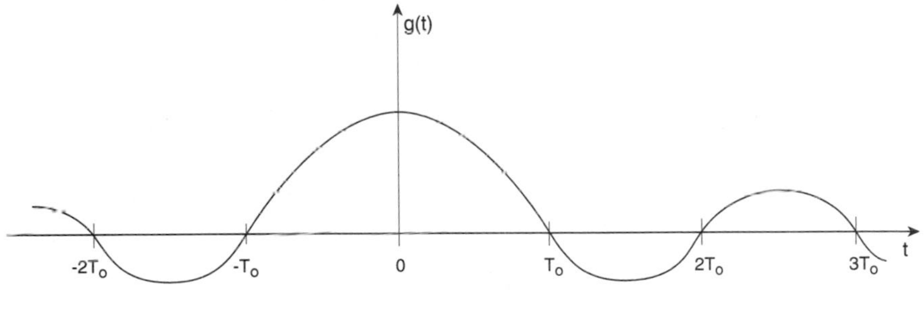

Fig. 2.62
Impulse response, g(t), that satisfies the condition for zero intersymbol interference

What is the equivalent of Eq. (2.7.5) in the frequency domain? To determine this condition, we first express the impulse response, g(t), in terms of the frequency response $G(f)$ via the inverse Fourier transform. In particular, the impulse response at the instants mT_0 is

$$g(mT_0) = \int_{-\infty}^{+\infty} G(f)\, e^{j2\pi m T_0 f}\, df \tag{2.7.6}$$

Since the exponential function, i.e., the complex sinusoid, in Eq. (2.7.6) is periodic, the expression for $g(mT_0)$ can be manipulated to read

$$g(mT_0) = \sum_{k} \int_{-\frac{1}{2}f_0}^{+\frac{1}{2}f_0} G(f - kf_0)\, e^{j2\pi m T_0 f}\, df \tag{2.7.7}$$

where $f_0 = 1/T_0$ is the symbol rate expressed in Hz and the sum is over all integer values of k.

Defining the function $R(f)$ by

$$R(f) = f_0 \sum_{k=-\infty}^{\infty} G(f - kf_0) \tag{2.7.8}$$

it can be seen that $R(f)$ is a periodic function of frequency with period f_0 and that Eq. (2.7.7) can be written as

$$g(mT_0) = \frac{1}{f_0} \int_{-\frac{1}{2}f_0}^{+\frac{1}{2}f_0} R(f)\, e^{j2\pi m T_0 f}\, df \tag{2.7.9}$$

and $g(mT_0)$ can be recognized as the (–m)th coefficient of the Fourier series of the periodic function $R(f)$.

The zero-**ISI** condition of Eq. (2.7.5) requires that all the Fourier series coefficients of $R(f)$, except the DC term, be zero. In other words,

$$R(f) = 1 \text{ for } f \in [-f_0, +f_0] \tag{2.7.10}$$

and the periodic nature of $R(f)$ requires that $R(f) = 1$ for all frequency. This provides us with a frequency domain counterpart for describing the zero-**ISI** condition, namely

$$\sum_{k=-\infty}^{\infty} G(f - kf_0) = \frac{1}{f_0} \tag{2.7.11}$$

Any $G(f)$ that satisfies Eq. (2.7.11) is a valid choice for achieving zero **ISI**. The choice is not unique. For comparison purposes the reader is urged to refer to Eqs. (2.3.25) and (2.4.66) and the associated discussions. The results there are quite similar to the notion of zero **ISI**. In fact the mathematical manipulations are identical, with the roles of f (frequency) and t (time) interchanged. There are several $G(f)$ that satisfy the zero-**ISI** condition, but only one will be bandlimited to f_0 (two-sided). This bandlimited choice of $G(f)$ is the ideal lowpass filter with cutoff frequency ($f_0/2$) shown in Fig. 2.63. With this choice of frequency response, the time domain condition can be easily verified since the inverse Fourier transform is given by

$$g(t) = \frac{\sin(\pi f_0 t)}{(\pi f_0 t)} \tag{2.7.12}$$

and will have zeros at values of t that are nonzero integer multiples of $T_0 = 1/f_0$. Examination of the zero-**ISI** criterion in the frequency domain yields the following conclusion. If the bandwidth of $G(f)$ is less than $(f_0/2)$ then the sum of shifted versions, $G(f + f_0)$ has "holes" in the sense that there are frequencies where the sum would be zero (or very small, anyway) and **ISI** cannot be avoided. The overlap of $G(f)$ and shifted versions thereof is depicted in Fig. 2.64. In the ideal case, when $G(f)$ satisfies the Nyquist criterion, the sum is a constant. If the bandwidth of $G(f)$ is greater than $(f_0/2)$ then there are overlaps between $G(f + nf_0)$ and $G(f + (n+1)f_0)$. In this case we could introduce bandlimiting behavior by reducing the bandwidth of either the transmit filter or receive filter, or both. Rather than bandlimiting, however, which requires ideal (or close to ideal) filters, we shape $G(f)$ by choosing $P(f)$ and $Q(f)$ such that the zero-**ISI** condition is satisfied and use as much bandwidth as the channel will allow.

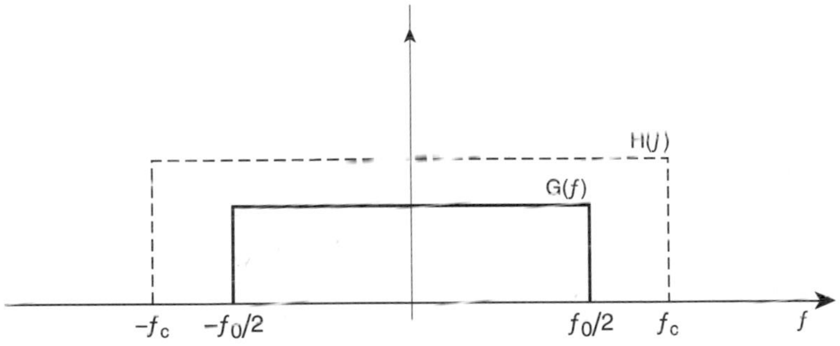

Fig. 2.63
Choosing G(f) as an ideal lowpass characteristic of bandwidth less than the bandwidth of the channel

Recognizing that $G(f)$ is the combination of the channel frequency response $H(f)$ and the transmit and receive filters, it is obvious that the cutoff frequency of the channel, f_c, should be greater than $f_0/2$. A possible representation of H(f) is also shown in Fig. 2.63. In fact, we can extend the combined response $G(f)$ upto the point where $f_0 = 2f_c$ and no further. Since f_0 is the data rate, we have provided a rationale for Nyquist's result, namely that a channel of cutoff frequency f_c can support a maximum symbol rate of $f_0 = 2f_c$ without **ISI**. Achieving this maximum data rate requires that the filters be **ideal**, and therefore not realizable. That is why we always interpret the result in terms of an upper bound on the symbol transmission rate of a practical system.

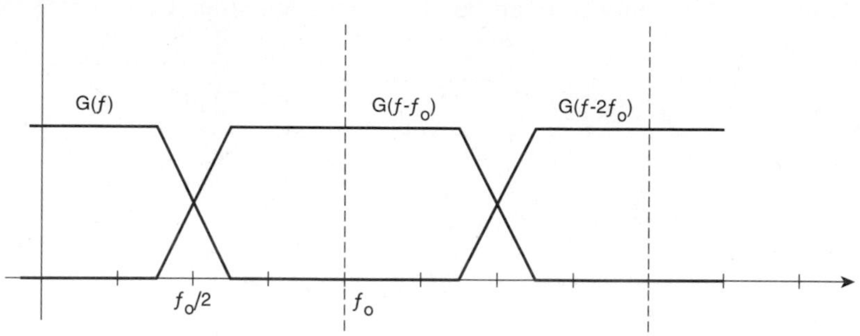

Fig. 2.64
Overlap of adjacent shifted versions of G(f)

The question arises as to whether there are choices for G(f) other than the ideal lowpass filter that will satisfy the zero-**ISI** criterion. There is a family of such frequency responses, generically called "raised cosine pulse(s) with X% excess bandwidth" where X is a parameter. The shape of a raised cosine pulse as a function of frequency is depicted in Fig. 2.65 (normalized to unity at f=0). It is flat up to a frequency ($f_0/2 - \Delta f$), zero beyond f_c, and rolls off as a cosine in between, with a complementary symmetry about $f_0/2$. The ratio of Δf to $f_0/2$, expressed as a percentage, is the parameter X. G(f) is symmetric about $f = 0$. For positive f, G(f) can be expressed mathematically as

$$G(f) = \frac{1}{2}\left\{1 + \cos\left[\left(\frac{\pi}{2\Delta f}\right)\left(f - \frac{f_0}{2} + \Delta f\right)\right]\right\} \text{ for } \left|f - (f_0/2)\right| \le \Delta f$$

$$G(f) = 1 \text{ for } 0 \le f \le \left(\frac{f_0}{2} - \Delta f\right)$$

$$G(f) = 0 \text{ for } f \ge \left(\frac{f_0}{2} + \Delta f\right) \qquad (2.7.13)$$

It can be verified that the equivalent impulse response, g(t) is given by

$$g(t) = f_0 \operatorname{sinc}(f_0 t) \frac{\cos(2\pi \Delta f\, t)}{1 - (4\Delta f\, t)^2} \qquad (2.7.14)$$

and that g(t) and G(f) do indeed satisfy the zero-**ISI** criteria of Eqs. (2.7.5) and (2.7.11)

A popular version of the raised cosine pulse is one with 100% excess bandwidth corresponding to $f_0 = f_c$. Thus in this case the symbol rate is half that of the upper bound set forth by Nyquist's result. The raised cosine pulse has an important advantage over the ideal rectangular pulse (in the frequency domain). This advantage relates to the impact of nonideal sampling. In practical systems it is not possible to guarantee the perfect synchronization between transmitter and receiver with respect to the sampling instant. In the **ideal** (rectangular) case, if the sampling instant deviates from

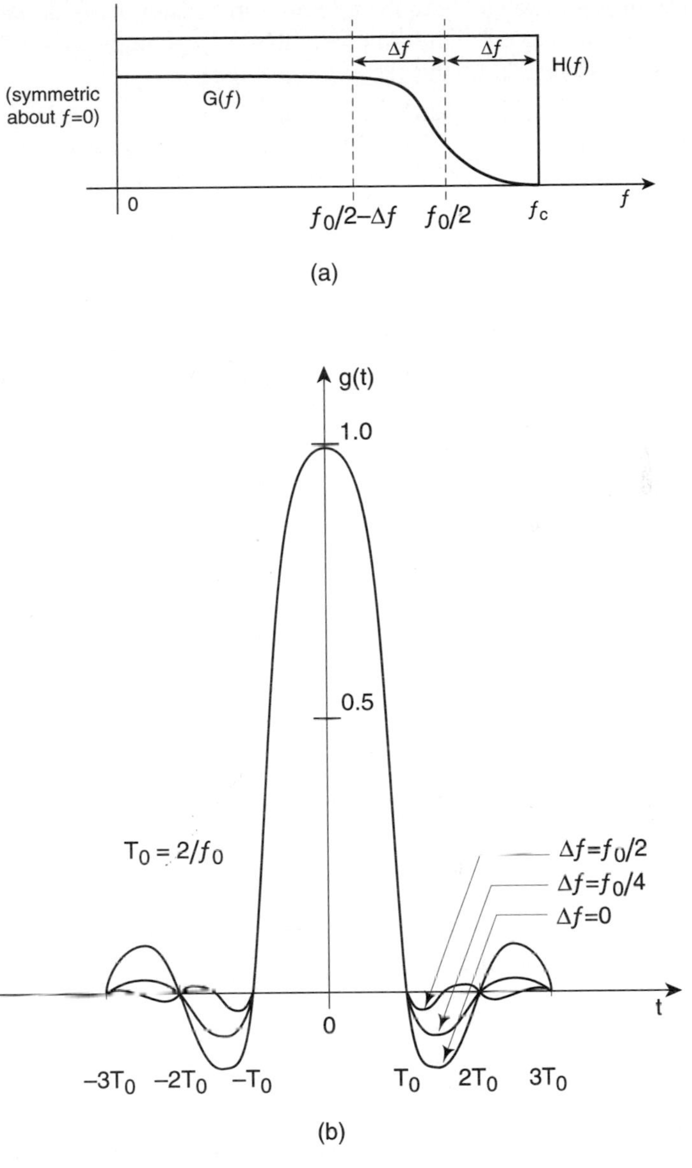

Fig. 2.65
The raised cosine pulse and associated impulse response

the ideal of $t=mT_0$ by a small value, ε, not only is zero-**ISI** no longer true, but the **ISI** is extremely large, regardless of how small the value of ε is. The raised cosine pulse shape is more tolerant of such *timing jitter*, as the deviation from ideal sampling instant is referred to, with the tolerance increasing with increasing excess bandwidth. This is tantamount to the statement that the eye in the case of the raised cosine pulse is wider than that of the rectangular pulse.

The eye opening is inversely proportional to the voltage contribution of **ISI** at the sampling instant. If we do have a timing error and sample the signal at $t = +\varepsilon$ instead of at $t=0$, then (for the case of binary data, say ± 1 on a normalized voltage scale) the **ISI** contribution in a worst-case scenario is

$$|e| = \sum_{n \neq 0} |g(nT+\varepsilon)| \tag{2.7.15}$$

If g(t) is the impulse response of the ideal (bandlimited) filter as given in Eq. (2.7.12), then

$$|e| = \sum_{n \neq 0} \frac{|\sin(\varepsilon)|}{|\pi(n+f_0\varepsilon)|} = \begin{cases} 0 \text{ for } \varepsilon = 0 \\ \infty \text{ for } \varepsilon \neq 0 \end{cases} \tag{2.7.16}$$

Thus a sampling error is devastating, primarily because g(t) decays as $(1/t)$. In the 100% excess bandwidth raised cosine pulse case, g(t) decays as $(1/t^3)$ and the impact of sampling phase error is considerably less.

The raised cosine pulse uses more bandwidth for transmission than that required by Nyquist's theorem but provides a more robust transmission scheme. This illustrates one of the principles of communication system design where bandwidth, usually a scarce resource, is traded-off for improved performance. Other pulse shapes that satisfy the zero-**ISI** criterion can be derived from pulse shapes that are known to satisfy it. For example, if either g(t) or h(t) satisfy the criterion then so will g(t)h(t). Since multiplication in the time domain is equivalent to convolution in the frequency domain, the transforms of these products are of the form $G(f)\,(*)\,H(f)$ and will be *wider* than either of the individual $G(f)$ or $H(f)$.

2.7.3 Data Transmission in the Presence of Noise

So far we have not considered the effect of noise. In practice the channel will have both bandwidth restrictions as well as additive noise. For explaining the impact, however, it is convenient to separate the two effects. Consequently, consider the situation depicted in Fig. 2.66 where x(t) is a binary waveform of the type indicated in Fig. 2.54. The channel is modeled solely as the addition of a noise signal n(t) that has a noise power of σ^2. The sampled signal in bit time n will thus be

$$y(nT) = x(nT) + \eta$$
$$x(nT) = \begin{cases} +V \text{ for } b(n) = +1 \\ -V \text{ for } b(n) = -1 \end{cases} \tag{2.7.17}$$

where the transmitted voltage is V if the data bit was **1** and -V if the data bit was **0**. The entity η represents the contribution of the noise signal at the sampling instant and, assuming the noise is an uncorrelated random process, the contribution at each sampling instant will be essentially independent from bit time to bit time. This noise contribution may cause errors. η can be modeled as a random variable with variance σ^2 (we always assume that such signals are zero-mean).

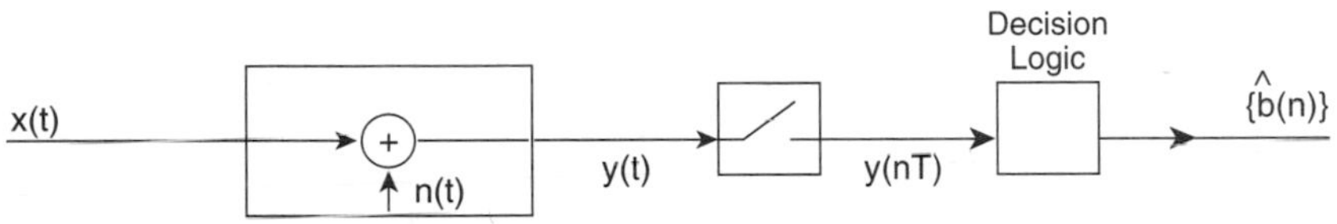

Fig. 2.66
Model for analyzing the impact of noise on data transmission

The data slicer or decision logic decides that a **1** is transmitted if the sampled value y(nT) is greater than a threshold, A, and that a **0** is transmitted otherwise. It is intuitively obvious that the threshold, A, is zero (midway between the voltage excursions expected). Thus if the transmitted bit is a **0** (–V) and the noise contribution exceeds V volts, the data slicer erroneously decodes a **1**. Clearly, the greater the strength of the noise signal, greater are the chances of such an error occurring. The efficacy of transmission is quantified, in a statistical sense, by the **probability of error, $\mathbf{p_e}$**. The probability of error is a useful measure of how well the data transmission scheme is performing—a low probability of error indicates that transmission is good whereas a large $\mathbf{p_e}$ constitutes poor transmission.

In order to compute the probability of error, we can define the following events:

$$\begin{aligned} \mathbf{E} &= \{\text{ Error Occurred}\} \\ \mathbf{A} &= \{\text{ Transmitter sent a } \mathbf{1}\} \\ \mathbf{B} &= \{\text{ Transmitter sent a } \mathbf{0}\} \end{aligned} \tag{2.7.18}$$

Clearly, the transmitter sent either a **1** or a **0**, so **A** and **B**, which are non intersecting sets, together constitute all possibilities. The event **E** is expressed in words as "the decoder decided the opposite from what was transmitted," and $\mathbf{p_e}$ is just the probability of the event **E**. Since **A** and **B** are disjoint, we can write

$$\mathbf{p_e} = \Pr\{\mathbf{E}\} = \Pr\{\mathbf{E|A}\}\Pr\{\mathbf{A}\} + \Pr\{\mathbf{E|B}\}\Pr\{\mathbf{B}\} \tag{2.7.19}$$

The reason for so doing is that it is easier to visualize the conditional events **E|A** and **E|B** than the event **E** directly. In words, the conditional event **E|A** translates to "the decoder decided a **0** was transmitted when actually a **1** was sent". Therefore Pr{**E|A**} is equivalent to $\Pr\{\eta < -V\}$ and Pr{**E|B**} is the same as $\Pr\{\eta > V\}$. Assuming that the data is random, the likelihood of a **1** or **0** would be the same, or Pr{**A**} = Pr{**B**} = 0.5, which yields

$$\begin{aligned} \mathbf{p_e} = \Pr\{\mathbf{E}\} &= [\Pr\{\mathbf{E|A}\} + \Pr\{\mathbf{E|B}\}]0.5 \\ &= [\Pr\{\eta < -V\} + \Pr\{\eta > V\}]0.5 \end{aligned} \tag{2.7.20}$$

The probability of error depends on the probability density function associated with the random variable η. The most common assumption is that η is Gaussian with **pdf** N(0, σ) and, since the Gaussian **pdf** is symmetric (about the mean, which is assumed to be 0), the probability of error is written in terms of the error function **erf**(x) defined in Section 2.5, $\mathbf{p}_e$ is given by

$$\mathbf{p}_e = 0.5\,[1 - \mathbf{erf}(\frac{V}{\sigma\sqrt{2}})] \tag{2.7.21}$$

The probability of error is related to the ratio (V/σ). Since **erf**(0)=0 and **erf**( )=1, Eq. (2.7.21) says quantitatively what we expect qualitatively, namely that Pr{**E**} is small when the signal is strong compared to the noise and worsens as the signal-to-noise ratio becomes small. Since the power of the signal component is V^2 and the noise power is σ^2, we define the signal-to-noise ratio (in dB) as

$$\text{SNR} = 10\log_{10}\left(\frac{V}{\sigma}\right)^2 = 20\log_{10}\left(\frac{V}{\sigma}\right) \tag{2.7.22}$$

and we can express the probability of error in terms of **SNR** using Eq. (2.7.21). Curves of $\mathbf{p}_e$ versus **SNR** are shown in Figs. 2.67 and 2.68. The probability of error is seen to drop rapidly as the **SNR** increases and for **SNR**s greater than about 14 dB, is less than one in a million if the noise is Gaussian.

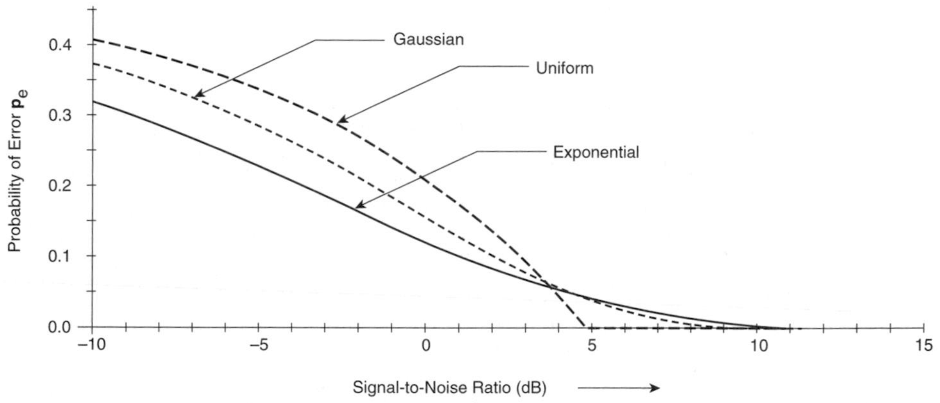

Fig. 2.67
Plot of probability of error versus signal-to-noise ratio

The probability of error is related to the notion of tail probability and thus to the manner in which the **pdf** decays for abscissas far removed from the mean. We have seen earlier that the exponential **pdf** was wider than the Gaussian **pdf** and consequently the probability of error drops more slowly in the former case as the **SNR** is increased. In the figures we also consider the case of the uniform distribution for the noise and in this case, as expected, the probability of error is zero for a large enough signal-to-noise ratio.

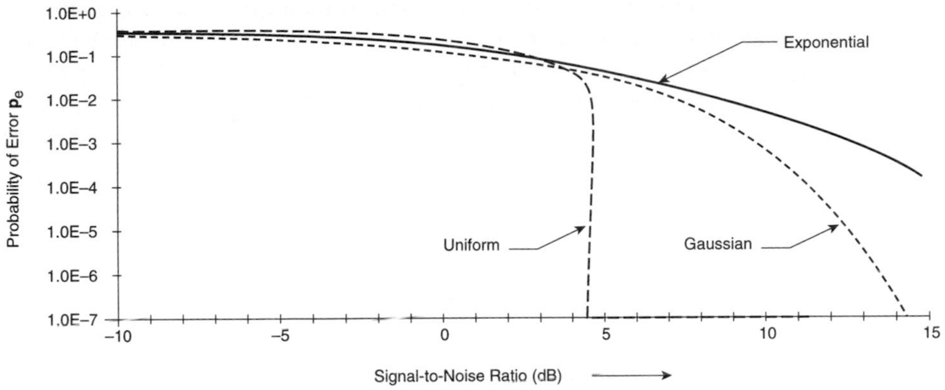

Fig. 2.68
Plot of probability of error versus signal-to-noise-ratio

This analysis of probability of error carries over to the case of multilevel symbols if we interpret V as one-half the "distance" between adjacent alphabet. If the voltage levels are of the form $\pm(2m+1)V$, then the distance between adjacent alphabet is 2V and an error occurs if the noise voltage is greater than V in magnitude, i.e., $|\eta| > V$, except at the "endpoints," where an error corresponds to $\eta > +V$ or $\eta < -V$.

2.7.4 The Matched Filter

The analysis of the impact of noise is most easily done when the symbols are binary, that is, the case of transmitting bits of information. We again consider the case depicted in Fig. 2.58 whereby a certain pulse shape is used for each symbol (bit) and in the binary case the two waveforms are +p(t) and –p(t) corresponding to a binary **1** or binary **0**. We will assume that the bandwidth of the channel is not the limiting factor—the bandwidth of the channel is essentially infinite. In this case the received signal will be equal to the transmitted signal with the addition of noise. This additive noise will be assumed, as usual, to be white Gaussian noise with power spectral density of $N_0/2$ (watts/Hz).

In the simplest case, the waveshape transmitted is a positive pulse for a **1** and a negative pulse for a **0**. The waveshape launched by the transmitter is then represented by Fig. 2.54 for a rectangular pulse. At the receiver the signal component of the received waveform is that of Fig. 2.54 but the actual waveform could be quite different because of the additive noise. One may be tempted to use the simple data slicer, wherein the received waveform is sampled at the symbol boundary and, depending on whether the sample value was positive or negative decide that a **1** or a **0** had been transmitted. The assumption of white noise (and very large bandwidth) renders this approach invalid since the noise power is infinite! How then can the signal be recovered?

The reason that the noise power is infinite is that there has been no limitation of the noise (and signal) bandwidth. With suitable filtering the bandwidth of noise and consequently the noise power can be reduced to a finite value. The generic scheme is depicted in Fig. 2.69. The receiver employs a filter, $Q(f)$, to limit the noise bandwidth. The output of the receive filter, s(t), has two components. One is the signal component comprising the filtered version of the transmitted waveform and the other a *noise* component equivalent to white noise filtered by $Q(f)$. The variance, or power, of the noise component is

$$\sigma_N^2 = N_0 \int_0^{+\infty} |Q(f)|^2 df \tag{2.7.23}$$

The sampled value of s(t) taken at time t=T is the signal component and will have contributions from the symbol launched at time t=0, i.e., b(0), and all prior symbols. We will make the assumption that the symbol waveform, p(t), is limited in time to one symbol interval. With this assumption, and assuming that the filter $Q(f)$ is causal with a finite duration (T long) impulse response q(t), the signal component at time t=T at the filter output can be written as

$$\begin{aligned} s_S(T) &= \int_{-\infty}^{+\infty} x(T-\tau)\,q(\tau)\,d\tau = \int_{-\infty}^{+\infty} q(T-\tau)\,x(\tau)\,d\tau \\ &= b(0)\int_0^T q(T-\tau)\,p(\tau)\,d\tau = b(0)\,V \end{aligned} \tag{2.7.24}$$

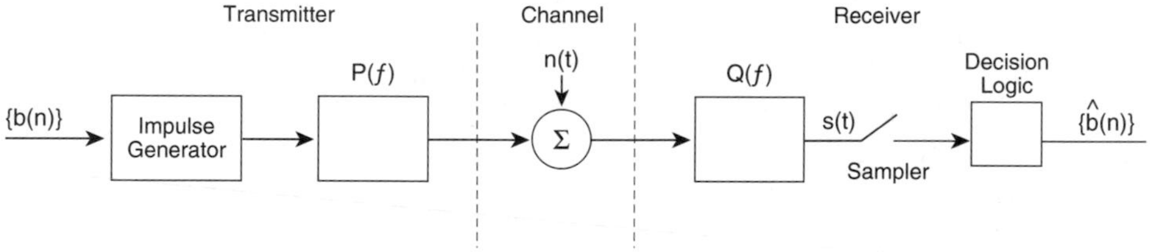

Fig. 2.69
Model for analyzing the impact of noise on data transmission

The assumption of finite duration of p(t) and q(t) is simply to eliminate any **ISI** effects so that the effect of the noise can be highlighted.

Eq. (2.7.24) provides the value of the signal component of the sample at time t=T. Hence we can describe the sampled signal as

$$s_S(T) = \begin{cases} +V + \eta & (\text{for } b(0) = +1) \\ -V + \eta & (\text{for } b(0) = -1) \end{cases} \tag{2.7.25}$$

where η is a random variable, i.e., noise, of variance σ_N^2 given by Eq. (2.7.23). This is quite similar to Eq. (2.7.17) and the derivation of the probability of error follows the same procedure described there. In particular, the probability of error is given by

$$\mathbf{p_e} = 0.5\,[1 - \mathbf{erf}(\frac{V}{\sigma_N \sqrt{2}})] \qquad (2.7.26)$$

where the quantity V is obtained from Eq. (2.7.24) as the filter output in the absence of noise.

Clearly, the probability of error depends on V, which in turn reflects the interaction of the p(t) and q(t), the choice of pulse shape and receive filter characteristic. Further, the probability of error can be minimized by maximizing (V/σ_N). This maximum is obtained when p(t) and q(t) are "matched" and the notion of matching is described below.

The optimal choice of q(t), for a given pulse shape p(t), can be derived by the application of the **Schwartz inequality**. This inequality states that if a(t) and b(t) are two finite energy functions, then

$$\left| \int a(t)\, b(t)\, dt \right|^2 \le [\int |a(t)|^2\, dt\,][\int |b(t)|^2\, dt\,] \qquad (2.7.27)$$

where the equality holds if and only if

$$a(t) = c\, b^*(t) \qquad (2.7.28)$$

for some constant c. As stated, the functions could be complex-valued. The condition for equality is that the two functions, a(t) and b(t), should be proportional to each other, and, if complex-valued, the proportionality should include conjugation.

Applying the Schwartz inequality to Eq. (2.7.24) we get

$$|V|^2 \le [\int_0^T |x(t)|^2\, dt\,][\int_0^T |q(T-t)|^2\, dt\,] \qquad (2.7.29)$$

and the equality condition, which maximizes |V|, is

$$q(T-t) = c\, p(t) \qquad (2.7.30)$$

The optimal choice for the impulse response q(t) of the receive filter is thus a time-reversed version of p(t) with an additional delay to make it causal. This is the notion of a matched filter. In the frequency domain, the equivalent condition is

$$Q(f) = c\, P^*(f) \exp(-j2\pi f T) \qquad (2.7.31)$$

and Q(f) is seen to be the complex conjugate of P(f) with the inclusion of a delay term and constant of proportionality.

With this optimal choice of q(t), the ratio (V/σ_N) can be expressed as

$$\left(\frac{V}{\sigma_N}\right)^2 = \frac{2}{N_0}\int_0^T |p(t)|^2 dt \tag{2.7.32}$$

by combining Eqs. (2.7.23), and (2.7.24) and recognizing that q(t) has been assumed to be of finite duration. The output of the matched filter is thus the autocorrelation function of the signal p(t) (delayed) such that the output at time T corresponds to the autocorrelation funcion for zero-lag, namely the energy of the pulse. Since

$$E_b = \int_0^T |p(t)|^2 dt \tag{2.7.33}$$

the probability of error is given by

$$\mathbf{p_e} = 0.5\,[1 - \mathbf{erf}(\frac{\sqrt{E_b}}{\sqrt{N_0}})] \tag{2.7.34}$$

The quantity E_b is just the energy of the pulse p(t) and is thus the **energy per bit**, considering we assumed binary transmission.

Some of the conclusions we could derive from the previous discussion are:

a) The probability of error is minimized when the receive filter is matched to the transmitter pulse shape.

b) The optimal probability of error is a function only of the noise spectral density, N_0, and the energy per bit, E_b.

Strictly speaking these conclusions are valid only when the channel has unlimited bandwidth *and* the additive noise is white and Gaussian *and* when the pulses used are of finite duration (T) *and* the sampling instants at the receiver are perfectly synchronized with the transmitter bit times. Since there are so many assumptions, it is common practice to "derate" the **SNR** by about 6 to 12 dB depending on the application.

What happens if the pulses are not perfectly time-limited to one symbol time? There are two effects than can be considered. First is that there could be intersymbol interference. **ISI** can be combated by choosing P(f) (and thus Q(f) via Eq. (2.7.31)) such that the combination G(f)=P(f)Q(f) satisfies the Nyquist criterion. The second is that, even if **ISI** has been eliminated by design of P(f), only a fraction of the total energy per symbol will be contained in the interval [0,T]. Clearly, the probability of error is minimized when this fraction is unity, i.e., the waveform is time-limited to [0,T]. In practice the channel does not have infinite bandwidth but is limited in bandwidth to some value, say B. Since bandwidth and time duration cannot be limited simultaneously, the problem of signal design is formulated as an optimization problem that chooses P(*f*) and Q(*f*) such that the product satisfies the Nyquist criterion and that the quantity C given by

$$C = \alpha \int_{|f|>B} |P(f)|^2 df + \beta \int_{|f|>T} |p(t)|^2 dt \qquad (2.7.35)$$

is minimized. The weights α and β reflect the weighting associated with the impact of finite bandwidth and fraction of symbol energy within one symbol time, respectively. The particular choice will be application specific.

2.7.5 M-ary Signaling

The matched filter as discussed earlier is the optimal receiver for deciding whether the transmitted waveshape was p(t) or –p(t) and thus well suited for the case of binary symbols and the analysis can be extended to multilevel signals (**PAM**) in a straightforward manner. One approach to extending the symbol alphabet to a size greater than 2 is the following. Suppose the alphabet size was M = 2m. For notational convenience we refer to the 2m symbols by {±1, ±2, ... , ±k, ... , ±N}. Since the size of the alphabet is greater than 2, such schemes are called "M-ary signaling" as opposed to "binary." It can be shown that using M-ary signaling is, in general, more efficient than binary (see, for example, Taub and Schilling [1.9], or Sklar [1.8]). That is, for a given additive noise power, and a specified signal strength, and for a fixed (overall) data rate (in bits/sec), M-ary schemes have a smaller probability of error than a binary scheme.

For a general M-ary transmission scheme we need to obtain m pulse shapes $\{p_i(t);\ i=1,2, ..., m\}$ which satisfy an orthogonality condition, namely

$$\int_0^T q_k(\tau)p_i(\tau)d\tau = \begin{cases} E_b & \text{for } i = k \\ 0 & \text{otherwise} \end{cases}$$

$$q_k(t) = p_k(T-t) \qquad (2.7.36)$$

The generation of the transmitted signal follow the rule that in the nth symbol interval we launch the pulse $p_k(t)$ if b(n)=+k and the pulse $-p_k(t)$ if b(n)=–k. This can be written as

$$x(t) = \pm p_k(t-nT) \text{ for } nT < t \leq (n+1)T \qquad (2.7.37)$$

The receiver for such a symbol transmission scheme comprises a bank of m matched filters, each matched to a particular pulse shape. This is shown in Fig. 2.70. The outputs of the m matched filters are sampled at the symbol boundary and fed to the decision logic. The decision logic then decides which of the symbols was the most likely to have been transmitted by examining all m outputs simultaneously.

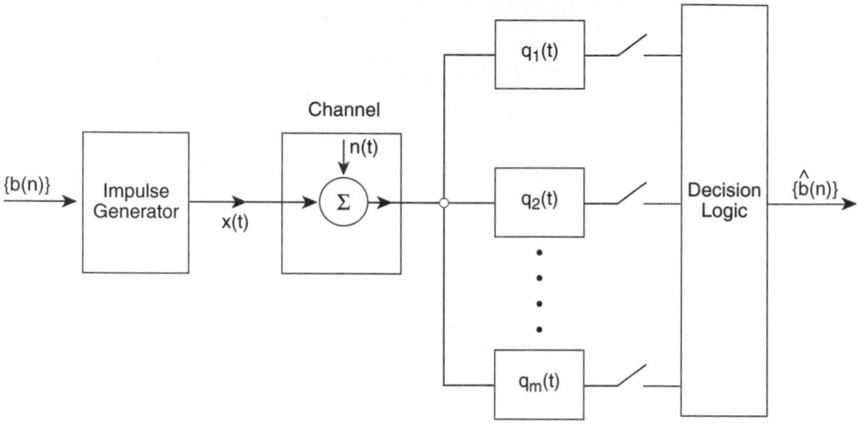

Fig. 2.70
Model for an M-ary signaling scheme

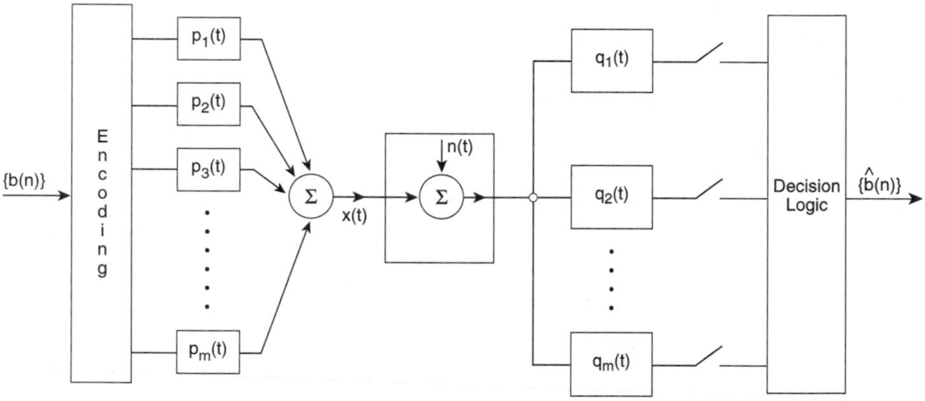

Fig. 2.71
A Code Division Multiplexing scheme that behaves like m independent subchannels

As shown in Fig. 2.70, the signal power transmitted is E_b/T since during any symbol time there is only one pulse of the set $\{p_i(t)\}$ is present. An alternative scheme, similar in form to Fig. 2.70, but that utilizes a transmitted signal power of mE_b/T and that can be used for an alphabet of 2^m symbols is shown in Fig. 2.71. The symbol to be transmitted, b(n), is represented by a m-bit code of the form [b(n,1) b(n,2) b(n,3) ... b(n,m)] where each of the b(n,j) can be viewed as binary symbols. The encoding logic block in the figure refers to the procedure for assigning this m-bit code to each symbol. Each binary signal is transmitted "independently" of the other. The orthogonality condition then guarantees that there would be no crosstalk between the separate symbol streams and thus they can be summed and transmitted simultaneously over the channel. This is the principle of ***Code Division Multiplexing*** (**CDM**), so called because the channel is being used simultaneously by m independent subchannels, and each subchannel distinguishes itself from the others because of the choice of pulse shape or "code." In fact from the viewpoint of each individual subchannel, all the other subchannels appear as "noise." The scheme could even be extended further by making each "bit" b(n,j) a symbol representing multiple bits. That is, each subchannel could be M-ary in nature.

2.7.6 Channel Capacity

For a given channel, characterized by a (finite) bandwidth and certain noise level, how much information can be carried across it? We shall use as the measure of information the bit rate. Thus we are looking for the relationship between maximum bit rate and the channel characteristics. This capacity is given by the celebrated Shannon formula (see Taub and Schilling [1.9], for example)

$$R = W \log_2(1+(S/N)) \tag{2.7.38}$$

where R is the maximum bit rate, W the (two-sided) bandwidth, and S/N the signal to noise ratio. Shannon's result states that it is possible to achieve this rate with a probability of error of zero! A rationale for this result is that we know from Nyquist's theorem that we can transmit symbols across a channel of bandwidth W Hz at a rate of W symbols/sec. Thus Shannon's formula can be "proved" if we show that each symbol can carry $\log_2(1+(S/N))$ bits of information. A rigorous proof is beyond the scope of this book but we could present a special case wherein the truth of the formula is evident. In particular we will consider the case of multilevel **PAM**.

Each symbol will be transmitted as a particular voltage level. The receiver (sampler) decides which level was transmitted based on slicing the received voltage at values midway between the ideal expected values as in Fig. 2.60. There will be no errors if the additive noise was smaller than half the difference in (voltage) value between two symbols. Thus if the symbol alphabet had M=2N entries, the associated voltage values are written as $\{(\pm)(2k+1)V;\ k = 0, 1, \ldots, (N-1)\}$, corresponding to a "symbol separation" of 2V. If the noise amplitude is less than V, then the receiver distinguishes, with zero probability of error, which symbol is transmitted. A measure of signal-to-noise ratio in this contrived example is computed by assuming that each symbol is equally likely and that the noise sample is uniformly distributed in [–V,+V]. The equiprobable assumption on the symbols yields

$$S = \frac{V^2}{N}\sum_{k=0}^{N-1}(2k+1)^2 \approx \frac{V^2}{3}(4N^2-1)^2 \tag{2.7.39}$$

for the signal power. The uniform distribution of the noise yields

$$\mathbf{N} = \frac{(2V)^2}{12} \tag{2.7.40}$$

for the noise power. Consequently

$$\left(1 + \frac{\mathbf{S}}{\mathbf{N}}\right) = 4N^2 = M^2 \Rightarrow \log_2\left(1 + \frac{\mathbf{S}}{\mathbf{N}}\right) = \log_2(M) \quad (2.7.41)$$

and we see that the Shannon formula is indeed correct in the simple case of uniform symbols and uniform noise as we have assumed. This computation is by no means a proof of Eq. (2.7.38) but does serve as an illustration as to why the formula is applicable.

2.7.7 Concluding Remarks on Data Transmission

In this section we covered several aspects of data transmission. The treatment was introductory and restricted to just a few topics. Several books are available on this subject, a few of which are listed in the bibliography. The topics that we did cover, namely Nyquist's theorem, M-ary signaling, and the discussion on channel capacity, are very similar to topics in digital signal processing covered in the next chapter. For example, the Nyquist result tells us the highest rate at which we can pump symbols through a channel of finite bandwidth; the sampling theorem addresses a similar problem where we try to assess the lowest rate at which we can sample a signal of finite bandwidth. The addition of noise to an M-ary signal is the logical dual of quantization, where the latter models an M-ary signal as the combination of the true signal value and additive noise. Our hypothetical noise signal in our discussion of Shannon's formula is identical to the noise model of quantization.

2.8 EXERCISES

2.1. A linear time-invariant system has input x(t) and output y(t). The relationship between input and output is y(t) = Kx(t) where K is a constant (the **LTI** is an ideal amplifier). What is the impulse response, h(t), of the system?

2.2. A linear time-invariant system has input x(t) and output y(t). The relationship between input and output is y(t) = Kx(t–τ) where K and τ are constants (the **LTI** is an ideal amplifier with delay). What is the impulse response, h(t), of the system?

2.3. A linear time-invariant system has a frequency response H(*f*) and corresponding impulse response h(t). The input, x(t), is an impulse train of strength I and period T_0, that is,

$$x(t) = \sum_{n=-\infty}^{\infty} I\delta(t - nT_0)$$

a) Derive an expression for the output, y(t), in terms of h(t).

b) Show that y(t) is periodic and derive an expression for the coefficients of the Fourier series of y(t) in terms of H(*f*).

2.4. Suppose x(t) is given by

$$x(t) = u(t)e^{-\alpha t}$$

a) Compute the Fourier transform $X(f)$.

b) Use this result and Parseval's relation to evaluate the integral

$$I = \int_{-\infty}^{+\infty} \frac{1}{1 + 4\pi^2 f^2} df$$

c) Evaluate the integral directly. Verify that we get the same result.

2.5. The impulse response of a system, h(t), is given by

$$h(t) = A\,[\,u(t) - u(t - T)\,]$$

a) Plot h(t).

b) Analyze the output when the input is periodic with period T. Express y(t) in terms of h(t) and x(t) (the convolution integral). Can it be simplified?

c) Express the Fourier series coefficients of y(t) in terms of those of x(t).

d) Compare this analysis with Nyquist's result on zero-**ISI**. Draw an analogy between the "frequency domain" result (Nyquist) and the "time domain" result of this exercise.

2.6. The Fourier transform, $X(f)$, of a time function, x(t), is such that the magnitude is given by

$$|X(f)| = B\,[\,u(f + f_0) - u(f - f_0)\,]$$

a) Plot $|X(f)|$.

b) Evaluate x(t) given that $\mathrm{Im}\{X(f)\} = 0$.

c) Evaluate x(t) given that the phase function of $X(f)$ is given by $\phi(f) = 2\pi f T_0$.

2.7. The autocorrelation function, $R_{xx}(\tau)$, of a periodic signal, x(t), is given by

$$R_{xx}(\tau) = \frac{1}{T}\int_0^T x(t)\,x(t+\tau)\,dt$$

where T is the period of x(t). Prove that $R_{xx}(0)$ $R_{xx}(\tau)$ for all values of τ. (Assume x(t) is real).

2.8. The impulse response, h(t), of an **LTI** is given by

$$h(t) = A\frac{\sin(2\pi f_0 t)}{2\pi f_0 t}$$

What is the output, y(t), when the input x(t) is periodic with fundamental frequency f_1 and described by a Fourier series with coefficients $\{a_n = (1/n^2);\ n = 1,2, \ldots\}$ and $f_1 = (2/3)f_0$.

2.9 Evaluate the following integrals:

(a) $$I = \int_{-\infty}^{+\infty} \frac{\sin(2\pi f_0 t)}{2\pi f_0 t}\, \delta(t+\tau)\, dt$$

(b) $$I = \int_{-\infty}^{+\infty} \frac{\sin(2\pi f_0 t)}{2\pi f_0 t}\, \delta(\tau - t)\, dt$$

(c) $$I = \int_{-\infty}^{+\infty} \frac{\sin(2\pi f_0 t)}{2\pi f_0 t}\, \frac{\sin(8\pi f_0 t)}{8\pi f_0 t}\, dt$$

Comment on the similarity, and difference, between (a) and (b).

2.10. Consider the following sequence of processes:

i) The input, x(t), is multiplied by w(t) to get $x_1(t)$.

ii) $x_1(t)$ is the input to an **LTI** with impulse response h(t) (and corresponding frequency response $H(f)$).

iii) The output of the **LTI**, $y_1(t)$, is multiplied by $w^*(t)$ to give y(t).

a) Draw a block diagram of the overall system.

b) Given that w(t) is the complex exponential $w(t) = \exp(j2\pi f_0 t)$, express the relationship between x(t) and y(t) in terms of the frequency response $H(f)$.

c) If $w(t) = \cos(2\pi f_0 t)$ can a similar result be derived?

d) Assuming h(t) is the same as from Exercise **2.8**, what additional manipulations are required to make case (c) "equivalent" to case (b)?

2.11. Suppose the frequency response of a system, $H(f)$, is given by

$$H(f) = \frac{G}{1 + 2\pi f \tau}$$

a) Plot $|H(f)|$ with τ as a parameter. What kind of filter is $H(f)$?

b) Denote the input to the filter as x(t) and the output as y(t). Analyze the output when the input is given by $x(t) = [A \sin(2\pi f_0 t)]^2$. In particular, what is the "average steady state output."

c) Show that the output y(t) contains a sinusoidal component. What is the frequency and amplitude of this component (as a function of τ)?

2.12. If x(t) is a finite energy signal with Fourier transform X(f), prove that (this is the Poisson Sum Formula)

$$\sum_{n=-\infty}^{\infty} x(t-nT) = \frac{1}{T}\sum_{m=-\infty}^{\infty} X\left(\frac{m}{T}\right) e^{j\frac{2\pi m t}{T}}$$

2.13. A sinusoid is modified by a nonlinear device whose input-output characteristic is defined below. The input is x(t) and the output is y(t).

If |x(t)| < A then y(t) = x(t); if x(t) > A then y(t) = A; if x(t) < –A then y(t) = –A.

a) Plot the input-output characteristic.

b) What will be the shape of the output, y(t), for different values of input amplitude.

c) The nonlinearity depicted is an ideal model for an amplifier with clipping (or saturation) that is symmetric (treats positive and negative inputs alike). Show that the Fourier series of the output will have components at odd multiples of the fundamental frequency provided that the input amplitude is large enough to experience clipping.

2.14. Consider the nonlinearity defined below. Again the input is a sinusoid.

If x(t) < 0 then y(t) = Ax(t); If x(t) > 0 then y(t) = A(1 + e)x(t).

a) Plot the input-output charateristic.

b) What will be the shape of the output, y(t).

c) The nonlinearity depicted is a model for a "push-pull" amplifier which is not matched. That is, the gain introduced is different for positive signal and negaive signal. Show that the Fourier series of the output includes terms corresponding to even multiples of the fundamental frequency.

2.15. Verify that multiplication in the time (frequency) domain is equivalent to convolution in the frequency (time) domain. That is, if x(t) and y(t) are time functions with Fourier transforms X(f) and Y(f), respectively, then

$$w(t) = x(t)\,y(t) \quad \text{implies that} \quad W(f) = X(f)\,(*)\,Y(f)$$

$$w(t) = x(t)\,(*)\,y(t) \quad \text{implies that} \quad W(f) = X(f)\,Y(f)$$

2.16. Consider a system **H** whose input-output relation is given by

$$y(t) = (1/T)\{\,x(t) - x(t-T)\,\}$$

a) Show that **H** is an **LTI** system.

b) Derive an expression for the frequency response H(f).

c) Compare H(f) with the frequency response of an ideal differentiator. How close is **H** to an ideal differentiator?

2.17. Suppose **x** and **y** are two independent random variables and we define a third random variable **w** by **w** = **x** + **y**. Show that the **pdf** of **w** is related to those of **x** and **y** by the convolution

$$\mathbf{p_w}(w) = \int_{-\infty}^{+\infty} \mathbf{p_x}(\alpha)\,\mathbf{p_y}(w-\alpha)\,d\alpha$$

2.18. In Exercise **2.17**, show that if **x** and **y** are Gaussian, then so is **w**. Extend the situation to the case where $\mathbf{w} = \alpha\,\mathbf{x} + \beta\,\mathbf{y}$. Express the mean and variance of **w** in terms of the means and variances of **x** and **y**.

2.19. If x(t) is a Gaussian pulse, show that the transform X(f) is also a Gaussian pulse. That is,

$$x(t) = A e^{-\alpha t^2} \Rightarrow X(f) = B e^{-\beta f^2}$$

Relate (A,α) and (B,β). Hint: **pdf** integrates to unity.

2.20. For a sinewave the peak-to-rms ratio is deterministic and can be computed easily. For random signals we will define the peak, parameterized by β, by saying that P is the peak value if Prob{|**x**| > P} < β. β is nominally a small number and we shall assume it is 0.01 (approximately). Compute the peak-to-rms ratio for the cases where **x** is a sinusoid, where **x** is Gaussian, and where **x** is exponentially distributed. What conclusion can you make regarding the use of the peak-to-rms ratio for distinguishing speech versus noise versus sinusoid?

2.9 REFERENCES AND BIBLIOGRAPHY

The references and bibliography have been split into sections. The first comprises those that are directly referenced in the body of Chapter 2. The others are included as examples of the rich literature available on the subject matter.

2.9.1 References

[1.1] Daniels, R. W., *Approximation Methods for Electronic Filter Design,* Bell Telephone Labs, McGraw-Hill Publishing Co., New York, 1974.

[1.2] Franks, L. E., *Signal Theory*, Prentice-Hall, Inc., Englewood Cliffs, NJ, 1969.

[1.3] Kailath, T., *Linear Systems*, Prentice-Hall, Inc., Englewood Cliffs, NJ, 1969.

[1.4] Korn, I., *Digital Communications*, Van Nostrand Reinhold Company, Inc.,New York, 1985.

[1.5] Nyquist, H., "Certain topics of telegraph transmission theory," *Trans. Am. Inst. Electr. Eng.*, vol. 47, April 1928.

[1.6] Papoulis, A., *Probability, Random Variables, and Stochastic Processes*, Second Ed., McGraw-Hill Publishing Co., New York, 1984.

[1.7] Papoulis, A., *Signal Analysis*, McGraw-Hill Publishing Co., New York, 1977.

[1.8] Sklar, B., *Digital Communications*, Prentice-Hall, Inc., Englewood Cliffs, NJ, 1988.

[1.9] Taub, H., and Schilling, D. L., *Principles of Communication Systems*, Second Edition, McGraw-Hill Publishing Co., New York, 1986.

[1.10] Vlach, J., *Computerized Approximation and Synthesis of Linear Networks*, John Wiley and Sons, New York, 1969.

[1.11] Recommendation **O.41**: Specification of a psophometer for use on telephone-type circuits, Specifications of Measuring Equipment, **CCITT** Series O Recommendations, **Blue Book**, 1988.

[1.12] Transmission Parameters Affecting Voiceband Data Transmission—Measuring Techniques. **Bell Pub. 41009**, *Bell System Data Communications Technical Reference*, May 1975.

[1.13] Recommendation **P.56**: Objective measurement of active speech level. Telephone Transmission Quality, **CCITT** Series P Recommendations, **Blue Book**, 1988.

2.9.2 Bibliography

There are several excellent books available on Ccommunication theory, linear systems, and other topics touched upon in this chapter. The following are those most commonly prescribed as text books for advanced undergraduate and graduate level courses or used as reference material by practicing engineers.

[2.1] Ash, R. B., *Information Theory*, Interscience Publishers, John Wiley and Sons, New York, 1965. Also available from Dover Publications, Mineola, NY , 1990.

[2.2] Bracewell, R. N., *The Fourier Transform and its Applications*, Second Ed., McGraw-Hill Publishing Co., New York, 1986.

[2.3] Cooper, G. R., and McGillem, C. D., *Modern Communications and Spread Spectrum*, McGraw-Hill Publishing Co., New York, NY, 1986.

[2.4] Cruz, J. B., and Van Valkenburg, M. E., *Introductory Signals and Circuits*, Ginn and Company, Waltham, 1967.

[2.5] Gallagher, R. G., *Information Theory and Reliable Communication*, John Wiley and Sons, New York, 1968.

[2.6] Lee, E. A., and Messerschmitt, D. G., *Digital Communications*, Kluwer Academic Publishers, Higham, MA, 1988.

[2.7] Oppenheim, A. W., and Willsky, A. S., with Young, I. T., *Signals and Systems*, Prentice-Hall, Inc., Englewood Cliffs, NJ, 1983.

[2.8] Proakis, J. G., and Salehi, M., *Communication Systems Engineering*, Prentice-Hall, Inc., Englewood Cliffs, NJ, 1994.

[2.9] Roden, M. S., *Analog and Digital Communication Systems*, Prentice-Hall, Inc., Englewood Cliffs, NJ, 1985.

[2.10] Schwartz, M., *Information Transmission, Modulation, and Noise*, Fourth Ed., McGraw-Hill Publishing Co., New York, 1990.

[2.11] Wozencraft, J. M., and Jacobs, I. M., *Principles of Communication Engineering*, John Wiley and Sons, New York, 1965.

FUNDAMENTALS OF DIGITAL SIGNAL PROCESSING

3

3.1 INTRODUCTION

The purpose of this chapter is to introduce the fundamentals of discrete-time and digital signal processing (**DSP**). The concepts and notions of signals, filters, and transforms are applicable in the case of discrete-time systems just as they were in the case of continuous-time, or analog, systems considered in Chapter 2. In that sense, this chapter extends those results to the case where the time axis is discretized and the values taken on by the signals discretized as well. There are certain operations and techniques that are available to designers of digital systems that analog filters cannot provide or cannot provide in a reliable, simple, fashion. These will be highlighted as features of **DSP**.

DSP, as a body of knowledge, is relatively young. The ability of digital computers to do arithmetic manipulations generated the first interest in **DSP** for purposes of control and simulation. Advances in semiconductor technology, especially in the Very Large Scale Integration (**VLSI**) of digital circuits, have been responsible for making complex algorithms economically practical and thus been instrumental in spurring the rapid advances in **DSP** theory. In the process **DSP** has acquired a certain aura of being capable of doing the impossible. Unfortunately **DSP** is not magic and has its fundamental limitations as do all techniques for problem solving. These limitations will be addressed, implicitly at times and explicitly at others.

The remaining sections of this chapter are organized in the following fashion. Section 3.2 introduces the notion of discrete-time signals and the associated theory. This theory covers the description and analysis of deterministic and random signals and the nature of the principal mathematical manipulations performed on discrete-time signals. Section 3.3 covers a cornerstone of discrete-time signal processing theory, the **Sampling Theorem**, and several of the derived properties. In particular, we provide a framework for transforming discrete-time signals from the time domain into the Fourier, that is frequency, domain. The peculiarities arising because of the discrete-time nature are emphasized.

Section 3.4 discusses the notion of Linear Time Invariance (**LTI**) as it applies to the discrete-time case. What becomes clear is that there is a very definite parallel between the concepts of signal processing covered in Chapter 2 to the concepts developed for manipulating, that is filtering, sequences. The **Z-transform**, covered in Section 3.5 is the principal tool for converting convolution into algebraic, polynomial, manipulation. The theory of the Z-transform is hardly new, having been studied in the field of Complex Analysis for a very long time.

Most information signals, such as speech, are available at the output of transducers, microphones, in an analog format. Applying **DSP** requires that these analog signals be converted into digital form. This is achieved by a device called an *Analog-to-Digital Converter* (**ADC**, or A/D converter). An A/D converter embodies two processes which make a signal "digital." These are the discretization of time, namely sampling, and the discretization of amplitude, namely quantization. The post processing conversion back to analog form is achieved by a *Digital-to-Analog converter* (**DAC**, D/A Converter). Section 3.6 describes how these affect the signal in terms of impairments and how to analyze these effects in terms of signal-to-noise ratio (**SNR**). The evaluation of **SNR**, for various wordlengths and quantization characteristics is the subject of Chapter 4.

Filters, from the viewpoint of **DSP**, fall into two broad categories, *Finite Impulse Response* (**FIR**) and *Infinite Impulse Response* (**IIR**) filters. The principal characteristics of these classes is covered in Section 3.7. **FIR** filters, in particular, are what distinguish **DSP** from analog processing. With **FIR** techniques we can implement filters of arbitrary (within certain limits) magnitude response while having a phase response that is perfectly linear phase. **FIR** filters are also unconditionally stable.

Two advanced topics in digital signal processing, namely *adaptive filters* and *interpolation and decimation*, are covered later in Chapters 6 and 7 respectively. Adaptive filters find applications in several areas other than telecommunications, for example in biomedical instrumentation, phased-array antenna design, and many more. In telecommunications the principal use is in *echo cancelers* for improving the quality of calls in long-distance telephone networks. The discussion of adaptive filters in Chapter 6 is slanted in this direction. Interpolation and decimation relate to changing the implicit or explicit sampling rate of a digital signal. In doing so we are able to achieve certain filtering operations in a more convenient manner than if the sampling rate was not altered. From a signal processing perspective, the study of interpolation and decimation is helpful in devising schemes that transfer the complexity from the analog to the digital domains. In Chapter 7 we provide a brief introduction to this subject, particularly in the light of bandpass filters and related applications. Additionally, interpolation and decimation play an important role in A/D and D/A conversion methods based on *Delta Sigma Modulation*, and these applications are described in Chapter 8.

3.2 DISCRETE-TIME SIGNALS

In Chapter 2, a signal was defined as a function of time. The independent variable, namely time, took on a continuum of values between – and + . For this reason we refer to such signals as continuous time. The notion of a discrete-time signal corresponds to the case where the independent variable takes on a discrete set of values. This set is countably infinite and we can make a one-to-one correspondence between this discrete set and the set of integers $\{n; ...-2, -1, 0, 1, 2, ...\}$. A discrete-time signal is therefore defined as a function that maps the set of integers into the real line, for real-valued signals, or the complex plane, for complex-valued signals. We thus create a distinction between signals that are functions of a real variable and signals that are functions of integers. The notion of a discrete-time signal is associated with the mathematical notion of a ***sequence***, which is considered a set of numbers, real or complex, that is indexed, with the set of indices being (a subset of) the set of all integers. The notation used for a sequence is $\{x_n\}$ or $\{x(n)\}$. For our purposes we make the association between indices and time by assuming that the sequence $\{x_n\}$ is related to a conventional continuous time signal, x(t), whereby the sequence is obtained by taking samples of x(t) at uniformly spaced instants of time. Thus we can write

$$x(n) = x(nT_s) \tag{3.2.1}$$

where the spacing in time, T_s, is the ***sampling interval*** and the reciprocal, $f_s = 1/T_s$, is called the ***sampling rate*** or ***sampling frequency***. The discrete-time signal $\{x_n\}$, corresponding to the continuous time signal, x(t), is thus the samples of x(t) taken at time instants $t = nT_s$.

Regarding notation, we will use different forms for representing the same discrete-time signal, the choice of form dictated by the application. These forms are $\{x_n\}$, $\{x(n)\}$, $\{x(nT_s)\}$. The third form, $\{x(nT_s)\}$, is cumbersome but does convey explicitly what the sampling interval is and this form is used when the sampling interval is not obvious from the context, such as when there is a sampling rate change. The second form, $\{x(n)\}$, is used, for example, if we need to provide names for several signals and have to resort to subscripts to provide the distinguishing feature. The first form, $\{x_n\}$, is what is conventionally used in mathematics for sequences.

3.2.1 Some Common Discrete-Time Signals

Most of the common continuous time signals described in the Chapter 2 have their discrete-time counterparts. Those which we use often are:

a) The unit **delta** function, δ(n). The discrete-time unit delta function is defined by

$$\delta(n) = \begin{cases} 1 & \text{for } n=0 \\ 0 & \text{for } n \neq 0 \end{cases} \tag{3.2.2}$$

That is, δ(n) is unity for time index 0 and is zero elsewhere. This is the same as the ***Kronecker Delta*** used in Mathematical Physics, δ_{nk}, which is defined as

$$\delta_{nk} = \begin{cases} 1 & \text{for } n=k \\ 0 & \text{for } n \neq k \end{cases} \tag{3.2.3}$$

b) The unit **step** function, u(n). The unit step function is defined by

$$u(n) = \begin{cases} 1 & \text{for } n \geq 0 \\ 0 & \text{for } n < 0 \end{cases} \tag{3.2.4}$$

As in the continuous time case, the step function is very useful in distinguishing positive indices (positive time) from negative ones.

c) The **sinusoidal** function. A discrete-time sinusoid can be expressed as

$$
\begin{aligned}
x(n) &= A\exp[j(2\pi f_0 n + \varphi)] = A\exp[j(\omega_0 n + \varphi)] \\
x(n) &= A\cos[2\pi f_0 n + \varphi] = A\cos[\omega_0 n + \varphi] \\
x(n) &= A\sin[2\pi f_0 n + \varphi] = A\sin[\omega_0 n + \varphi]
\end{aligned}
\qquad (3.2.5)
$$

where the three common forms, complex, real-cosine, and real-sine are shown. The parameter f_0 is the ***frequency*** of the sinusoid and, in the discrete-time case, is expressed in **normalized** units as a fraction of the underlying sampling rate. Similarly, the term ω_0 is the normalized radian frequency. The expression in Eq. (3.2.5) does not explicitly show the (implied) sampling interval. If we insist on expressing the frequency in the usual units of Hz (or kHz, MHz, etc.) then we explicitly include the notion of a time interval by writing

$$x(n) = A\exp[j(2\pi f_0 n T_s + \varphi)] = A\exp[j(2\pi(\frac{f_0}{f_s})n + \varphi)] \qquad (3.2.6)$$

3.2.2 Operations on Discrete-Time Signals

The mathematical operations performed on discrete-time signals closely parallel those described for continuous-time signals. Of particular significance are:

i) **Correlation and Autocorrelation.** The correlation between two sequences comprises a third sequence that is a measure of how closely the two sequences resemble each other for a variety of (differential) delays (advances). Specifically, if $\{x(n)\}$ and $\{y(n)\}$ are two discrete-time signals, the correlation between the two is defined as

$$R_{xy}(k) = \sum_{n} x(n)\, y^*(n+k) \qquad (3.2.7)$$

where, to accommodate complex-valued sequences, the second signal is complex-conjugated. The summation is carried out over all integer values, n.

The autocorrelation of a discrete-time signal is defined as

$$R_{xx}(k) = \sum_{n} x(n)\, x^*(n+k) \qquad (3.2.8)$$

and is a measure of how closely a signal resembles a delayed (or advanced) version of itself. Strictly speaking, the autocorrelation, as defined above, is valid only for signals that have finite energy.

ii) **Convolution.** In dealing with linear time-invariant systems we saw that the output of a system could be expressed as a convolution integral between the input signal and an entity called the impulse response. A similar operation can be defined in the discrete-time case. Specifically, the convolution of two discrete-time signals is defined as

$$w(n) = \sum_{k} x(k)\,y(n-k) = \sum_{k} x(n-k)\,y(k) \tag{3.2.9}$$

where the signal $\{w(n)\}$ is the convolution of the signals $\{x(n)\}$ and $\{y(n)\}$.

Examination of the definitions of correlation and convolution would indicate that there is a close parallel between discrete-time and continuous-time operations on signals. In the latter the principal mathematical operation is integration over the independent variable, namely time, and in the latter case we use the notion of a sum over the index set, which constitutes discrete-time.

3.2.3 Finite Length and Periodic Sequences

As in the continuous-time case, there are several properties that make the mathematical machinations less complicated. Two of the most important ones are the notions of periodicity and the notion of duration or finiteness over time.

i) **Finite Length Sequences**. The notion of finite sequence length is equivalent to finite signal duration. Thus $\{x(n)\}$ is considered finite (length or duration) if

$$x(n) = 0 \quad \text{if } n < 0 \quad \text{or if } n > M \tag{3.2.10a}$$

or

$$x(n) = 0 \quad \text{if } n < -N \quad \text{or if } n > +N \tag{3.2.10b}$$

In the first definition, the length of the sequence is (M+1), indicating that other than (M+1) indices, the signal is zero. It may happen that some of the (M+1) values are also zero. In the second definition the length is (2N+1) samples and is usually employed when we intend to use some other property such as symmetry about the "origin." The two definitions are equivalent in the sense that if $\{x(n)\}$ is finite duration based on one definition then we could construct a delayed (or advanced) version $\{y(n-K)\}$ for some delay (advance) K that would satisfy the other definition. Sometimes, in order to reinforce the notion that a sequence is of finite duration we use the nomenclature $\{x(n);\ n = 0,1,2, \dots, (M-1)\}$ for a signal of duration M samples. The converse, namely infinite length sequences, consist of those sequences for which a finite limit cannot be given for the indices of nonzero sequence values.

ii) **Periodic Sequences.** A sequence {y(n)} is said to be periodic with period of N (samples) if

$$y(n) = y(n+kN) \qquad (3.2.11)$$

where the relationship must be satisfied for ***all*** possible indices n and ***all*** possible offsets kN where k is any integer. The period is the least value of N {y(n)} cannot be of finite duration. A second consequence is that knowledge of {y(n)} for any N consecutive samples (indices) uniquely specifies the values at all other time instants (indices). The latter observation allows us to describe a periodic signal {y(n)}, of period N samples, in terms of a finite duration signal {x(n); n = 0,1,2, ... , (N–1)} as

$$y(n) = \sum_{k=-\infty}^{\infty} x(n-kN) \qquad (3.2.12)$$

The expression in Eq. (3.2.12) expresses directly the notion that a periodic signal is repetitive and that {y(n)} delayed by N samples will be exactly the same as {y(n)}. All periodic signals can be expressed in this manner. In fact the finite duration sequence {x(n); n = 0,1, ... , (N–1)} can be generated from {y(n)} as

$$x(n) = \begin{cases} y(n) & \text{for } 0 \le n \le (N-1) \\ 0 & \text{otherwise} \end{cases} \qquad (3.2.13)$$

where k is any integer. What if {w(n)} was not a finite duration signal and we created {y(n)} by summing delayed versions of {w(n)} in the following manner:

$$y(n) = \sum_{k=-\infty}^{\infty} w(n+kN) \qquad (3.2.14)$$

Then, because of the nature of the infinite sum, a delayed version {y(n–N)}, for a delay of N samples, is indistinguishable from {y(n)}. In effect we have created a finite duration signal {x(n); n=0,1, ... , (N–1)} from {w(n)} by

$$x(n) = [u(n) - u(N)] \sum_{k=-\infty}^{\infty} w(n+kN) \qquad (3.2.15)$$

from which the periodic signal {y(n)} is described by Eq. (3.2.12). An important conclusion is that while there may be an infinite number of sequences {w(n)} which when delayed and summed according to Eq. (3.2.14) would give rise to a specific periodic signal {y(n)}, there is only one finite duration signal {x(n); n=0,1, ... , (N–1)}, which would yield {y(n)} via Eq. (3.2.14) (or (3.2.12)).

A mathematical concept that is particularly useful in dealing with periodic sequences is the notion of modulo–N arithmetic on the index. By definition, the value of an integer modulo–N is represented and defined by

$$((n))_N = n - kN \tag{3.2.16}$$

where k is some integer chosen such that

$$0 \quad ((n))_N \quad (N-1) \tag{3.2.17}$$

Thus $((n))_N$, verbalized as "n modulo N," is a value between 0 and (N–1) obtained by subtracting from (or adding to) n a suitable integer multiple of N. With this definition of modulo–N, equality modulo–N of two indices n_1 and n_2 implies that the difference $(n_1 - n_2)$ is an integer multiple of N. The concept of index manipulation modulo–N allows us to describe periodicity, with a period of N samples, in the following compact manner:

$$y(n) = y(((n))_N) \tag{3.2.18}$$

Eq. (3.2.18) states that $\{y(n)\}$ is periodic and its period is N or an integer submultiple thereof.

3.2.4 Discrete-Time Random Processes

A discrete-time random process can be defined as a sequence of random variables $\{\mathbf{x}_n\}$ that satisfy certain consistency conditions. We shall assume that all processes we deal with are stationary. For our purposes, the principal measures associated with a random process are:

a) **Mean value.** The mean value of a process would be $\mathbf{E}\{\mathbf{x}_n\}$ which is, in general, a function of the index n. For stationary processes the mean value is independent of the time index. That is,

$$\mathbf{mean}(n) = \mathbf{E}\{\mathbf{x}_n\} = \mathbf{m} = \text{constant} \tag{3.2.19}$$

The constant **m** is thus the mean value of the process. Unless otherwise specified, we shall assume that this mean value is zero.

b) **Autocorrelation.** The autocorrelation of the process is a function of a single time index and is given by

$$R_{\mathbf{xx}}(m) = \mathbf{E}\{\mathbf{x}_n \mathbf{x}_{n+m}\} \tag{3.2.20}$$

Of special interest is the autocorrelation at *zero lag*, $R_{xx}(0)$, which is

$$R_{\mathbf{xx}}(0) = \mathbf{E}\{\mathbf{x}_n^2\} \tag{3.2.21}$$

and is interpreted as the ***power*** of the process expressed in terms of *per unit sample* (power is normally expressed as *per unit time*). The independence of the mean and autocorrelation with respect to the specific time index is equivalent to assuming that the process is ***wide-sense stationary***. We shall actually assume a more stringent restriction on the random processes considered, namely the property of **ergodicity**. As with the continuous-time stochastic processes considered in Chapter 2, the notion of ergodicity implies that a time average computed over a single instantiation of the random process, that is, the observed sequence of sample values, is equivalent to the ensemble average defined in Eq.(3.2.19) and (3.2.20). In particular, if $\{x_n\}$ is the sequence of observed values of the process, then

$$\mathbf{m} = \mathbf{E}\{\mathbf{x}_n\} = \lim_{N \to \infty} \frac{1}{N} \sum_{N} x_n \tag{3.2.22}$$

and

$$R_{\mathbf{xx}}(m) = \mathbf{E}\{\mathbf{x}_n \mathbf{x}_{n+m}\} = \lim_{N \to \infty} \frac{1}{N} \sum_{N} x_n x_{n+m} \tag{3.2.23}$$

In Eq. (3.2.22) and (3.2.23) the nomenclature associated with the summation means that there are N terms in the summation, corresponding to N distinct time-indices "n."

3.2.5 The Notion of the Strength of Discrete-Time Signals

The strength of a discrete-time signal can be expressed in a variety of ways. The intention is to provide a measure that conveys the information related to the values we would expect to see, larger numerical values would correspond to stronger signals. In much the same way as we considered the strength of continuous-time signals, we could define strength in terms of ***peak value***, ***energy***, and ***power***.

a) **Peak value.** The peak value of deterministic signals is easy to define. The largest sample value is the peak or maximum of the discrete-time signal.

$$M_x = \max_{n} \{|x(n)|\} \tag{3.2.24}$$

For random signals the peak value must be expressed in terms of probability to account for the fact that the probability density function may be nonzero for arbitrarily large sample values. Thus we temper the maximum by defining it in terms of probability. With β as a parameter, where β will be close to zero, we define

$$M_{\mathbf{x}}(\beta) = \text{minimum}\left[B;\ \text{such that} \Pr\{|x(n)| > B\} \le \beta\right] \tag{3.2.25}$$

implying that the peak value is that number such that the probability of observing a sample value greater than the peak is less than β.

b) **Energy.** The energy of a discrete-time signal takes into account the duration, in terms of the number of samples, that are substantially nonzero. Thus

$$E_x = \sum_n |x(n)|^2 \tag{3.2.26}$$

Signals that are of strictly finite duration, where only N samples are nonzero for some N, are always finite-energy. Signals that are not strictly of finite duration, but have finite energy, will be characterized by sample values that, for large indices $|n|$, will tend to zero

$$E_x < \infty \Rightarrow |x(n)| \to 0 \text{ as } |n| \to \infty \tag{3.2.27}$$

and in fact

$$E_x < \infty \Rightarrow |n|\,|x(n)|^2 \to 0 \text{ as } |n| \to \infty \tag{3.2.28}$$

c) Power. Some discrete-time signals, notably those that are periodic, or almost periodic (see Example 2), or random processes, have infinite energy. For such signals an appropriate definition of strength is ***power***, defined by

$$P_x = \lim_{N\to\infty} \frac{1}{2N+1} \sum_{k=-N}^{N} |x(n)|^2 \tag{3.2.29}$$

The units of power are related to energy in that power is energy per unit sample. If $\{x(n)\}$ is periodic with period N, then since any one period of N samples completely defines $\{x(n)\}$, we can write

$$P_x = \frac{1}{N} \sum_{k=0}^{N-1} |x(n)|^2 \tag{3.2.30}$$

If $\{\mathbf{x}(n)\}$ is a random sequence, then the assumption of ergodicity implies that

$$P_x = \mathbf{E}\{|\mathbf{x}(n)|^2\} = \sigma_{\mathbf{x}}^2 + \mathbf{m}_{\mathbf{x}}^2 \tag{3.2.31}$$

where we consider the random process at any time index n as a random variable with mean $\mathbf{m}_{\mathbf{x}}$ and variance $\sigma_{\mathbf{x}}^2$.

All signals that we will deal with come under the classification of finite energy or finite power. For these signals we define the autocorrelation in the following way:

$$R_{xx}(m) = \sum_{k=-\infty}^{\infty} x(k)\,x(k+m) \quad \{\text{ for } E_x < \infty\} \tag{3.2.32}$$

$$R_{xx}(m) = \lim_{N\to\infty} \frac{1}{2N+1} \sum_{k=-N}^{N} x(k)\,x(k+m) \quad \{ \text{for } P_x < \infty \} \tag{3.2.33}$$

With this definition, $R_{xx}(0)$ is equivalent to the strength of the signal, energy or power, whichever is appropriate.

Some examples of the computation of energy and power are discussed below.

Example 1. Consider the signal {x(n)} that is given by

$$x(n) = Ae^{-\alpha n} u(n) \quad \{ \text{with } \alpha > 0 \} \tag{3.2.34}$$

The maximum value of {x(n)} is clearly $M_x = A$ since the sample value at $n = 0$ is A and, since $\alpha > 0$, the sample values for $n > 0$ decay exponentially to zero. {x(n)} is a finite energy signal and E_x can be computed by recognizing that the terms $|x(n)|^2$ follow a geometric sequence.

$$E_x = A^2 \sum_{k=0}^{\infty} e^{-2\alpha n} = \frac{A^2}{1 - e^{-2\alpha}} \tag{3.2.35}$$

Example 2. Suppose {x(n)} was generated by sampling the complex exponential $A\exp(j2\pi f_0 t)$ at a sampling rate $f_s = 1/T_s$. Then

$$x(n) = A\exp(j2\pi f_0 n T_s) \tag{3.2.36}$$

The continuous-time signal is periodic with period T_0 (sec) = $1/f_0$ (Hz). However, the discrete-time signal {x(n)} may not be periodic! That is, it may not be possible to find an integer N such that x(n) = x(n+N). This situation arises if the sinusoid frequency f_0 and the sampling frequency f_s are not "nicely" related to each other. If the ratio f_0/f_s (assuming $f_s > f_0$) is not a rational number then {x(n)} would not be periodic. If the ratio (f_0/f_s) is a rational number, say M/N, then {x(n)} may periodic with period N samples. Since, given any irrational number **c**, we can find a rational number (K/N) that is close to **c**, the signal is "almost periodic." {x(n)} is clearly not finite energy. The power, P_x, can be computed easily since

$$x(n)\,x^*(n) = |A|^2 \tag{3.2.37}$$

and thus the power is $|A|^2$. The peak value is also straightforward to evaluate and is equal to |A|. In some cases, depending on the relationship between the sinusoid and the sampling frequency, and the initial phase, the maximum value of |A| may not be actually observed over any finite interval of observation.

Example 3. If {x(n)} is composed of two complex sinusoids,

$$x(n) = A_1\exp(j2\pi f_1 n T_s) + A_2\exp(j2\pi f_2 n T_s) \tag{3.2.38}$$

Then the magnitude-squared value at time index n is

$$x(n)\,x^*(n) = |A_1|^2 + A_1A_2^*\exp[j2\pi(f_1 - f_2)nT_s] + |A_2|^2 + A_2A_1^*\exp[j2\pi(f_2 - f_1)nT_s] \tag{3.2.39}$$

It can be shown, with some difficulty, that

$$\frac{1}{N}\sum_{n=0}^{N-1}\exp[j2\pi(f_1 - f_2)nT_s] = \begin{cases} 1 & \text{if } f_1 = f_2 \\ \approx 0 & \text{if } f_1 \neq f_2 \end{cases} \tag{3.2.40}$$

where the notion of being approximately equal to zero for f_1 unequal to f_2 is exact as N tends to infinity. In Eq. (3.2.40) the range of indices is shown as going from 0 through (N–1) but the result is true when the summation is done over any N consecutive indices. Using this result it follows that

$$P_x = |A_1|^2 + |A_2|^2 \tag{3.2.41}$$

This results can be generalized to the sum of M distinct complex sinusoids as

$$P_x = \sum_{k=0}^{M-1}|A_k|^2 \tag{3.2.42}$$

Example 4. If {x(n)} is periodic with period N (samples) and is the sum of sinusoids, then the frequencies of the component sinusoids will be of the form

$$f_k = k\left(\frac{f_s}{N}\right) \quad \text{for} \quad 0 \leq k \leq (N-1) \tag{3.2.43}$$

where it is assumed that the frequencies are less than the sampling frequency, f_s. That is, each component will have a period that is an integer submultiple of N. The most general form of {x(n)} as a sum of sinusoids is

$$x(n) = \sum_{k=0}^{N-1} A_k\exp\left[j2\pi\left(\frac{f_s}{N}\right)kT_s\right] = \sum_{k=0}^{N-1} A_k e^{j2\pi\frac{nk}{N}} \tag{3.2.44}$$

from which the squared-magnitude at time index n is

$$x(n)\,x^*(n) = \sum_{k=0}^{N-1}\sum_{m=0}^{N-1} A_k A_m^* e^{j2\pi\frac{n(k-m)}{N}} \tag{3.2.45}$$

The power of {x(n)} from Eq. (3.2.30) is, therefore,

$$P_x = \frac{1}{N}\sum_{n=0}^{N-1}\sum_{k=0}^{N-1}\sum_{m=0}^{N-1} A_k A_m^* e^{j2\pi\frac{n(k-m)}{N}} \tag{3.2.46}$$

The triple summation of Eq. (3.2.46) can be simplified by using the equivalent of Eq. (3.2.40), namely

$$S = \frac{1}{N} \sum_{n=0}^{N-1} e^{j2\pi \frac{n(k-m)}{N}} = \delta_{km} \tag{3.2.47}$$

where δ_{km} is the Kronecker delta. While proving Eq. (3.2.40) is not easy, proving Eq. (3.2.47) is. To see why this is true, recognize S as the sum of N terms of a geometric series from which we ascertain that

$$S = \frac{1}{N} \frac{1 - \left(e^{j2\pi\frac{\alpha}{N}}\right)^N}{1 - \left(e^{j2\pi\frac{\alpha}{N}}\right)} \tag{3.2.48}$$

and since α, which is the difference (k–m), is an integer, the numerator is zero (exp(j2π*integer) = 1) for α not equal to zero. When $\alpha = 0$, S is clearly equal to 1 since each term in the summation in Eq. (3.2.47) is unity in this case. Rearranging the order of summation in Eq. (3.2.46) and applying the result from Eq. (3.2.47) we get

$$P_x = \sum_{k=0}^{N-1} \sum_{m=0}^{N-1} A_k A_m^* \delta_{km} = \sum_{k=0}^{N-1} |A_k|^2 \tag{3.2.49}$$

Example 5. If {x(n)} is derived as the samples of a real-valued sinusoid, say

$$x(n) = A\cos(2\pi f_0 nT_s + \varphi) \tag{3.2.50}$$

then we can express x(n) as the sum of two complex exponentials as

$$x(n) = \frac{Ae^{j\varphi}}{2} e^{j2\pi f_0 nT_s} + \frac{Ae^{-j\varphi}}{2} e^{-j2\pi f_0 nT_s} \tag{3.2.51}$$

Applying Eq. (3.2.42) for the power of the sum of two complex sinusoids yields

$$P_x = \frac{A^2}{4} + \frac{A^2}{4} = \frac{A^2}{2} \tag{3.2.52}$$

3.3 THE SAMPLING THEOREM AND ITS IMPLICATIONS

The sampling theorem provides a basis for relating a continuous time signal x(t) with the sequence {x(n)} obtained from the values of x(t) taken T units of time apart. It also provides the underlying theory for relating operations performed on the sequence with equivalent operations on the signal x(t) directly. In this context, the sampling theorem is the cornerstone of discrete-time signal processing.

3.3.1 Statement of the Sampling Theorem

The sampling theorem states that a continuous-time signal x(t) can be recovered from its samples, $\{x(nT_s)\}$, taken T_s seconds apart, provided the sampling frequency, $f_s = 1/T_s$, is greater than the bandwidth of the signal x(t). In the most commonly encountered case, called "baseband sampling," the sampling frequency must be greater than twice the highest signal frequency.

The equivalence of the sequence and the time function thus requires that the signal be bandlimited and the granularity of sampling be fine enough to capture the time-varying characteristics of the signal. For example, if x(t) were a constant, or *DC*, with no significant components at frequencies other than zero, a single sample would be sufficient to describe the signal; if x(t) were a sinusoidal function, a few values per cycle of the sinusoid would enable us to establish the nature, i.e., amplitude and frequency, of the signal. The sampling theorem extends this notion to general signals.

3.3.1.1 Validation of the Sampling Theorem

A rationale for the sampling theorem is provided next. One of the principal benefits of the discussion is that first it establishes the conditions such that a continuous time signal can be reconstructed from its samples and second it provides a measure of the distortion introduced when the said conditions are not met.

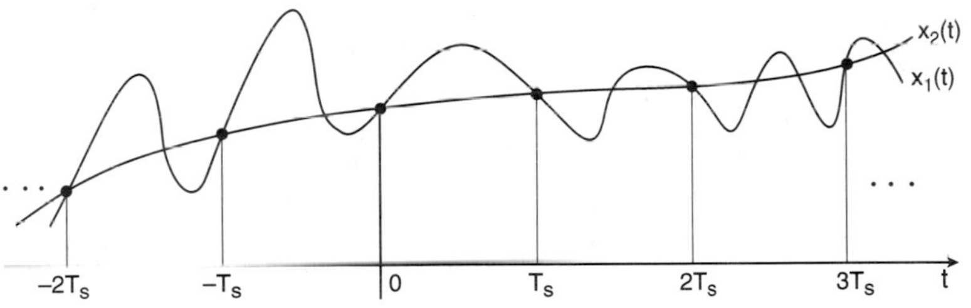

Fig. 3.1
Illustrating the ambiguity introduced by sampling

Fig. 3.1 depicts the process of sampling. It shows two signals, $x_1(t)$ and $x_2(t)$ that are sampled at a rate of $1/T_s$. Both signals provide the same sample values on this grid of sampling instants. How can we distinguish between the two signals $x_1(t)$ and $x_2(t)$ based on the samples? How are $x_1(t)$ and $x_2(t)$ related, if at all? Is there a third signal $x_3(t)$ that will have exactly the same sample values on the sampling grid? The following discussion of the sampling theorem will provide the answers to these and other questions.

Consider the special function $\Theta(t)$ given by

$$\Theta(t) = \sum_{k=-\infty}^{\infty} \delta(t - kT_s) \tag{3.3.1}$$

$\Theta(t)$ is a train of impulses of unit strength separated by T_s seconds in time as shown in Fig. 2.19 in the previous chapter. Clearly, $\Theta(t)$ is not a signal that we could generate physically but is a useful mathematical abstraction when dealing with discrete-time signals and is called the ***sampling function*** since it is zero at all instants of time that do not correspond to sampling instants. $\Theta(t)$ is periodic and can be expressed in terms of a Fourier series as (as shown in Chapter 2)

$$\Theta(t) = \frac{1}{T_s} \sum_{k=-\infty}^{\infty} e^{j2\pi f_s kt} \quad \{\text{where } f_s = \frac{1}{T_s}\} \tag{3.3.2}$$

$\Theta(t)$ does not satisfy the usual conditions we impose on functions before taking the Fourier series, namely finite power, but we shall nevertheless assume that the Fourier expansion given in Eq. (3.3.2) is valid. Eq. (3.3.2) indicates that $\Theta(t)$ is equivalent to the combination of sinusoids of frequency kf_s and every integer multiple of the sampling frequency is represented in the expansion. The Fourier transform of $\Theta(t)$ is obtained from Eq. (3.3.2) by recognizing that the Fourier transform of a complex exponential is a delta function in the frequency domain and so

$$\Xi(f) = f_s \sum_{k=-\infty}^{\infty} \delta(f - kf_s) \tag{3.3.3}$$

The sampling function is quite special in the sense that the representation in both the time and frequency domains is a periodic stream of delta functions, the period being T_s in the time domain and f_s in the frequency domain. In fact we could represent the frequency domain description of the sampling function, $\Xi(f)$, by a Fourier series with frequency as the independent variable, as

$$\Xi(f) = \sum_{k=-\infty}^{\infty} e^{j2\pi T_s kf} \tag{3.3.4}$$

Eqs. (3.3.1),(3.3.2), (3.3.3), and (3.3.4) are all equivalent descriptions of the sampling function in the time and frequency domains.

The notion of sampling is the process of ascertaining what the values of a signal are at certain time instants that are separated by T_s, the sampling interval. Consider the situation in Fig. 3.2, which involves multiplying (amplitude modulation) the signal x(t) by the sampling function (carrier) $\Theta(t)$. The resulting signal y(t) is given by

$$y(t) = \sum_{k=-\infty}^{\infty} x(t)\,\delta(t - kT_s) = \sum_{k=-\infty}^{\infty} x(kT_s)\,\delta(t - kT_s) \tag{3.3.5}$$

Since the delta functions are zero except at the sampling instants kT_s, it is clear that the signal y(t) depends **only** on the sampled values of x(t) taken at the time instants kT_s. Fig. 3.3 depicts the action of this modulation. The continuous-time signal x(t) is represented as the smooth curve and y(t) is comprised of the impulse train. Purely for visual clarity, the height of the arrow that represents a delta function is made "equal" to the value of x(t) at the sampling instants (remember that a unit delta function is infinite in value with an "area" or strength equal to unity).

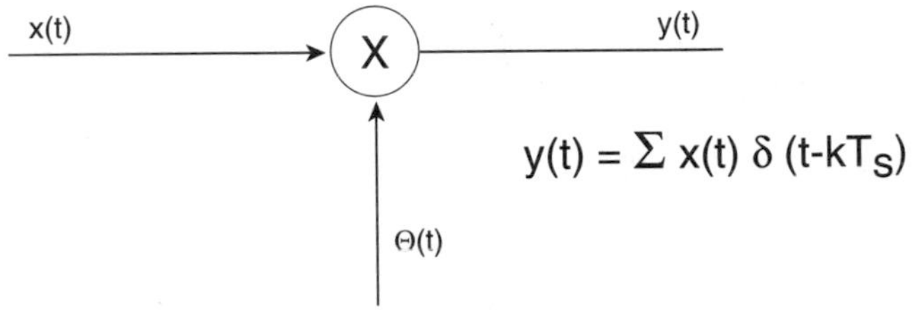

Fig. 3.2
Model of the sampling process as the modulation of the sampling function Θ(t) by the information signal x(t)

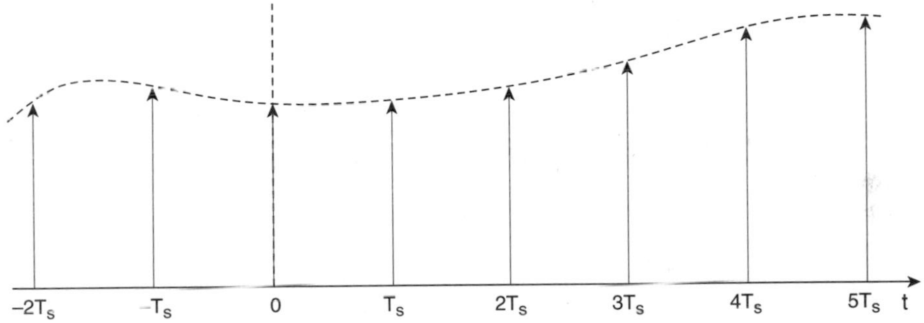

Fig. 3.3
Representation of the discrete-time signal as the composite signal obtained by modulating the sampling function Θ(t) by x(t)

The signal y(t) is an impulse train with impulses, at the sampling epochs kT_s, of strength, or "area," equal to the sampled value $x(kT_s)$. We could also consider the signal y(t) as the continuous time signal derived from a discrete-time sequence by defining an ***ideal discrete-time to continuous-time converter*** via Fig. 3.4, which depicts the creation of a stream of impulses of strength equal to the values of the discrete-time sequence. Based *solely* on y(t) it is impossible to discern whether y(t) was obtained by modulating a continuous-time signal x(t) or was generated by creating an impulse stream from a discrete-time sequence {w(n)} if the sequence values were the same as the sampled values $x(nT_s)$. Thus y(t) is representative of **all** continuous time signals that have the **same sampled values** at $t=nT_s$.

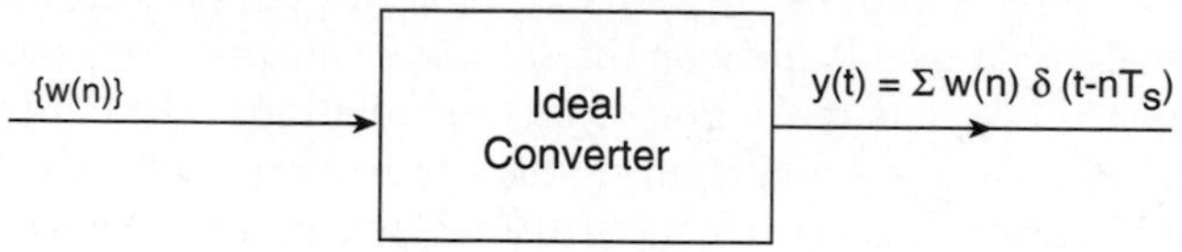

Fig. 3.4
Generation of the signal y(t) via an ideal discrete-time to continuous-time converter from the sequence $\{w(nT_s)\}$

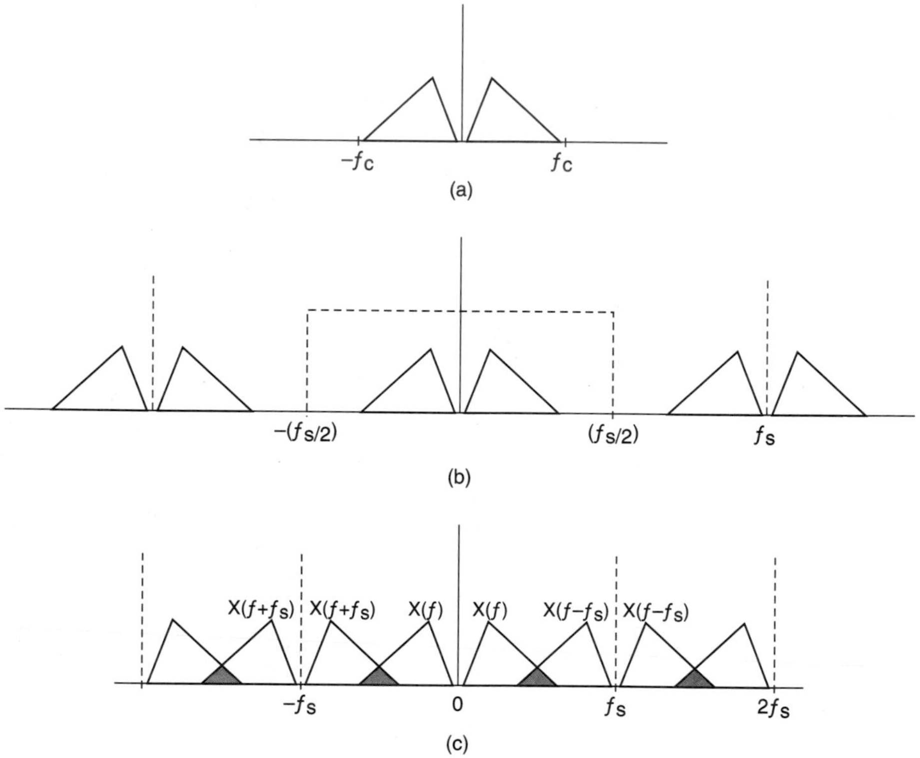

Fig. 3.5
Illustration of the phenomena of replication and aliasing. (b) and (c) show the resulting spectrum for two different sampling frequencies.

The relationship between x(t) and y(t) in the frequency domain provides the clue as to the uniqueness property relating x(t), a continuous-time signal and $\{x(nT_s)\}$, a discrete-time signal that happens to be equal to x(t) at the sampling instants. Since multiplication in the time domain is equivalent to convolution in the frequency domain, we can write

$$Y(f) = X(f)\ (*)\ \Xi(f) = \int_{-\infty}^{+\infty} X(f-\eta)\,\Xi(\eta)\,d\eta \qquad (3.3.6)$$

and using Eq. (3.3.3) and the fundamental property of delta functions we get

$$Y(f) = f_s \sum_{k=-\infty}^{\infty} X(f - kf_s) \qquad (3.3.7)$$

The spectrum of y(t), Y(f), is a combination of the spectrum X(f) together will **all** shifted versions thereof, with the shifts being all integer multiples of f_s. This is depicted graphically in Fig. 3.5. Fig. 3.5(a) depicts the spectrum of the signal x(t), i.e., X(f), and conveys the notion of the signal being bandlimited to $[-f_c, f_c]$. That is,

$$|X(f)| = 0 \text{ for } |f| > f_c \qquad (3.3.8)$$

Fig. 3.5(b) illustrates the case when $f_c < (f_s/2)$; Fig. 3.5(c) the case where f_c is just greater than $(f_s/2)$; the case when f_c is much greater than $(f_s/2)$ is difficult to draw without destroying the identity of X(f) and its shifted versions.

Now suppose that y(t) is filtered by an ideal lowpass filter with cutoff frequency $(f_s/2)$ as depicted in Fig. 3.6.

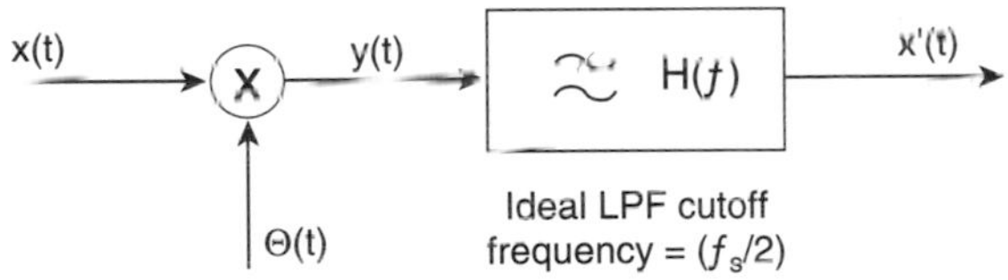

Fig. 3.6
Recovering x(t) from y(t) using an ideal lowpass filter

It is clear from Fig. 3.5 that if $(f_s/2) > f_c$ then the output of the lowpass filter, x'(t), will have a spectrum that is equal to X(f). That is,

$$f_s \geq 2f_c \;\Rightarrow\; X'(f) = X(f) \;\Rightarrow\; x'(t) = x(t) \qquad (3.3.9)$$

where the equality of x'(t) and x(t) is to be interpreted as equality *almost everywhere*; x'(t) will, for all practical purposes, be equal to x(t). Since y(t) is uniquely specified by the samples of x(t), this means that samples taken T_s seconds apart completely represent the continuous-time signal, provided that $(1/T_s) = f_s > 2f_c$. Thus the notion of modulation and (ideal) lowpass filtering provide the mathematical basis for the sampling theorem. This result should be compared with Nyquist's result for data transmission.

3.3.1.2 An Uncertainty Principle and the Notion of Aliasing

In the frequency domain the sampled signal y(t) is given by $Y(f)$, which is composed of $X(f)$, the spectrum of the underlying continuous-time signal, and shifted versions, $X(f-kf_s)$, thereof. These shifted versions, $X(f-kf_s)$, are called the ***spectral replicates*** (or just ***replicates***) of $X(f)$.

There are several (actually an infinite number of) signals that, when sampled, provide the same values at the sampling instants. For example, assuming x(t) has the spectral content depicted in Fig. 3.5(a), we could define $x_1(t)$ by

$$x_1(t) = x(t)\, e^{j2\pi f_s t} \tag{3.3.10}$$

and $X_1(f)$ comprises the shifted version of $X(f)$

$$X_1(f) = X(f - f_s) \tag{3.3.11}$$

The periodic nature of the complex exponential ensures that

$$x_1(nT_s) = x(nT_s)\, e^{j2\pi f_s nT_s} = x(nT_s) \tag{3.3.12}$$

and thus $x_1(t)$ and x(t), as discrete-time signals $\{x_1(nT_s)\}$ and $\{x(nT_s)\}$, are indistinguishable. In order to discriminate between the signals $x_1(t)$ and x(t) it is necessary to have some additional information above and beyond the sampled values. This information is equivalent to the center frequency of the spectrum. This is depicted in Fig. 3.7.

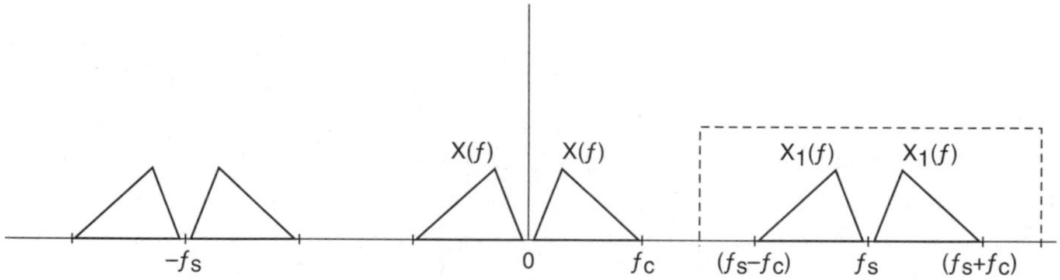

Fig. 3.7
Spectral replicates constitute an uncertainty regarding the passband of the original analog signal prior to sampling

$X(f)$ and $X_1(f)$ are both bandlimited in the sense that their spectra are nonzero over a frequency range of $2f_c$; $X(f)$ is centered around $f=0$ whereas $X_1(f)$ is centered around $f=f_s$. This permits the recovery of the signal from its sampled values by applying the suitable filter as shown in Fig. 3.8.

As shown in Fig. 3.8, the signal x(t) or $x_1(t)$ could be sampled by $\Theta(t)$ to yield (the same) y(t). Recovery of x(t) from y(t) is achieved by applying an ideal lowpass filter; $x_1(t)$ can be recovered from y(t) by applying an ideal bandpass filter centered around f_s and with passband extending over a range greater than $2f_c$.

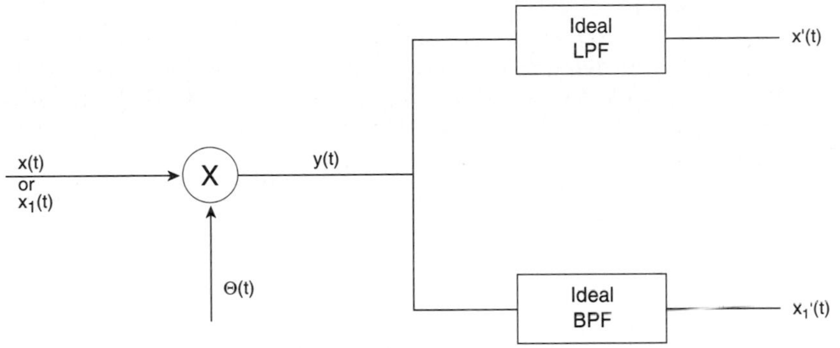

Fig. 3.8
Extraction of different replicates from the same discrete-time signal

This notion of discrete-time equivalence of signals can be extended easily. There are an infinite number of signals, bandlimited in the sense that their spectra are nonzero over a frequency range $2f_c$, which yield the same discrete-time signal upon sampling. The bandlimited nature is required to ensure that the replicates do not overlap in the frequency domain. The uncertainty as to which of these signals is the source of the samples is equivalent to the uncertainty as to which of the spectral replicates is the main one. Any one of the replicates can be treated as the primary since the corresponding continuous-time signal can be extracted by applying a suitable (ideal) bandpass filter.

What if the signal x(t) is not bandlimited? The frequency content of the (sampled) signal y(t), i.e., $Y(f)$, is still given by the expression in Eq. (3.3.7). In this case it is not possible to extract $X(f)$, or any individual replicate $X(f-kf_s)$ from $Y(f)$. How then can we categorize the relationship between y(t) and x(t)? In order to do this we first note that $Y(f)$ is periodic in the frequency domain with period f_s. That is, specification of $Y(f)$ for f in the range $[-f_s/2, f_s/2]$ completely specifies $Y(f)$ for all values of f. Based on this notion we define $X^{\sim}(f)$ by

$$X^{\sim}(f) = \begin{cases} \frac{1}{f_s} Y(f) & \text{for } |f| \le \frac{1}{2} f_s \\ 0 & \text{for } |f| > \frac{1}{2} f_s \end{cases} \tag{3.3.13}$$

Clearly $X^{\sim}(f)$ corresponds to a bandlimited signal which, in the frequency range $[-f_s/2, f_s/2]$, has the same (frequency) content as $Y(f)$ (up to a constant factor). With this definition we can write

$$Y(f) = f_s \sum_{k=-\infty}^{\infty} X^{\sim}(f - kf_s) \tag{3.3.14}$$

which is just a restatement of the periodic nature of $Y(f)$—$Y(f)$ is the same function in each period.

We can recognize $\tilde{X}(f)$ as the Fourier transform of a bandlimited signal, $\tilde{x}(t)$, which when sampled would yield $y(t)$. If the signal $x(t)$ were indeed bandlimited to $f_c < f_s/2$ then clearly $\tilde{X}(f) = X(f)$ or $\tilde{x}(t) = x(t)$; if $x(t)$ were bandlimited (such as the signal $x_1(t)$) but $X(f)$ was not localized to the frequency band around DC then $\tilde{X}(f)$ would be a replicate of $X(f)$. This notion can be extended to the case where $x(t)$ is not bandlimited as

$$
\begin{aligned}
\tilde{X}(f) &= \sum_{k=-\infty}^{\infty} X(f - kf_s) \text{ for } |f| \le \frac{1}{2}f_s \text{ and} \\
\tilde{X}(f) &= 0 \text{ for } |f| > \frac{1}{2}f_s
\end{aligned}
\tag{3.3.15}
$$

Note the similarity between Eq. (3.3.15), which extracts a one-period portion of a periodic function in the frequency domain, and its counterpart in the time domain described in Eq. (2.3.23). The composition of $\tilde{X}(f)$ from $X(f)$ is depicted graphically in Fig. 3.9. $\tilde{X}(f)$ is created by adding the frequency components of $X(f)$ in the following fashion. Since

$$
\begin{aligned}
\tilde{X}(f) &= X(f) + X(f - f_s) + X(f - 2f_s) + \ldots \\
&\quad + X(f + f_s) + X(f + 2f_s) + \ldots
\end{aligned}
\tag{3.3.16}
$$

a) $\tilde{X}(f)$ contains that portion of $X(f)$ for $-f_s/2 < f < f_s/2$ from $X(f)$

b) $\tilde{X}(f)$ contains that portion of $X(f)$ for $f_s/2 < f < 3f_s/2$ from $X(f+f_s)$

c) $\tilde{X}(f)$ contains that portion of $X(f)$ for $-3f_s/2 < f < -f_s/2$ from $X(f-f_s)$ and so on.

It can be seen from Fig. 3.9 that $\tilde{X}(f)$ contains the entire spectral content of $x(t)$ because of the effect of shifts. This phenomenon is called ***aliasing*** or ***folding***. The spectral content of $\tilde{X}(f)$ at any frequency f_α within $[-f_s/2, f_s/2]$ depends on the spectral content $X(f_\alpha)$, $X(f_\alpha+f_s)$, $X(f_\alpha-f_s)$, ... and it is not possible, based solely on $\tilde{X}(f)$, to determine which replicate $X(f+kf_s)$ determined $\tilde{X}(f_\alpha)$.

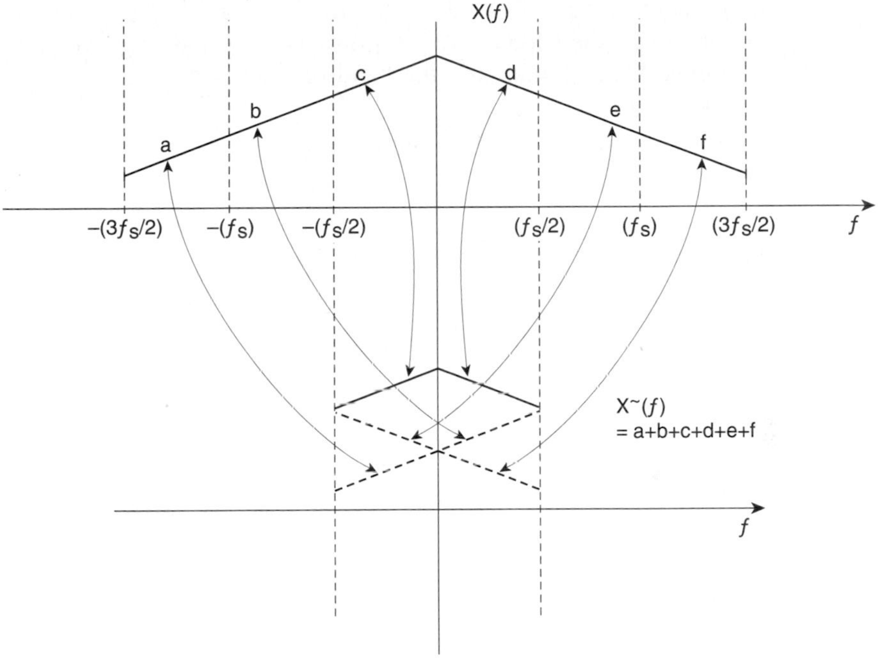

Fig. 3.9
Upon sampling, ***all*** signal components alias into baseband

To alleviate the problem of folding, or aliasing, the signal is usually filtered prior to sampling. If the desired signal is lowpass, that is, concentrated around $f = 0$, then the appropriate filter is lowpass; if the desired signal is bandpass, that is, concentrated around some center frequency (usually a multiple of the sampling frequency) then the proper filter is bandpass. The intent of the filter is to remove signal components that may, because of aliasing, corrupt desired signal components and thus the nomenclature used for the filter is ***anti-aliasing***. This is depicted in Fig. 3.10, which shows the desired signal x(t), which is bandlimited, corrupted by interference, n(t). n(t) represents all the signal power outside the bandwidth limit as well as any noise that may be present. To illustrate the effects of aliasing, the desired signal is recovered using an ideal lowpass filter to yield the signal labeled x~(t).

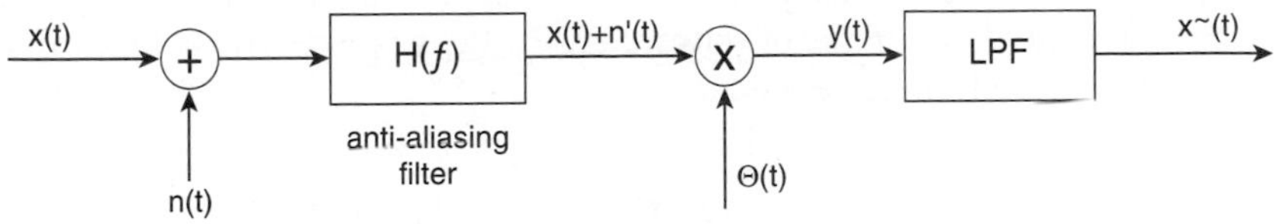

Fig. 3.10
Pre-sampling filter attenuates out-of-band signals that would alias back in-band

Now $\tilde{x}(t)$ is different from x(t) because of two effects. One is the impact of filtering x(t) by the anti-aliasing filter $H(f)$; the response of $H(f)$ is nominally unity in the frequency range of interest and will be ignored. The second is the result of signal components outside $[-f_s/2, f_s/2]$ that pass through the (non-ideal) filter $H(f)$, albeit attenuated, and alias into the band of interest. The form of this aliasing is discussed below.

Suppose n(t) consisted of a sinusoid of frequency f_1:

$$n(t) = A\cos(2\pi f_1 t) \text{ where } |f_1| > \frac{f_s}{2} \tag{3.3.17}$$

This gives rise to a component at the output of the filter given by

$$n_1(t) = A|H(f_1)|\cos(2\pi f_1 t + \varphi_1) \tag{3.3.18}$$

where $|H(f_1)|$ and $\varphi_1 = \Phi(f_1)$ are the magnitude and phase responses of the anti-aliasing filter at frequency f_1. This component aliases into the band of interest as a sinusoidal signal of frequency f_0, where $|f_0| < f_s/2$, given by

$$f_0 = f_1 - k f_s \text{ where k is an integer such that } |f_0| \le \frac{f_s}{2} \tag{3.3.19}$$

and $\tilde{x}(t)$ is composed of

$$\tilde{x}(t) = x(t) + A|H(f_1)|\cos(2\pi f_0 t + \varphi_1) \tag{3.3.20}$$

This is especially detrimental if the desired signal had a sinusoidal component at frequency f_0. For example, if

$$x(t) = B\cos(2\pi f_0 t + \varphi_0) \tag{3.3.21}$$

then, depending on the relative phase difference $(\varphi_1 - \varphi_0)$, the equivalent component in $\tilde{x}(t)$ could have an amplitude, C, between two extremes:

$$B - A|H(f_1)| \le C \le B + A|H(f_1)| \tag{3.3.22}$$

corresponding to the extremes of destructive interference to constructive interference. Since the relative phase difference is usually not known, we quantify the effect of the aliasing as the *power* of $n_1(t)$, given by $(A^2/2)|H(f_1)|^2$. Constructive (or destructive) interference is regarded as "coherent" addition since we are adding (subtracting) in amplitude; adding in power is called "incoherent" addition and is akin to assuming that the two signals are uncorrelated with each other. This notion can be extended to the case where the interfering signal had several (sinusoidal) components at frequencies $f_1, f_2, f_3, \ldots$ of powers $P_1, P_2, P_3, \ldots$ respectively by evaluating the power of the aliased interference as

$$P_I = \sum_{i} |H(f_i)|^2 P_i \tag{3.3.23}$$

If the interfering signal is noise with power spectral density $S_{nn}(f)$ then the total noise power that corrupts $\tilde{x}(t)$ is evaluated as

$$\sigma_I^2 = \int_{-\infty}^{+\infty} S_{nn}(f)\left|H(f)\right|^2 df \tag{3.3.24}$$

Eqs. (3.3.23) and (3.3.24) are the mathematical equivalent of saying that **all the interference present at the sampler will alias, or fold back, into the band of interest**. The anti-aliasing filter thus represents our **only** means of defense from interfering components that lie outside $[-f_s/2,\ f_s/2]$.

To summarize:

a) The sampling theorem provides a minimum sampling rate that must be employed in order to represent a (bandlimited) signal.

b) Any signal components outside $[-f_s/2,\ f_s/2]$ will alias back and become indistinguishable from components within $[-f_s/2,\ f_s/2]$.

c) The power of the interference in the signal, post-sampling, will be equal to the total power of the interference, pre-sampling.

d) After sampling a signal will lose its frequency-location identity in the sense that a signal becomes indistinguishable from a replicate.

3.3.2 The Fourier Transform of a Discrete-Time Signal

Given a discrete-time signal, $\{x(n)\}$, the Fourier transform is defined in terms of the transform of the (continuous-time) signal generated as the output of an ideal converter (see Fig. 3.4), namely $x(n)\ \delta(t-nT_s)$. Thus the Fourier transform of y(t), $Y(f)$, is, by definition, the Fourier transform of the discrete-time signal $\{x(n)\}$ $(= \{x(nT_s)\})$. That is,

$$X_s(f) = \mathbf{F}\{\sum_n x(n)\,\delta(t-nT_s)\} = \sum_n x(n)\,e^{-2\pi n f T_s} \tag{3.3.25}$$

We shall refer to the function $X_s(f)$ as the ***Discrete-Time Fourier Transform*** or **DTFT** of the discrete-time signal $\{x(n)\}$. The subscript has been used to distinguish the Fourier transforms of the underlying continuous time signal x(t) and the sampled signal.

3.3.2.1 Periodicity of the Fourier Transform of Discrete-Time Signals

Examination of Eq. (3.3.25) indicates that the spectral content of the discrete-time signal can be written as a function of the **normalized frequency** (f/f_s) or the **normalized radian frequency** $\omega = 2\pi(f/f_s)$. This allows us to consider an alternate notation for the Fourier Transform of the discrete-time signal, i.e., the **DTFT**, as

$$X(e^{j\omega}) = \sum_n x(n)\,e^{jn\omega} \tag{3.3.26}$$

Both Eqs. (3.3.25) and (3.3.26) provide a representation of the signal in terms of a frequency variable. The former is preferable when the notion of frequency needs to be expressed absolutely, in Hz, kHz, etc. The latter is convenient when frequency is normalized such that the sampling frequency is unity. The frequency range $[-f_s/2, f_s/2]$ in f translates to a range $[-\pi, +\pi]$ in ω, which has units of radians/sec.

As shown before, $Y(f)$, and thus $X_s(f)$, are periodic in the frequency domain with period f_s. Thus $X_s(f)$ is completely specified by its values over one period. This period is usually taken as the frequency band $[-f_s/2, f_s/2]$ or $[0, f_s]$, all other frequencies being implied by periodic continuation. Assuming the former we have

$$X_s(f) = f_s \tilde{X}(f) \text{ for } |f| \le \frac{f_s}{2} \tag{3.3.27}$$

which relates the spectral representation of the discrete-time signal with a bandlimited continuous-time signal $\tilde{x}(t)$. The relationship between $\tilde{x}(t)$ and $x(t)$ was described in the previous section. The scaling factor f_s is often ignored since it is a known constant though its function is to indicate a scale factor between quantities that are expressed either in terms of *per unit time* or *per sample.*

Since $X_s(f)$ is periodic, it can be expressed as a Fourier series as

$$X_s(f) = \sum_n \chi_n e^{-j2\pi n f T_s} \tag{3.3.28}$$

where

$$\chi_n = \frac{1}{f_s} \int_{-\frac{1}{2}f_s}^{+\frac{1}{2}f_s} X_s(f) e^{-j2\pi n f T_s} df \tag{3.3.29}$$

The coefficients of the Fourier series, $\{\chi_k\}$ are related to the sample values of the discrete-time signal $\{x(nT_s)\}$. To see this, χ_k can be written as

$$\chi_n = \int_{-\frac{1}{2}f_s}^{+\frac{1}{2}f_s} \tilde{X}(f) e^{-j2\pi n f T_s} df = x(-kT_s) \tag{3.3.30}$$

which shows that the Fourier series coefficients of the spectral representation of the discrete-time signal are none other than the samples of the continuous-time signal (taken in reverse order). If $\{\chi_k\}$ is a finite energy sequence then Fourier theory tells us that $X_s(f)$ is continuous and differentiable (with derivatives of all orders).

3.3.2.2 Fourier Transforms of Periodic Discrete-Time Signals

The function $X_s(f)$ will be continuous if $\{x(n)\}$ is a finite energy signal. If $\{x(n)\}$ is not finite energy, but finite power, then $X_s(f)$ defined by Eq. (3.3.25) will contain delta functions. This mathematical implication of periodicity is exactly analogous to the continuous-time situation where we coined the term **line spectrum** to describe transforms containing delta functions.

Two cases are considered. The first case is when $\{x(n)\}$ is almost periodic, that is, when the discrete-time signal is generated as samples of a periodic continuous-time signal whose period is not an integer or rational multiple of the sampling interval. The second case is when the sequence is periodic with period of N samples.

Case 1: Suppose $\{x(n)\}$ is generated from the periodic signal x(t) and we wish to calculate $X_s(f)$ over the range $[0,f_s)$. Since x(t) is periodic, it can be expressed as a Fourier series with coefficients $\{c_k\}$ and the Fourier transform of x(t) will be of the form

$$X(f) = \sum_{k=-\infty}^{\infty} c_k \delta(f - kf_0) \tag{3.3.31}$$

where $f_0 = 1/T_0$ is the fundamental frequency. x(t) may not be bandlimited to $[-f_s/2,f_s/2]$ even if $f_0 < f_s/2$. Because of aliasing, however, **every** component of $X(f)$ is represented in $X_s(f)$, which can thus be expressed as

$$X_s(f) = \sum_{k=-\infty}^{\infty} c_k \delta(f - ((kf_0))) \quad \text{for } 0 \le f < f_s \tag{3.3.32}$$

where the notation $((f))$ involves shifting the frequency to lie within the band of interest, namely $[0,f_s)$ by the following operation

$$((f)) = (f - kf_s) \text{ where k is an integer such that } 0 \le (f - kf_s) < f_s \tag{3.3.33}$$

Thus $X_s(f)$ is a line spectrum and over the frequency range $[0,f_s)$ will have one line for each harmonic component of x(t).

If x(t) is, in addition to being periodic, bandlimited to $[-f_s/2,f_s/2]$, then $X(f)$ is comprised of (2N+1) delta functions occurring at frequencies kf_0, where k=–N, –(N–1), ... ,–1,0,1, ... , (N–1), N as expressed in Eq. (3.3.34).

$$X(f) = \sum_{k=-N}^{N} c_k \delta(f - ((kf_0))) \tag{3.3.34}$$

In this case $X_s(f)$ is given by

$$X_s(f) = \sum_{k=0}^{N} c_k \delta(f - kf_0) + \sum_{k=1}^{N} c_k \delta(f + kf_0 - f_s) \tag{3.3.35}$$

where the delta functions in the frequency range $[-f_s/2,0)$ have been shifted intothe range $[0, f_s)$.

If x(t) is periodic with period T_0, which is a rational multiple of T_s, the sampling interval, then

$$T_0 = \frac{M}{K}T_s \text{ and } f_0 = \frac{K}{M}f_s \tag{3.3.36}$$

where K and M are some integers. The process of computing $((f))$ for the harmonic components of x(t) then shows that $X_s(f)$ will have at most M distinct lines and these lines correspond to frequencies that are a multiple of (f_s/M):

$$((k\frac{K}{M}f_s)) = \frac{\kappa}{M}f_s \text{ where } 0 \le \kappa \le (M-1) \tag{3.3.37}$$

By writing x(t) by its Fourier series representation, it can be observed that {x(n)} is periodic with period M samples, since

$$x(t) = \sum c_k e^{j2\pi(\frac{K}{M}f_s)kt} \Rightarrow x(n) = \sum c_k e^{j2\pi(\frac{Knk}{M})} \tag{3.3.38}$$

and replacing n by (n+mM), for any integer m, does not affect the complex exponential terms. Thus if the period of x(t) is rationally related to the sampling interval, the spectrum of {x(n)} is a line spectrum with at most M lines in the frequency band $[0, f_s)$. Since {x(n)} is periodic, it follows that such periodic signals can be represented by exactly M samples. This leads to the following result: If x(t) is bandlimited and periodic, it can be represented completely by M samples taken at a rate Mf_0.

Case 2: Suppose {x(n)} is periodic with a period of N samples. Then the Fourier transform can be written formally as

$$X_s(f) = \sum_{n} x(n)e^{-j2\pi nfT_s} \tag{3.3.39}$$

and by setting

$$n = k + rN \; ; \; 0 \le k \le (N-1) \tag{3.3.40}$$

we get the following double summation for $X_s(f)$:

$$\begin{aligned} X_s(f) &= \sum_{r=-\infty}^{\infty}\sum_{k=0}^{N-1} x(k+rN)e^{-j2\pi(k+rN)fT_s} \\ &= \sum_{k=0}^{N-1} x(k)e^{-j2\pi kfT_s}\left[\sum_{r=-\infty}^{\infty} e^{-j2\pi rNfT_s}\right] \end{aligned} \tag{3.3.41}$$

In Eq. (3.3.41), the periodic nature of x(n), which implies x(n) = x(n+rN), has been utilized. Using Eqs (3.3.3) and (3.3.4), the term in parentheses can be written as

$$\sum_{r=-\infty}^{\infty} e^{-j2\pi rNfT_s} = \frac{1}{NT_s}\sum_{r=-\infty}^{\infty} \delta(f - \frac{rf_s}{N}) \qquad (3.3.42)$$

Thus a periodic discrete-time signal will have a line spectrum with components at frequencies $r(f_s/N)$ and, further, the periodic behavior in the frequency domain implies that the components at $r = 0,1, \ldots, (N-1)$ would completely specify $X_s(f)$. At frequency $f = r(f_s/N)$, $X_s(f)$ has a line, i.e., delta function, of strength

$$\frac{1}{NT_s}\sum_{k=0}^{N-1} x(k)e^{-j2\pi rk(\frac{f_s}{N})T_s} = \frac{1}{NT_s}\sum_{k=0}^{N-1} x(k)e^{-j\frac{2\pi rk}{N}} = \frac{X(r)}{NT_s} \qquad (3.3.43)$$

Other than a constant of proportionality, the strength of the components at frequencies $f_r = r(f_s/N)$ is given by the N-point sequence $\{X(r), r = 0,1, \ldots,(N-1)\}$ where

$$X(r) = \sum_{k=0}^{N-1} x(k)e^{-j\frac{2\pi rk}{N}} \qquad (3.3.44)$$

The expression in Eq. (3.3.44) is the defining expression for the ***Discrete Fourier Transform*** or **DFT**, which is quite different from the discrete-time Fourier Transform (**DTFT**). Strictly speaking, the **DFT** is a mapping between an N-point sequence in the "time" domain and an N-point sequence in the "frequency" domain that has a superficial resemblance to the discrete-time Fourier transform. The **DFT** is, however, applicable in the computation of the **DTFT** of a periodic sequence as we have just seen and also for computing the **DTFT** of finite length sequences, as we shall see next.

3.3.2.3 Fourier Transforms of Finite-Length Sequences

If $\{x(n); n=0,1, \ldots ,(N-1)\}$ is a sequence of finite length, then its **DTFT** is given by

$$X_s(f) = \sum_{k=0}^{N-1} x(k)e^{-j2\pi kfT_s} \qquad (3.3.45)$$

Note the similarity between Eq. (3.3.45) and Eq. (3.3.44), which defined the **DFT**. In fact, calculation of the **DTFT** of $\{x(n)\}$ on the finite grid of equispaced frequencies corresponding to $r(f_s/N)$ corresponds to the **DFT**.

This gives rise to the following concept that is important in cases where we are trying to calculate the spectral content of a signal based on a finite record. That is, given N samples of a signal we are trying to compute its Fourier transform. The behavior at sample indices other than $\{0,1, \ldots , (N-1)\}$ can only be assumed. Two possible choices are first that the signal is zero outside this interval, and second that the signal is periodic with period N samples. In both cases the **DFT** expression of Eq. (3.3.44) can be calculated. The interpretation is, however, quite different. In the first case (zero outside) the **DTFT** is continuous and the sequence {X(r); r=0,1, ... , (N−1)} is interpreted as samples of $X_s(f)$ using a uniform frequency grid. In the second case (periodic) the **DTFT** is of the form of a line spectrum and the sequence {X(r); r=0, 1, ... , (N−1)} is interpreted as the strength of the individual (complex sinusoidal) components.

3.3.2.4 Inversion of the Discrete-Time Fourier Transform

Inverting a **DTFT** is the process whereby from $X_s(f)$ we obtain the sequence $\{x(nT_s)\}$. The inversion is based on the observation that the time samples are the Fourier coefficients of the expansion of $X_s(f)$, which is periodic with period f_s (the sampling frequency). Consequently,

$$x(nT_s) = \frac{1}{f_s}\int_0^{f_s} X_s(f)\, e^{j2\pi n f T_s}\, df \qquad (3.3.46)$$

When we are dealing with Fourier transforms expressed in terms of the normalized radian frequency ω, the inversion operation can be written as

$$x(nT_s) = \frac{1}{2\pi}\int_0^{2\pi} X_s(e^{j\omega})\, e^{jn\omega}\, d\omega \qquad (3.3.47)$$

The notation used in Eqs. (3.3.46) and (3.3.47) is consistent with that defined in Eqs. (3.3.25) and (3.3.26).

3.4 LINEAR TIME-INVARIANT DISCRETE-TIME SYSTEMS

The notions of linearity and time invariance as applied to discrete-time systems are exactly the same as for continuous-time systems. The results described in Chapter 2 are directly applicable once we recognize that a discrete-time signal is a sequence whereas a continuous-time signal is a time function. A LTI system operates on an input sequence {x(n)} to produce an output sequence {y(n)}. Consequently, we can write

$$y(n) = \mathbf{H}\{\, x(n) \,\} \qquad (3.4.1)$$

Linearity, or superposition, is that property of a linear time-invariant (**LTI**) system that preserves linear combinations. Thus if the response of the system to the input signal $\{x_i(n)\}$ is $\{y_i(n)\}$, for i=1,2, ... , taken independently, then if a new signal {x(n)} is constructed as a linear combination of the $\{x_i(n)\}$, the output of the system for this combination input will be the same combination of the separate outputs. That is, if

$$y_i(n) = \mathbf{H}\{\, x_i(n) \,\} \qquad (3.4.2)$$

and

$$x(n) = a_i\, x_i(n) \tag{3.4.3}$$

then

$$y(n) = a_i y_i(n) \tag{3.4.4}$$

Time invariance is that property of an **LTI** that preserves time shifts; if the input is delayed by m samples, then the output will be delayed by m samples. Thus if $\{x_1(n)\}$ and $\{y_1(n)\}$ form an input-output pair, the output $\{y_2(n)\}$, for an input $\{x_2(n)\}$, which is a delayed version of $\{x_1(n)\}$, will be a delayed version of $\{y_1(n)\}$. That is,

$$\begin{aligned} y_1(n) &= \mathbf{H}\{x_1(n)\} \text{ and } x_2(n) = x_1(n-m) \\ \Rightarrow y_2(n) &= \mathbf{H}\{x_2(n)\} = y_1(n-m) \end{aligned} \tag{3.4.5}$$

One of the principal implications of linearity and time-invariance is that the output, $\{y(n)\}$, can be expressed as a convolution of the input $\{x(n)\}$ and a signal $\{h(n)\}$ which is characteristic of the system and is called the ***impulse response***. This implication is of the *if and only if* variety; if the output can be expressed as the convolution of the input and a (given) signal then the system is **LTI**. Consider the case where the input is a discrete-time delta function, $\{\delta(n)\}$. The response is then, by definition, the impulse response $\{h(n)\}$:

$$h(n) = \mathbf{H}\{\delta(n)\} \tag{3.4.6}$$

To see that the output is the convolution of the input and the impulse response, we can first express $\{x(n)\}$ as a combination of impulses by writing

$$x(n) = \underset{k}{\ } x(k)\delta(n-k) \tag{3.4.7}$$

and then the **LTI** properties imply that

$$\begin{aligned} y(n) &= \mathbf{H}\{y(n)\} = \mathbf{H}\{\underset{k}{\ } x(k)\delta(n-k)\} \\ &= \underset{k}{\ } x(k)\mathbf{H}\{\delta(n-k)\} = \underset{k}{\ } x(k)h(n-k) \end{aligned} \tag{3.4.8}$$

In Eqs. (3.4.7) and (3.4.8) the limits of the summation correspond to all possible values of k, namely k = – to k = + . Because of the nature of the limits we can employ a change of variable (index) and obtain

$$y(n) = \underset{k}{\ } h(k)x(n-k) \tag{3.4.9}$$

The notion of a **causal** system is one that does not provide a response prior to the application of input. In terms of the impulse response, causality can be expressed by saying that

$$\text{causality} \Longleftrightarrow h(n) = 0 \;\; \text{for } n < 0 \tag{3.4.10}$$

which implies that if the input is zero prior to n=0 then the output will, likewise, be zero for n < 0. One implication of causality is that the convolution expressions for the output can be expressed as

$$y(n) = \sum_{k=0}^{\infty} h(k)x(n-k) = \sum_{k=-\infty}^{n} x(k)h(n-k) \tag{3.4.11}$$

which is what Eq. (3.4.8) or (3.4.9) reduce to when Eq. (3.4.10) is considered.

A second implication of the **LTI** properties is the notion of frequency response. If the input to the system is a complex exponential, then the (steady state) output will be a complex exponential of the same frequency and the relative change in magnitude and phase between input and output will be a function of the frequency of the sinusoid. Thus if {x(n)} is a unit complex sinusoid with normalized frequency f_0, (sampling frequency normalized to unity)

$$x(n) = e^{j2\pi f_0 n} = e^{j\omega_0 n} \tag{3.4.12}$$

then the output will be given by

$$y(n) = H(f_0)\, e^{j2\pi f_0 n} \tag{3.4.13}$$

The entity $H(f_0)$ is a complex function of the real variable f_0 and provides the magnitude and phase of {y(n)} relative to {x(n)}. If we choose to use the normalized radian frequency as the independent variable then, since $\omega = 2\pi f$, we can write

$$y(n) = H(e^{j\omega_0})\, e^{j\omega_0 n} \tag{3.4.14}$$

Notice the difference in nomenclature between Eqs. (3.4.13) and (3.4.14) for the same entity, namely frequency response. The notation $H(f_0)$ specifically identifies the function as a function of frequency; the notation $H(e^{j\omega_0})$ identifies the function as one of frequency but indirectly. The latter notation will be seen to be less illogical when we relate the **DTFT** and **Z**-transform in the next section.

For causal systems, a third implication of linearity and time invariance is that the input-output relationship can be expressed as a **linear difference equation** (compare linear differential equation for continuous-time systems) of the form

$$y(n) = \sum_{k=1}^{D} b_k y(n-k) + \sum_{k=0}^{N} a_k x(n-k) \tag{3.4.15}$$

That is, the output can be constructed as a linear combination of the current and past values of the input and past values of the output. Two structures can be perceived in the difference equation. One is the contribution from past outputs and is called the Auto-Regressive (**AR**) or recursive part; the other involves a linear combination of inputs and is called the Moving-Average (**MA**) or nonrecursive part. A general system is thus **ARMA** (for auto-regressive, moving-average); when $b_k = 0$ for k=1,2, ... the system output depends only on the input and the system is considered **MA** or nonrecursive ; and when $a_k = 0$ for k=1,2, ... the system output depends only on the current input and past outputs the system is considered **AR** or recursive.

In the case of continuous-time systems, relating the underlying differential equation to an actual circuit implementation is difficult if not impossible. In sharp contrast, the difference equation defining an **LTI** system is an essential description of its implementation. The difference equation succinctly and completely describes the mathematical operations that must be performed on the sequence(s) to achieve the action of the **LTI** system. It is the basis upon which the code is developed, if the **LTI** system is implemented on a programmable processor, or the basis of sequencing the operation of arithmetic logic units, if the **LTI** system is implemented via (special-purpose) digital hardware.

In summary, linear time-invariant systems can be specified in three, equivalent, ways: by the impulse response $\{h(n)\}$, or the frequency response $H(e^{j\omega})$, or the (N+D+1) coefficients $\{a_k,\ k = 0, 1, \dots, N;\ b_k,\ k = 1, 2, \dots, D\}$.

In discrete-time systems the impulse response plays a very major role. In fact systems can be characterized in terms of the impulse response. One such characterization is based on the length of the impulse response. For instance, suppose that the impulse response $\{h(n)\}$ is of finite length. That is, if

$$h(n) = 0 \text{ for } n < 0 \text{ and } n \geq N \tag{3.4.16}$$

then the impulse response length is N samples. Such a system is called ***Finite Impulse Response*** or **FIR**. **FIR** systems can be described by a difference equation of the form

$$y(n) = \sum_{k=0}^{N-1} h(k)x(n-k) \tag{3.4.17}$$

which in one fell swoop describes the difference equation **and** the convolution sum expressing the output in terms of the input. Nonrecursive systems are always **FIR**. If the impulse response is not a finite-length sequence, then the system is ***Infinite Impulse Response*** or **IIR**.

The stability of a system can also be characterized by the impulse response. One form of the definition of stability can be expressed in words as "the system does not oscillate" implying that if the input is removed (x(n) = 0 for all indices n greater than some value, say 0), then the output must die out. Since oscillatory behavior is characterized by infinite energy, this stability requirement will be satisfied if

$$\sum_{k} |h(k)|^2 = G_N < \infty \qquad (3.4.18)$$

3.5 THE Z-TRANSFORM

The Z-transform is a technique that permits us to manipulate sequences, that is discrete-time signals, in a simple fashion. The underlying mathematical theories associated with the Z -transform are by no means new and have been used in Complex Analysis for several decades. Complex Analysis, that is, the study of complex functions of a complex variable, provides us with several theorems and results that make dealing with Z-transforms quite straightforward. Several excellent text books on Complex Analysis are available, such as Churchill [1.1].

3.5.1 The Basis for the Z-Transform

The basis for the Z-transform can be found in Complex Analysis, and in particular, a concept called the ***Laurent Series***. If F(**z**) is a complex function of a complex variable **z**, the Laurent series expansion of F(**z**) can be written formally as

$$F(\mathbf{z}) = \sum_{k=-\infty}^{\infty} a_k \mathbf{z}^{-k} \qquad (3.5.1)$$

That is, the function can be expressed as a power series involving all (integer) powers of **z**. This expansion is valid only for those values of **z** for which the power series converges; where it does converge, the series in the right-hand side of Eq. (3.5.1) will converge to F(**z**). If the series does not converge at any point **z**, then the association with F(**z**) is not valid for that **z**.

The region of convergence, that is the values of **z** for which the series expansion is valid, is usually depicted as an annular region, as shown in Fig. 3.11, contained between two circles of radius R_1 and R_2. The significance of the annular region can be explained by breaking up the power series into two series, one with positive powers of **z** and the other with negative powers of **z**:

$$F(\mathbf{z}) = \sum_{k=0}^{\infty} \alpha_k \mathbf{z}^{-k} + \sum_{k=0}^{\infty} \beta_k \mathbf{z}^{+k} = F_1(\mathbf{z}) + F_2(\mathbf{z}) \qquad (3.5.2)$$

The series comprising the negative powers of **z** can be viewed as a one-sided Laurent series of a function $F_1(\mathbf{z})$ and similarly the series comprising the positive powers of **z** can be viewed as the one-sided Laurent series of a function $F_2(\mathbf{z})$. The Laurent series for $F_1(\mathbf{z})$ will converge outside a circle radius R_1 provided all the singularities of $F_1(\mathbf{z})$ lie inside the said circle. The Laurent series for $F_2(\mathbf{z})$ will converge inside a circle of radius R_2 provided that all the singularities of $F_2(\mathbf{z})$ lie outside the said circle. In the annular region between the two circles both the series converge and hence the sum converges to F(**z**). For values of **z** within the region of convergence, the function F(**z**) and the series are, for all intent and purposes, the

same and for any operation performed on the series, an equivalent operation may as well be performed on the function F(**z**) and vice versa.

Given a sequence (discrete-time signal) {x(n)}, we define the Z-transform by the association

$$X(\mathbf{z}) = \sum_{n} x(n)\mathbf{z}^{-n} \tag{3.5.3}$$

where the index n ranges over the set for which x(n) is nonzero, possibly – to + . Provided the variable **z** is within the region of convergence, X(**z**) and the sequence {x(n)} are equivalent; X(**z**) serves not only as a representation for the sequence, but any manipulation of the sequence we may choose to do will have a direct (and sometimes simpler) counterpart as a modification of X(**z**).

For one-sided sequences the Z-transform takes the form

$$X(\mathbf{z}) = \sum_{n=0}^{\infty} x(n)\mathbf{z}^{-n} \tag{3.5.4}$$

Based on our prior discussion, the region of convergence is outside a circle of radius R_1 which encompasses all the singularities of X(**z**). In discrete-time signal processing applications the unit circle, that, is the circle with unit radius and centered at the origin of the complex **z**-plane, is of special interest. We will require, as will be clear shortly, that R_1 be less than one, implying that the unit circle is within the region of convergence of the Z-transform.

For example, consider the causal sequence defined as

$$x(n) = a^{n}u(n) \tag{3.5.5}$$

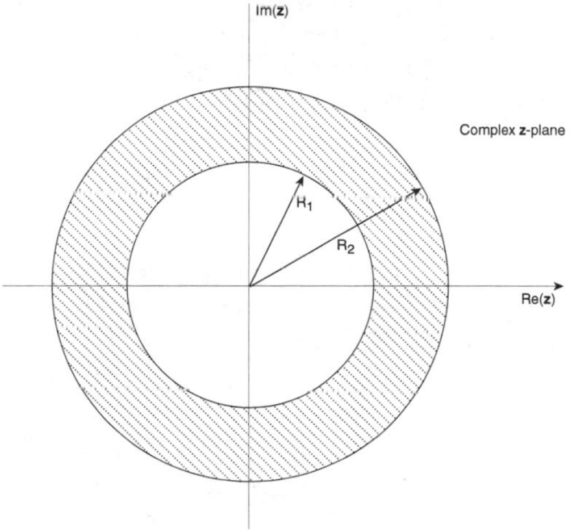

Fig. 3.11
Annulus in **z**-plane showing region of convergence of the Laurent series

From Eq. (3.5.4), the defining relationship, the Z-transform of the sequence is given by

$$X(\mathbf{z}) = \sum_{n=0}^{\infty} a^{n} \mathbf{z}^{-n} = \frac{1}{1 - a\mathbf{z}^{-1}} \tag{3.5.6}$$

The second part of Eq. (3.5.6) comes from recognizing that the terms of the sum followed a geometric progression with common ratio of $a\mathbf{z}^{-1}$. The compact expression for X(**z**) is valid only if the geometric progression dies out, which implies that Eq. (3.5.6) is valid only if

$$\left|a\mathbf{z}^{-1}\right| < 1 \Rightarrow |\mathbf{z}| > |a| \tag{3.5.7}$$

and from which we can discern the region of convergence as the outside of a circle of radius R_1 $|a|$. Examination of X(**z**) indicates that it has only one singularity, at $\mathbf{z} = a$. When $\mathbf{z} = a$, X(**z**) is of the form 1/0 which is ill-defined (infinity). With the said choice of R_1 it is then true that all singularities of X(**z**) are inside the circle of radius R_1.

The Z-transform of a finite-length sequence is a polynomial in $\mathbf{z}^{-1}$. This is obvious since if {h(n)} has a finite length of N samples,

$$H(\mathbf{z}) = \sum_{n=0}^{N-1} h(n)\mathbf{z}^{-n} \tag{3.5.8}$$

Such transforms are very well-behaved functions of **z** and the region of convergence is the whole complex plane, except for the origin **z**=0.

3.5.2 Inversion of the Z-Transform

The process of inverting the Z-transform refers to the procedure used to obtain the sequence {x(n)} given the function X(**z**). The formula for the inverse transform can be derived using the following result from Complex Analysis: If C is a closed contour that encompasses the origin, then

$$\frac{1}{2\pi j}\oint_C \mathbf{z}^{k} \frac{d\mathbf{z}}{\mathbf{z}} = \delta(k) \;\; \{= 1 \text{ for } k = 0;\, 0 \text{ otherwise }\} \tag{3.5.9}$$

where the contour is traversed in a counterclockwise direction. For our purposes, the closed contour is almost always the unit circle in the **z**-plane ($|\mathbf{z}| = 1$). Applying this operation to both sides of the definition Eq. (3.5.3) yields

$$\begin{aligned} \frac{1}{2\pi j}\oint_C \mathbf{z}^{k} X(\mathbf{z})\frac{d\mathbf{z}}{\mathbf{z}} &= \sum_{n} x(n)\frac{1}{2\pi j}\oint_C \mathbf{z}^{k-n}\frac{d\mathbf{z}}{\mathbf{z}} \\ &= \sum_{n} x(n)\,\delta(k-n) = x(k) \end{aligned} \tag{3.5.10}$$

It should be kept in mind that the series and X(**z**) are equivalent only in the region of convergence and hence the closed contour must be in this region. By requiring that $R_1 < 1$ we are assured that the unit circle will indeed be in the region of convergence.

Computing the contour integral is not quite as difficult as it may first appear. Evaluation of the integral uses the following result (see , for example, Churchill [1.1]). Let F(**z**) be analytic inside the closed contour C except at a finite number of points z_i, i = 1, 2, ... , P, where the singularities are "poles." A singularity at z_i is considered a pole if $F_i(\mathbf{z})$ given by

$$F_i(\mathbf{z}) = (\mathbf{z} - z_i)^m F(\mathbf{z}) \tag{3.5.11}$$

is analytic at z_i and *m* is the order of the pole. *m* = 1 corresponds to a simple pole. If all the singularities of F(**z**) inside the contour C are poles then

$$\frac{1}{2\pi j}\oint_C F(\mathbf{z})\, d\mathbf{z} = \sum_i \left[\text{residue of } F(\mathbf{z}) \text{ at } z_i\right] \tag{3.5.12}$$

The residue at z_i, $\mathbf{R}_i$, is given by:

$$\mathbf{R}_i = \frac{1}{(m-1)!}\frac{d^{(m-1)}}{d\mathbf{z}^{(m-1)}}(\mathbf{z} - z_i)^m F(\mathbf{z}) \text{ evaluated at } \mathbf{z} = z_i \tag{3.5.13}$$

In most, if not all, cases, the Z-transform that we will invert will be of the form

$$X(\mathbf{z}) = \sum_{n=0}^{N-1} \alpha_n \mathbf{z}^{-n} + \frac{\sum_{n=0}^{D-1} a_n \mathbf{z}^{-n}}{1 - \sum_{n=1}^{D} b_n \mathbf{z}^{-n}} = P(\mathbf{z}) + \frac{N(\mathbf{z})}{D(\mathbf{z})} \tag{3.5.14}$$

where we have shown X(**z**) as a combination of a polynomial in $\mathbf{z}^{-1}$ plus a ratio of polynomials. All the singularities of X(**z**) will occur at the roots of the denominator polynomial D(**z**). These roots are the poles of X(**z**). Inversion of the Z-transform requires the evaluation of

$$x(n) = \frac{1}{2\pi j}\oint_C \mathbf{z}^n \left[P(\mathbf{z}) + \frac{N(\mathbf{z})}{D(\mathbf{z})}\right]\frac{d\mathbf{z}}{\mathbf{z}} \tag{3.5.15}$$

and the singularities of the integrand are clearly the poles of X(**z**) and, possibly, the origin **z**=0.

Consider the example where X(**z**) is given by

$$X(\mathbf{z}) = \frac{1}{1 - a\mathbf{z}^{-1}} = \frac{\mathbf{z}}{\mathbf{z} - a} \tag{3.5.16}$$

where $|a| < 1$. The inversion of the Z-transform is, therefore,

$$x(n) = \frac{1}{2\pi j}\oint_C \mathbf{z}^n \frac{d\mathbf{z}}{\mathbf{z} - a} \qquad (3.5.17)$$

and the choice of the unit circle for the contour is valid. The integrand has a single pole at $\mathbf{z} = a$ that is inside the contour (the unit circle) and application of Eqs. (3.5.12) and (3.5.13) yields

$$x(n) = a^n u(n) \qquad (3.5.18)$$

as we would expect. It can be shown that for $n < 0$, x(n) evaluated using Eq. (3.5.17) is indeed zero.

3.5.3 Some Properties of the Z-Transform

The Z-domain representation of a sequence would not be useful if the typical operations on sequences did not have a corresponding operation on the transform. To this end we shall consider what the equivalent operation on the transform is for the common operations performed on sequences.

a) Delay. If {y(n)} is a delayed version of {x(n)}, by m samples, the Z-transform Y(**z**) can be related to X(**z**) in the following manner.

$$Y(\mathbf{z}) = \sum_n y(n)\mathbf{z}^{-n} = \sum_n x(n-m)\mathbf{z}^{-(n-m)-m} = \mathbf{z}^{-m} X(\mathbf{z}) \qquad (3.5.19)$$

Thus a delay in the time domain corresponds to multiplication by $\mathbf{z}^{-m}$ in the Z-domain. In terms of a block diagram, a delay of m samples can be represented in the manner depicted in Fig. 3.12. It can be shown that the region of convergence of Y(**z**) is the same as that of X(**z**).

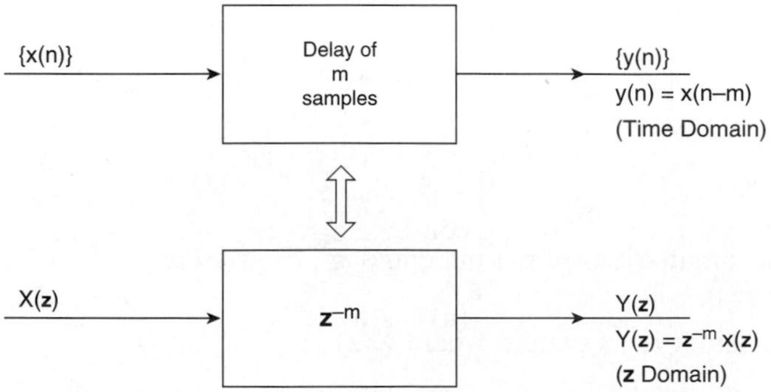

Fig. 3.12
Block diagram for representing a delay of m samples

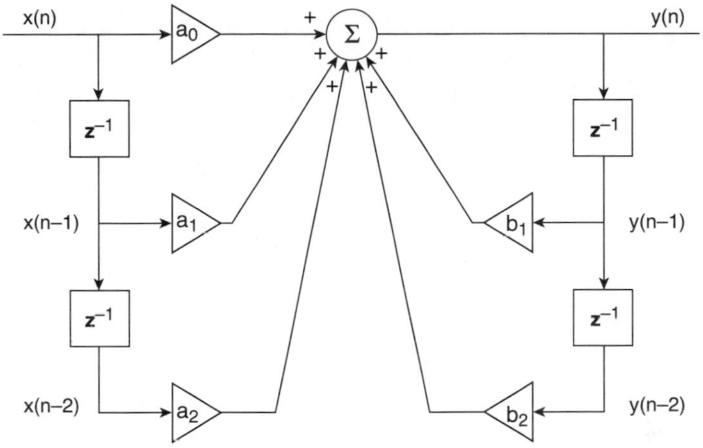

Fig. 3.13
Block diagram of second-order difference equation

Associating a delay of one sample with $\mathbf{z}^{-1}$ allows us to pictorially represent difference equations. For example, consider the difference equation

$$\begin{aligned} y(n) &= a_0 x(n-0) + a_1 x(n-1) + a_2 x(n-2) \\ &\quad + b_1 y(n-1) + b_2 y(n-2) \end{aligned} \tag{3.5.20}$$

The equivalent block diagram, using the notion that a delay of one sample is equivalent to $\mathbf{z}^{-1}$, is shown in Fig. 3.13. The operation of multiplication is represented by the triangle (the symbol for an amplifier).

b) **Time reversal.** If {y(n)} is the time-reversed version of {x(n)} then its Z-transform can be evaluated as

$$Y(\mathbf{z}) = \sum_{n} y(n)\mathbf{z}^{-n} = \sum_{n} x(-n)\mathbf{z}^{-n} = X(\mathbf{z}^{-1}) \tag{3.5.21}$$

Time reversal is equivalent to replacing $\mathbf{z}$ by $(1/\mathbf{z})$ or $\mathbf{z}^{-1}$. The region of convergence is important. If $X(\mathbf{z})$ converges in the annulus defined by radii R_1 and R_2 ($R_2 > R_1$), then $Y(\mathbf{z})$ converges in the annulus defined by radii $(1/R_1)$ and $(1/R_2)$. In all cases of interest the radii R_1 and R_2 will satisfy $0 < R_1 < 1 < R_2$ and consequently the unit circle, $|z|=1$, will be in the region of convergence of the $X(\mathbf{z})$ as well as $X(\mathbf{z}^{-1})$.

c) **Convolution.** If {y(n)} is the convolution of two sequences {x(n)} and {h(n)}, the Z–transform allows us to represent $Y(\mathbf{z})$ as the product of $X(\mathbf{z})$ and $H(\mathbf{z})$. That is, convolution in the time domain is equivalent to multiplication in the transform domain. This can be derived in the following way. Since

$$y(n) = \sum_{k} x(k)h(n-k) \tag{3.5.22}$$

we can express the Z-transform of {y(n)} as

$$Y(\mathbf{z}) = \sum_{n} y(n)\mathbf{z}^{-n} = \sum_{n}\sum_{k} x(k)h(n-k)\mathbf{z}^{-n}$$

$$= \sum_{k} x(k)\{\sum_{n} h(n-k)\mathbf{z}^{k-n}\}\mathbf{z}^{-k} = X(\mathbf{z})H(\mathbf{z}) \tag{3.5.23}$$

In the derivation we have employed two techniques that are often used. First, since the range of indices is nominally – to + , the change of variable m = (n–k) will retain the infinite limits. Second is the interchange of the order in which the summations are performed. This interchange is valid provided that the variable **z** lies within the region where both X(**z**) and H(**z**) are valid. The region of convergence of Y(**z**) is thus the intersection of those of X(**z**) and H(**z**).

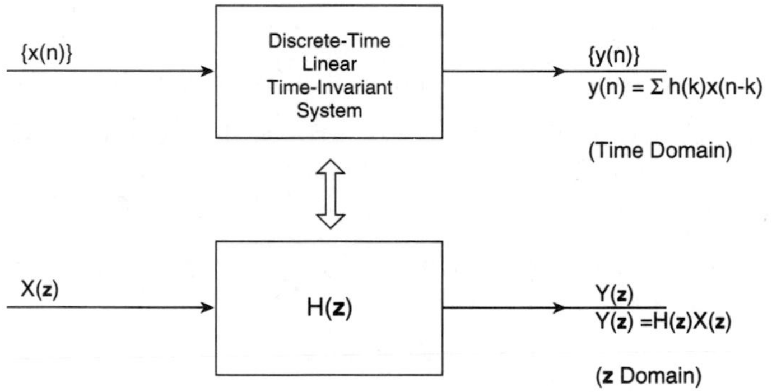

Fig. 3.14
Equivalence of time domain and Z-domain representations of an **LTI** system

The transform H(**z**) of the impulse response, {h(n)}, is called the ***transfer function*** of the system. In fact, there is an equivalence between the time domain and Z-domain representations, as depicted in Fig. 3.14. The transfer function can be established from the difference equation of the **LTI** system in the following way. If

$$y(n) = \sum_{k=1}^{D} b_k\, y(n-k) + \sum_{k=0}^{N} a_k\, x(n-k) \tag{3.5.24}$$

and we take the Z-transform of both sides, then

$$Y(\mathbf{z}) = \sum_{k=1}^{D} b_k \mathbf{z}^{-k} Y(\mathbf{z}) + \sum_{k=0}^{N} a_k \mathbf{z}^{-k} X(\mathbf{z}) \tag{3.5.25}$$

If $Y(\mathbf{z})$ is expressed as the product of $X(\mathbf{z})$ and the transfer function $H(\mathbf{z})$ as in Eq. (3.5.23), then it follows that

$$H(\mathbf{z}) = \frac{Y(\mathbf{z})}{X(\mathbf{z})} = \frac{\sum_{k=0}^{N} a_k \mathbf{z}^{-k}}{1 - \sum_{k=1}^{D} b_k \mathbf{z}^{-k}} \tag{3.5.26}$$

d) **Product of sequences.** Suppose a sequence $\{y(n)\}$ is created by taking the term-by-term product of two sequences $\{x(n)\}$ and $\{w(n)\}$. That is,

$$y(n) = x(n)\,w(n) \tag{3.5.27}$$

The Z-transform $Y(\mathbf{z})$ can be expressed in terms of $X(\mathbf{z})$ and $W(\mathbf{z})$ in the following manner. First we write the formal expression for $Y(\mathbf{z})$ as

$$Y(\mathbf{z}) = \sum_{n} x(n)\,w(n)\,\mathbf{z}^{-n} \tag{3.5.28}$$

and substitute the inversion formula for either $x(n)$ or $w(n)$, say $w(n)$, to get

$$\begin{aligned} Y(\mathbf{z}) &= \sum_{n} x(n)\mathbf{z}^{-n} \frac{1}{2\pi j}\oint_C \eta^{n} W(\eta)\frac{d\eta}{\eta} \\ &= \frac{1}{2\pi j}\oint_C W(\eta)\Big[\sum_{n} x(n)\Big(\frac{\mathbf{z}}{\eta}\Big)^{-n}\Big]\frac{d\eta}{\eta} \\ &= \frac{1}{2\pi j}\oint_C W(\eta)X(\mathbf{z}\eta^{-1})\frac{d\eta}{\eta} \end{aligned} \tag{3.5.29}$$

The last portion of Eq. (3.5.29) is the Z-domain equivalent of convolution and thus multiplication in the time domain is equivalent to convolution in the transform domain. A suitable choice of the contour C would be the unit circle since we would expect all transforms $X(\mathbf{z})$, $X(\mathbf{z}^{-1})$, $W(\mathbf{z})$, $W(\mathbf{z}^{-1})$ to be convergent on $|\mathbf{z}|=1$.

e) **Correlation of sequences.** The correlation of two sequences $\{x(n)\}$ and $\{w(n)\}$ is a third sequence $\{y(n)\}$ given by

$$y(m) = \sum_{n} x(n)\, w^*(n-m) \quad \text{(Correlation)}$$

$$y(m) = \sum_{n} x(n)\, x^*(n-m) \quad \text{(Autocorrelation)} \tag{3.5.30}$$

When the sequences are the same, we get the autocorrelation $\{R_{xx}(m)\}$. This can be manipulated as

$$Y(\mathbf{z}) = \sum_{m}\sum_{n} x(n)\, w^*(n-m)\, \mathbf{z}^{-m}$$

$$= \sum_{n} x(n)\, \mathbf{z}^{-n} \sum_{m} w^*(n-m)\, \mathbf{z}^{-m}\, \mathbf{z}^{+n} = X(\mathbf{z})\, W^*(\mathbf{z}^{-1}) \tag{3.5.31}$$

Consequently

$$S_{xx}(\mathbf{z}) = X(\mathbf{z})\, X^*(\mathbf{z}^{-1}) \tag{3.5.32}$$

The quantity $S_{xx}(\mathbf{z})$ is the ***squared-magnitude function*** of the sequence $\{x(n)\}$. In the case of real-valued sequences, the complex conjugation denoted by the * superscript can be dropped altogether.

3.5.4 Relating the Z-Transform and the Fourier Transform

The relationship between the Z-transform and the discrete-time Fourier transform can be obtained by inspection. By definition, if $\{x(n)\}$ is a discrete-time signal (sequence) then

$$X(\mathbf{z}) = \sum_{n} x(n)\, \mathbf{z}^{-n} \quad \text{(Z-transform)} \tag{3.5.33a}$$

$$X(e^{j\omega}) = \sum_{n} x(n)\, e^{-jn\omega} \quad (\text{DTFT}(\omega)) \tag{3.5.33b}$$

$$X(f) = \sum_{n} x(n)\, e^{-j2\pi n f T_s} \quad (\text{DTFT}(f)) \tag{3.5.33c}$$

Consequently, the **DTFT** can be obtained by setting $\mathbf{z} = e^{j\omega}$ if the **DTFT** is considered a function of the normalized radian frequency ω.

$$X(e^{j\omega}) = [X(\mathbf{z})]_{\mathbf{z}=e^{j\omega}} \quad \text{or} \quad X(f) = [X(\mathbf{z})]_{\mathbf{z}=e^{j2\pi f T_s}} \tag{3.5.34}$$

This provides the rationale for our notation of $X(e^{j\omega})$. If the sampling frequency is known, $f_s = 1/T_s$, then the **DTFT** expressed as a function of f can be obtained by setting $\mathbf{z} = \exp(j2\pi(f/f_s)) = \exp(j2\pi f T_s)$. $\mathbf{z} = e^{j\omega}$ is the parametric form for expressing values of $\mathbf{z}$ on the unit circle $|z|=1$. Thus in Fig. 3.15, which shows the complex Z-plane, we can identify frequency as an angle. The point $\mathbf{z} = 1$ corresponds to 0 frequency (DC). The point $\mathbf{z} = -1$ corresponds to a radian frequency of $+\pi$ or $-\pi$ depending on whether we approach $\mathbf{z} = -1$ from above or below. Thus values of $\mathbf{z}$ on the unit circle represent frequencies in the range $[-\pi, +\pi]$, which is exactly one period of the **DTFT** that we know is periodic with period 2π when considered a function of the normalized frequency ω.

The frequency response and transfer function are related by

$$H(e^{j\omega}) = [H(\mathbf{z})]_{\mathbf{z}=e^{j\omega}} \quad \text{or} \quad H(f) = [H(\mathbf{z})]_{\mathbf{z}=e^{j2\pi f T_s}} \tag{3.5.35}$$

which is what we would expect from Eq. (3.5.34).

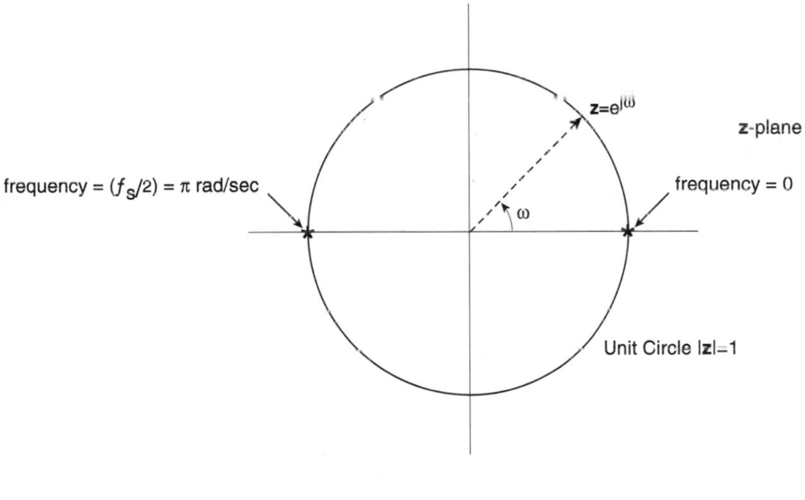

Fig. 3.15
The unit circle in the **z**-plane and its association with frequency.

The equivalence of **DTFT** and Z-transform based on Eq. (3.5.33) implicitly assumes that the unit circle is in the region of convergence. This is a requirement that will, in all cases of interest, be satisfied. A consequence of this equivalence on the unit circle allows us to draw a parallel with Eq. (3.5.29), which defined the notion of convolution in the Z-domain and the "normal" view of convolution in the frequency domain. Choosing the contour C as the unit circle we can substitute $\mathbf{z} = e^{j\omega}$ and $\eta = e^{j\zeta}$ and Eq. (3.5.29) reduces to

$$Y(f) = \left[Y(\mathbf{z})\right]_{\mathbf{z} = e^{j2\pi f T_s}} = \frac{1}{f_s}\int_0^{f_s} W(\zeta)X(f-\zeta)\,d\zeta \qquad (3.5.36a)$$

$$Y(e^{j\omega}) = \left[Y(\mathbf{z})\right]_{\mathbf{z} = e^{j\omega}} = \frac{1}{2\pi}\int_{-\pi}^{-\pi} Y(e^{j\zeta})X(e^{j(\omega-\zeta)})\,d\zeta \qquad (3.5.36b)$$

where the two cases correspond to the choice of frequency variable, one normalized and the other explicitly including the sampling frequency. Thus multiplication in the time domain is, as would be expected, equivalent to convolution in the frequency domain. The limits of integration accentuate the fact that sampled data systems exhibit a periodicity in the frequency domain.

The squared-magnitude function $S_{xx}(\mathbf{z})$ can be related to an equivalent function of (real) frequency by defining

$$S_{xx}(e^{j\omega}) = \left[S_{xx}(\mathbf{z})\right]_{\mathbf{z} = e^{j\omega}} \text{ or } S_{xx}(f) = \left[S_{xx}(\mathbf{z})\right]_{\mathbf{z} = e^{j2\pi f T_s}} \qquad (3.5.37)$$

Recognizing that

$$\left[X^*(\mathbf{z}^{-1})\right]_{\mathbf{z} = e^{j\omega}} = X^*(e^{j\omega}) \qquad (3.5.38)$$

we can combine the definition of Eq. (3.5.37) with Eq. (3.5.32) to get

$$S_{xx}(e^{j\omega}) = \left|X(e^{j\omega})\right|^2 \text{ or } S_{xx}(f) = \left|X(f)\right|^2 \qquad (3.5.39)$$

3.5.5 Parseval's Relation

Suppose $\{x(n)\}$ was a finite energy sequence. That is,

$$E_x = R_{xx}(0) = \sum_n |x(n)|^2 < \infty \qquad (3.5.40)$$

Expressing $R_{xx}(0)$ via the inverse Z-transform of $S_{xx}(\mathbf{z})$ we get

$$E_x = \sum_n |x(n)|^2 = \frac{1}{2\pi j}\oint_C X(\eta)X(\eta^{-1})\frac{d\eta}{\eta} \qquad (3.5.41)$$

where C is a closed contour lying in the common region of convergence of $X(\mathbf{z})$ and $X(1/\mathbf{z})$. If the unit circle is an appropriate contour, then by expressing $\mathbf{z}$ on C as $\mathbf{z}=e^{j\omega}$ (or $\mathbf{z}=j2\pi f T_s$) we get

$$E_x = \sum_{n} |x(n)|^2 = \frac{1}{2\pi}\int_{-\pi}^{+\pi} \left|X(e^{j\omega})\right|^2 d\omega \qquad (3.5.42a)$$

$$E_x = \sum_{n} |x(n)|^2 = \frac{1}{f_s}\int_{0}^{f_s} |X(f)|^2 df \qquad (3.5.42b)$$

Since X(f) is a periodic function with period f_s, and since we have established that {x(n)} are the coefficients of the Fourier series expansion of X(f), Eq. (3.5.42) is a restatement of Parseval's relation established in Chapter 3 (Section 3.4).

3.5.6 Stability of Linear Time-Invariant Systems

We have established that **LTI** systems can be characterized by the impulse response {h(n)} or, equivalently, by the transfer function H(**z**), which is the Z-transform of {h(n)}. Examination of H(**z**) can provide us with an indication of the stability of the **LTI** system.

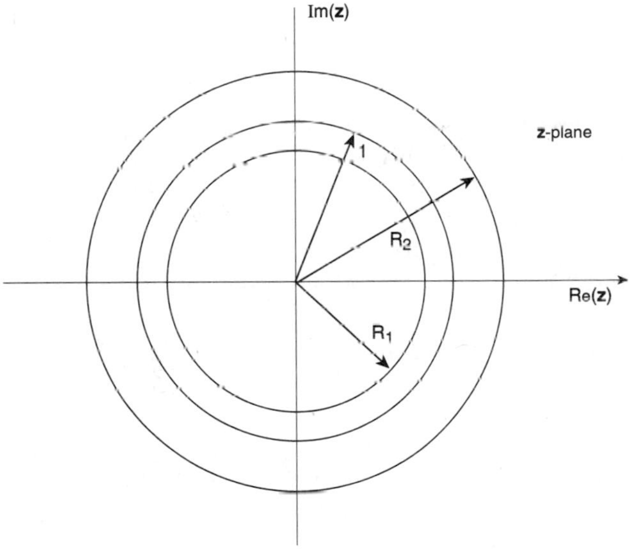

Fig. 3.16
Regions of convergence for H(**z**) and H(1/**z**)

Let {h(n)} be the impulse response of a causal system. Then H(**z**) will converge outside a circle of radius R_1 in the Z-plane that encompasses all the poles of H(**z**). Similarly H(1/**z**) will converge inside a circle of radius $R_2 = 1/R_1$. These are depicted in Fig. 3.16. The system will be stable if and only if

$$E_h = \sum_n |h(n)|^2 < \infty \qquad (3.5.43)$$

and the discussion on Parseval's relation would lead to the conclusion that the system is stable if and only if a closed contour (such as the unit circle) can be placed within the annulus. This is possible if and only if

$$\frac{1}{R_1} > R_1 \Rightarrow R_1 < 1 \qquad (3.5.44)$$

This leads to the following stability condition: A (causal) **LTI** system is stable if and only if all the poles of the transfer function H(**z**) lie entirely within the unit circle.

3.6 CERTAIN ASPECTS OF A/D AND D/A CONVERSION

In the prior sections we implicitly assumed that sample values, x(n), were representable exactly, with infinite precision. An obvious constraint of physically realizable systems is that sample values, that is, numbers, cannot be represented with infinite precision. Digital filters and systems will invariably be implemented on general purpose digital computers or application specific digital hardware where representation of numerical values will have an intrinsic constraint of finite wordlength. The fundamental distinction between discrete-time signal processing and digital signal processing is that of wordlength. The former assumes that signal values can take on a continuum of values, while the latter does not. However, we use the same acronym, **DSP**, in both cases. When the wordlength is large, for instance when the computations are done on a general purpose computer with multiple-precision floating point representation of numbers and arithmetic, the distinction between discrete-time and digital is not pronounced.

If the (finite) wordlength of the machine is B bits, then only $N = 2^B$ different numbers can be represented exactly. Let us denote this set of numbers that can be represented exactly by the set $\mathbf{Y} = \{ y_k ; k=0,1, \ldots , (N-1)\}$. Then in order to represent a value x, which is a real number (infinite precision), then the best we can do is to choose from the set **Y** that value y_k that is closest to x. This process is called ***quantization*** and the error $(x - y_k)$ in the representation is referred to as ***quantization error***. The notion of a ***quantizer***, from the viewpoint of digital signal processing is a device that converts a sequence {x(n)}, characterized by having great (infinite) precision in representing numerical values into a quantized sequence $\{x_Q(n)\}$ that is characterized by having a finite precision representation of numerical values. This is depicted in Fig. 3.17. The quantizer **Q** is considered as an entity that knows the N possible output numerical values, embodied by a set $\mathbf{Y} = \{y_k; k=0,1, \ldots , (N-1)\}$, and for a given input sample x(n) provides an output sample $x_Q(n)$ that takes on values from the set Y. The quantizing error can be viewed as a sequence {e(n)} defined by the difference

$$e(n) = x(n) - x_Q(n) \tag{3.6.1}$$

The sequence $\{x_Q(n)\}$ is considered a ***digital*** signal. The distinction between a discrete-time signal and a digital signal is that the former is a sequence of real numbers and thus a mathematical entity; the latter is such that each sample value can be represented by a finite number of bits and thus can be stored and processed by actual, physically realizable, digital hardware (such as a computer).

Sometimes we have a digital signal $\{w(n)\}$ that has been represented using B_1 bits of precision, but for the purposes of storage, or transmission, we wish to reduce this precision further, to B_2 bits. If B_1 is substantially larger than B_2 the quantizer concept depicted in Fig. 3.17 is still applicable. In general, quantization is the process whereby the precision of numerical representation is reduced; the error so introduced is, generically, quantization error.

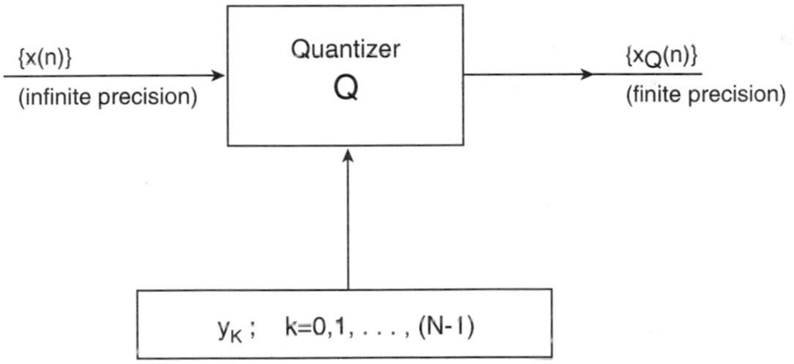

Fig. 3.17
Illustrating the principle of quantization

3.6.1 A/D Conversion.

In order to process an analog signal digitally, it is necessary to first convert it into a digital form. Conversion from analog to digital, called ***A/D conversion***, is performed by an ***A/D converter***. The process has two distinct aspects to it that are depicted in Fig. 3.18. First is the process of sampling, whereby the time scale is discretized; the second is quantization, whereby the amplitude scale is discretized. The reason for making this distinction is that each process introduces distortion—the distortion introduced by the sampling process is aliasing or folding; restriction to a finite set of values introduces quantization noise.

In Chapter 4 we cover in detail the analysis of quantizers from the viewpoint of signal-to-noise ratio. In this section we will address some of the other aspects of A/D conversion.

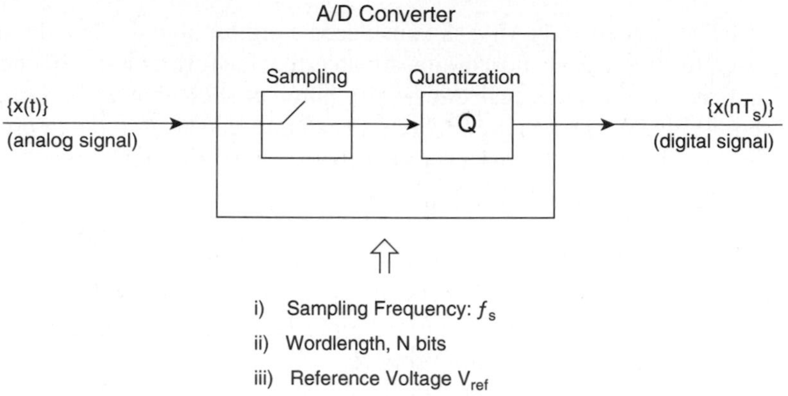

Fig. 3.18
A/D conversion as the combination of sampling and quantization

For the purposes of analyzing the impact of A/D conversion from a system perspective, the quantizer is modeled as an additive component as depicted in Fig. 3.19.

$$x'(n) = x(n) + e(n) \tag{3.6.2}$$

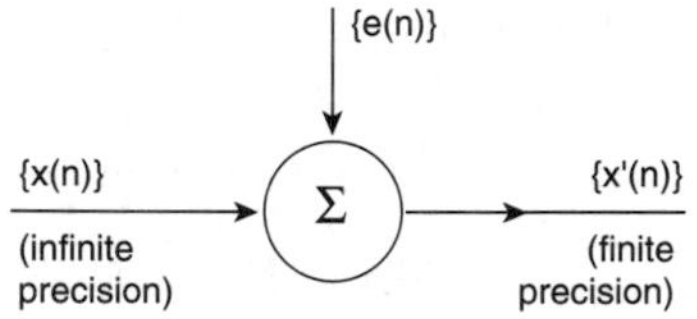

Fig. 3.19
Model of quantization as additive noise

In addition, we make the following assumptions:

i) The sequence {**e**(n)} is independent of (or at least uncorrelated with) the sequence {**x**(n)}. This quite a drastic assumption considering that the sequence {**e**(n)} can be computed exactly, given the sequence {**x**(n)}. The validity of this assumption is borne out by practice. System performance measures computed based on this assumption closely track experimental results. One instance where this assumption is not valid is when the signal {x(n)} is periodic. In such a case the quantization error will also be periodic and cannot be modeled as being uncorrelated with the input. For this reason when we generate sinusoidal signals for test purposes, the frequency of the sinusoid is chosen such that it is not an integer submultiple of the sampling rate. In telephony applications, for example, the sampling rate used to convert speech signals is 8 kHz; test signals are usually 1.01 kHz or 1.004 kHz, rather than exactly 1.0 kHz.

ii) The sequence {**e**(n)} is a white noise sequence. That is, the autocorrelation and squared-magnitude function of {**e**(n)} are of the form

$$R_{ee}(n) = \sigma_e^2 \delta(n) \text{ and } S_{ee}(z) = \sigma_e^2 \tag{3.6.3}$$

Again, the proof of the pudding is in the eating. If $\{x(n)\}$ is sufficiently *random*, that is, sample values from time epoch to time epoch change by an amount commensurate with the step size, then this assumption is valid. The assumption *breaks down* when the signal $\{x(n)\}$ is constant, or nearly so. The assumption is also not quite accurate if the signal $\{x(n)\}$ is periodic with a period of N samples where N is "small" (DC is a special case with N=1).

iii) The number representation implicitly assumes a normalization of the actual signal voltage. This normalization is via the reference voltage V_{ref}. An input voltage of V_{ref} will be converted to the largest number representable and constitutes the "saturation" or "clipping" level.

The output of the quantizer, being limited to a finite number of distinct values, can be encoded into a finite wordlength. We can still consider this as a "real" number and use the mathematics developed for discrete-time signal processing by treating the quantizer as an additive noise. This leads to the interpretation of A/D conversion, based on the model of Fig. 3.19, that the quantizer error is such that by adding it to the "real number" x(n) we get a "real number" x'(n) that just happens to be one of the finite set of values that can be encoded exactly.

As mentioned before, A/D conversion introduces two forms of impairment. One, because of the finite wordlength, is modeled as an additive quantization noise signal. The other impairment is related to the sampling process. If the analog signal being sampled has components outside $[-f_s/2, f_s/2]$ then these components will alias into the band of interest. To remove these out-of-band frequency components, an A/D converter is always preceded by an ***anti-aliasing*** filter as shown in Fig. 3.20.

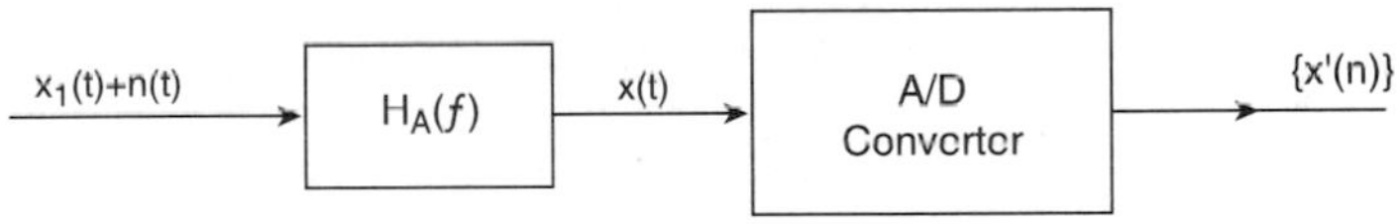

Fig. 3.20
Preceding an A/D converter by an anti-aliasing filter to remove out-of-band signals

The analog signal is assumed to be the sum of the desired signal, $x_1(t)$, that is suitably bandlimited, and an additive component $\mathbf{n}(t)$ that is not bandlimited. The requirement of the anti-aliasing filter is to attenuate (ideally annihilate) this noise component. The conversion is performed on the signal x(t) given by

$$x(t) = x_2(t) + \mathbf{s}(t) \tag{3.6.4}$$

where $\mathbf{s}(t)$ and $x_2(t)$ are the filtered versions of $\mathbf{n}(t)$ and $x_1(t)$, respectively. Since the sampling process is accompanied by aliasing, all the signal components fold into the band of interest and the sampled signal $\{x(n)\}$ can be written as

$$x(n) = x_2(n) + \mathbf{s}(n) \tag{3.6.5}$$

where $\{\mathbf{s}(n)\}$ is the aliased distortion component. The power of this component is equal to the power of $\mathbf{s}(t)$ prior to sampling and is given by

$$\sigma_s^2 = \int_{-\infty}^{+\infty} |H_A(f)|^2 S_{nn}(f)\,df \tag{3.6.6}$$

where $S_{nn}(f)$ is the power spectral density of $\mathbf{n}(t)$.

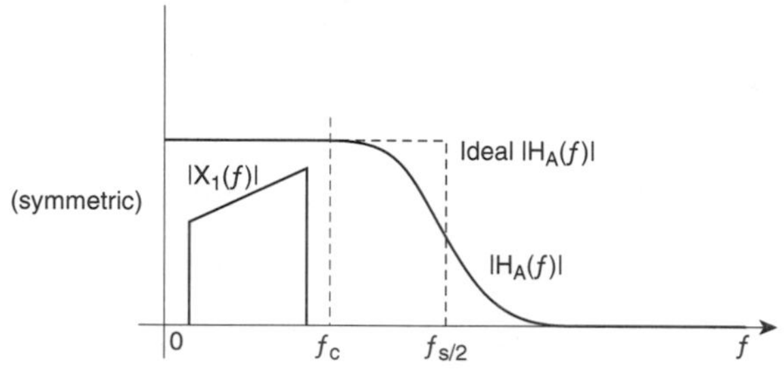

Fig. 3.21
Typical anti-aliasing filter frequency response

If we assume that the bandwidth of $x_1(t)$ is $f_c < f_s/2$, the frequency response of the anti-aliasing filter should be of the form shown in Fig. 3.21. Ideally it would have the brickwall shape of an ideal filter. In practice we try to implement a response that is characterized by

$$\begin{aligned} |H_A(f)| &\approx 1 \text{ for } |f| \le f_c \\ E_h &= \int_{-\infty}^{+\infty} |H_A(f)|^2 df \text{ (small)} \\ |H_A(f)| &\approx 0 \text{ for } |f| > f_c \end{aligned} \tag{3.6.7}$$

indicating that the filter does not affect $x_1(t)$ but reduces the impact of aliasing. Further, if possible, the phase response of $H_A(f)$ within the band $|f| < f_c$ should be as "linear" as possible. Optimizing the response, $H_A(f)$, requires some knowledge of the interfering signal $\mathbf{n}(t)$. If $\mathbf{n}(t)$ consisted of a few sinusoids of known frequency then it is advantageous for the filter to have zeros at these frequencies. In the absence of any knowledge of the characteristics of the interference, we usually assume that $\mathbf{n}(t)$ is white noise. If $S_{nn}(f) = N_0/2$, then using Eq. (3.6.6) the **SNR** (in dB) at the input of the A/D converter is

$$\mathbf{SNR}_{in} = 10 \log_{10}\left(\frac{2\sigma_x^2}{E_h N_0}\right) \tag{3.6.8}$$

The process of sampling, when implemented by realizable electronic components, is not quite as simple as assumed. Practical sampling or ***sample-and-hold*** circuits involve a filtering described in terms of an aperture or window. This can be modeled as in Fig. 3.22. The analog signal is sampled by charging a capacitor by opening a gate for a short duration called the aperture. Assuming the capacitor is discharged between samples, the actual sample value, $x(nT_s)$, will be given by

$$x(nT_s) \approx \int_{nT_s}^{nT_s+T_\alpha} \frac{1}{RC} x(t)\, dt \qquad (3.6.9)$$

which is a filter with impulse response $h_s(t)$ where

$$h_s(t) = \frac{T_\alpha}{RC}\left[u(t) - u(t-T_\alpha) \right] \qquad (3.6.10)$$

and corresponds to a spectral shaping given by

$$\left| H_s(f) \right| = \text{constant} \left| \text{sinc}(f T_\alpha) \right| \qquad (3.6.11)$$

In practice T_α is kept as small as is practical and thus should not impact the (lowpass) signal $x_1(t)$ substantially.

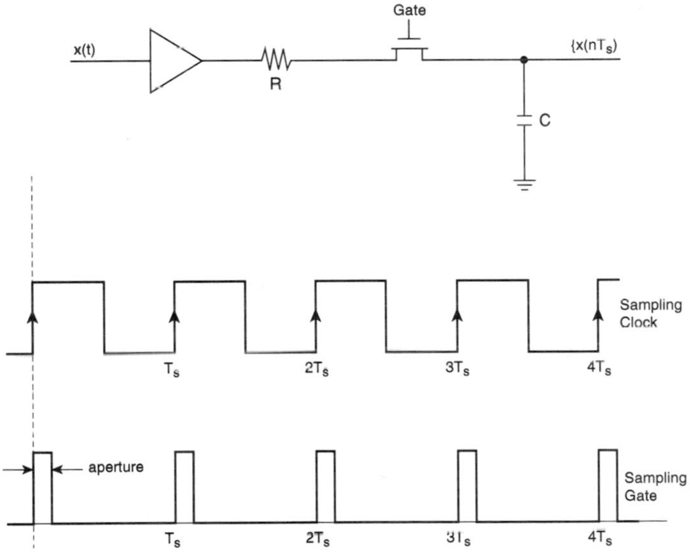

Fig. 3.22
Model for analyzing the spectral shaping introduced by the sampling process

The overall model of the A/D conversion process can thus be diagrammed as in Fig. 3.23, which shows the ideal signal $\{x_1(n)\}$ being corrupted by two impairments, $\{\mathbf{s}(n)\}$ and $\{\mathbf{e}(n)\}$, comprising the aliasing and quantization components, respectively. In addition, the signal is modified by a discrete-time filter, $H_1(f)$, which is the effective combination of the anti-aliasing filter $H_A(f)$ and the effect of sampling, $H_s(f)$, over the frequency range of interest.

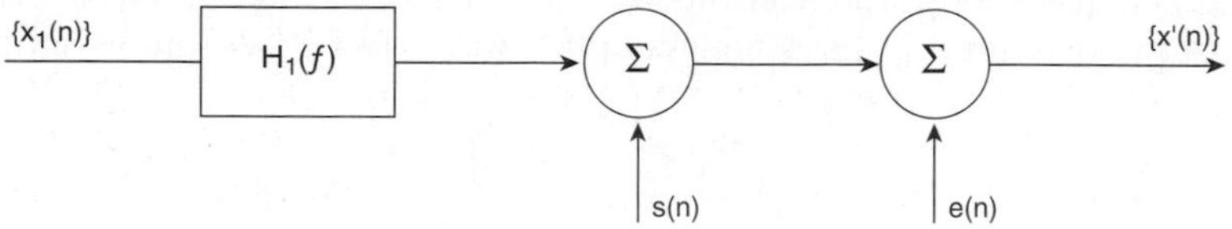

Fig. 3.23
Aliasing and quantization modeled as additive noise signals; anti-aliasing and sampling modeled as a frequency shaping

3.6.2 D/A Conversion

The conversion between a digital format and the corresponding analog signal is accomplished by a device called a ***Digital to Analog (D/A) converter***. The principal function of a D/A converter is to accept a digital signal, $\{x(nT_s)\}$, as input and provide an output analog signal, usually embodied as a train of pulses, with each pulse representative in amplitude to the value represented by the corresponding digital input sample. The D/A converter will be followed by an ***antireplication*** or ***replicate-rejection*** filter, which will choose the appropriate replicate in the frequency domain. This is depicted in Fig. 3.24.

The D/A converter is distinguished from the ideal discrete-time to analog converter defined in Section 3.2 by two fundamental characteristics:

i) A D/A converter can output pulses of amplitudes from a **discrete** set of values. This is the notion of *digital*, corresponding to finite wordlength. A B-bit D/A converter will have a repertoire of at most 2^B (usually $2^B - 1$) pulse heights. These pulse heights will correspond to values of the form $b_k V$, akin to the output values of a quantizer. If, for the most part, the differences $(b_k - b_{k-1})$ are a constant (step size) then the D/A converter is considered ***uniform***. For convenience we shall choose the range of values of b_k such that $|b_k| < 1$. Thus the output voltage of the D/A converter will be limited to V (volts) in magnitude. The voltage V, which is a constant of proportionality linking the representation of numbers and an absolute voltage (or current) is called the reference voltage of the D/A converter. The ideal converter on the other hand, can output pulses from an uncountably infinite set, representative of the real numbers.

ii) An ideal converter provides output pulses in the form of impulses or delta functions. A realizable D/A converter will have pulse shapes that are, usually, rectangular in shape. These pulses have a width less than the sampling interval. The subsequent replicate-rejection filter smooths the waveform to give it the continuity expected of an analog signal.

For purposes of analysis, it is useful to define an ideal D/A converter as one that provides output pulses that are delta functions. With this definition, realizable D/A converters can be modeled as a combination of an ideal converter followed by a filter whose impulse response defines the shape of the overall D/A pulse. This is depicted in Fig. 3.25. This notion is related to the representation of signals as the output of an **LTI** excited by a train of delta functions as discussed in Section 2.3.

Unlike its counterpart, the A/D converter, the D/A converter may not add any quantization noise to the signal. The noise it might add would be that associated with circuit imperfections, thermal noise, and so forth. If the wordlength used to represent the samples at the D/A input was B_1, then the D/A would technically need to generate only 2^{B_1} (usually $2^{B_1}-1$) output levels. If the number of permissible output levels of the D/A exceeded this number then the D/A would not actually add any quantizing noise. If the number of permissible output levels was less, implying that the wordlength of the D/A converter was less than B_1, then a wordlength reduction, or quantization, phase would be required, implying the addition of quantization noise.

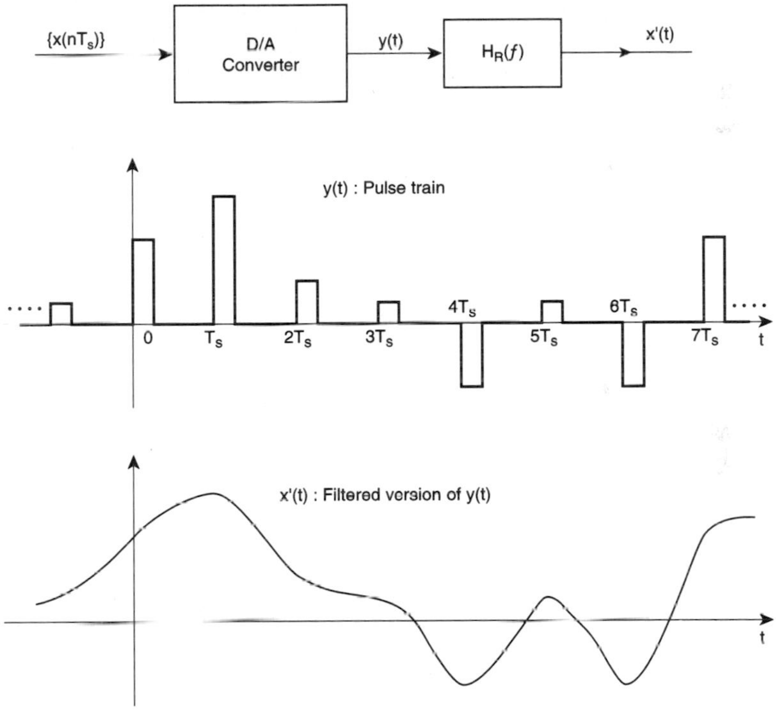

Fig. 3.24
Waveshapes associated with a D/A converter

The general model for analyzing the performance of a D/A converter is shown in Fig. 3.26. The quantizer is required only if the D/A wordlength B is less than B_1, the wordlength used to represent the input samples. From a system level viewpoint, the impairments introduced by the D/A converter are the quantization noise and the frequency shaping introduced by the pulse shaping. The output of the D/A, embodied in $y_1(t)$ or $y_2(t)$ will contain all the spectral replicates associated with the discrete-time signal $\{x(nT_s)\}$. The intent of the filter $H_R(f)$ is to attenuate these replicates.

The inability of the replicate reduction filter to delete the unwanted replicates can be viewed as an impairment with the following caveat—the additive distortion associated with these (attenuated) replicates is at frequencies outside the band of interest and can, at least in principle, be eliminated using an appropriate lowpass/bandpass filter. There are situations where the out-of-band signal, the replicates, could be harmful and in those situations the replicate-rejection filter is the only means of defense. A case where this may arise is if the output of the D/A is used as a control voltage for some device (e.g., in a phase locked loop the voltage could modify, or control, the frequency of a voltage-controlled oscillator).

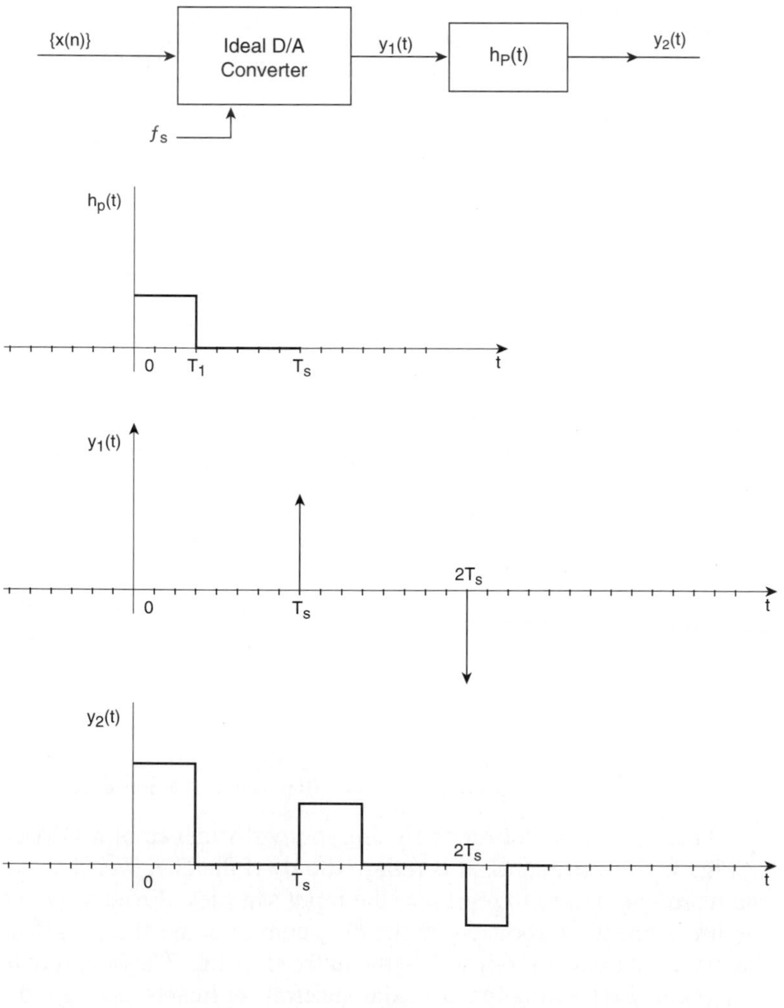

Fig. 3.25
Model of a D/A converter using a filter to mimic the pulse shape

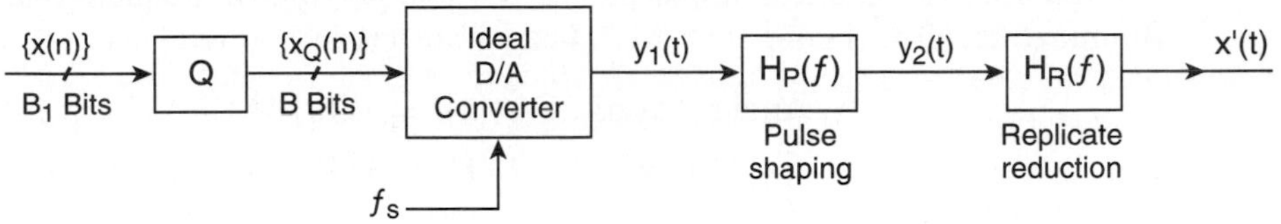

Fig. 3.26
Model for analyzing performance of a D/A converter

3.6.3 Combined Effects of A/D and D/A Conversion in Realizations of Spectral Shaping Filters

Assessing the deleterious impact of the impairments introduced by A/D and D/A conversions is application dependent. To this end consider the example depicted in Fig. 3.27 where it is desired to implement the digital equivalent of an analog filter. Thus the desired, or target, system is described by the transfer function $G(f)$, which may or may not be realizable using traditional analog filter synthesis techniques. The overall impairments associated with the digital implementation are discussed below. An impairment will be defined as any contribution that makes y'(t) different from y(t).

First it should be appreciated that the sampled-data nature of the digital implementation introduces a very fundamental constraint on the type of $G(f)$ that can be approximated. The nominal frequency range over which a digital signal processing system can assert control is limited to f_s (Hz). This frequency band may be "positioned" in frequency by appropriately choosing the passband of the anti-aliasing and replicate-removal filters. The (two-sided) frequency response $H(f)$ of the digital filter is necessarily periodic with period f_s and thus the implemented response must be a replicate of $H(f)$.

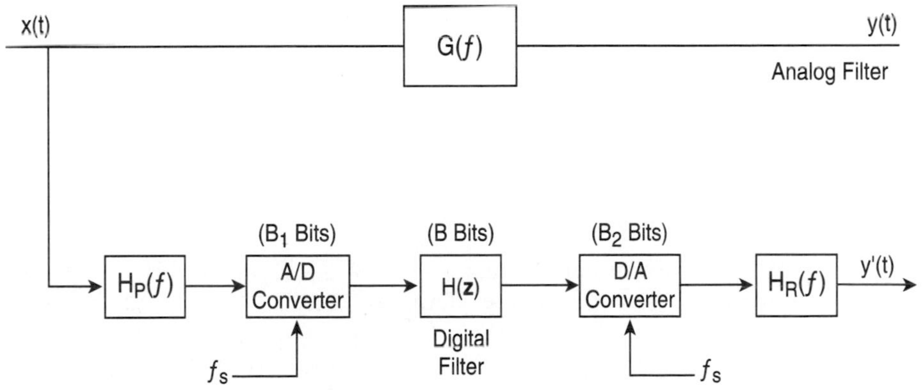

Fig. 3.27
Block diagram of model used to analyze the impairments introduced when an analog response is implemented using a digital filter

Ignoring the additive impairments, the comparison of frequency responses is illuminating. The analog and digital cases are given below.

$$\text{Analog case:}\quad G_A(f) = G(f)$$
$$\text{Digital case:}\quad G_D(f) = H_A(f)H_s(f)H(f)H_p(f)H_R(f) \tag{3.6.12}$$

The digital case is seemingly complicated. The overall response is affected by the anti-aliasing filter, the A/D aperture effect, the digital filter itself, $H(f)$, the effect of the D/A hold circuit, and the replicate-rejection filter. Of these the A/D aperture effect is usually the least problematic and will be neglected. That is, we assume that the aperture time is small enough for $H_s(f)$ 1.

The most common case of desired filter response is when $G(f)$ is substantially nonzero around DC (baseband) and can be assumed to be practically zero for $|f| > f_s/2$. That is,

$$|G(f)| \approx 0 \text{ for } |f| > \frac{f_s}{2} \tag{3.6.13}$$

Since $H(f)$, the digital filter frequency response, is periodic, the approximation to $G(f)$ for $|f| > f_s/2$ is the responsibility of the anti-aliasing (analog) filter $H_A(f)$ in combination with the (analog) $H_R(f)$ and, to a limited extent, $H_p(f)$ that describes the pulse shaping in the D/A converter. This is appropriate since, to avoid aliasing, $H_A(f)$ is lowpass, with

$$|H_A(f)| \approx 0 \text{ for } |f| > \frac{f_s}{2} \tag{3.6.14}$$

and, to reject the replicates other than at baseband, $H_R(f)$ is also lowpass with

$$|H_R(f)| \approx 0 \text{ for } |f| > \frac{f_s}{2} \tag{3.6.15}$$

Assuming that the D/A converter hold circuit corresponds to a pulse-width of T (sec) (the most common implementation, for simplicity and cost), $H_p(f)$ is of the form

$$|H_p(f)| \approx |\text{sinc}(fT)| \tag{3.6.16}$$

which is also lowpass with its first transmission zero at $f_0 = 1/T$ (Hz).

Within the band of interest, $[-f_s/2, f_s/2]$, $H_A(f)$ and $H_R(f)$ should nominally not introduce any frequency shaping, i.e.,

$$|f| \le \frac{f_s}{2} \Rightarrow |H_R(f)| \approx 1 \text{ and } |H_A(f)| \approx 1 \tag{3.6.17}$$

and the overall frequency response of the digital system is dominated by $H(f)$, the digital filter, and to some extent is affected by the pulse shape of the D/A converter.

For the digital system to be an approximation of the analog filter $G(f)$, the design criterion for the digital filter is thus

$$H(f)H_p(f) \cong G(f) \quad \text{for} \quad |f| < \frac{f_s}{2} \qquad (3.6.18)$$

The construction of $G_D(f)$ is depicted in Fig. 3.28 (only positive frequencies shown for convenience).

If $G(f)$ is a bandpass type of filter, located close to a frequency that is an integer multiple of the sampling frequency, then we use $H_A(f)$ and $H_R(f)$ to choose a particular replicate of $H(f)$. This is depicted in Fig. 3.29. The desired $G(f)$ is located around kf_s; this means that $H_A(f)$ and $H_R(f)$ must be bandpass, and must be of the appropriate bandwidth to pick off the appropriate replicate of $H(f)$. The effect of the D/A converter pulse shape can be quite detrimental in cases where the passband is at a frequency of the order of $f_0 = 1/T$; this passband approach is valid only if f_0 is greater than the passband center frequency.

To analyze the impact of noise, consider Fig. 3.30, which shows where noise injection points can be modeled. The noise components are:

a) the effect of aliasing, $\eta_1(n)$;

b) quantization in the A/D converter, $\eta_2(n)$;

c) additive noise in the digital filter implementation caused by the finite wordlength, called roundoff noise, $\eta_3(n)$;

d) quantization in the D/A converter, $\eta_4(n)$.

Denoting by $S_1(f)$ the power spectral density of $\{\eta_1(n)\}$, the noise power contribution due to $\{\eta_1(n)\}$ in the output is given by

$$\sigma_1^2 = \frac{1}{f_s}\int_0^{f_s} |H(f)|^2 S_1(f)\,df \qquad (3.6.19)$$

A similar expression can be derived for the contribution of the quantization in the A/D converter, σ_2^2. Denoting by σ_3^2 and σ_4^2 the powers of the roundoff noise and the quantization in the D/A converter, the total noise power contribution in the output signal $\{y(n)\}$ would be σ^2, where

$$\sigma^2 = \sigma_1^2 + \sigma_2^2 + \sigma_3^2 + \sigma_4^2 \qquad (3.6.20)$$

Thus the overall action of the digital system in Fig. 3.27 could be shown as in Fig. 3.31, where the "equivalence" is appropriate for the frequency range of the principal replicate, usually the frequency band $[-f_s/2, f_s/2]$.

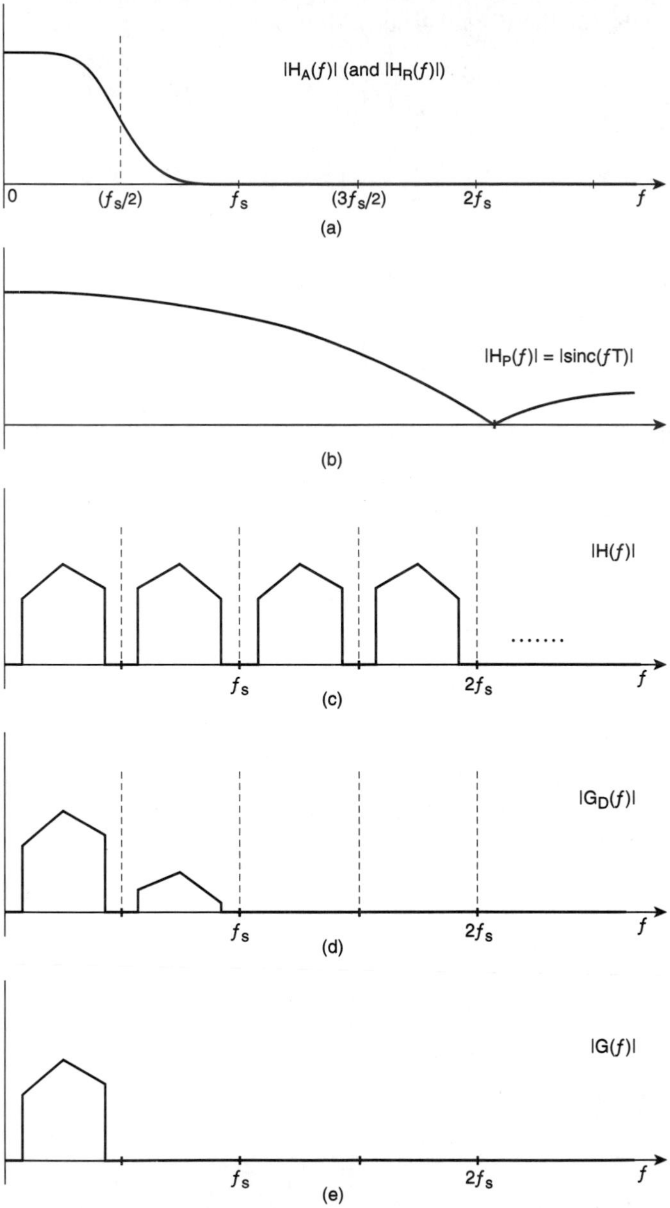

Fig. 3.28
Development of the combined frequency response of the digital filter together with anti-aliasing, replicate-rejection, and converter effects

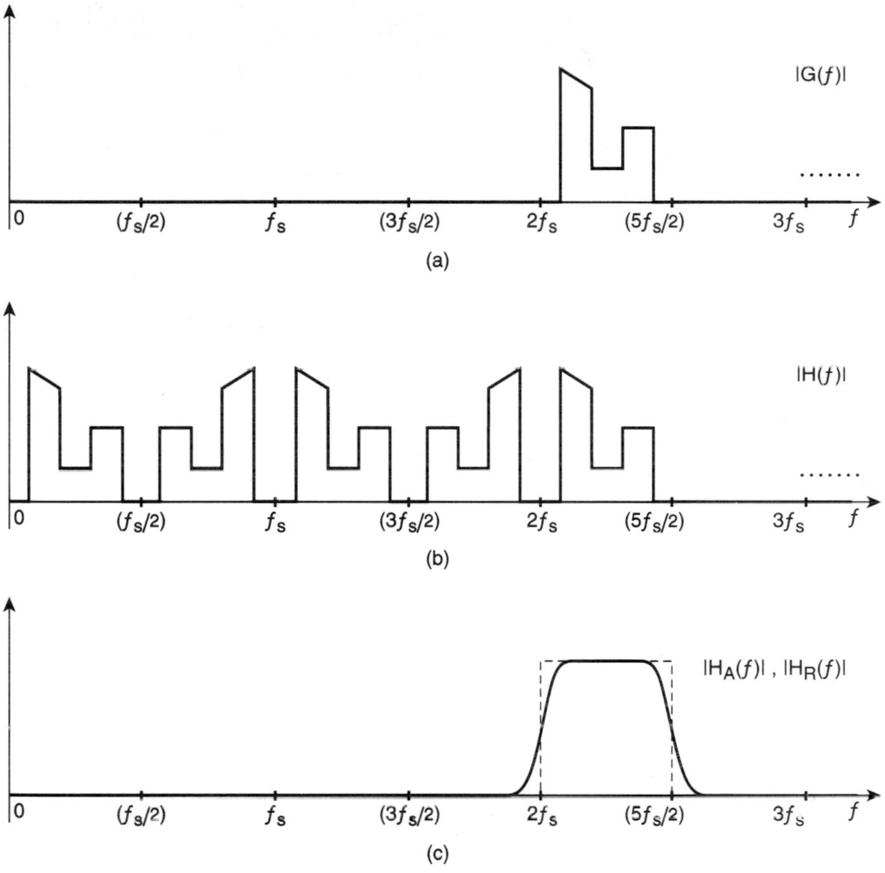

Fig. 3.29
Using passband anti-aliasing and replicate-rejection so that a digital filter can implement a passband response centered at a multiple of the sampling frequency

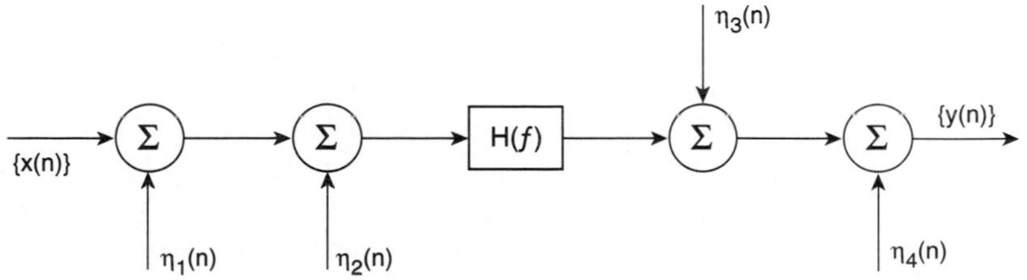

Fig. 3.30
Injection of noise arising from the conversion processes and roundoff noise in the implementation of the digital filter

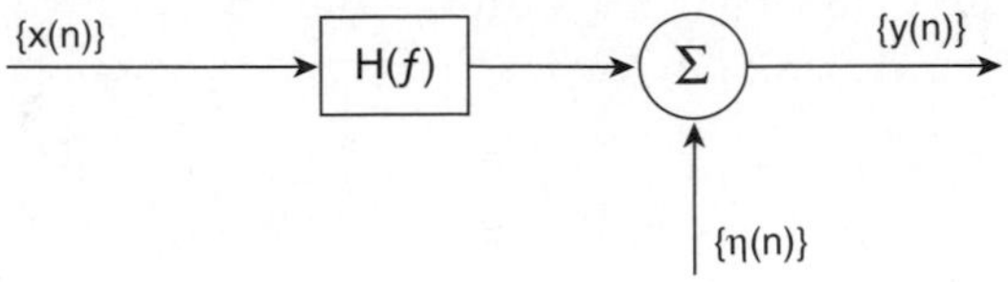

Fig. 3.31
Model applicable when a filter is implemented in a digital manner

3.7 DIGITAL FILTERS

The operation of a discrete-time filter is specified completely by the underlying difference equation. For example, the transfer function

$$H(\mathbf{z}) = \frac{\sum_{k=0}^{N} a_k \mathbf{z}^{-k}}{1 - \sum_{k=1}^{D} b_k \mathbf{z}^{-k}} \tag{3.7.1}$$

and the difference equation

$$y(n) = \sum_{k=1}^{D} b_k y(n-k) + \sum_{k=0}^{N} a_k x(n-k) \tag{3.7.2}$$

are equivalent. The difference equation provides us a procedure for determining the current output in terms of the present and past inputs as well as past outputs. A digital filter is nothing other than implementation of the difference equation either by using special-purpose arithmetic hardware or by firmware on a *Digital Signal Processor* (**DSP**), or software on a general-purpose computer. The distinction between a digital filter and a discrete-time filter is solely the impact of finite wordlength. Consequently we consider digital filters to be "equivalent" to discrete-time filters, the difference being modeled as "roundoff noise."

3.7.1 Flow Diagram of a Digital Filter

The difference equation can be represented by a ***flow*** diagram as in Fig. 3.32. The structure is referred to as a direct form implementation since it tries to mimic the difference equation. A distinguishing characteristic of direct form implementations is that the coefficients of the difference equation, $\{a_k;\ k= 0, 1, \ldots, N\}$ and $\{b_k;\ k=1, 2, \ldots, D\}$ appear directly as multipliers. When implemented in **DSP** chips or by digital hardware, the *adders* shown in Fig. 3.32 are implemented using an accumulator and the intermediate quantity, $\{w(n)\}$, may not be available unless the **DSP** code is specifically written to extract this quantity "part-way" through the process of implementing the (N+D+1) "multiply-accumulates."

The operation of the structure in Fig. 3.32 can be described by the following pseudo-code:

```
/* FOR EACH INPUT SAMPLE (DENOTED BY "X_IN") : */
/* BUMP THE "X" DELAY LINE : */
   FOR I=N,(N-1), ... ,1
       { X(I) = X(I-1); }
   X(0) = X_IN;
/* PERFORM THE MULTIPLY-ACCUMULATES: */
   TERM=0;
   FOR I=0,1, ... ,N
       { TERM = TERM + A(I)*X(I); }
   W = TERM;                  (OPTIONAL, IF W(N) IS REQUIRED)
   FOR I=1, ... ,D
       { TERM = TERM + B(I)*Y(I); }
   Y = TERM;                  ("Y" IS THE OUTPUT SAMPLE)
/* BUMP THE "Y" DELAY LINE : */
   FOR I=D,(D-1), ... ,2
       { Y(I) = Y(I-1); }
   Y(1) = Y;
```

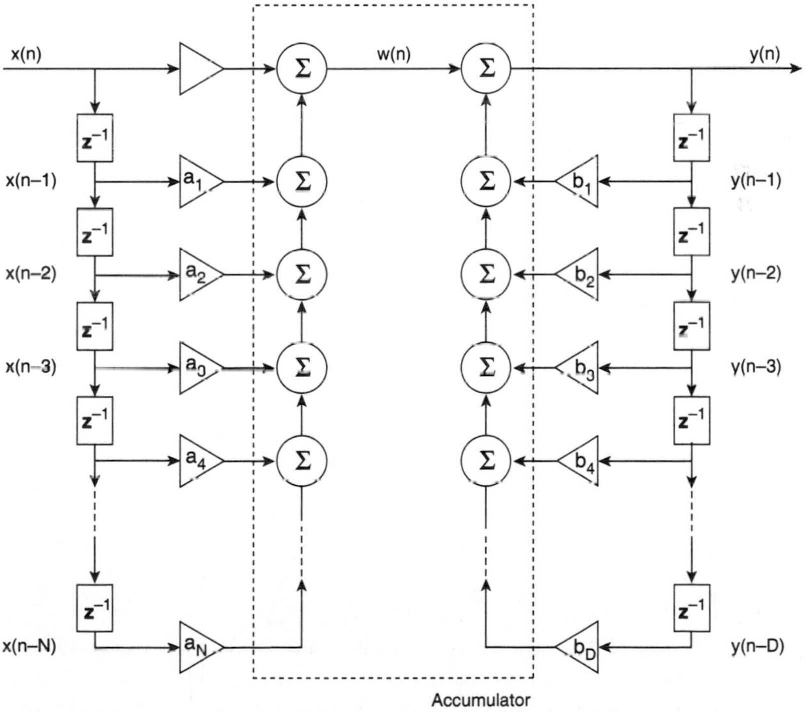

Fig. 3.32
Block diagram of a general difference equation

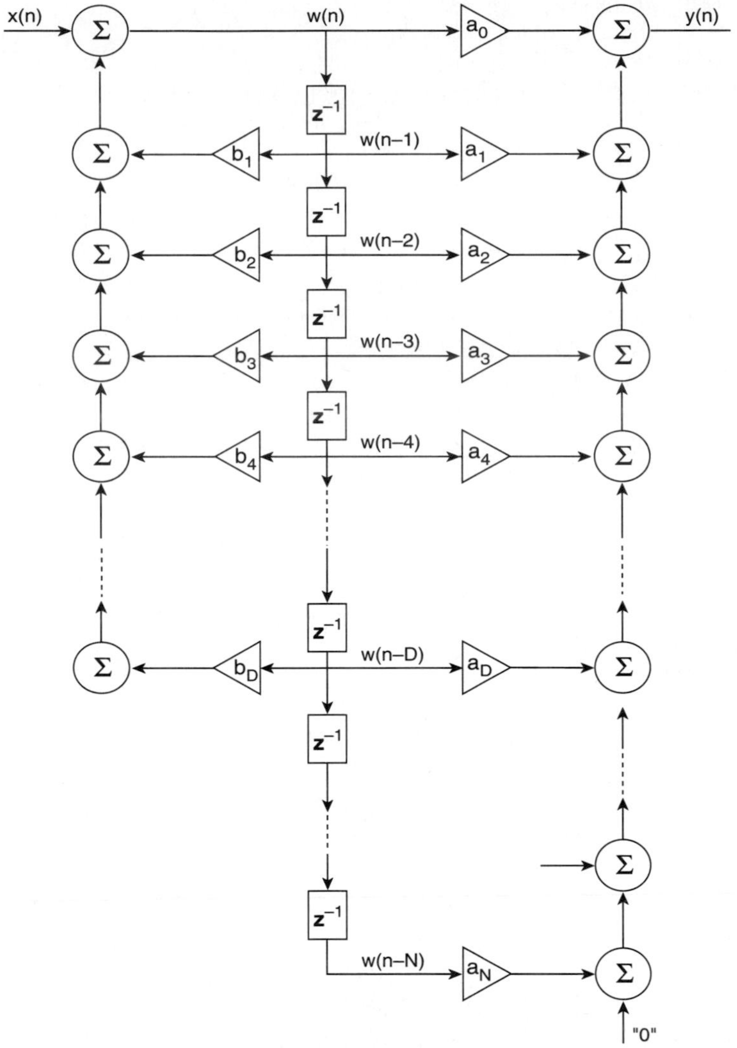

Fig. 3.33
Canonical representation of a difference equation (N>D). An alternate form of Fig. 3.32

Another way of representing the difference equation is depicted in Fig. 3.33. The difference between the two structures is the order in which the computations are performed. Fig. 3.32 depicts the nonrecursive portion of the difference equation being performed on the input, to yield an intermediate signal $\{w(n)\}$ followed by the recursive section that uses this intermediate signal as an input. In Fig. 3.33 the order is reversed with the recursive section being performed directly on the input followed by the nonrecursive section, which has the intermediate signal as input.

Are the two structures equivalent? That is, for the same input sequence $\{x(n)\}$ will the two structures provide identical output sequences? The answer is a qualified yes. Since realizable implementations imply a finite wordlength constraint, and the two structures have different noise behavior, the outputs would be different. A second reason relates to initial conditions. Assuming that the underlying difference equation, or transfer function, is **stable** and the input sequence is applied starting from time – , then the two outputs coincide, other than because of finite wordlength effects. If the inputs to the two are equal from time = 0 (some arbitrary starting point) but are different prior to that point, the outputs will be different, the difference related to the difference in "initial conditions" corresponding to the stored values at time = 0. If the filter is stable, these transient conditions "die out" and the two structures provide equivalent steady state outputs.

Examination of Fig. 3.32 indicates that (N+D) memory locations are required to store past history for the nonrecursive and recursive sections but that (assuming N > D) only N (the larger of N and D) are actually required as shown in what is known as a ***canonical*** structure depicted in Fig. 3.33 (assuming N > D). The notion of canonical is that the structure uses the minimal number of storage elements.

The structures can be viewed in terms of Z-transforms, i.e., transfer functions, by defining

$$H_N(\mathbf{z}) = \sum_{k=0}^{N} a_k \mathbf{z}^{-k} \tag{3.7.3}$$

as the transfer function associated with the numerator, or nonrecursive section, and

$$H_D(\mathbf{z}) = \frac{1}{1 - \sum_{k=1}^{D} b_k \mathbf{z}^{-k}} \tag{3.7.4}$$

as the transfer function associated with the denominator, or recursive section. $H_N(\mathbf{z})$ corresponds to the **FIR** part and $H_D(\mathbf{z})$ is responsible for the **IIR** part of the general transfer function $H(\mathbf{z})$. Clearly,

$$H(\mathbf{z}) = H_N(\mathbf{z})\, H_D(\mathbf{z}) = H_D(\mathbf{z})\, H_N(\mathbf{z}) \tag{3.7.5}$$

The order in which $H(\mathbf{z})$ is constructed from $H_N(\mathbf{z})$ and $H_D(\mathbf{z})$ corresponds to the structures in Fig. 3.32 and 3.33. That is,

$$W(\mathbf{z}) = H_N(\mathbf{z})X(\mathbf{z}) \quad \text{and} \quad Y(\mathbf{z}) = H_D(\mathbf{z})W(\mathbf{z}) \tag{3.7.6}$$

describes the structure in Fig. 3.32 and

$$W(\mathbf{z}) = H_D(\mathbf{z})X(\mathbf{z}) \text{ and } Y(\mathbf{z}) = H_N(\mathbf{z})W(\mathbf{z}) \tag{3.7.7}$$

describes the structure in Fig. 3.33.

When $D > 2$, that is, the order of the recursive section is greater than 2, the filter is rarely, if ever, implemented using the direct form structure shown in Figs. 3.32 or 3.33. The approach taken in such a case is to "divide and conquer." The transfer function $H(\mathbf{z})$, a ratio of polynomials, is split into smaller filter sections in the following way:

$$H(\mathbf{z}) = H_1(\mathbf{z})H_2(\mathbf{z})\ldots H_M(\mathbf{z}) \tag{3.7.8}$$

provides a ***cascade*** factorization while

$$H(\mathbf{z}) = G_1(\mathbf{z}) + G_2(\mathbf{z}) + \ldots + G_M(\mathbf{z}) \tag{3.7.9}$$

provides a ***parallel*** factorization. The particular terminology is natural when the two forms are shown in the block flow diagram of Fig. 3.34.

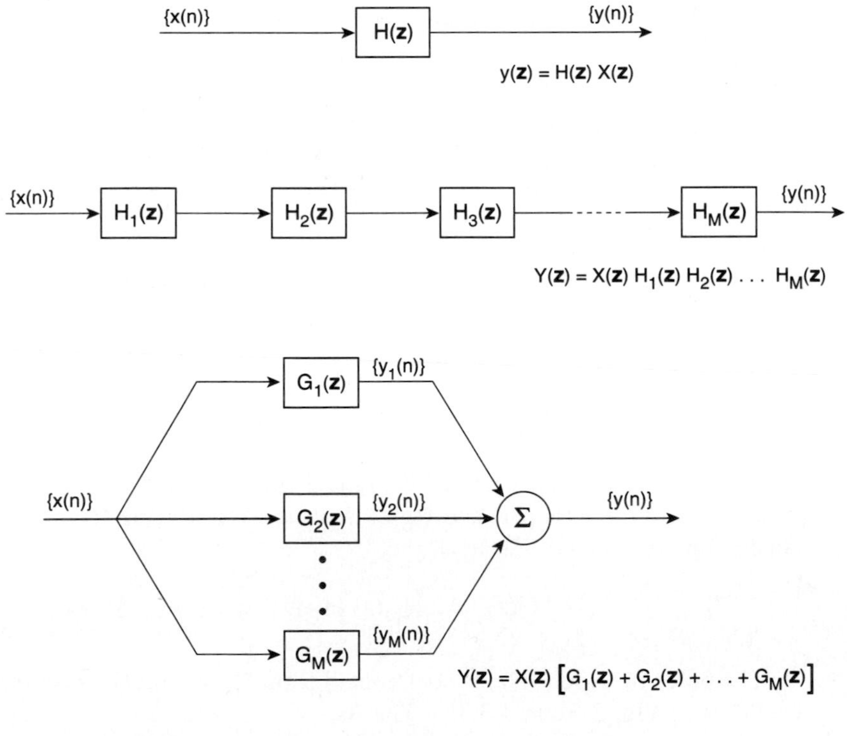

Fig. 3.34
Cascade and parallel factorizations for implementing a digital filter

The notion of designing a digital filter is the procedure or technique employed to obtain the coefficients $\{a_k, b_k\}$ of the difference equation Eq. (3.7.2) or the equivalent for each individual factor $H_i(\mathbf{z})$ or $G_i(\mathbf{z})$ according to whether the implementation will be of the cascade or parallel form, respectively. The notion of designing an implementation of a digital filter is to establish the structure of choice and determine the particular impact of the finite wordlength. The remainder of Section 3.7 considers the various structures employed, the equivalent frequency response, and the implication of finite wordlength.

Note: Unless otherwise specified, we shall assume that the coefficients of the filters we deal with are **real valued**. The complex filters (complex coefficients) that we come across are derived from, or associated with, filters with real coefficients.

3.7.2 FIR Filters

FIR filters are characterized by transfer functions that are polynomials, such as

$$H(\mathbf{z}) = \sum_{k=0}^{N-1} h(k)\mathbf{z}^{-k} \tag{3.7.10}$$

where the coefficients are directly the impulse response of the filter, $\{h(n)\}$. The most common form of implementing an **FIR** filter is the direct form of Fig. 3.32 with the recursive section removed. This is diagramed in Fig. 3.35. The form of the block diagram gives rise to the terminology of ***tapped delay line*** and the coefficients as ***tap weights***. The length of an **FIR** filter is the number of taps, N (and thus our convention of using indices from 0 through (N–1) for the coefficients).

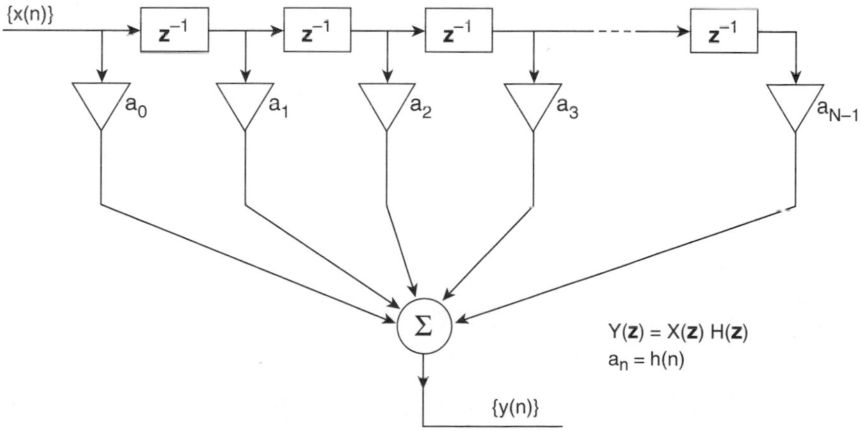

Fig. 3.35
Direct form N-point **FIR** filter

FIR filters have two principal advantages. First is that **FIR** filters are inherently stable when implemented in a manner such as shown in Fig. 3.35. The finite length of the impulse response guarantees that the output will go to zero within N samples from the epoch the input goes to zero. Second is that filters with a precisely linear-phase characteristic can be designed and implemented. The disadvantage of **FIR** filters is the computational complexity; for each output sample, N multiply-accumulate ("MAC") operations need to be performed. For the most part a target (magnitude) frequency response can be approximated using **IIR** filters of less computational complexity, expressed in MACs per input sample, than **FIR** filters. If, however, there is a sampling rate change then **FIR** filters can be constructed to compute only those output samples that are really necessary, thus reducing the computational burden. If we allow the **FIR** filter to operate on data in blocks, rather than sequentially on a sample-by-sample basis, there are other techniques to reduce the number of MACs per output sample. Some of these are covered in Chapter 7. Since **FIR** filters operate only on the input samples, they are a logical candidate for adaptive filters. The notions of adaptive operation and sampling-rate changes will be discussed in Chapter 6 and Chapter 7, respectively.

3.7.2.1 Frequency Response of FIR Filters

The frequency response of a causal **FIR** filter can be obtained from the impulse response, by substituting $\mathbf{z} = \exp(j\omega)$:

$$H(e^{j\omega}) = \sum_{k=0}^{N} h(k)e^{-jk\omega}$$

$$= \sum_{k=0}^{N} h(k)\cos(k\omega) - j\sum_{k=0}^{N} h(k)\sin(k\omega) \qquad (3.7.11)$$

The notion of causality as applied to **FIR** filters can be relaxed somewhat. Consider the noncausal **FIR** filter given by $\{g(n); n=-M, -(M-1), \ldots, -1, 0, 1, \ldots, (M-1), M\}$. Thus **G** is a noncausal **FIR** filter of length 2M+1, with transfer function $G(\mathbf{z})$ given by

$$G(\mathbf{z}) = \sum_{k=-M}^{k=+M} g(k)\mathbf{z}^{-k} \qquad (3.7.12)$$

The corresponding causal **FIR** filter is constructed from **G** by taking

$$H(\mathbf{z}) = \mathbf{z}^{-M} G(\mathbf{z}) \qquad (3.7.13)$$

from which we see that $\{h(n)\}$ is the impulse response of a causal filter and the coefficients of **G** and **H** are related by

$$h(k) = g(k-M) \text{ for } k=0, 1, \ldots, (2M+1) \qquad (3.7.14)$$

The frequency responses $G(e^{j\omega})$ and $H(e^{j\omega})$ are related by

$$H(e^{j\omega}) = e^{-jM\omega} G(e^{j\omega}) \qquad (3.7.15)$$

and the distinction is just a linear-phase term $e^{jM\omega}$, which is equivalent to a fixed delay of M samples. Except in some special cases that are inherently delay sensitive (such as in a feedback loop), a fixed delay is normally considered inconsequential. Thus **G** and **H** are considered the "same." Since noncausal **FIR** filters can be made causal by the introduction of an appropriate delay, we usually do not insist on design procedures providing causal designs.

A principal advantage of **FIR** filters is the ability to design filters that exhibit a linear-phase (fixed delay) characteristic. The linear-phase characteristic is possible because we can force the impulse response coefficients to exhibit symmetry. Four cases are considered corresponding to whether the symmetry is odd or even and to whether the filter length is even or odd.

i) Case A. Odd length and even symmetry. This situation is shown in Fig. 3.36. Since N is odd, (N–1)/2 is an integer. Shifting the response $\{h(n)\}$ by (N–1)/2 samples creates a (noncausal) impulse response $\{g(n)\}$ that is an even sequence. That is,

$$g(n) = g(-n) \qquad (3.7.16)$$

and thus the transfer function $G(\mathbf{z})$ is given by

$$G(\mathbf{z}) = g(0) + \sum_{k=1}^{\frac{(N-1)}{2}} g(k)[\mathbf{z}^{+k} + \mathbf{z}^{-k}] \qquad (3.7.17)$$

Substituting $\mathbf{z}=e^{j\omega}$ to get the frequency response in terms of the normalized radian frequency ω, we obtain

$$G(e^{j\omega}) = g(0) + 2\sum_{k=1}^{\frac{(N-1)}{2}} g(k)\cos(k\omega) \qquad (3.7.18)$$

which is a purely real function of frequency. That is, the frequency response has no imaginary part. The phase angle is either zero or π. We shall call such filters ***zero-phase***. The phase angle of π is obtained when the response is negative, a condition that we have only when the desired response at and around that particular frequency is null and in which case the phase response is moot. Since H(**z**) is a delayed version of G(**z**), it follows that $H(e^{j\omega})$ will correspond to a ***linear-phase*** filter.

$$H(e^{j\omega}) = e^{-j\frac{(N-1)}{2}\omega}\, G(e^{j\omega}) \tag{3.7.19}$$

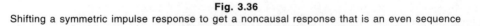

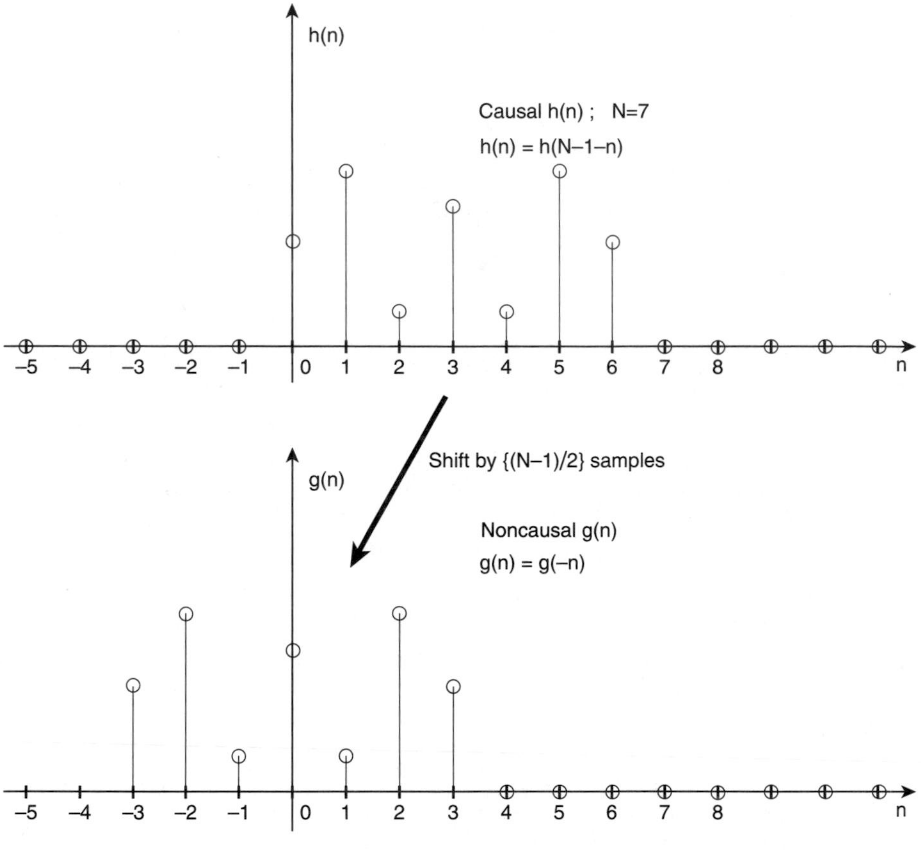

Fig. 3.36
Shifting a symmetric impulse response to get a noncausal response that is an even sequence

A typical lowpass frequency response $G(e^{j\omega})$ is shown in Fig. 3.37. Since $G(e^{j\omega})$ is real, it is shown directly, rather than $|G(e^{j\omega})|$. Also, since a phase angle of π (radians) is indistinguishable from a phase angle of $-\pi$ (radians), we have chosen the sign to provide a phase response that is an odd function of frequency. Note that when the phase is π ($-\pi$) the frequency is in the stopband of the filter and the phase is not that meaningful. The symmetry of $G(e^{j\omega})$ about $\omega = 0$ is a consequence of the filter coefficients being real (the real part of the **DTFT** of a real-valued sequence is an even function of frequency). Since discrete-time frequency responses are periodic, only one period (of 2π in normalized radian frequency) is shown.

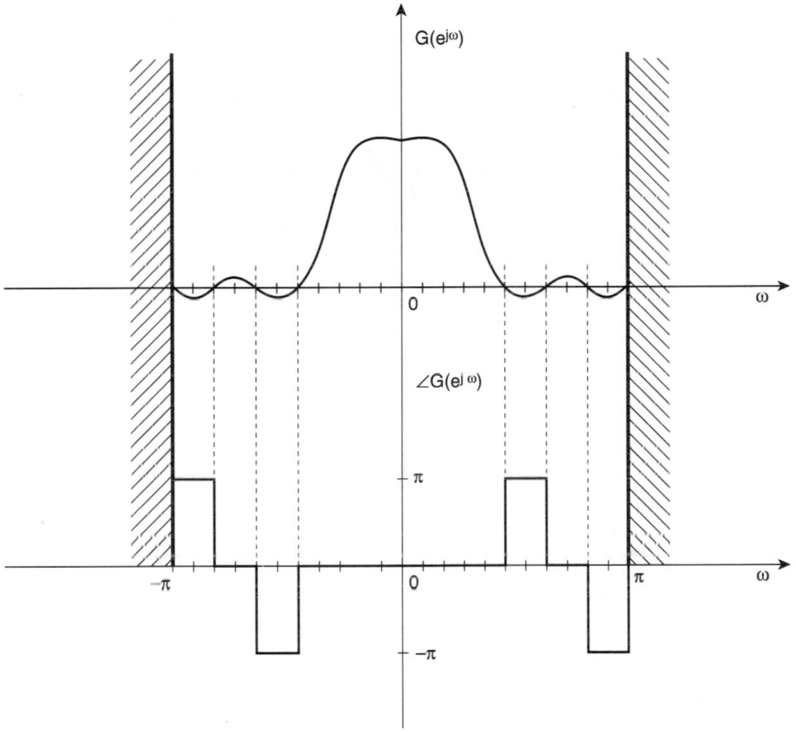

Fig. 3.37
Typical frequency response of an **FIR** filter of odd length with even symmetry

Thus for real-coefficient **FIR** filters of odd length and even symmetry, the frequency response is even-symmetric in the frequency domain, is described completely by [(N–1)/2 + 1] coefficients {g(k); k=0, 1, ... , (N–1)/2}, and is comprised of the product of a linear-phase term and a zero-phase response .

ii) Case B. Odd length and odd symmetry. An odd-length, odd-symmetry causal **FIR** filter is depicted in Fig. 3.38. A shift of (N–1)/2 samples creates a noncausal response {g(n)} that is an odd sequence.

Therefore, since

$$g(n) = -g(-n) \tag{3.7.20}$$

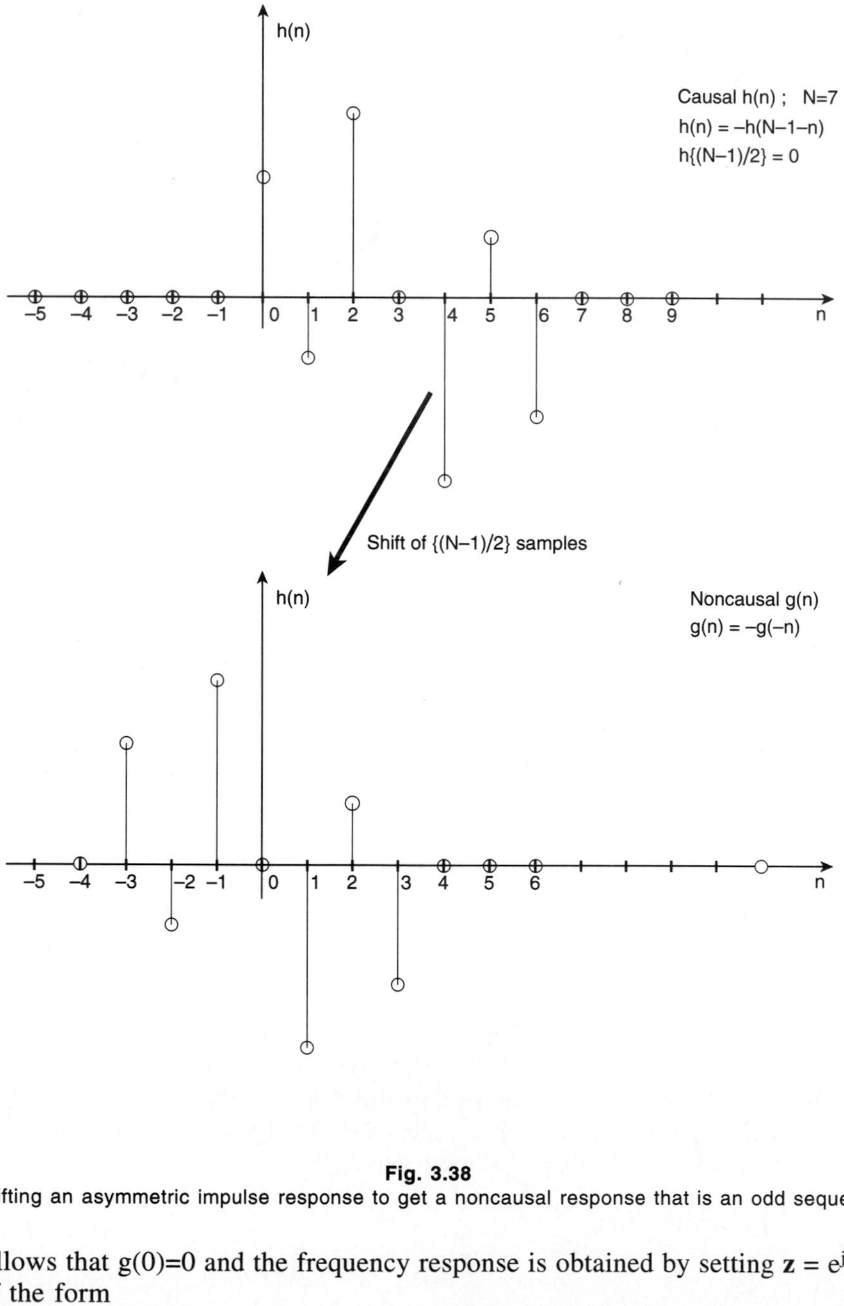

Fig. 3.38
Shifting an asymmetric impulse response to get a noncausal response that is an odd sequence

it follows that g(0)=0 and the frequency response is obtained by setting $\mathbf{z} = e^{j\omega}$ and is of the form

$$G(e^{j\omega}) = 2j \sum_{k=1}^{\frac{(N-1)}{2}} g(k)\sin(k\omega) \qquad (3.7.21)$$

which is a purely imaginary function of frequency—the real part is zero. This is equivalent to a phase shift of $(\pi/2)$ radians across the entire range of positive frequencies $0 < \omega < \pi$ {and $-(\pi/2)$ radians for negative frequencies}. We refer to such filters as ***Hilbert Transformers*** (see the discussion in Chapter 2). If we ignore the regions of frequency where the filter introduces considerable attenuation, we can

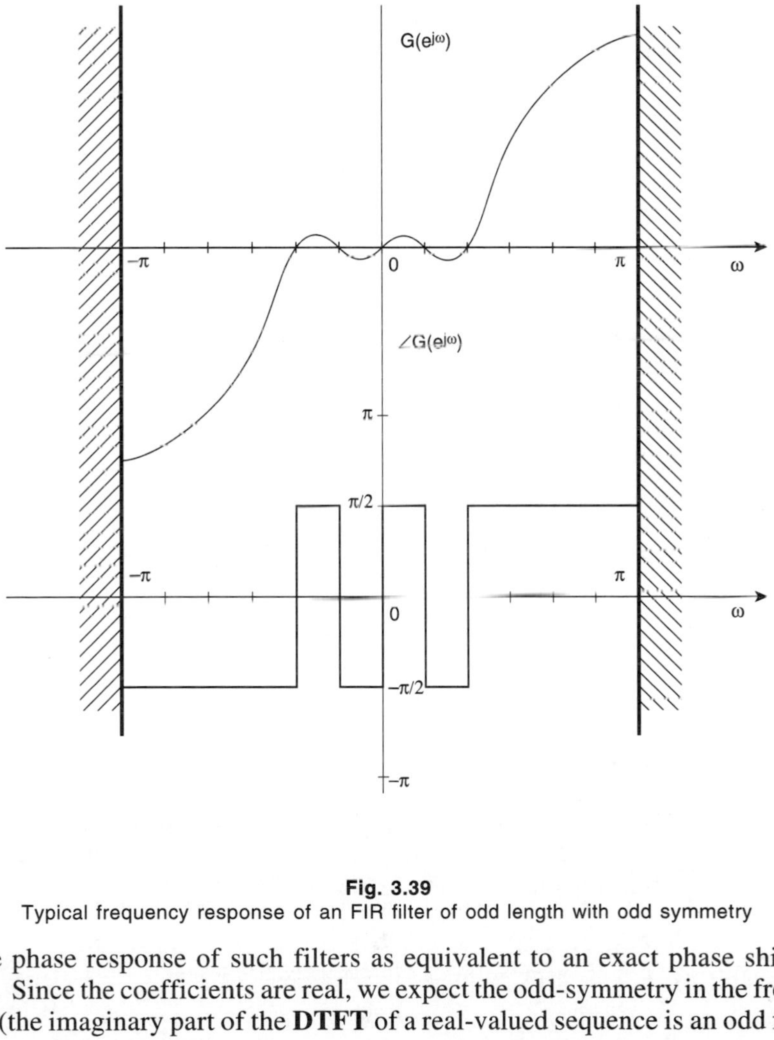

Fig. 3.39
Typical frequency response of an FIR filter of odd length with odd symmetry

treat the phase response of such filters as equivalent to an exact phase shift of 90 degrees. Since the coefficients are real, we expect the odd-symmetry in the frequency domain (the imaginary part of the **DTFT** of a real-valued sequence is an odd function of frequency). A sample response is graphed in Fig. 3.39. For $\omega = 0$ the response is zero, regardless of the coefficients. Such symmetry is obviously not appropriate for lowpass filters.

Thus for real-coefficient **FIR** filters of odd length and odd symmetry, the response is odd-symmetric in the frequency domain, is described completely by $[(N-1)/2 + 1]$ coefficients $\{g(k); k=0, 1, \ldots, (N-1)/2\}$, and is comprised of the product of a linear-phase term and a $(\pi/2)$-phase response.

iii) Case C. Even length and even symmetry. A causal **FIR** response of even length and even symmetry is depicted in Fig. 3.40. The primary distinction between the even-length and odd-length cases is that in the latter case there is the notion of a mid-point or center sample. In the even-length case the midpoint lies between two samples. The response cannot be shifted by an integer number of samples to obtain an even sequence.

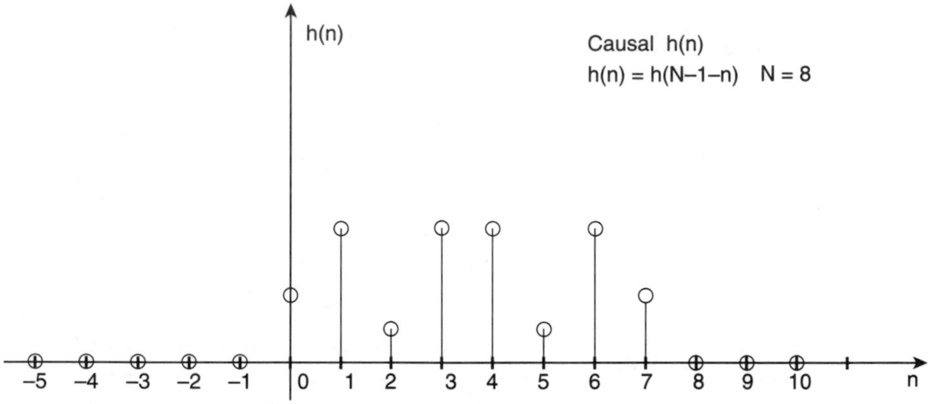

Fig. 3.40
Even-length, even-symmetry finite impulse response

The frequency response of H(**z**) is obtained in the following manner. Since the filter does have symmetry,

$$h(n) = h(N-1-n) \text{ for } n = 0, 1, 2, \ldots, (N-1) \quad (3.7.22)$$

from which we define $\{g(k); k=0, 1, \ldots, (N/2-1)\}$ via

$$g(k) = h(\tfrac{N}{2} - 1 - k) = h(\tfrac{N}{2} + k) \text{ for } k = 0, 1, \ldots, (\tfrac{N}{2} - 1) \quad (3.7.23)$$

The filter transfer function H(**z**) is written as

$$H(\mathbf{z}) = \sum_{k=0}^{N-1} h(k)\mathbf{z}^{-k} = \sum_{k=0}^{\frac{N}{2}-1} h(k)\mathbf{z}^{-k} + \mathbf{z}^{-\frac{N}{2}} \sum_{k=0}^{\frac{N}{2}-1} h(k+\tfrac{N}{2})\mathbf{z}^{-k} \quad (3.7.24)$$

and by a simple manipulation of indices we get

$$H(\mathbf{z}) = \mathbf{z}^{-\frac{N}{2}} \sum_{k=0}^{\frac{N}{2}-1} \{h(\frac{N}{2}-1-k)\mathbf{z}^{k+1} + h(\frac{N}{2}+k)\mathbf{z}^{-k}\} \qquad (3.7.25)$$

We have associated $\mathbf{z}^{-n}$ with a delay of n samples, where n is an integer, or an advance if n is negative. If n is not an integer then we still treat the term as a delay, albeit of a noninteger number of samples. This is valid in a mathematical framework even though it is not "physical." With this in mind, we derive from Eq. (3.7.25) the following expression for $H(\mathbf{z})$:

$$H(\mathbf{z}) = \mathbf{z}^{-\frac{(N-1)}{2}} \sum_{k=0}^{\frac{N}{2}-1} g(k)\{ \mathbf{z}^{(k+\frac{1}{2})} + \mathbf{z}^{-(k+\frac{1}{2})}\} \qquad (3.7.26)$$

where, since (N–1) is odd, (N–1)/2 corresponds to the linear-phase term of a noninteger delay. Setting $\mathbf{z} = e^{j\omega}$ provides the frequency response as

$$H(e^{j\omega}) = 2e^{-j\frac{(N-1)}{2}\omega} \sum_{k=0}^{\frac{N}{2}-1} g(k)\cos\{(k+\frac{1}{2})\omega\} = e^{-j\frac{(N-1)}{2}\omega} A(\omega) \qquad (3.7.27)$$

where $A(\omega)$ is a real-valued even function of ω. Thus the filter can be viewed as a linear-phase filter where the equivalent delay is (N–1)/2.

Thus for real-coefficient **FIR** filters of even length and even symmetry, the frequency response is even-symmetric in the frequency domain, is comprised of the product of a linear-phase term and a zero-phase response, and is described completely by N/2 coefficients $\{g(k);\ k=0, 1, \ldots, (N/2-1)\}$. Note that at $\omega = \pi$, the response is zero, regardless of the coefficients $\{g(k)\}$, making such symmetry inappropriate for highpass filters.

iv) **Case D. Even length and odd symmetry.** The derivation of frequency response in this case follows the previous case. First the odd symmetry means that

$$h(n) = -h(N-1-n) \quad \text{for} \quad n = 0, 1, \ldots, (N-1) \qquad (3.7.28)$$

from which we define $\{g(k)\}$ via

$$g(k) = h(\frac{N}{2}-1-k) = -h(\frac{N}{2}+k) \quad \text{for } k=0, 1, \ldots, (\frac{N}{2}-1) \qquad (3.7.29)$$

Following Case C, we get the following form for $H(\mathbf{z})$:

$$H(\mathbf{z}) = \mathbf{z}^{-\frac{(N-1)}{2}} \sum_{k=0}^{\frac{N}{2}-1} g(k)\{ \mathbf{z}^{(k+\frac{1}{2})} - \mathbf{z}^{-(k+\frac{1}{2})}\} \qquad (3.7.30)$$

from which the frequency response $H(e^{j\omega})$ is given by

$$H(e^{j\omega}) = 2je^{-j\frac{(N-1)}{2}\omega} \sum_{k=0}^{\frac{N}{2}-1} g(k) \sin\{(k+\tfrac{1}{2})\omega\} = je^{-j\frac{(N-1)}{2}\omega} B(\omega) \quad (3.7.31)$$

Thus for real-coefficient **FIR** filters of even length and odd symmetry, the frequency response is odd-symmetric in the frequency domain, is comprised of the product of a linear-phase term and a ($\pi/2$)-phase response, and is described completely by N/2 coefficients $\{g(k);\ k=0,\ 1,\ \ldots\ ,\ (N/2-1)\}$.

3.7.2.3 Impact of Finite Wordlength on the Frequency Response

The procedure of designing an **FIR** filter involves the determination of the N coefficients $\{h_I(n);\ n=0,\ 1,\ \ldots\ ,\ (N-1)\}$ such that the frequency response $H_I(e^{j\omega})$ approximates a given function of frequency, $G(\omega)$. That is,

$$H_I(e^{j\omega}) = \sum_{k=0}^{N-1} h_I(k) e^{-jk\omega} \cong G(\omega) \quad (3.7.32)$$

This procedure is usually implemented on a digital computer using very high precision arithmetic, usually multiple-precision floating point. Compared to the wordlengths used in an implementation of the filter, usually 32 bits or less, this multiple-precision is, for all practical purposes, "infinite" and hence the subscript I. For implementation, the coefficients need to be quantized to the appropriate wordlength and the actual filter is thus described by coefficients $\{h(n)\}$ where

$$h(n) = \mathbf{Q}\{h_I(n)\} \quad (3.7.33)$$

This introduces an error that can be expressed as

$$E(e^{j\omega}) = H(e^{j\omega}) - H_I(e^{j\omega}) = \sum_{n=0}^{N-1} \Delta h(n) e^{-jn\omega} \quad (3.7.34)$$

where

$$\Delta h(n) = h(n) - h_I(n) \quad (3.7.35)$$

We shall assume that the ideal coefficients can be normalized to have a maximum value of unity. Then if the wordlength used to represent the fractional part is B bits (total wordlength = (B+1) to accommodate the sign) then the error coefficients are limited by

$$|\Delta h(n)| \le 2^{-B} \quad (3.7.36)$$

In practice both $\{h_I(n)\}$ and $\{h(n)\}$ are known and the error can be computed in any desired form to provide a measure of the impact of how "good" the approximation is. For example,

$$E_\infty = \max_\omega \left| H(e^{j\omega}) - H_I(e^{j\omega}) \right| W(\omega) \tag{3.7.37}$$

defines the measure as the maximum deviation of the implemented response from the infinite precision version. This deviation can be weighted by a (non-negative) function $W(\omega)$ which represents the notion that certain frequencies are more important than others. We shall assume that this weighting function is unity. Other measures are E_1 and E_2 where

$$E_1 = \sum_{k=0}^{N-1} |\Delta h(k)| \tag{3.7.38a}$$

$$E_2 = \sum_{k=0}^{N-1} |\Delta h(k)|^2 \tag{3.7.39b}$$

There is a relation between these measures. Application of the **triangle inequality** yields

$$\left|E(e^{j\omega})\right| \le \sum_{k=0}^{N-1} |\Delta h(k)| \Rightarrow E_\infty \le E_1 \tag{3.7.40}$$

Since $|h(n)|$ and by implication $|\Delta h(n)|$ are less than unity,

$$|\Delta h(k)| \le 1 \Rightarrow |\Delta h(k)|^2 \le |\Delta h(k)| \Rightarrow E_2 \le E_1 \tag{3.7.41}$$

The three measures described above have certain physical interpretation that can be explained with respect to Fig. 3.42.

Suppose that the input, $\{x(n)\}$, to the filter is a sinusoid. Then the output, $\{y(n)\}$ is also a sinusoid of the same frequency. Assuming that the only finite-wordlength effect being considered is the quantization of the coefficients, then

$$\begin{aligned} x(n) &= A\, e^{jn\omega_0} \\ y(n) &= A\, H(e^{j\omega_0})\, e^{jn\omega_0} \\ y_I(n) &= A\, H_I(e^{j\omega_0})\, e^{jn\omega_0} \end{aligned} \tag{3.7.42}$$

Consequently, the error can be bounded in the following manner:

$$\left| y(n) - y_I(n) \right| \le |A| \left| H(e^{j\omega}) - H_I(e^{j\omega}) \right| \le |A| E_\infty \tag{3.7.43}$$

Thus E is a measure of how well, or badly, the implemented filter will be with respect to the infinite-precision version, when the signals of interest are nominally sinusoidal in nature.

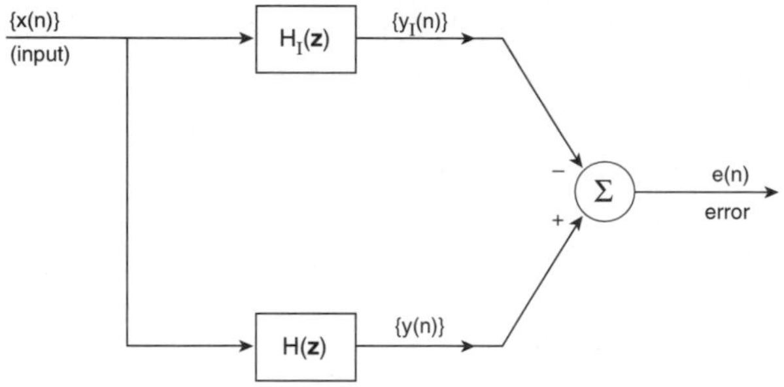

Fig. 3.42
Model for analyzing the error introduced by coefficient quantization

Suppose now that the input is random, say white noise. Then the error is a noise sequence as well. The power of the error sequence can be computed in terms of the power of the input as

$$e(k) = x(n)(*)\Delta h(n) \Rightarrow \sigma_e^2 = \sigma_x^2 \sum_{k=0}^{N-1} |\Delta h(k)|^2 = \sigma_x^2 E_2 \qquad (3.7.44)$$

Thus E_2 is a suitable measure of "goodness" when the signals of interest are random.

If the only information available on the input is that $|x(n)| < M_x$, i.e., the input is bounded, then an similar bound on the error can be obtained as

$$|e(k)| \leq \sum_{n=0}^{N-1} |\Delta h(n)\, x(k-n)| \leq M_x \sum_{k=0}^{N-1} |\Delta h(k)| = M_x E_1 \qquad (3.7.45)$$

These measures can be computed since, in principle, both the infinite-precision and quantized versions of the filter are known once the filter design procedure and coefficient wordlength have been established. Prior to the design process we can "guesstimate" a bound for the error measures in terms of the expected wordlength as

$$E_1 = \sum_{k=0}^{N-1} |\Delta h(k)| \leq N 2^{-B} \qquad (3.7.46)$$

This bound indicates the dependence of the **goodness of fit** between the two filters is a function of the **length** of the filter as well as the **wordlength** used to represent the coefficients. In considering fixed (predefined) **FIR** filters the bound in Eq. (3.7.46) is not very useful but in the case of adaptive filters, as discussed in Chapter 6, does provide a guideline for the wordlength required in representing coefficients.

3.7.3 IIR Filters

In dealing with **IIR** filters, we can assume that the transfer function is of the form

$$H(\mathbf{z}) = \frac{\sum_{k=0}^{N} a_k \mathbf{z}^{-k}}{1 + \sum_{k=1}^{D} b_k \mathbf{z}^{-k}} \tag{3.7.47}$$

where the order of the numerator, N, is not greater than the order of the numerator, D. The procedure for designing the filter will provide the coefficients $\{a_i\}$ and $\{b_i\}$ with "infinite" precision and hence we use the subscript I for the transfer function.

The direct-form implementation of an **IIR** filter is (almost) never used. Rather, the transfer function is factored into **first-order** and **second-order** sections of the form

$$H_1(\mathbf{z}) = \frac{a_0 + a_1 \mathbf{z}^{-1}}{1 + b_1 \mathbf{z}^{-1}} \quad \text{(first-order)} \tag{3.7.48a}$$

$$H_2(\mathbf{z}) = \frac{a_0 + a_1 \mathbf{z}^{-1} + a_2 \mathbf{z}^{-2}}{1 + b_1 \mathbf{z}^{-1} + b_2 \mathbf{z}^{-2}} \quad \text{(second-order)} \tag{3.7.48b}$$

and the composite transfer function is given by

$$H_I(\mathbf{z}) = \prod_i H_i(\mathbf{z}) \ \text{(cascade)} \quad \text{or} \quad H_I(\mathbf{z}) = \sum_i H_i(\mathbf{z}) \ \text{(parallel)} \tag{3.7.49}$$

where each $H_i(\mathbf{z})$ is either a first- or second-order section. Needless to say, Eq. (3.7.49) describes the form of the implementation; the $H_i(\mathbf{z})$ are different for the cascade and parallel forms. Of these, the cascade form is more popular and our discussion will assume that this is the architecture of choice.

The cascade and parallel forms both necessitate the extraction of the roots of the polynomial D(**z**). For convenience we will assume that the roots are distinct. Since the coefficients are real-valued, these roots will occur in complex conjugate pairs or will be real. It is known that for polynomials of order less than four these roots can be obtained by formulas. Roots of higher-order polynomials cannot be obtained by formulas but must be obtained by employing numerical methods or symbolic division. This implies that, for high-order filters, the "roots" computed are actually approximations of the true roots. Usually, but not always, the larger the order of the polynomial D(**z**), the less is the accuracy with which the roots can be computed (it is possible though to have high-order polynomials whose roots can be computed accurately but lower-order polynomials whose roots cannot be computed with the same accuracy). Therefore, for **IIR** filters we are starting out with an approximation of H(**z**), which in turn is an approximation of the "target" or "desired" frequency response, even before we apply the finite-wordlength constraint.

The roots of D(**z**) can be grouped in pairs; if D is odd then an additional "group" of a single real root needs to be considered as well. With this grouping we can write

$$D(\mathbf{z}) = (1 + \beta_1 \mathbf{z}^{-1} + \gamma_1 \mathbf{z}^{-2})(1 + \beta_2 \mathbf{z}^{-1} + \gamma_2 \mathbf{z}^{-2}) \ldots \text{(even)} \quad (3.7.50a)$$

$$D(\mathbf{z}) = (1 + \beta_0 \mathbf{z}^{-1})(1 + \beta_1 \mathbf{z}^{-1} + \gamma_1 \mathbf{z}^{-2}) \ldots \text{(odd)} \quad (3.7.50b)$$

The coefficients (β,γ) of each second order section of the denominator are related to the roots of D(**z**), which are the poles of the transfer function, in the following way.

$$\text{Complex roots: } u_1 + j\,v_1 \text{ and } u_1 - j\,v_1$$
$$\beta_1 = -2u_1 \text{ and } \gamma_1 = u_1^2 + v_1^2 \quad (3.7.51a)$$

$$\text{Real roots: } r_1 \text{ and } s_1$$
$$\beta_1 = -(r_1 + s_1) \text{ and } \gamma_1 = r_1 s_1 \quad (3.7.51b)$$

Assume now that N=D, the order of the numerator was equal to the denominator. Then we can find the roots of N(**z**) and organize them in much the same way as for D(**z**):

$$N(\mathbf{z}) = G(1 + \xi_1 \mathbf{z}^{-1} + \eta_1 \mathbf{z}^{-2})(1 + \xi_2 \mathbf{z}^{-1} + \eta_2 \mathbf{z}^{-2}) \ldots \text{(even)} \quad (3.7.52a)$$

$$N(\mathbf{z}) = G(1 + \xi_0 \mathbf{z}^{-1})(1 + \xi_1 \mathbf{z}^{-1} + \eta_1 \mathbf{z}^{-2}) \ldots \text{(odd)} \quad (3.7.52b)$$

where G is a constant "gain" term. If the order of the numerator were less than the denominator, then some of the quantities $\{\xi_i, \eta_i\}$ would be zero. If N(**z**) was a constant then all these coefficients would be zero. The roots of N(**z**) are associated with the *zeros* of the transfer function.

The cascade form implementation of the transfer function $H_I(\mathbf{z})$ can be expressed as

$$H_I(\mathbf{z}) = \frac{G(1+\xi_0 \mathbf{z}^{-1})(1+\xi_1 \mathbf{z}^{-1}+\eta_1 \mathbf{z}^{-2})\dots}{(1+\beta_0 \mathbf{z}^{-1})(1+\beta_1 \mathbf{z}^{-1}+\gamma_1 \mathbf{z}^{-2})\dots} \tag{3.7.53}$$

and is depicted in Fig. 3.42.

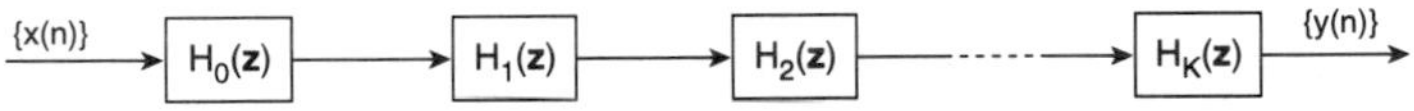

Fig. 3.42
Cascade implementation of an **IIR** filter

If D is even then $H_0(\mathbf{z})$ is not present. The number of second-order sections, K, is D/2 (D even) or (D–1)/2 (D odd). The variety of such structures for implementing a given $H(\mathbf{z})$ can be quite large since we have freedom in associating poles with zeros as well as freedom in ordering the chosen pole-zero pairs.

The basic second-order section is fundamental to the implementation of **IIR** filters and so deserves some discussion. Consider the recursive section $G(\mathbf{z})$ corresponding to a second-order section with complex poles

$$G(\mathbf{z}) = \frac{1}{(1+\beta \mathbf{z}^{-1}+\gamma \mathbf{z}^{-2})} \tag{3.7.54}$$

The complex conjugate pair of poles of $G(\mathbf{z})$ are related to the coefficients in a simple manner, especially if the pole locations are expressed in polar notation.

$$\text{Poles at } re^{\pm j\theta} \Rightarrow \beta = -2r\cos(\theta) \text{ and } \gamma = r^2 \tag{3.7.55}$$

The locations of the poles in the **z**-plane is depicted in Fig. 3.43. For the filter to be stable, the poles must lie inside the unit circle. Thus, for a complex pole pair, stability of the filter is guaranteed if $\gamma < 1$. For a pair of real poles, the condition $\gamma < 1$ is necessary but not sufficient.

The magnitude response and group delay response of $G(\mathbf{z})$ are depicted in Fig. 3.44. The figures are not to scale and only positive frequencies are considered since the responses will be symmetric about zero frequency. The shape of the response is important. First, the peak values of the magnitude response, and the group delay, will occur at, approximately, the normalized frequency θ, the "angle" of the pole. The peak values will be proportional to $1/(1-r)$. That is, closer the pole is to the unit circle, the greater is the peak of the response. This is important to realize since a pole on the unit circle corresponds to an unstable system that will oscillate at the frequency corresponding to θ. The closeness to the unit circle is thus a measure of the instability of the filter.

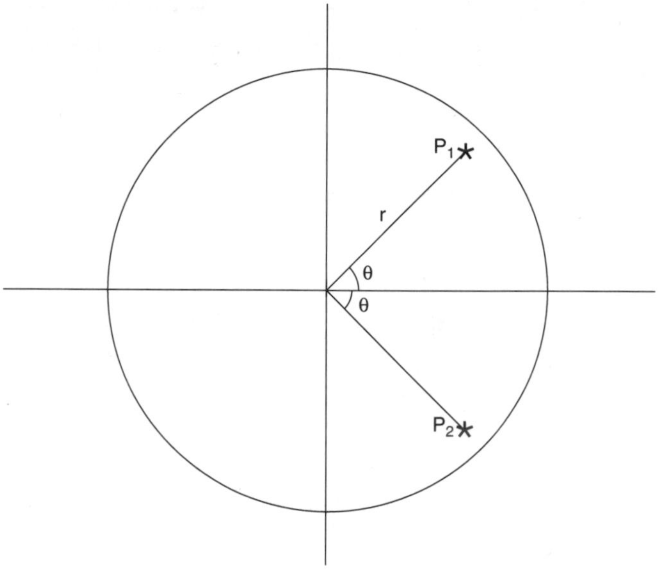

Fig. 3.43
Location of the poles of a typical second-order recursive section

The numerator of a typical second-order system is of the form

$$G(\mathbf{z}) = (1 + \xi \mathbf{z}^{-1} + \eta \mathbf{z}^{-2}) \tag{3.7.56}$$

In most cases **IIR** filters are designed to achieve a bandpass (or lowpass or highpass) or a bandstop (such as a notch) response. Consequently, it helps if the zero lay on the unit circle as depicted in Fig. 3.45. In this case the coefficients can be written in terms of the (normalized radian) frequency of the transmission zero as

$$\text{Zeros at } e^{\pm j\varphi} \Rightarrow \xi = -2\cos(\varphi) \text{ and } \eta = 1 \tag{3.7.57}$$

Thus, based on the symmetry, the numerator has a linear-phase characteristic and a magnitude characteristic depicted in Fig. 3.46. At a frequency corresponding to the angle of the zero there is an absolute null. Clearly, if the zero is not on the unit circle there is just a minimum (approximately) at that frequency rather than a null. In Chapter 9 we describe a design procedure that permits the positioning of transmission zeros (on the unit circle) at prescribed frequencies while maintaining an equiripple passband behavior.

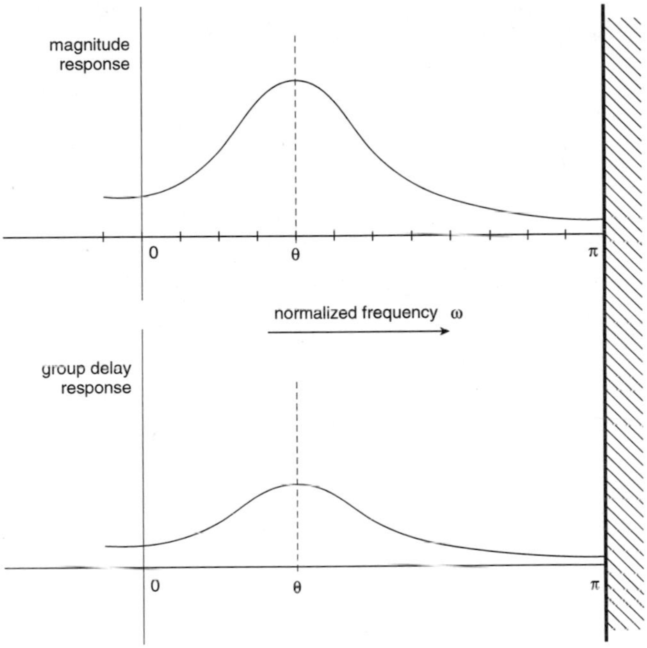

Fig. 3.44
Frequency response of a typical second-order recursive section

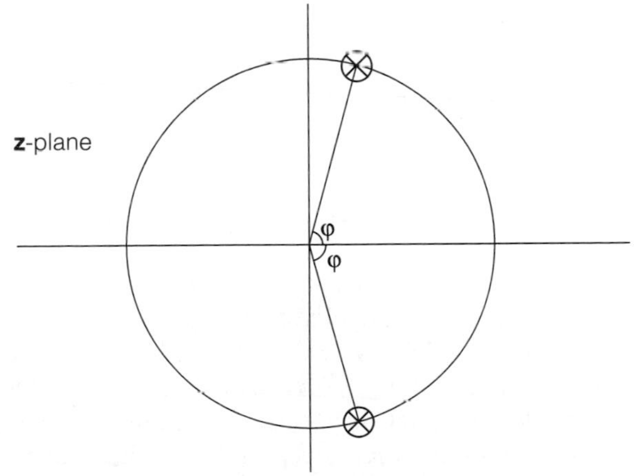

Fig 3.45
Zero on the unit circle associated with the typical second-order nonrecursive section achieving a bandpass/bandstop response

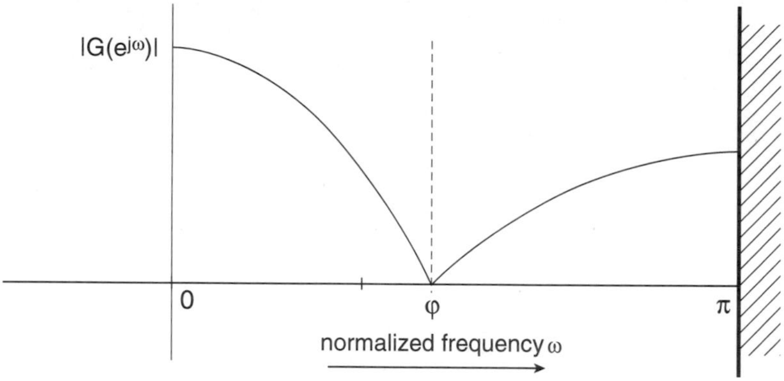

Fig. 3.46
Typical frequency response associated with a zero on the unit circle

The impact of finite wordlength on the representation of the coefficients of an **IIR** filter is somewhat more complicated than for an **FIR** filter. First note that a coefficient of unity does not involve any quantization error. Thus quantization of the numerator coefficients changes only the frequency of the transmission zero. Quantization of the denominator coefficients moves the pole from its ideal position and care must be taken to ensure that because of quantization a pole close to the unit circle does not move onto (or outside) the unit circle. It is possible in principle to establish the impact of quantization as a "frequency response error" in much the same way as E (ω) was defined in the **FIR** case as well as the error measures E_1 and E_2. Unfortunately, a description of this deviation as an "error signal" as in the **FIR** case is not so straightforward.

3.7.4 Allpass Filters

There is a class of **IIR** filters that have a frequency response that is special. The magnitude response is unity for all frequencies and thus all frequencies pass through unattenuated. This property gives rise to the name ***allpass*** for this class of filters.

Consider the transfer function H(**z**) given by

$$H(\mathbf{z}) = \mathbf{z}^{-D} \frac{1 + \sum_{k=1}^{D} b_k \mathbf{z}^{+k}}{1 + \sum_{k=1}^{D} b_k \mathbf{z}^{-k}} \qquad (3.7.58)$$

There is a definite relation between the numerator and denominator. Specifically the numerator coefficients can be viewed as the same as the denominator coefficients taken in reverse order. The expression in Eq.(3.7.58) is written in such a way as to emphasize the relationship. Thus if the denominator is considered a polynomial in $\mathbf{z}^{-1}$ then the numerator, with an appropriate delay term, is the same polynomial in

z^{+1}. Hence the pole-zero pattern of such a filter will follow the pattern depicted in Fig. 3.47. Each pole, and its complex conjugate, will be accompanied by a zero, and its complex conjugate, and the magnitude of the pole and zero would be reciprocal. Depending which pole we call **p** and which is **p***, the angles of the pole and zero would be the same (or negative). That is,

$$|\mathbf{p}| = \frac{1}{|\mathbf{q}|} \text{ and } \text{angle}(\mathbf{p}) = \pm\,\text{angle}(\mathbf{q}) \tag{3.7.59}$$

The frequency response of the filter, $H(e^{j\omega})$, can be obtained by setting $\mathbf{z} = e^{j\omega}$ in the expression for $H(\mathbf{z})$. It can be shown very easily that the relationship of the numerator and denominator coefficients implies that $H(e^{j\omega})$ has the form

$$H(e^{j\omega}) = \frac{A_N(\omega)\,e^{-j\beta_N(\omega)}}{A_D(\omega)\,e^{-j\beta_D(\omega)}} = e^{-j2\beta_N(\omega)} \tag{3.7.60}$$

which exhibits the constant magnitude-versus-frequency characteristic discussed before. The phase response is (twice) that of the denominator and can be expressed in terms of the coefficients as

$$\beta_D(\omega) = \arctan\left(\frac{\sum_{k=1}^{D} b_k \sin(k\omega)}{1 + \sum_{k=1}^{D} b_k \cos(k\omega)}\right) \tag{3.7.61}$$

Allpass filters are used for providing phase equalization. If a desired phase response can be expressed mathematically, the coefficients $\{b_k\}$ can be obtained by approximation techniques to match, as closely as possible, this desired phase response.

Allpass filters can also be used as phase-splitters, where the actual phase response is not as important as having a 90-degree phase shift between two (filtered) versions of the same signal. 90-degree phase-splitters are also called ***Hilbert Transformers***. In our discussion on **FIR** filters we saw that we could get this phase response by having appropriate symmetry. Unfortunately, with **FIR** filters, we cannot get the allpass magnitude characteristic exactly. Designing **FIR** Hilbert transformers is equivalent to approximating the "allpass" characteristic, usually over certain bands of frequency rather than all frequencies. With **IIR** filters on the other hand, the magnitude response is just right, unity (constant) over the entire frequency band, but the phase response is only an approximation to the 90-degree phase shift.

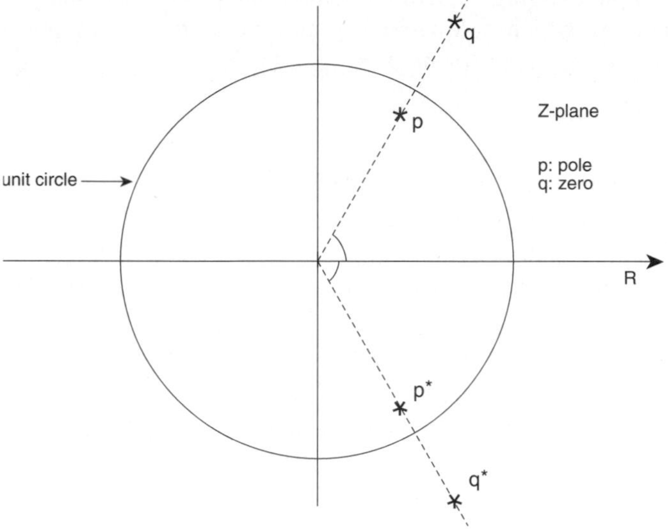

Fig. 3.47
Pole-zero relationship in allpass filters.

3.8 EXERCISES

3.1. a) Suppose that the Fourier transform of the signal x(t) is $X(f)$. Express $Y(f)$ in terms of $X(f)$ where y(t) is given by $y(t) = x(t)\cos(2\pi f_0 t)$. b) Suppose the **DTFT** of {x(n)} is $X(f)$. Express $Y(f)$ in terms of $X(f)$ where {y(n)} is given by $y(n) = x(n)\cos(2\pi f_0 n)$.

3.2. Suppose {e(n)} is an exponential sequence, i.e., $e(n) = a^n$ for some a. If {x(n)} and {y(n)} are two arbitrary finite-energy sequences, show that

$$[\ e(n)\ x(n)\ ]\ (*)\ [\ e(n)\ y(n)\ ] \quad = \quad e(n)\ [\ x(n)\ (*)\ y(n)\ ]$$

3.3. Suppose that {x(nT)} is a discrete-time signal with underlying sampling rate $f_s = (1/T)$. A new signal {y(nT)} is generated as $y(nT) = (-1)^n\, x(nT)$. Relate the spectrum of {y(nT)} to the spectrum of {x(nT)}. That is, express $Y(f)$ in terms of $X(f)$. Also provide a graphical representation of the solution.

3.4. {x(n)} is a discrete-time signal whose amplitude is bounded That is, |x(n)| B. This signal is the input to an **FIR** filter with impulse response {h(n); n = 0,1, ... , (N–1)}. Show that the output sequence {y(n)} is bounded and satisfies

$$|y(n)| \leq B \sum_{k=0}^{N-1} |h(k)|$$

and further that there actually exists a sequence {x(n)} that is bounded and for which the equality holds.

3.5. If {x(n); n = 0,1, ... , (N–1)} and {X(k); k = 0, 1, ... , (N–1)} are related by the **DFT** sum then show that a Parseval's relation can be developed. That is,

$$\sum_{n=0}^{N-1} |x(n)|^2 = \frac{1}{N} \sum_{k=0}^{N-1} |X(k)|^2$$

3.6. Consider the following signal processing chain. The input {x(n)} is multiplied by a signal {c(n)} to yield {v(n)}. {v(n)} is applied to a discrete-time **LTI** with impulse response {h(n)} and associated frequency response $H(f)$. The output of the **LTI**, {w(n)} is multiplied by the signal {c*(n)} (the complex conjugate of c(n)) to yield the output {y(n)}.

a) Draw a block diagram indicating the various signals and their relationship.

b) Assume that $c(n) = \exp(j2\pi f_0 n)$. Express the overall frequency response between {x(n)} and {y(n)} in terms of $H(f)$ and f_0. Express the impulse response in terms of h(n) and f_0.

c) Repeat part **b)** assuming $c(n) = \cos(2\pi f_0 n)$. State any assumptions made. If necessary, assume that $H(f)$ is the frequency response of an ideal lowpass filter (bandpass filter).

3.7. {**x**(n)} is a random signal with autocorrelation function $R_{xx}(n)$ and power spectrum $S_{xx}(f)$. This is applied to a digital filter with transfer function H(**z**) (and frequency response $H(f)$ and impulse response {h(n)}) and the output is {**y**(n)}. Then

a) Express the autocorrelation $R_{yy}(n)$ in terms of $R_{xx}(n)$ and the filter (impulse response).

b) Express the squared magnitude function $S_{yy}(\mathbf{z})$ in terms of $S_{xx}(\mathbf{z})$ and the transfer function H(**z**).

c) Express $S_{yy}(f)$ in terms of $S_{xx}(f)$ and $H(f)$.

d) If {**x**(n)} is a white noise sequence of power σ_x^2, derive an expression for the power of {y(n)}, σ_y^2, in terms of the frequency response of the filter (the expression is in the form of an integral).

e) If {**x**(n)} is a white noise sequence of power σ_x^2, derive an expression for the power of {y(n)}, σ_y^2, in terms of the impulse response of the filter (the expression is in the form of an summation).

3.8. With reference to the previous exercise, do parts **d)** and **e)** assuming that $H(\mathbf{z}) = (1 - \mathbf{z}^{-1})$. Assuming that the underlying sampling frequency, f_s, is much greater than f_c, (which defines the "frequency band of interest"), what fraction of the power of {y(n)} lies in the frequency band $[-f_c, +f_c]$?

3.9. The transfer function of a filter is given by $H(z) = (1 - a)(1 - a\mathbf{z}^{-1})^{-1}$. The input to the filter is the signal {x(n)}, where $x(n) = [A\cos(2\pi f_0 n)]^2$. The output is {y(n)}. Assume that $|a| < 1$.

a) Plot the frequency response of the filter for different values of a. Comment on the type of filter (lowpass/highpass/bandpass, etc.).

b) Show that the output {y(n)} contains a "constant" component and a sinusoidal component (in steady state, after the transients have died down). What is the amplitude and frequency of this steady state sinusoidal component. What is the value of the constant component?

c) Show that the impulse response length (defined appropriately) increases as a is closer to 1. Comment on the frequency selectivity of the filter in the same situation.

d) Convince yourself that if $x(n) = [v(n)]^2$, then the constant component of the output is related to the power of {v(n)}. What is a reasonable compromise for the coefficient a in order to use this scheme for power measurement?

e) Simulate the action of the filter as applied to power estimation using for x(n) a random noise signal.

3.10. A signal {x(n)} is observed and a "snapshot" of N samples, {x(n), n = 0, 1, 2, ... , (N–1)} taken. Based on this snapshot we wish to determine whether the signal was a sinusoid, random (Gaussian) noise, or speech (exponentially distributed random process). How can this be achieved? In particular, we can compute the following: max $\{|x(n)|\,;\, n = 0,1,\ldots,(N-1)\}$ and the sum of the square of $|x(n)|$ over the N samples. Can the ratio be used to discriminate between the three types of signals? What is the impact of the size, N, of the observation interval?

3.11. A signal x(t) is bandlimited to $[-f_c, +f_c]$. An A/D converter operating at a sampling rate $f_s > 2f_c$ is used to convert the signal to digital format. Unfortunately, x(t) is corrupted by white noise, **n**(t), with a power spectral density $S_{\mathbf{nn}}(f) = (N_0/2)$ watts/Hz. A pre-conversion (anti-aliasing) lowpass filter is applied prior to sampling, $H(f)$, with

$$H(f) \approx 1 \text{ for } f \in [-f_c, +f_c] \text{ and } \int_{-\infty}^{+\infty} |H(f)|^2 df = \alpha$$

What is a reasonable number of bits, B, for the A/D converter, given that the maximum value for x(t) is V volts? That is, $|x(t)| < V$. Explain your reasoning.

3.12. Write a program to compute and plot the frequency response of the filter H(**z**) given by

$$H(\mathbf{z}) = \frac{(1+2\cos(\theta)\mathbf{z}^{-1}+\mathbf{z}^{-2})}{(1+2r\cos(\theta+\Delta\theta)\mathbf{z}^{-1}+r^2\mathbf{z}^{-2})}\,\frac{(1+2\cos(\theta)\mathbf{z}^{-1}+\mathbf{z}^{-2})}{(1+2r\cos(\theta-\Delta\theta)\mathbf{z}^{-1}+r^2\mathbf{z}^{-2})}$$

Choose various values for r (in the range of 0.9 to 0.99, for example), and various value for $\Delta\theta$ (approximately 0.01π, for example). Use this to design a fourth order "notch filter" that removes the components at frequency θ (normalized radian frequency). If the input to the filter is a combination of a tone (at θ) and white noise of power σ_n^2, what is the power at the output. Describe how such a filter can be used to estimate the signal-to-noise ratio when the desired signal is a tone.

3.13. Show that the second-order all-pole section whose denominator is given by $B(\mathbf{z}) = 1 + b_1\mathbf{z}^{-1} + b_2\mathbf{z}^{-2}$ is stable if

$$1 + b_1 + b_2 > 0 \ \underline{\text{and}}\ 1 - b_1 + b_2 > 0\ \underline{\text{and}}\ |b_2| < 1.$$

3.9 REFERENCES AND BIBLIOGRAPHY

This section is split into two parts. The first part lists those publications that are referenced directly in this chapter. The second part is a small subset of the various publications available that deal with discrete-time and digital signal processing.

3.9.1 References

[1.1] Churchill, R.V., Brown, J.W., and Verhey, R.F., *Complex Variables and Applications,* Third Edition, McGraw-Hill Publishing Co., New York, NY, 1974.

3.9.2 Bibliography: Digital Signal Processing

[2.1] Antoniou, A., *Digital Filters: Analysis and Design,* McGraw-Hill Publishing Co., New York, 1979.

[2.2] Bellanger, M., *Digital Signal Processing, Theory and Practice,* John Wiley and Sons, New York, NY, 1989.

[2.3] Elliott, D. F. (Ed.), *Handbook of Digital Signal Processing*, Academic Press, San Diego, 1987.

[2.4] Hamming, R. W., *Digital Filters*, Prentice-Hall, Inc., Englewood Cliffs, NJ, 1977.

[2.5] Lane, J., and Hillman, G., "Implementing IIR/FIR Filters with Motorola's DSP56000/DSP56001," *Motorola Application Note* APR7/D, 1990. [This is but one of several excellent Application Notes generated by the Digital Signal Processing Group at Motorola.]

[2.6] Oppenheim, A. V., *Applications of Digital Signal Processing*, Prentice-Hall Inc., Englewood Cliffs, NJ, 1978.

[2.7] Oppenheim, A. V., and Schafer, R. W., *Discrete-Time Signal Processing*, Prentice-Hall, Inc., Englewood Cliffs, NJ, 1989.

[2.8] Rabiner, L. R., and Gold, B., *Theory and Application of Digital Signal Processing*, Prentice-Hall, Inc., Englewood Cliffs, NJ, 1975.

References [2.9] and [2.10] are collections of key papers in the development of digital signal processing and include papers from a variety of journals including those published by the IEEE. A subset of these papers have been listed as being most pertinent relative the material covered in this book.

[2.9] Rabiner, L.R., and Rader, C.M., Editors, *Digital Signal Processing*, IEEE Press, New York, 1972.

[2.9.1] J.F. Kaiser, "Design methods for sampled data filters," *Proc. 1st Annual Allerton Conf. on Circuits and System Theory*, 1963.

[2.9.2] C.M. Rader and B. Gold, "Digital filter design techniques in the frequency domain," *Proc. IEEE*, Feb. 1967.

[2.9.3] A.G. Constantinides, "Spectral transformations for digital filters," *Proc. IEEE*, Aug. 1970.

[2.9.4] L.R. Rabiner, "Techniques for designing finite duration impulse response digital filters," *IEEE Trans. Comm. Tech.*, Apr. 1971.

[2.9.5] J.W. Cooley and J.W. Tukey, "An algorithm for the machine calculation of complex Fourier series," *Mathematics of Computation,* 1965.

[2.9.6] G.D. Bergland, "A guided tour of the Fast Fourier Transform," *IEEE Spectrum*, July 1969.

[2.9.7] IEEE Group on Audio and Electroacoustics Subcommittee on Measurement Concepts, "What is the FFT?," *IEEE Trans. Audio and Electroacoustics*, June 1967.

[2.9.8] Liu, B., "Effect of finite wordlength on the accuracy of digital filters—a review," *IEEE Trans. Circuit Theory*, Nov. 1971.

[2.9.9] Rader, C.M., and Gold, B., "Effects of parameter quantization on the poles of a digital filter," *Proc. IEEE*, May 1967.

[2.9.10] Jackson, L.B., "On the interaction of roundoff noise and dynamic range in digital filters," *Bell System Tech. Journal*, Feb. 1970.

[2.9.11] Ebert, P.M., Mazo, J.E., and Taylor, M.G., "Overflow oscillations in digital filters ," *Bell System Tech. Journal*, Nov. 1969.

[2.10] Digital Signal Processing Committee, IEEE Acoustics, Speech, and Signal Processing Society, Editors, *Digital Signal Processing, II*, IEEE Press, New York, 1976.

[2.10.1] J. Makhoul, "Linear Prediction: A tutorial review ," *Proc. IEEE*, April 1975.

[2.10.2] J.H. McClellan, T.W. Parks, and L.R. Rabiner, "A computer program for designing optimum FIR linear-phase digital filters," *IEEE Trans. on Audio and Electroacoustics*, Dec. 1973.

[2.10.3] L.R. Rabiner, J.H. McClellan, and T.W. Parks, "FIR digital filter design using weighted Chebyshev approximation," *Proc. IEEE*, Apr. 1975.

[2.10.4] J.F. Kaiser, "Nonrecursive digital filter design using the I_0-sinh window function," *Proc. 1974 IEEE International Symposium on Circuits and Systems,* Apr. 1974.

[2.10.5] A.G. Deczky, "Synthesis of recursive digital filters using the minimum p-error criterion," *IEEE Trans. Audio and Electroacoustics*, Oct. 1972.

[2.10.6] R.W. Schafer and L.R. Rabiner, "A digital signal processing approach to interpolation," *Proc. IEEE*, June 1973.

[2.10.7] J.B. Knowles and E.M. Olcayto, "Coefficient accuracy and digital filter response," *IEEE Trans. Circuit Theory*, Mar. 1968.

[2.10.8] D.S.K. Chan and L.R. Rabiner, "Analysis of quantization errors in the direct form for finite impulse response digital filters," *IEEE Trans. on Audio and Electroacoustics*, Aug. 1973.

[2.10.9] I.W. Sandberg and J.F. Kaiser, "A bound on limit cycles in fixed-point implementations of digital filters," *IEEE Trans. Audio and Electroacoustics*, June 1972.

[2.10.10] J. L. Long and T. N. Trick, "An absolute bound on limit cycles due to roundoff errors in digital filters," *IEEE Trans. Audio and Electroacoustics*, Feb. 1973.

[2.10.11] B. Eckhardt and W. Winkelnkemper, "Implementation of a second order digital filter section with stable overflow behavior ," *Nachrichtentechnische Zeitschrift*, June 1973.

[2.11] The Applications Engineering Staff of Analog Devices, DSP Division, Mar, A., Ed., *Digital Signal Processing Applications using the ADSP-2100 Family*, Prentice-Hall, Inc., Englewood Cliffs, NJ, 1990.

[2.12] Edited by the Digital Signal Processing Committee of the IEEE Acoustics, Speech, and Signal Processing Society, *Programs for Digital Signal Processing*, IEEE Press, New York, 1979.

[2.13] *Digital Signal Processing Applications with the TMS320 Family*, Texas Instruments, 1986; Prentice-Hall, Inc., Englewood Cliffs, NJ, 1987.

[2.14] *PC-MATLAB*: Software package from The MathWorks, Inc., Also, the *Signal Processing Toolbox*, which can be used with *MATLAB*.

QUANTIZATION AND FINITE-WORDLENGTH EFFECTS IN DIGITAL FILTERS

4

4.1 INTRODUCTION

Rapid advancements in Very Large-Scale Integrated (**VLSI**) semiconductor technology, the steep decline in the cost of digital hardware, the availability of high-performance Digital Signal Processors (**DSP**s), the ease of encryption, the repeatability and stability (over time) of digital systems, and for a myriad other reasons, there has been a transition from analog to digital techniques for signal processing, transmission, and switching. The cornerstone of digital systems is the means for converting signals between analog and digital formats. The devices that accomplish this are the A/D converter (**co**der) and D/A converter (**dec**oder), the combination of which is referred to as a **codec**.

For the purposes of switching and transmission, standards have evolved that specify the nature of the codec. These devices adhere to *companding laws*, μ-law in North America and parts of Northeast Asia, and A-law in Europe and throughout most of the rest of the world. For signal processing purposes, the implication of arithmetic operations is that the digital signals are represented in a uniform format. Consequently, it is necessary to study what the relationship is between these companded laws and a uniform code. This relationship has consequences when viewed from two different angles. In telecommunications, use of a uniform coder has implications from the viewpoint of conversion from uniform to companded formats. In general signal processing the implication is just the reverse. The availability of inexpensive companded codecs justifies their use as *front-end* devices for a digital system (not necessarily for telecommunications) and the nature of distortion introduced must be known *a priori*. The study of companded codecs will be done by first introducing the notion of an ***optimal*** codec, an optimal codec being one that minimizes the power of the quantization noise. This notion will be extended to the notion of "optimality" from a slightly different viewpoint, that of "constancy" of the signal-to-noise ratio.

In dealing with discrete-time systems we assume that sample values, x(n), are representable exactly, with infinite precision. An obvious constraint of physically realizable systems is that sample values, that is, numbers, are quantized. In this chapter we consider some of the effects of finite wordlength. As usual, we quantify the effect of quantization noise in terms of signal-to-noise ratio.

In Section 4.2 we treat the case of uniform quantization and examine the cases of quantizing Gaussian and exponential random signals (variables). The key message is the relationship between wordlength of the quantizer and **SNR** and we provide a rationale for the thumb rule of "6 dB per bit." Quantization noise can be quite large if saturation or overload occurs and we see that the exponential distribution is more likely to saturate a quantizer than a Gaussian distribution of the same power (variance). This is in keeping with the notion of tail probability and "width" of distributions discussed in Chapter 2 (Section 2.5).

The notion of an "optimal" quantizer is introduced in Section 4.3. An optimal quantizer is one that maximizes the signal-to-noise ratio by suitably choosing the parameters of the quantizer as an (implicit) function of the **pdf** of the signal being quantized. The theory of optimal quantizers was first published by Max [1.2.1], who showed that that the calculation of the parameters of the optimal quantizer is, in general, a difficult one. In the case of a uniform distribution, however, the optimal quantizer is easy to find. Consequently, we propose a method for obtaining a near-optimal quantizer by establishing a function that would map the signal being quantized into a uniformly distributed signal.

In telephony the notion of optimal quantizer is somewhat different. Rather than try to minimize the noise power, the optimal quantizer is that which maximizes the range of input signal power over which the **SNR** is greater than some prescribed value. Fixed (not adaptive) coders striving to be optimal in this sense utilize a nonlinear function that "compresses" the range of the signal before applying a uniform quantizer. At the decoder the reverse or "expanding" function is employed. Two such "companding" codecs are the μ-law and A-law, which are used in North America and Europe, respectively. These are treated in Section 4.4.

The notion of finite wordlength (**FWL**) is implicit in all digital signal processing. Physical hardware, by its very nature, cannot provide the infinite wordlength required to represent all real numbers. This restriction to a finite set of numerical values introduces quantization noise. The impact of this noise is analyzed in Section 4.5. Implementing **IIR** filters, because of the impact of finite wordlength, can be problematic because the errors can accumulate. The technique for optimizing the signal-to-noise ratio by arranging the order of processing and maintaining the proper signal level is called *scaling* and is introduced in Section 4.5.

4.2 INTRODUCTION TO QUANTIZATION

The notion of a ***quantizer*** was developed in Chapter 3 and is elaborated upon here and analyzed in terms of signal-to-noise ratio (**SNR**). A quantizer is a device that converts a sequence $\{x(n)\}$, characterized by having great (infinite) precision in representing numerical values into a quantized sequence $\{x_Q(n)\}$ that is characterized by having a finite precision representation of numerical values. The quantizing error is the sequence $\{e(n)\}$ defined by the difference

$$e(n) = x(n) - x_Q(n) \tag{4.2.1}$$

A quantizer is modeled as a device with an input-output relation described by the staircase function shown in Fig. 4.1.

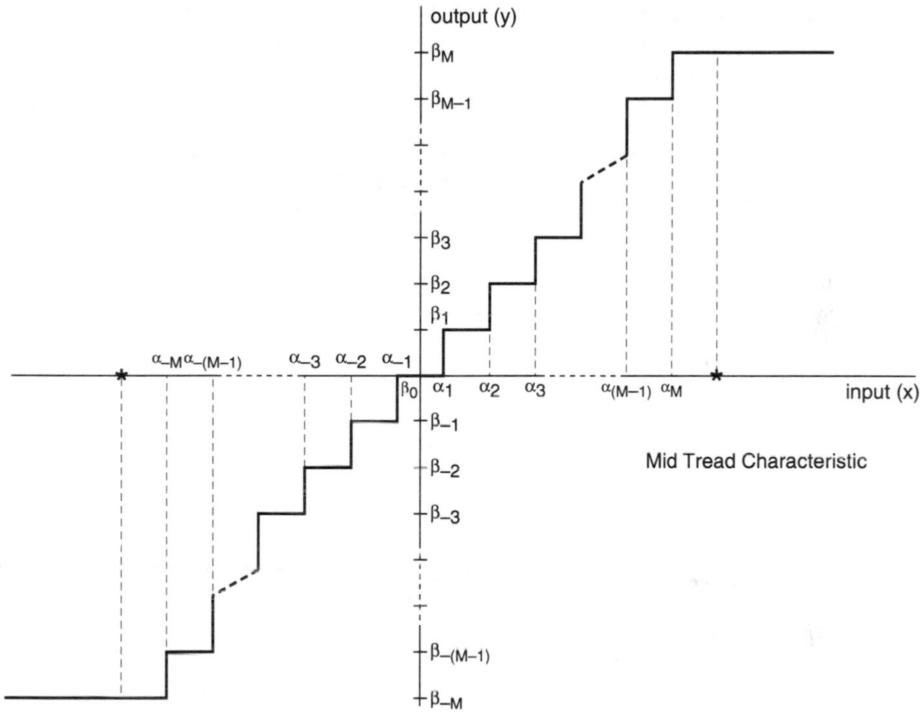

Fig. 4.1
Staircase representation of the input-output characteristic of a quantizer.

The permissible values of the output are described by the set $\{\beta_k;\ k = -M, \dots, -1, 0, 1, 2, \dots, M\}$. The grid along the input axis has points labeled α_k. These are called the decision points or breakpoints. The function of the quantizer can thus be expressed in words as follows. If the input value lies between decision points α_k and $\alpha_{(k+1)}$ then the output will be assigned the value β_k (for k >0); if the input is between α_{-k} and $\alpha_{-(k+1)}$ then the output is assigned the value β_{-k} (for negative input values). This can be written as

$$\alpha_k \le x < \alpha_{k+1} \Rightarrow y = \beta_k \text{ and } \alpha_M \le x \Rightarrow y = \beta_M \quad (4.2.2a)$$

$$\alpha_{-(k+1)} \le x < \alpha_{-k} \Rightarrow y = \beta_{-k} \text{ and } \alpha_M \le x \Rightarrow y = \beta_M \quad (4.2.2b)$$

Since the number of output values is finite, β_M is obviously a finite number. Regardless of the input, the output cannot be larger than this value. Thus A/D converters will exhibit a saturation or clipping phenomenon. That is, if the input sample has a voltage greater than α_M then the output is assigned the code corresponding to the largest representable value.

The nomenclature associated with quantizers borrows from conventional staircase construction. The horizontal lines are ***treads*** and the vertical lines are ***risers***. The characteristic of Fig. 4.1, since it has a tread at the origin is called a ***mid-tread*** characteristic. A ***mid-riser*** characteristic has a riser at the origin as depicted in Fig. 4.2. In the following treatment we shall assume we are dealing with a mid-tread characteristic unless otherwise specified. We will also assume that the quantizer is symmetric, that is

$$\alpha_k = -\alpha_{-k} \text{ and } \beta_k = -\beta_{-k} \tag{4.2.3}$$

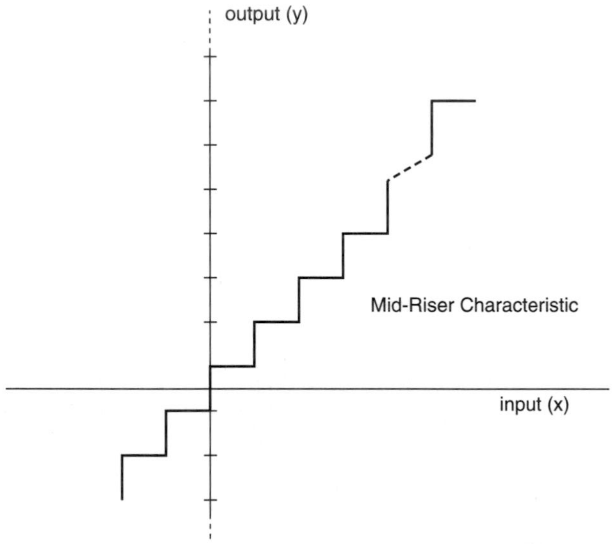

Fig. 4.2
Input-output characteristic of a mid-riser quantizer

The difference between adjacent output values is called the output step size; the difference between the decision points is called the input step size. Note that the values of the step size could vary across the quantizer characteristic. In the special case where the output step size is equal to (or is a known fixed multiple of) the input step size and this step is constant across the characteristic (except, possibly, at the endpoints around α_M and α_{-M}), the quantizer is considered ***uniform***. We will usually denote step size by Δ. Thus for a uniform quantizer, we can write

$$\begin{aligned} \alpha_k &= (2k+1)\frac{\Delta}{2} \text{ for } k = \pm 1, \pm 2, \ldots, \pm M \\ \beta_k &= k\Delta \text{ for } k = \pm 1, \pm 2, \ldots, \pm M \end{aligned} \tag{4.2.4}$$

The error introduced, e, is the difference between the input and the output and for input values in the interior of the characteristic

$$|e| = |y - x| \leq \frac{\Delta}{2} \tag{4.2.5}$$

For input values greater than $|\alpha_M|$, the quantizer error can become quite large, of the same magnitude as the input if it is large enough (infinite). We define a fictitious decision point $\alpha_{(M+1)}$ (and its counterpart $\alpha_{-(M+1)}$ for the negative input axis) as

$$\alpha_{M+1} = \alpha_M + \Delta \tag{4.2.6}$$

and then we can state that

$$|x| \leq \alpha_{M+1} \Rightarrow |e| \leq \frac{\Delta}{2} \tag{4.2.7}$$

This fictitious decision point is called the ***crashpoint*** of the quantizer characteristic. As long as the input is less than this value, the quantizer error is bounded by $\Delta/2$; for larger inputs the quantizer error is greater than $\Delta/2$. Other names for this decision value are "overload point," "saturation level," "peak level," and "virtual decision level."

The specific output level is encoded into B bits where B is the wordlength of the A/D converter or quantizer. Strictly speaking, what is encoded into the B bits is the index, k, of the output value β_k. With B bits available for representation, up to 2^B possible output levels can be considered. In mid-tread quantizers, the desire to both have symmetry and a output level of zero, implies the "loss" of one level. Mid-tread quantizers use (2M+1) levels where $M = 2^{(B-1)} - 1$. Therefore, for a uniform mid-tread quantizer,

$$\alpha_{M+1} = V \text{ and } \Delta = \frac{2V}{2^B + 1} \approx \frac{V}{2^{B-1}} \tag{4.2.8}$$

where the crashpoint is assumed to be equal to V (volts). If the signal level is less than V (volts) then the error voltage will be less than $\Delta/2$.

The quantization error is obviously a function of the input; if the input value is known, the quantization error can be computed using the staircase characterization of the quantizer. If the input, **x**, is treated as a random variable with probability density function $p_{\mathbf{xx}}(x)$, a **pdf** for the error, **e**, namely, $p_{\mathbf{ee}}(e)$ can be generated in the following manner (c.f. Section 2.5). Let the events $\mathbf{A}_k$, k = –M, ... , 0, 1, 2, ... , M, be defined as

$$\begin{gathered}\mathbf{A}_0 = \{\alpha_{-1} < \mathbf{x} < \alpha_{+1}\} \\ \mathbf{A}_k = \{\alpha_k < \mathbf{x} < \alpha_{k+1}\} \text{ for } k = 1, 2, \ldots, (M-1) \\ \mathbf{A}_M = \{\alpha_M < \mathbf{x} < \infty\} \\ \{\mathbf{A}_{-k} \text{ for } k = 1, 2, \ldots, M\} \text{ by symmetry}\end{gathered} \tag{4.2.9}$$

Then the collection $\{\mathbf{A}_k\}$ is a disjoint union that spans the space of possible input values, and we can express the **pdf** of the error in terms of the conditional probabilities. When the event $\mathbf{A}_k$ is true, then the quantization error is the difference between the input, $\mathbf{x}$, and y_k, the output value corresponding to $\mathbf{A}_k$.

$$p_{\mathbf{e}}(e) = \sum_n p_{\mathbf{x}\,|\,\mathbf{A}_n}((e+y_n)\,|\,\mathbf{A}_n)\,\Pr\{\mathbf{A}_n\} \tag{4.2.10}$$

The expression in Eq. (4.2.10) is applicable for all ranges of input values. If the input is restricted to values less than the crashpoint, then the error is restricted in value to less than $\Delta/2$ and the following approximation is used:

$$p_{\mathbf{e}}(e) = \begin{cases} \frac{1}{\Delta} & \text{for } |e| \le \frac{\Delta}{2} \\ 0 & \text{for } |e| > \frac{\Delta}{2} \end{cases} \tag{4.2.11}$$

That is, the quantization error is modeled as a uniform random variable. With this approximation, the power, or variance, of the quantization error is given by

$$\sigma_{\mathbf{e}}^2 = \frac{\Delta^2}{12} \tag{4.2.12}$$

Eq. (4.2.12) indicates that the quantization noise power is the same, regardless of the input, and thus the model may be suspect. In turns out that the model expressed in Eq. (4.2.11) is indeed quite accurate, especially if the number of quantization levels is large and the input values span a wide range of quantization steps.

The noise power is one of the principal measures of *goodness* associated with an A/D converter. However, to disassociate the performance measure from the absolute units used (volts, amps, etc.) it is conventional to define a "figure of merit" (**FOM**) based on the signal-to-noise ratio. The signal-to-noise ratio, **SNR**, is defined as the ratio of the input power to the power of the quantization error,

$$\mathrm{SNR} = \frac{\sigma_{\mathbf{x}}^2}{\sigma_{\mathbf{e}}^2} \quad \text{(linear units)}$$

$$\mathrm{SNR} = 10\log_{10}\left(\frac{\sigma_{\mathbf{x}}^2}{\sigma_{\mathbf{e}}^2}\right)\ \mathrm{dB} \tag{4.2.13}$$

From Eq. (4.2.12) the quantization error variance is a "constant," so the **SNR** is linearly related to the input signal power. The peak signal-to-noise ratio corresponds to the **SNR** value obtained for peak or maximum signal level. The intention in defining a maximum signal level is to ensure that the quantizer is not in an overload condition. We have related overload to signal amplitude but not signal power; for random signals the maximum amplitude, relative to the power, or variance, is not quite straightforward but for sinusoidal signals the peak value and the power have a definite relation. Thus for purposes of obtaining a figure of merit, we assume that the signal being quantized is sinusoidal with amplitude equal to the crashpoint of the quantizer. Thus the Figure of Merit (**FOM**) for a B-bit quantizer is given by

$$\mathrm{FOM} = \mathrm{SNR}_{\mathrm{peak}} = 10 \log_{10}\left(\frac{\frac{1}{2}V^2}{\frac{1}{12}\Delta^2}\right) \mathrm{dB}$$

$$\approx 10 \log_{10}(6 \times 2^{2B-2}) \approx (6B + 1.76)\,\mathrm{dB} \tag{4.2.14}$$

(as usual we approximate the logarithm of 2 by 0.3).

The **FOM** is directly related to the number of bits and Eq. (4.2.14) is the basis of the rule of thumb "the **SNR** improves by 6 dB for each additional bit."

If we plot the **SNR** versus input power we get a straight line, as depicted in Fig. 4.3, for nominal levels of input signal power. As the signal power is increased, we observe the effect of saturation, or overload, whereby the **SNR** flattens out and then starts decreasing.

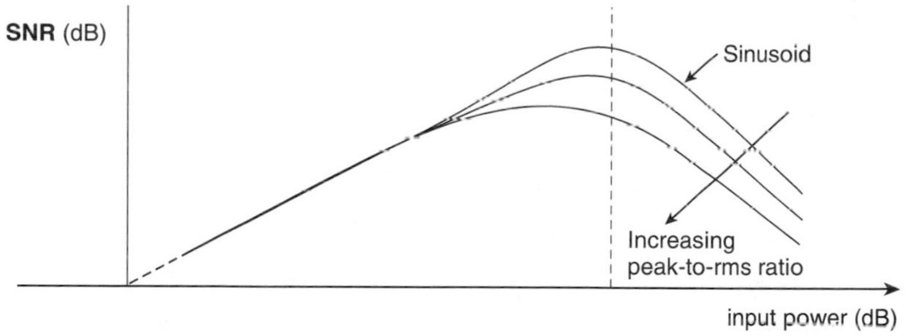

Fig. 4.3
Typical graph of signal-to-noise ratio versus signal power

By convention, we express the power of the input signal in dB relative to a suitable reference level. This reference level is defined as the power of a sinusoidal signal of amplitude equal to the crashpoint (V volts). This power level is, by definition, +3 dBm0 (actually 3.17 dBm0 or 3.14 dBm0 for North American and European and rest-of-the-world applications, respectively). That is, the peak sinusoid has a power of 3 dBm at a 0 Transmission Level Point (0 **TLP**). Consequently, for a signal power of X dBm0, the expected **SNR**, for $X < 3$, is

$$\mathrm{SNR} = \mathrm{SNR}_{\mathrm{peak}} + (X-3) = X + 6B - 1.24 \tag{4.2.15}$$

For signals other than sinusoids, we could see input values greater than the crashpoint even for signal powers less than 3 dBm0. If the peak-to-rms ratio is R dB then for signal powers greater than (3–R) dB we expect sample values of greater than the crashpoint with nonnegligible probability. Random signals will exhibit clipping even at a moderate signal power and the onset of clipping is experienced at lower signal power for signals that have a higher peak-to-rms ratio. The impact of clipping, or saturation, on the **SNR** curve will be a flattening out, followed by a decrease, as shown in Fig. 4.3.

The mathematically correct way of obtaining the quantization error power is to take into account the **pdf** of the input **x** by using Eq. (4.2.10). If only the variance of **e** is to be computed, the computation can be simplified by writing (assuming the **pdf** of **x** is symmetric)

$$\begin{aligned}\sigma_{\mathbf{e}}^2 &= \mathbf{E}\{(\mathbf{x}-\mathbf{y})^2\} = \\ &\int_{-\alpha_1}^{+\alpha_1} x^2 p_{\mathbf{x}}(x)\,dx + 2\sum_{n=1}^{M}\int_{\alpha_n}^{\alpha_{n+1}} (x-y_n)^2 p_{\mathbf{x}}(x)\,dx \\ &= \sigma_{\mathbf{x}}^2 - 4\sum_{n=1}^{M}\int_{\alpha_n}^{\alpha_{n+1}} (xy_n)^2 p_{\mathbf{x}}(x)\,dx + \sigma_{\mathbf{y}}^2\end{aligned} \tag{4.2.16}$$

where

$$\sigma_{\mathbf{y}}^2 = 2\sum_{n=1}^{M} y_n^2 \Pr\{\mathbf{A}_n\} \tag{4.2.17}$$

is the power of the quantized signal (we have assumed a mid-tread quantizer so $y_0 = 0$). By letting α_{M+1} be infinite we can incorporate the effect of clipping.

A program to compute the noise power for a given signal power is simple to generate if the **pdf** is known or assumed. The usual choices are the Gaussian and exponential **pdfs**. For these choices, the following results are useful. In the Gaussian case

$$\int_a^b t\exp\left(-\frac{t^2}{2}\right)dt = e^{-a} - e^{-b} \tag{4.2.18}$$

and the corresponding result for the exponential case is

$$\int_a^b t e^{-t}\,dt = (a+1)e^{-a} - (b+1)e^{-b} \tag{4.2.19}$$

The variation of **SNR** as a function of signal power is shown in Fig. 4.4 assuming that the **pdf** is Gaussian. Curves are presented for B = 14, 13, 12, 10, and 8 bits. Also shown in Fig. 4.4 is the quantization mask depicting the minimum **SNR** (see Chapter 1, Section 1.3) as specified for telephony applications. Similar calculations assuming an exponential **pdf,** are graphed in Fig. 4.5

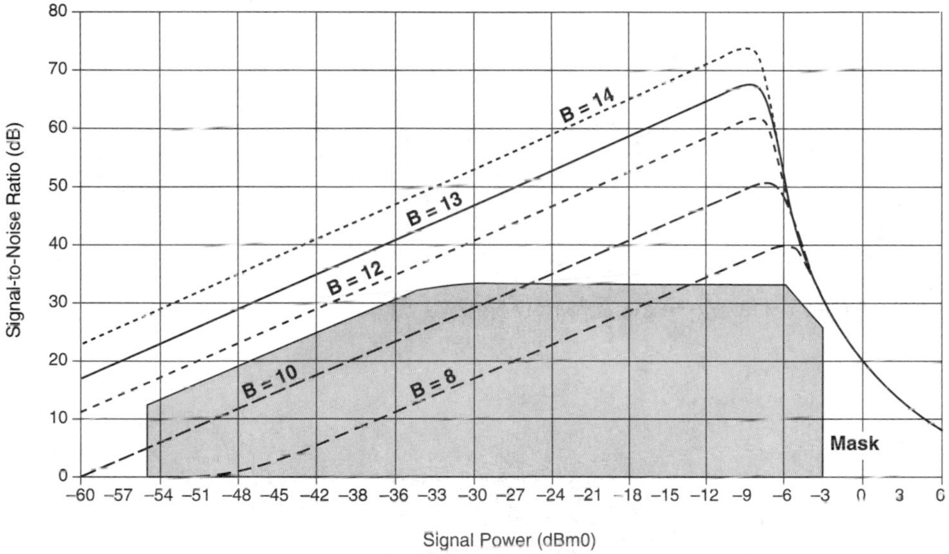

Fig. 4.4
SNR curves for 14-, 13-, 12-, 10-, and 8-bit uniform A/D converters assuming Gaussian signals. Also shown is the quantization noise mask discussed in Section 1.3 (Chapter 1)

From the graphs in Fig. 4.4 and Fig. 4.5 we can draw the following conclusions. First is that the application of Eq. (4.2.15) will provide an **SNR** estimate that will be within one-half dB of the correct value over a wide range of input power. Second is that the crashpoint of a quantizer becomes evident at about –9 dBm0 for the Gaussian **pdf** and at about –15 dBmo for the exponential **pdf**. The latter observation indicates that, for the Gaussian **pdf**, the probability of experiencing sample values greater than 12 dB above rhe rms is rare, with negligible impact on system performance. The corresponding level for the exponential **pdf** is about 18 dB.

In practice the observed, or measured, **SNR** is worse than the values shown in the graphs. The principal reason for this deviation is the presence of imperfections in the circuit implementations of the A/D converter. A common rule of thumb is that the measured **SNR** will be about 3 to 6 dB worse than the computed value based on a mathematical analysis.

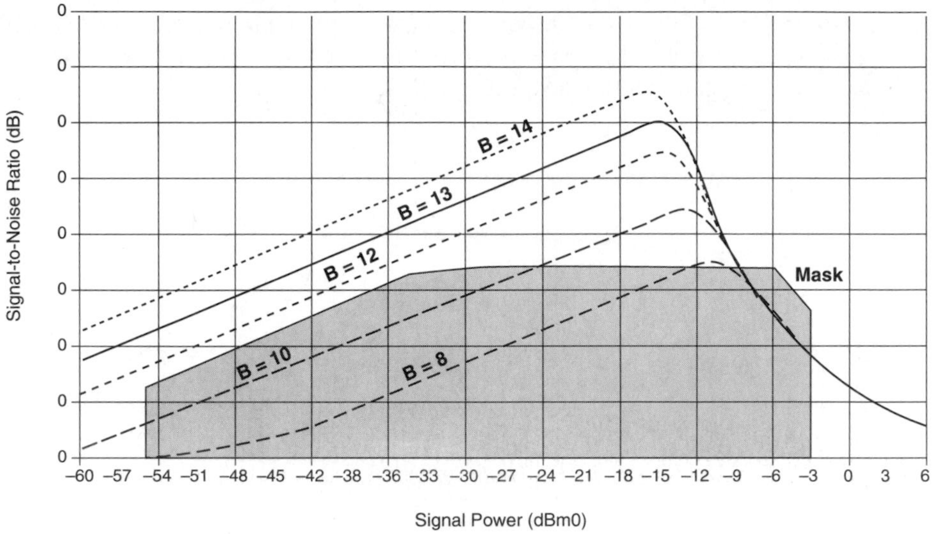

Fig. 4.5
SNR curves for 14-, 13-, 12-, 10-, and 8-bit uniform A/D converters assuming the signal has an exponential **pdf**. Also shown is the quantization noise mask discussed in Section 1.3 (Chapter 1)

4.3 QUANTIZATION FOR MINIMUM DISTORTION

The uniform quantizer, while well suited for general signal processing applications, does not take into account the nature of the signal being quantized. In 1960, Max published a paper [1.2.1], with the same title as that of this section, which showed how to organize the decision points and output levels of a quantizer so as to achieve a minimum for the power of the quantization noise.

The input-output characteristic of a general quantizer is pictured in Fig. 4.6 where we have taken the liberty to assume that the quantizer is of the mid-tread variety and that the characteristic is symmetric about the origin. This means that zero is a possible output value and that $x_i = -x_{-i}$. Also shown is a contrived decision value x_{N+1} which, as we shall see, is a ***virtual decision level***.

The quantizing noise power, **Q**, can be expressed in terms of the quantizer characteristic and the **pdf** of the input, $p_{\mathbf{x}}(x)$, as

$$\mathbf{Q} = \int_{x_{-1}}^{x_1} x^2 p_{\mathbf{x}}(x)\,dx + 2\sum_{n=1}^{N-1}\int_{x_n}^{x_{n+1}} (x-y_n)^2 p_{\mathbf{x}}(x)\,dx + 2\int_{x_N}^{\infty} (x-y_N)^2 p_{\mathbf{x}}(x)\,dx \tag{4.3.1}$$

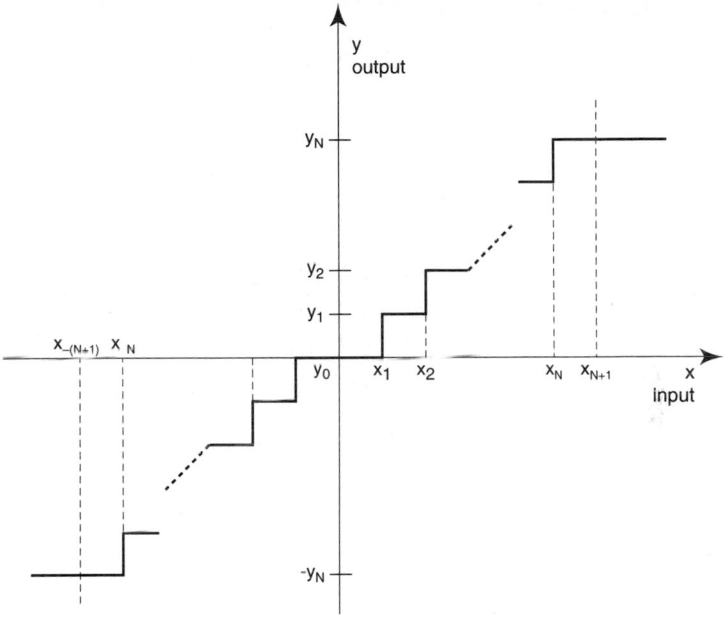

Fig. 4.6
Staircase function describing the input-output characteristic of a quantizer.

For a given **pdf**, the optimal decision levels $\{x_i\}$ and the optimal output levels $\{y_i\}$ can be obtained by setting the derivative of **Q** with respect to each of the parameters to zero. As shown by Max [1.2.1], the optimal decision and output values are given by

$$y_i = \frac{1}{P_i}\int_{x_i}^{x_{i+1}} x\, p_{\mathbf{x}}(x)\, dx = \mathbf{E}\{\,\mathbf{x} \mid x_i < \mathbf{x} < x_{i+1}\,\} \tag{4.3.2}$$

$$x_i = \frac{1}{2}(y_i + y_{i-1}) \quad \text{for } i = 1, 2, \ldots, N \tag{4.3.3}$$

That is, y_i is the expected value of **x**, given that **x** lies in the range $x_i < \mathbf{x} < x_{i+1}$. In our discussion on A/D conversion this event was called $\mathbf{A}_i$ and the probability of this event was termed P_i. The decision values are midway between the output values. Plugging these optimal choices of quantizer parameters into the expression for **Q** in Eq. (4.3.1) yields the following ***optimal*** (i.e., minimum) quantizing noise power:

$$\mathbf{Q}_{\text{opt}} = \int_{-\infty}^{+\infty} x^2 p_{\mathbf{x}}(x)\,dx - 2\sum_{i=1}^{N-1} y_i^2 \Pr\{ x_i < \mathbf{x} < x_{i+1} \}$$
$$= \mathbf{E}\{\mathbf{x}^2\} - \mathbf{E}\{\mathbf{y}^2\} \quad (= \text{input power} - \text{output power}) \tag{4.3.4}$$

Obtaining the optimal quantizer parameters is not a straightforward exercise. The equations (4.3.2) and (4.3.3) must be solved simultaneously, a task complicated by the fact that the **pdf** of the input, $p_{\mathbf{x}}(x)$, is implicitly involved. Max has shown that these equations can be solved using iterative techniques and has solved the equations for certain typical cases.

There is one case where the equations can be solved easily. This is the case where the input is a uniform random variable. Here we can express the **pdf** of **x** as

$$p_{\mathbf{x}}(x) = \begin{cases} \frac{1}{2W} & \text{for } |x| \le W \\ 0 & \text{for } |x| > W \end{cases} \tag{4.3.5}$$

For the uniform **pdf** Eq. (4.3.2) reduces to

$$y_i = \frac{1}{2}(x_i + x_{i+1}) \text{ for } i = 1, 2, \ldots, N \tag{4.3.6}$$

and we can derive the following:

$$y_0 = 0;\ x_1 = \frac{1}{2}y_1;\ x_2 = \frac{3}{2}y_1;\ldots;\ x_n = \frac{2n-1}{2}y_1;\ldots \tag{4.3.7a}$$

$$y_2 = \frac{1}{2}(x_1 + x_2) = 2y_1;\ldots;\ y_n = ny_1;\ldots \tag{4.3.7b}$$

Thus for the case of a uniform random variable the optimal quantizer is the uniform quantizer described by the step size Δ where

$$\text{stepsize } \Delta = y_1 \tag{4.3.8}$$

There is a certain **endpoint** effect. To take this into account we define the virtual decision level x_{N+1} by

$$x_{N+1} = W + \frac{\Delta}{2} \tag{4.3.9}$$

and with this definition, we can guarantee that the quantization error, ε, which is the difference between the input, **x**, and the output, **y**, is bounded by

$$|\varepsilon| = |\mathbf{x} - \mathbf{y}| \le \frac{\Delta}{2} \tag{4.3.10}$$

This virtual decision level, x_{N+1}, is the crashpoint of the quantizer.

If the **pdf** of the input is not uniform, then the uniform quantizer is not optimal. That is, in general, the uniform quantizer does not provide the least possible quantization noise power. One approach to determining a nonuniform quantizer that is "better" than a uniform one is to consider manipulating, or compressing the input prior to quantization. For example, consider the transformation

$$\eta = g(\mathbf{x}) \qquad (4.3.11)$$

that converts the variable **x**, into the variable η. We assume that the mapping is one-to-one, symmetric, and maps the input range of (– , +) into the range (–1/2, +1/2)

$$g(-x) = g(x);\ g(0) = 0;\ g(\infty) = \frac{1}{2} \qquad (4.3.12)$$

The **pdf** of η can be obtained as

$$p_\eta(\eta) = \left[\frac{p_\mathbf{x}(x)}{\left|\partial g(x)/\partial x\right|}\right]_{x = g^{-1}(\eta)} \qquad (4.3.13)$$

The intent of this mapping is that if η is a uniform random variable then we know that the optimal quantizer for η will be a uniform quantizer. By mapping the decision levels and output values for this uniform quantizer back to the variable **x** via $\mathbf{x} = g^{-1}(\eta)$ we obtain a nonuniform quantizer for **x** that will be, hopefully, better than a uniform one.

Consider the example where **x** has a **pdf** of the exponential form given by

$$p_\mathbf{x}(x) = \frac{\lambda}{2} e^{-\lambda|x|} \qquad (4.3.14)$$

This particular **pdf** is very close to the **pdf** for speech signals. If η is to be uniform, then

$$\frac{\partial g(x)}{\partial x} = p_\mathbf{x}(x);\ x > 0 \qquad (4.3.15)$$

which implies that

$$g(x) = \frac{1}{2} e^{-\lambda x} + \text{constant};\ x > 0 \qquad (4.3.16)$$

From endpoint considerations we can evaluate the constant and get the forward mapping

$$\eta = g(x) = \frac{1}{2}(1 - e^{-\lambda x});\ x > 0 \qquad (4.3.17)$$

and the inverse mapping

$$x = g^{-1}(\eta) = -\frac{1}{\lambda}\ln[1 - 2\eta];\ 0 \le \eta < \frac{1}{2} \tag{4.3.18}$$

for positive values of **x** and η. Symmetry is used to obtain the equivalent for negative values of **x** and η.

The mapping between **x** and η for the case N = 7 is shown in Fig. 4.7. The optimal (uniform) quantizer for η is defined by the breakpoints $\{\eta_i\}$ and the output values $\{\mu_i\}$. These map into breakpoints $\{x_i\}$ and output values $\{y_i\}$ for the nonuniform quantizer for **x**. Though x_8 is actually infinite, it is shown as a finite value for convenience (to make the curve resemble a compressing characteristic discussed in the next section)

It is tempting to assume that this procedure provides an optimal quantizer for **x**. Unfortunately this is not strictly true; the breakpoints $\{x_i\}$ and output values $\{y_i\}$ do not satisfy the conditions for optimality laid down in Eqs. (4.3.2) and (4.3.3). With some difficulty it is possible to show that the inverse mapping provides a nearly optimal quantizer if the function $g^{-1}(\eta)$ is (approximately) linear in the region of each step. When we have several output levels, corresponding to a large number of bits in the encoding process, this condition will be satisfied.

The reason the optimal quantizer, or the near-optimal quantizer obtained by the mapping approach, keeps the quantizing error small can be explained in the following manner. Consider the size of the step $(x_{i+1} - x_i)$. In the uniform quantizer this step size is the same regardless of where the input sample value lies within the quantizer input range. Since the quantizing error is somewhat proportional to this step size, the quantizer introduces error "uniformly" over the input range. In the case of the optimal quantizer this step is not constant. In fact, writing

$$(\eta_{i+1} - \eta_i) \approx (x_{i+1} - x_i)\left[\frac{\partial g(x)}{\partial x}\right]_{x = y_i} \tag{4.3.19a}$$

$$(x_{i+1} - x_i) \approx (\eta_{i+1} - \eta_i)\left[\frac{1}{p_{\mathbf{x}}(x)}\right]_{x = y_i} \tag{4.3.19b}$$

we see that the step size is, loosely speaking, inversely proportional to the probability of seeing an input sample in the range of the step in question. Thus the optimal quantizer uses a larger number of output values in the region where the probability of seeing an input sample is high, and, conversely, uses a sparse number of output values where the probability of seeing an input sample is low.

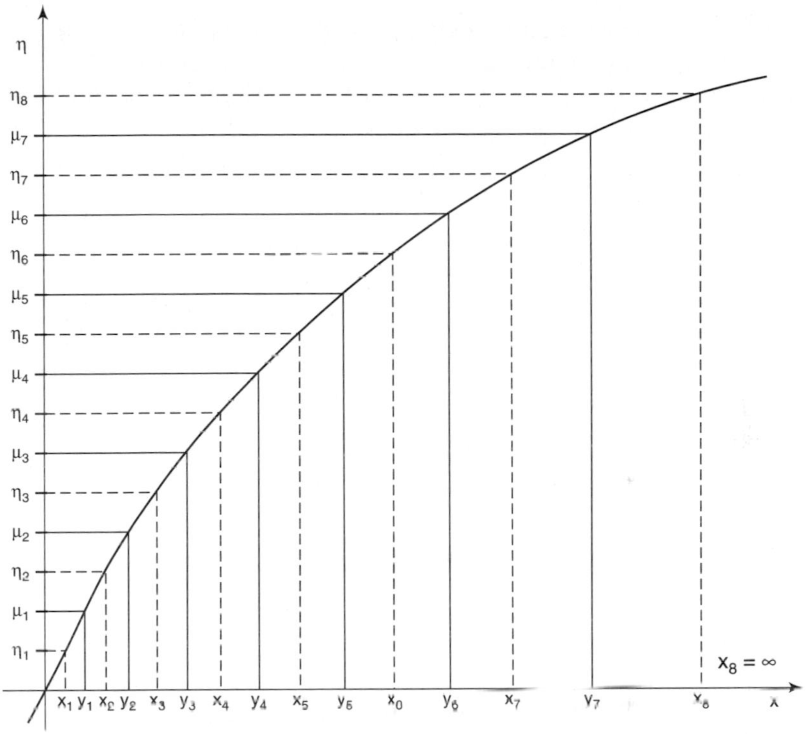

Fig. 4.7
Mapping method for obtaining a near-optimal characteristic from a uniform quantizer

The optimal quantizer obtained either by solving the determining equations Eqs. (4.3.2) and (4.3.3), or by using the inverse-mapping technique, suffers one principal drawback. The quantizer parameters are tightly coupled to the shape of the **pdf** as well as the width of the **pdf** (the variance of **x**). Thus we need to know the shape of the **pdf** as well as the variance of **x** to determine what the optimal breakpoints and output values should be. In the case of single-parameter **pdf**s, such as the Gaussian or exponential (we always assume zero mean), it is possible to devise an optimal quantizer for the ***normalized*** case, which scales up or down in a proportional fashion with respect to the standard deviation (square root of the variance). Such a change of quantizer characteristic is associated with adaptive quantizers that alter the scale of the quantizer, preserving the ratios of the parameters, in accordance to some rule that takes into account the power of the signal being quantized. Such a technique cannot be used if the quantizer is to be fixed (nonadaptive).

4.4 COMPANDING CODECS

In telephony, the notion of "optimal" is not related to maximizing the signal-to-noise ratio but, rather, maximizing the range of signal power over which the **SNR** is greater than some prescribed value. This is achieved by using a nonuniform transformation ("compressing") on the signal prior to applying a uniform quantizer characteristic. Conversion back to analog requires that the signal be "expanded." The technique is called companding. The concept can be explained with reference to Fig. 4.8.

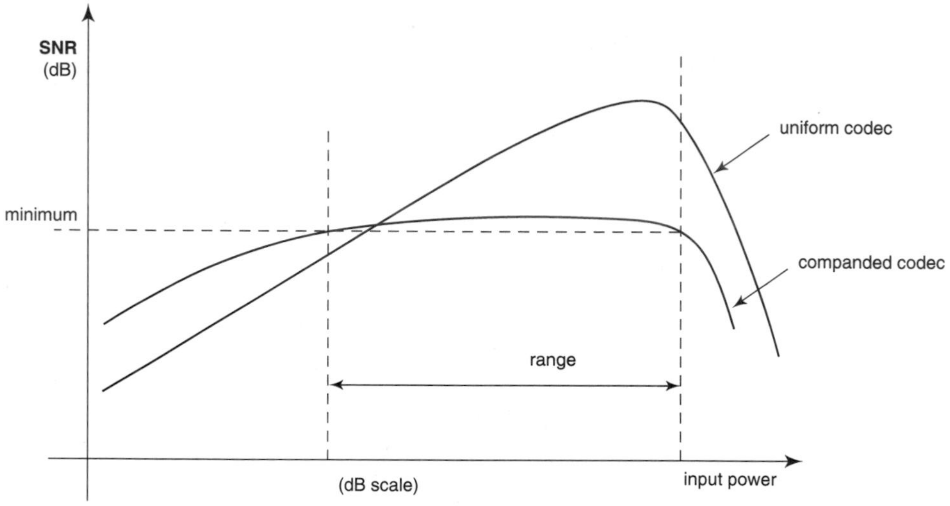

Fig. 4.8
SNR as a function of signal power for uniform and companded codecs.

The signal-to-noise ratio (**SNR**) associated with a uniform codec appears as a line with some fixed slope. This is because the noise power is fixed, determined by the uniform step size. The desired behavior is a curve that is "flat," with a nominally constant **SNR** over a wide range of input signal power. This is the notion of "optimal" as applied to companding codecs.

Both the optimal companding codec and the optimal (minimum noise) codec share the following property. The number of output levels is large where the majority of input samples are "expected" and a sparse number of output values is provided where the probability of seeing input samples is small. The shape of the compression characteristic (compare with g(x)) typically takes the form shown in Fig. 4.9. One interpretation of the underlying principle in companding codecs is that the power of the compressed signal is reasonably constant over a wide range of powers of the uncompressed signals. The compressed signal is quantized in a uniform manner and thus, considering that the quantizing power is constant, we expect the signal-to-noise ratio to follow the same constant behavior. This desired property of the compression characteristic is evident from Fig. 4.8.

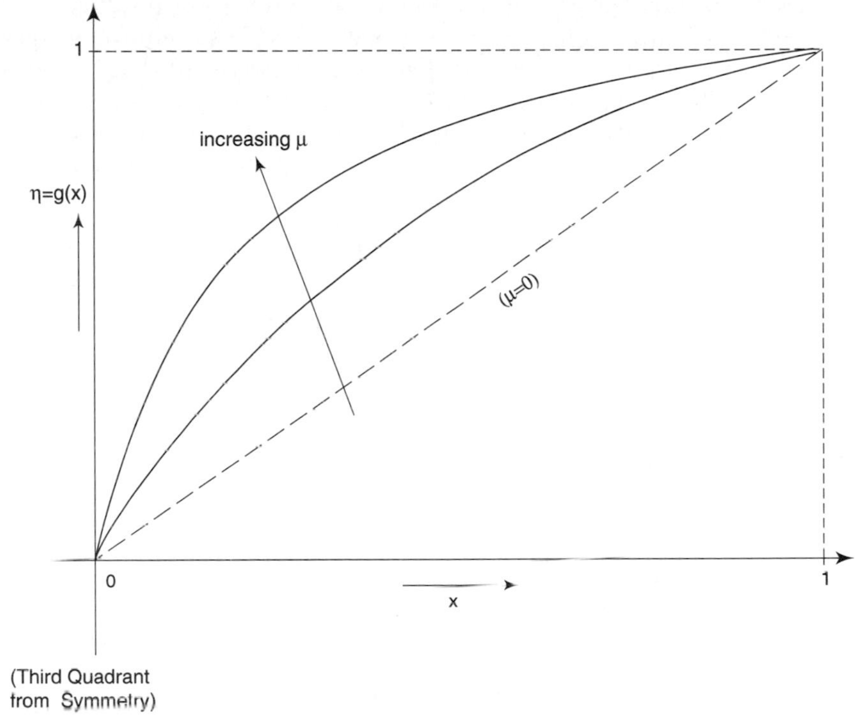

Fig. 4.9
Compression characteristic following Eq. (4.4.1).

Fig. 4.9 depicts a compression characteristic that follows

$$\eta = g(x) = \frac{\ln[1 + \mu|x|]}{\ln[1 + \mu]} \tag{4.4.1}$$

Such a characteristic assumes that there is a maximum associated with the input signal, normalized to unity. The μ-law codec characteristics employed in telephony applications is an approximation to this characteristic; specifically, a piece-wise-linear approximation to the logarithmic function of Eq. (4.4.1). The A-law characteristic is also a piece-wise-linear approximation to a logarithmic compression curve similar to Eq. (4.4.1). The maximum value of the input signal is assumed to be (approximately) 3 dBm0. That is, a sinusoidal signal of maximum amplitude will have a power of 3 dBm0 (3 dBm measured at a 0 **TLP**). The μ-law and A-law differ very slightly in the specification of maximum value; the μ-law assumes 3.17 dBm0 and the A-law 3.14 dBm0.

The piece-wise-linear form of the compression characteristic is depicted in Fig. 4.10 (not to scale). Each region where the curve is linear is called a ***segment***. The width of the segments on the η axis is uniform, corresponding to a uniform codec, while the width of the segments on the x axis is not uniform, increasing by a factor of 2 from segment to segment. For the A-law segments 0 and 1 have the same slope, i.e., width. If one considers the negative axes as well, segment 0 has a corresponding segment –0, which is contiguous in slope and is thus considered the "same" segment. Thus the μ-law is often called a 15-segment companding coder while the A-law is called a 13-segment companding coder.

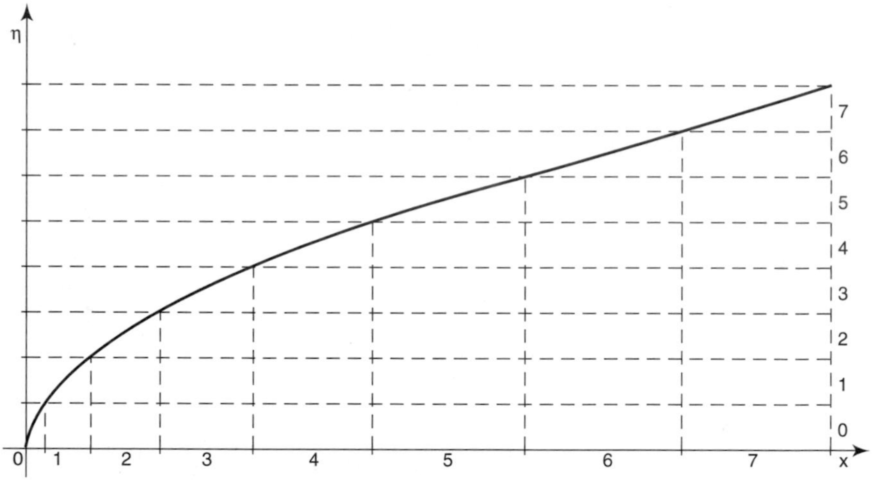

Fig. 4.10
Approximating the compression characteristic of Fig. 4.9 by straight lines in each of 15 segments (not to scale).

4.4.1 The μ-Law Coder

The μ-law coder employs a mid-tread characteristic, implying that zero is one of the allowed output values. From the viewpoint of symmetry, it is convenient to distinguish between "minus zero" and "plus zero," an artificial distinction that arises because the μ-law uses a sign-magnitude representation of the output values. Within each segment the quantizer characteristic is uniform and each segment has the same number of steps, 16. Consequently the step-size increases from segment 0 through segment 7, by a factor of two from segment to segment.

The number of bits used in the encoding of the output value into a digital code is 8. The eight bits can be viewed in the manner depicted in Fig. 4.11.

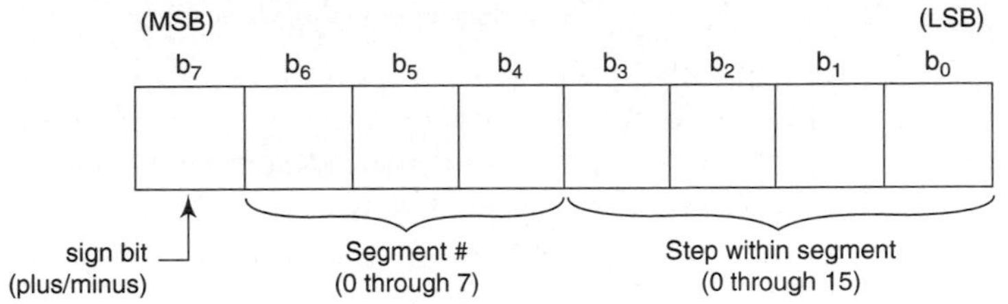

Fig. 4.11
Bit allocation in μ-law codeword

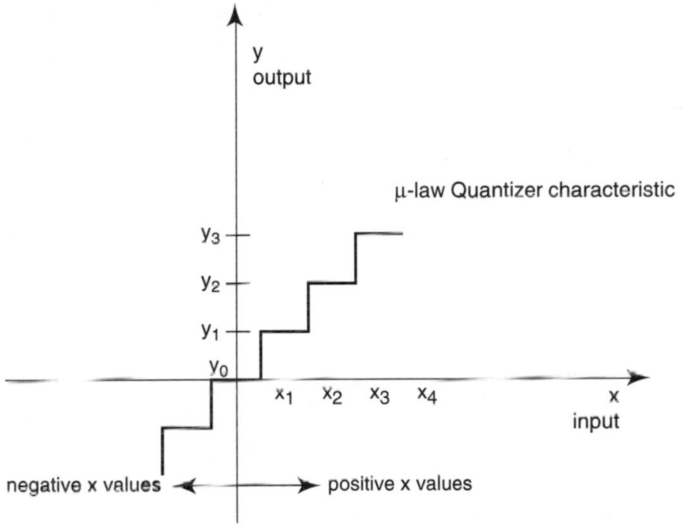

Fig. 4.12
μ-law quantizing characteristic for small values of input signal (mid-tread).

The most-significant bit indicates the sign of the sample value, the next 3 bits point to the segment, and the 4 least-significant bits identify one of 16 possible output values within that segment.

The μ-law codec is an industry standard and has been spelled out in complete detail in several publications such as **CCITT** Recommendation **G.711** [1.1]. The manner in which the codec is described warrants some explanation. Fig. 4.12 shows the quantizer input-output characteristic for a few steps around the origin. The mid-tread nature is evident. The codec is specified by providing numerical values for each of the decision levels, $\{x_i\}$ and each of the output values $\{y_i\}$. The convention followed forces these values to be integers. That is, the values are normalized. This is depicted in Table 4.1.

Table 4.1
Specification of the μ-law quantizer

μ-law Characteristic
Positive Values

Segment 0		Segment 1		•••	Segment 7	
Break Points	Output Values	Break Points	Output Values		Break Points	Output Values
	$y_0 = 0$		$y_{16} = 33$	•••		$y_{112} = 4191$
$x_1 = 1$		$x_{17} = 35$			$x_{113} = 4319$	
	$y_1 = 2$		$y_{17} = 37$			$y_{113} = 4447$
$x_2 = 3$		$x_{18} = 39$			$x_{114} = 4575$	
	$y_2 = 4$		$y_{18} = 41$	•••		$y_{114} = 4703$
$x_3 = 5$		$x_{19} = 43$			$x_{115} = 4831$	
	$y_3 = 6$		$y_{19} = 45$			$y_{115} = 4959$
$x_4 = 7$		$x_{20} = 47$			$x_{116} = 5087$	
•	•	•	•		•	•
•	•	•	•		•	•
•	•	•	•		•	•
$x_{15} = 29$		$x_{31} = 91$			$x_{127} = 7903$	
	$y_{15} = 28$		$y_{31} = 93$	•••		$y_{127} = 8031$
$x_{16} = 31$		$x_{32} = 95$			$(x_{128} = 8159)$	
Step Size = 2		Step Size = 4			Step Size = 256 x_{128} is a virtual decision value	

The breakpoints $\{x_i\}$ can be obtained using the following algorithm for positive input values. The breakpoints for negative input values are obtained by applying symmetry.

For segment **I**, **I**=0, 1, 2, 3, 4, 5, 6, 7:

The 16 break points are x(16**I**+k); k = 1, 2, 3, ... , 16, where

$x(16\mathbf{I}+k) = T(\mathbf{I}) + kD(\mathbf{I})$; k=0, 1, 2, ... , 16

The initial value and step size for each segment are:

$T(0) = 1$; $D(0) = 2$
$T(1) = 35$; $D(1) = 4$
$T(2) = 103$; $D(2) = 8$
$T(3) = 239$; $D(3) = 16$
$T(4) = 511$; $D(4) = 32$
$T(5) = 1055$; $D(5) = 64$
$T(6) = 2143$; $D(6) = 128$
$T(7) = 4319$; $D(7) = 256$

The maximum x-value, $x_{128} = 8159$, is a virtual decision value and corresponds to the amplitude of a 3.17 dBm0 sinusoid. That is, a sinewave of amplitude 8159 (normalized units) will have a power of 3.17 dBm0 (3.17 dBm at a 0 **TLP**).

The output values $\{y_i\}$ are computed from the breakpoints in the following manner for positive inputs. The negative output values are obtained using symmetry.

y(0) = 0;

For **I** = 1, 2, 3, ... , 127,

y(**I**) = 0.5(x(**I**) + x(**I**+1));

With this definition, the μ-law code word is constructed by first encoding the index of y_k, i.e., k, as a binary number between 0 and 127 and appending the most significant bit to match the sign (0 for positive). This yields a code word as per Fig. 4.13.

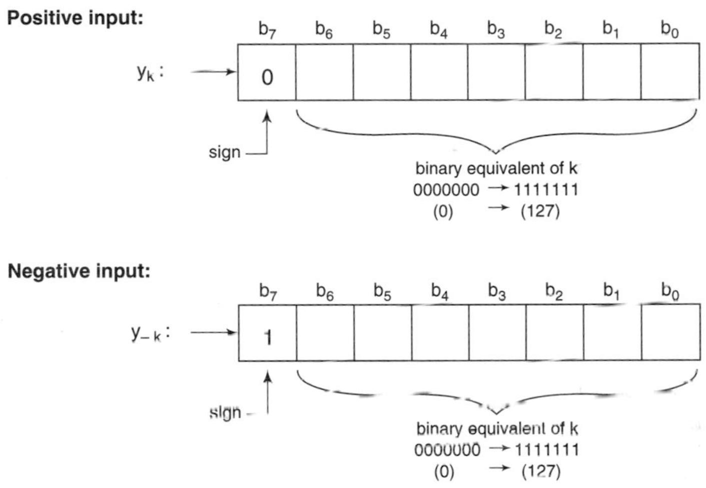

Fig. 4.13
Method for associating a μ-law codeword with a numeric level

The actual μ-law code word is obtained by inverting all eight bits of the pattern obtained in the intermediate step of Fig. 4.13.

4.4.2 The A-Law Coder

The A-law coder has a fundamental difference from the μ-law coder in that the A-law quantizer characteristic is of the mid-riser variety. The input-output relation around the origin is shown in Fig. 4.14.

Fig. 4.14 shows the quantizer input-output characteristic for a few steps around the origin. The mid-riser nature is evident. The codec is specified by providing numerical values for each of the decision levels $\{x_i\}$ and each of the output values $\{y_i\}$. The convention followed forces these values to be integers. That is, the values are normalized. This is depicted in Table 4.2. The description in Table 4.2 is somewhat unconventional but is chosen to show a parallel between the μ-law and A-law characteristics.

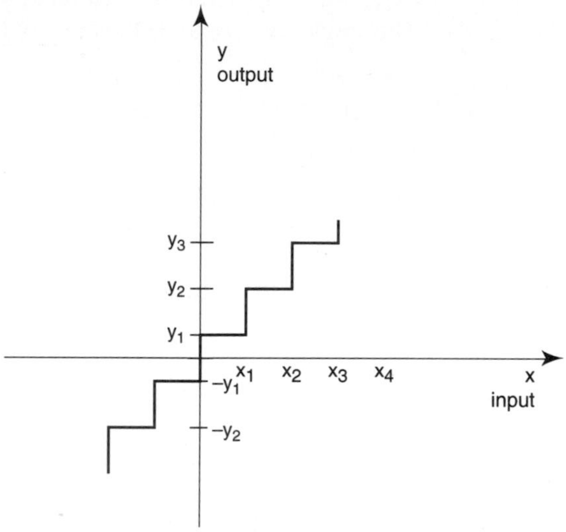

Fig. 4.14
A-law quantizer characteristic for small values of input signal (mid-riser)

The breakpoints $\{x_i\}$ can be obtained using the following algorithm for positive input values. The breakpoints for negative input values are obtained by symmetry.

For segment **I**, **I**=0, 1, 2, 3, 4, 5, 6, 7:
The 16 break points are x(16**I**+k); k = 0, 1, 2, 3, ... , 15, where
x(16**I**+k) = T(**I**) + kD(**I**); k=0, 1, 2, ... , 15.

The initial value and step size for each segment are:

T(0) = 0 ;	D(0) = 2
T(1) = 32 ;	D(1) = 2
T(2) = 64 ;	D(2) = 4
T(3) = 128 ;	D(3) = 8
T(4) = 256 ;	D(4) = 16
T(5) = 512 ;	D(5) = 32
T(6) = 1024 ;	D(6) = 64
T(7) = 2048 ;	D(7) = 128

The maximum x-value, x_{128} = 4096, is a virtual decision value and corresponds to the amplitude of a 3.14 dBm0 sinusoid. That is, a sinewave of amplitude 4096 (normalized units) will have a power of 3.14 dBm0 (3.14 dBm at a 0 **TLP**).

The output values $\{y_i\}$ are computed from the breakpoints in the following manner for positive inputs. The negative output values are obtained using symmetry.

y(1) = 1;
For **I** = 2, 3, ... , 128,
y(**I**) = 0.5(x(**I**) + x(**I**–1));

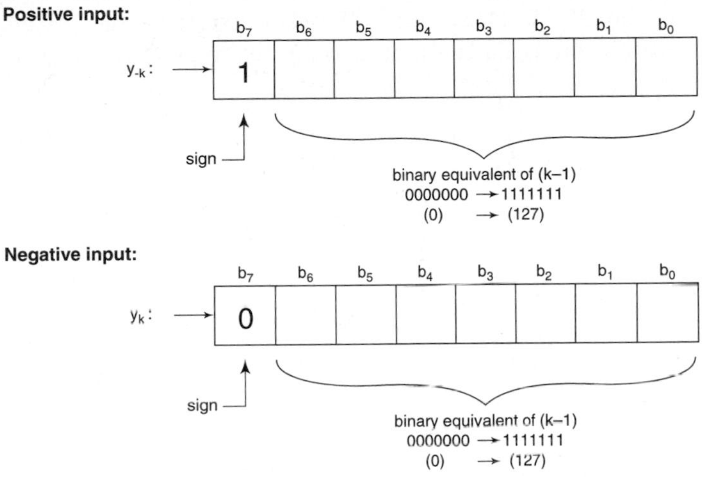

Fig. 4.15
Method for associating an A-law codeword with a numeric level

Table 4.2
Specification of the A-law quantizer characteristic

A-Law Characteristic
Positive Values

Segment 0		Segment 1		Segment 2		•••	Segment 7	
Break Points	Output Values	Break Points	Output Values	Break Points	Output Values		Break Points	Output Values
$x_0 = 0$		$x_{16} = 32$		$x_{32} = 64$			$x_{112} = 2048$	
	$y_1 = 1$		$y_{17} = 33$		$y_{33} = 66$	•••		$y_{113} = 2112$
$x_1 = 2$		$x_{17} = 34$		$x_{33} = 68$			$x_{113} = 2176$	
	$y_2 = 3$		$y_{18} = 35$		$y_{34} = 70$			$y_{114} = 2240$
$x_2 = 4$		$x_{18} = 36$		$x_{34} = 72$			$x_{114} = 2304$	
	$y_3 = 5$		$y_{19} = 37$		$y_{36} = 74$			$y_{115} = 2368$
$x_3 = 6$		$x_{19} = 38$		$x_{35} = 76$			$x_{115} = 2432$	
•	$y_4 = 7$	•	$y_{20} = 39$	•	$y_{38} = 78$		•	$y_{116} = 2496$
						•••		
•	•	•	•	•	•		•	•
•	•	•	•	•	•		•	•
	•		•		•			•
$x_{15} = 30$		$x_{31} = 62$		$x_{47} = 124$			$(x_{128} = 4096)$	
	$y_{16} = 31$		$y_{32} = 63$		$y_{48} = 126$	•••		$y_{127} = 4032$
Segment 1 Step Size = 2				Segment 2 Step Size = 4			Segment 7 Step Size = 128 x_{128} is a virtual decision value	

With this definition, the A-law codeword is constructed by first encoding the index of y_k, actually (k–1), as a binary number between 0 and 127 and appending the most-significant bit to match the sign (0 for positive). This yields a code word as per Fig. 4.15.

The actual codeword used for transmission purposes is obtained by inverting the alternate bits of the 8-bit word obtained in the intermediate step of Fig. 4.15.

4.4.3 Performance of μ-Law and A-Law Coders

The degradations suffered due to quantization will be quantified, as usual, in terms of signal-to-noise ratio (**SNR**). We shall make the artificial distinction that signal levels above –60 dBm0 are "active" and that signal powers below –60 dBm0 correspond to an "idle channel" where the signal is probably background noise (no speech present).

Generating a computer program to compute the **SNR** as a function of signal level for a given **pdf** is quite straightforward. Fig. 4.16 and Fig. 4.17 show the variation of **SNR** for the Gaussian and exponential cases, respectively. The figures also include the **CCITT** quantization noise mask (specified for Gaussian inputs). It is clear from the figures that both the μ-law and A-law codecs meet the requirement with some margin (Gaussian case). The assumption of an exponential **pdf**, which has a very high peak-to-rms ratio indicates a violation of the mask at high signal levels. For this reason most transmission systems are engineered to maintain a nominal maximum power of –16 to –10 dBm0 for speech circuits.

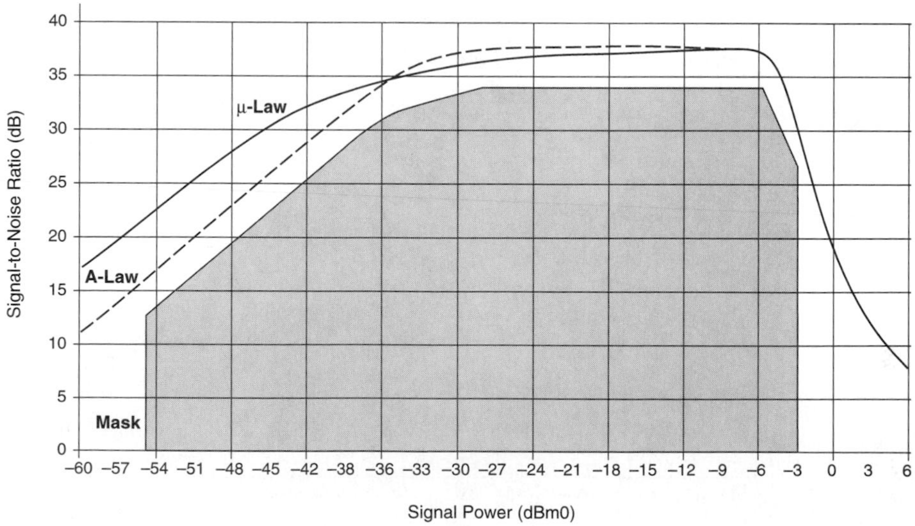

Fig. 4.16
SNR versus signal level for the Gaussian **pdf**

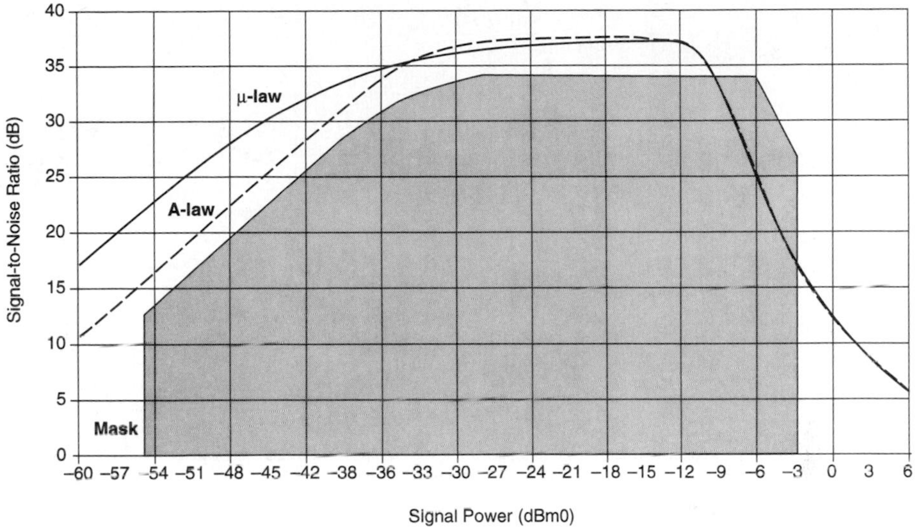

Fig. 4.17
SNR versus signal level for the exponential **pdf**

It is instructive to establish an equivalence between the 8-bit companding codecs and certain uniform codecs. Fig. 4.18 shows, on the same chart, the **SNR** behavior, assuming a Gaussian **pdf**, of the μ-law and 13-bit uniform codecs. Fig. 4.19 does the same for the A-law and 12-bit uniform codecs.

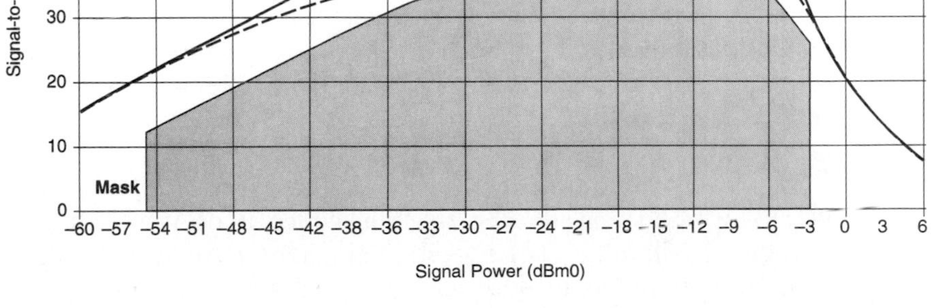

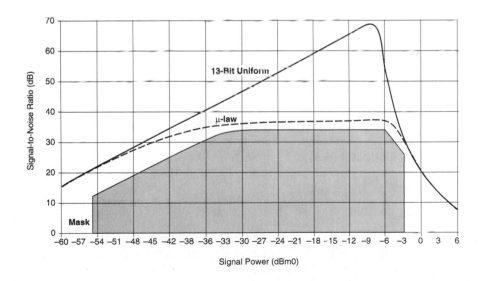

Fig. 4.18
μ-law and 13-bit uniform codec performance assuming a Gaussian **pdf**

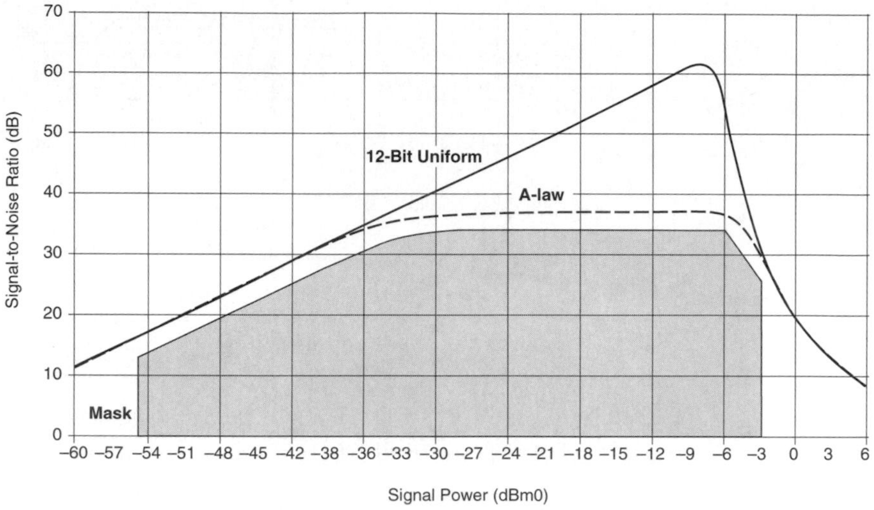

Fig. 4.19
A-law and 12-bit uniform codec performance assuming a Gaussian **pdf**

Note that at high signal levels all four quantizers exhibit the same **SNR** since in this region the distortion is dominated by the error due to clipping (saturation). At medium signal levels, from about –10 dBm0 to about –40 dBm0, the companded codecs are approximately equivalent and the **SNR** is approximately flat at about 35 dB. At low signal levels the μ-law coder and the 13-bit uniform coder are about the same, whereas the A-law coder mimics the performance of a 12-bit uniform coder.

It appears that the companding codecs are a subset of a 13-bit uniform codec. This statement can be "verified" by analyzing the hypothetical cases depicted in Fig. 4.20, which show a code conversion being performed. That is, the analog-to-digital conversion is performed by one codec and the output digital code mapped into the possible output values of the other codec by assigning the closest available choice. The **SNR** curves for the configuration of Fig. 4.20(a) are practically identical to the curves in Fig. 4.16 and Fig. 4.17 for the Gaussian and exponential **pdf**s. An immediate inference is that if code conversions are performed, the overall **SNR** is typically that of the companding codec. Using a 14-bit uniform coder instead of 13-bits in the scheme of Fig. 4.20(a) yields a very small improvement in overall **SNR**. This improvement is less than 1.7 dB or 0.8 dB for the μ-law or A-law intermediate codes, respectively.

The conversion error defined in Fig. 4.20(b) is zero. That is, for either uniform codec (13- or 14-bit) and either companding codec (μ- or A-law), the companded code is reproduced faithfully. The overall degradation introduced by the chain analog-companded-uniform-companded-analog is that of the companding codec itself. The intermediate conversion to uniform is not apparent.

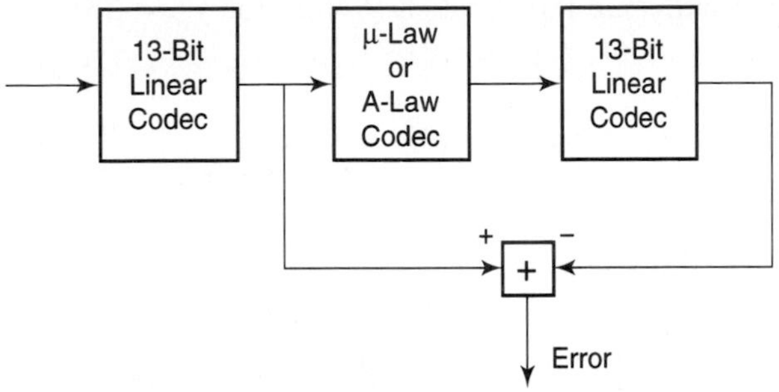

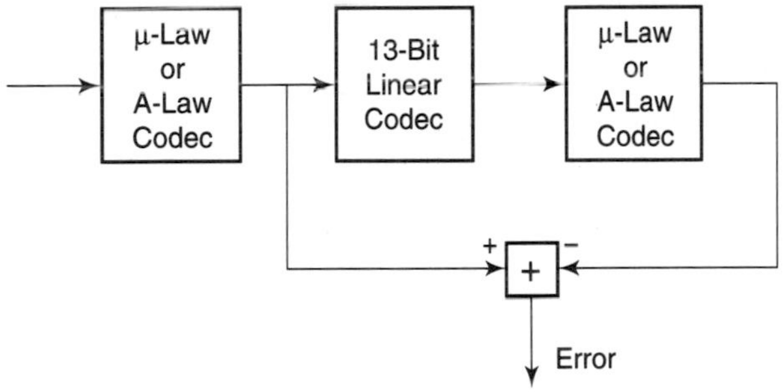

Fig. 4.20
Depicting the error associated with mapping between uniform and companded coding laws

An explanation for these observations relating the companding codecs to the 13-bit uniform coder is apparent from Fig. 4.21, which shows the input-output characteristics of the three codecs in question for small values of input. In the first segment, where the companding codecs have the greatest density of output levels, the A-law is a subset of the 13-bit uniform coder and the μ-law and 13-bit codecs are equivalent.

In the second (third, fourth, etc.) segment, the μ-law has one-half (one-fourth, one-eighth, etc.) as many output levels as the 13-bit codec. For each 13-bit output level, however, there is a unique "closest" μ-law output level. Although the mapping from 13-bit to μ-law is many-to-one, it is unambiguous. The mapping of 13-bit output levels into A-law output levels is ambiguous for those 13-bit output levels that lie exactly midway between A-law output levels (a consequence of the A-law being mid-riser). This "ambiguity" is equivalent to a DC offset of one-half of the smallest A-law step.

Notice that certain μ-law output values, such as $y_{16} = 33$ normalized units do not have an exact counterpart in the case of the 13-bit uniform coder (which will have output values of 32 and 34 in the same normalized units). This should not be construed as an incompatibility since every "missing" μ-law output level is flanked by two 13-

bit levels. If the 13-bit codes are mapped into the μ-law codes, both these flanking levels will map into the "missing" level (both output values 32 and 34 for the 13-bit codec will map into the μ-law code for $y_{16} = 33$).

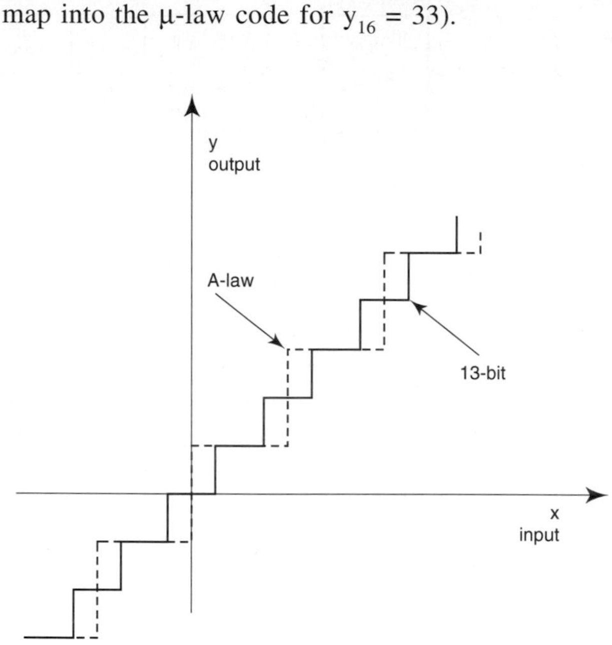

Fig. 4.21 (a)
Comparison of A-law and 13-bit uniform quantizers for small values of input signal

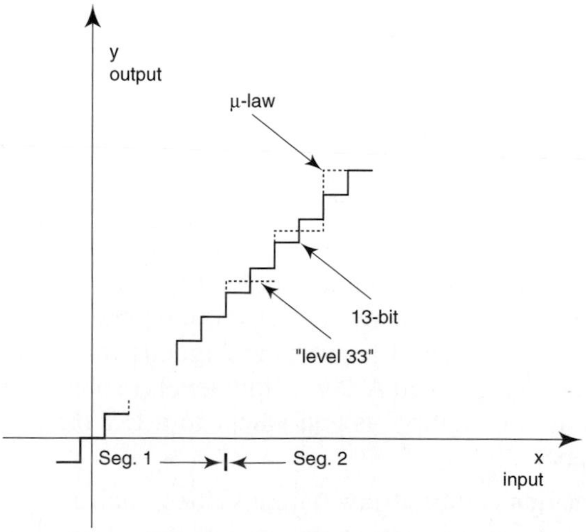

Fig. 4.21 (b)
Comparison of μ-law and 13-bit uniform quantizers for small values of input signal

4.4.4 Idle Channel Behavior of μ-Law and A-Law Codecs

When the signal is absent at the input, the codec is said to be in an ***idle state***. In this state the actual input is a weak interference signal such as noise or crosstalk and we do not expect the coder to quantize the signal with great accuracy. However, we do require that the codec not amplify this noise. In order to analyze the behavior of a coder in the idle state we shall model the signal as a Gaussian white noise process of variance (power) σ^2.

The dependence of the output noise power on the input power is shown in Fig. 4.22. Only input powers in the range less than –60 dBm0 are considered as idle states. It is seen that for the A-law the output power does not drop below –66 dBm0 as the input power is reduced. This phenomenon is due to the mid-riser nature of the quantizer characteristic. In contrast, the mid-tread quantizers demonstrate the property of squelching low level signals. A sharp drop in output noise level is observed when the input noise level drops below –78 dBm0 for the 13-bit (and μ-law) case and –84 dBm0 for the 14-bit case. Just prior to the drop, the output level can be greater than the input level. This amplification, or enhancement, is noticeable only when the noise power σ^2 lies in the range

$$-\frac{\delta^2}{16} \le \sigma^2 \le \frac{\delta^2}{4} \tag{4.4.2}$$

where δ is the step size in that region. Heuristically, when $\sigma < \delta/4$, most samples of a Gaussian distribution lie within $[-\delta/2, \delta/2]$, the dead zone of a mid-tread quantizer for which the output is zero. Samples between $\delta/2$ and δ are mapped into δ, experiencing a gain. When $\delta/2 > \sigma > \delta/4$, a reasonable number of samples experience this amplification, enough to influence the average.

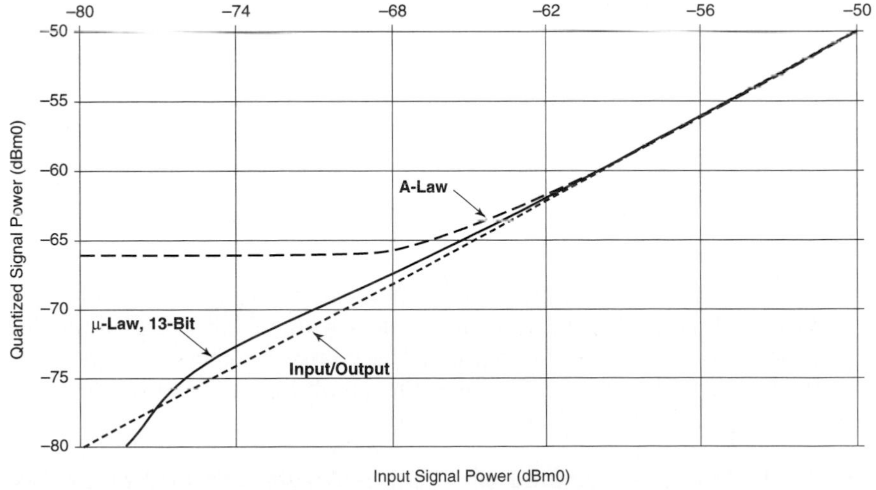

Fig. 4.22
Behavior of various codecs for low input power (idle channel)

4.4.5 Concluding Remarks on μ-Law and A-Law Codecs

Based on our discussion of companded codecs it is clear that if we start with a uniform coder and then use mapping techniques to mimic μ-law and A-law codecs then we require an effective (uniform) quantization level of 13- or 12-bits, respectively. Conversely, μ-law and A-law codecs may be used in applications calling for a quantization level of 13- or 12-bits, respectively, over a wide range of signal power.

Most commercially available codecs, especially those produced in large volume, build companded codecs in a "conventional" manner, by employing a bank of comparators to determine which output level is closest to the input sample. These codecs also include the anti-aliasing and replicate-rejection filters. Though specified to operate at a sampling rate of 8 kHz (nominal), they could probably operate at higher (or lower) sampling rates. If the built-in filters utilize switched-capacitor technology then the effective frequency response will scale proportionally to the sampling rate. Consequently, with just an adjustment to the clock rate (i.e., the master clock that is internally divided down to the sampling rate of 8 kHz, nominally) we could use a μ-law or A-law in other applications. Clearly, such a codec will be vastly superior to a uniform codec of the same word size, namely 8 bits. The companding characteristic, however, is matched to signals that have a large peak-to-rms ratio, such as speech, and would be "nonlinear." The impact of this nonlinearity must be considered before deciding whether a companded codec is applicable in any particular (nontelephony) application. Considering that a conventional μ-law or A-law codec filter ("combo" as they are often called when the unit has both the codec and the lowpass filters) costs but a fraction of that of a uniform 13/12 bit codec, such a study is well worth the effort.

4.5 FINITE-WORDLENGTH EFFECTS IN DIGITAL FILTERS

The impact of finite wordlength can be divided into two categories. First, since the coefficients of the filter are quantized, the implemented filter will be different from the target filter and the achieved frequency response will be different from the desired one. Second, the arithmetic operations would not be exact; each arithmetic operation performed will have an associated error, called roundoff *noise*. The first effect is deterministic. Since we know what the quantized coefficients are, we can, in principle at least, evaluate how different the achieved frequency response is from the target or desired response. The second effect is necessarily dependent on the values of the signal being processed and is described in a statistical manner.

In this section we treat roundoff noise as the (small) error introduced by the inability to represent very small numbers with a finite wordlength. The companion problem of the inability to represent very large numbers is beyond the scope of this book. We will attempt to minimize the possibility of overflow by appropriate scaling and assume that overflow does not occur.

4.5.1 Noise Associated with Finite-Wordlength Arithmetic Operations

All digital filter operations are necessarily implemented by ***adders*** and ***multipliers*** and ***storage elements*** that are of finite wordlength (**FWL**). This gives rise to distortions or inaccuracies that are generically referred to as ***roundoff noise*** when the imperfections are introduced because of lack of precision. That is, roundoff noise is related to the inability of a physical implementation to represent small numbers, those that correspond to a significance level ***less*** than that of the least-significant bit (**LSB**). Finite wordlength also implies that there is a maximum value representable. The error involved with the inability to represent a quantity that is too large is called ***overflow*** and the deleterious impact of overflow can be catastrophic. Roundoff noise is akin to the quantization noise associated with A/D conversion and wordlength reduction is troublesome but not usually catastrophic. The science of ***scaling*** is the attempt by the designer to achieve a compromise—minimize the roundoff noise, relative to the signal level, while maintaining a very, very, low probability of overflow, for a given finite wordlength.

Scaling is usually considered a problem with fixed-point , or integer, arithmetic. Floating-point processors have a built-in safeguard against overflow (up to a point) at the expense of having roundoff noise that varies with signal level. From a standpoint of economics, fixed-point implementations are less expensive but require greater ingenuity on the part of the designer. Consequently, we shall assume that the implementations we discuss use fixed-point arithmetic.

To establish the impact of **FWL** it is necessary to have a consistent convention for representing numbers. For our discussion we will assume the ***2-s complement*** method for number representation defined below via Eq. (4.5.1). Further, all signal samples will be considered as integers and all coefficients will be considered as fractions. This convention is purely for convenience. Any consistent convention will yield the same analytical results. The advantage in representing sample values as integers is that all roundoff effects have a maximum value of "1." An alternate convention is to consider all values as fractions; the roundoff effects then have a maximum value of 2^{-B}, where B is the wordlength used to represent the fractional part. This alternate convention is perfectly applicable when the wordlength is known. For a general discussion, the integer convention for sample values is more convenient.

Let N be the wordlength associated with signals. Then all memory locations, or registers storing sample values, will be N bits wide. The implication is that if intermediate computations are done using a larger wordlength, then these would have to be quantized, that is, the least significant bits discarded, in order to store these values. It is this discarding of bits, albeit in the least-significant positions, that gives rise to roundoff noise. The N bits associated with signal samples are interpreted in the manner shown in Fig. 4.23. Each bit, b_i, is either 0 or 1. Our convention of assuming signal samples as integers is equivalent to placing the binary point as shown in the figure. The numerical value represented by the N bits is given by (the 2-s complement system)

$$X = -b_{N-1}2^{N-1} + \sum_{i=0}^{N-2} b_i 2^i \qquad (4.5.1)$$

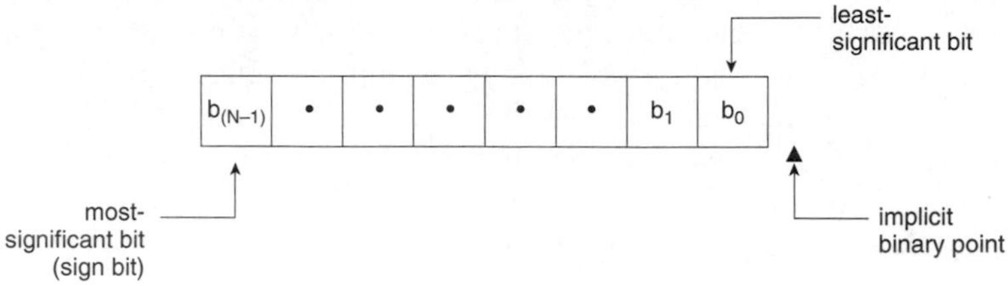

Fig. 4.23
Interpretation of an N-bit word as an integer

The largest positive value that can be represented is $(2^N - 1)$ and will correspond to the code 011 ... 111; the largest negative value that can be represented is -2^N. corresponding to the code 100 ... 000. Attempting to represent numbers outside this range results in an overflow condition. This overflow condition can create severe problems. For example if we add 1 to the most positive number, and retain the same number of bits, we get a code that represents the most negative number!

If we wish to increase the wordlength for signals, without changing the numerical value, we can do so in two ways. One is to pad zero bits to the right of the implied binary point to yield a mixed integer/fraction. Alternatively we "sign-extend" the word by replicating the sign bit to the left (more significant bits). The latter method maintains the notion of an integer and allows for (later) growth in numerical value.

Suppose the coefficient wordlength is M. In most hardware implementations using **DSP** processors, the same wordlength is used for both signal and coefficient representation. In specialized implementations built out of separate arithmetic and storage elements, it is not uncommon for the wordlengths to be different. The M bits associated with a coefficient are interpreted as shown in Fig. 4.24 and the corresponding numerical value given by

$$C = -c_0 + \sum_{i=1}^{M-1} c_i 2^{-i} \quad \text{with } |C| \le 1 \qquad (4.5.2)$$

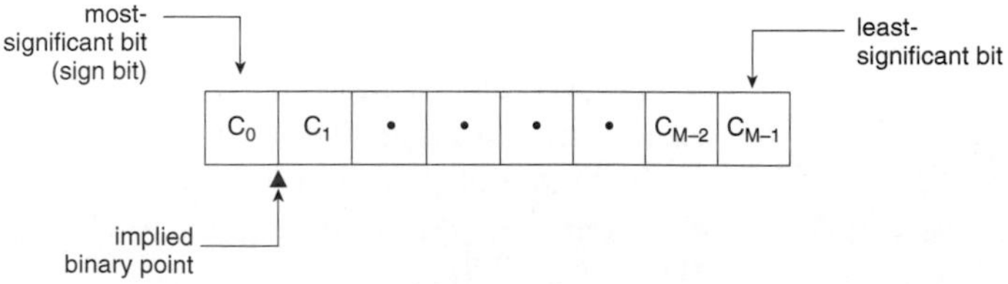

Fig. 4.24
Interpretation of an M-bit word as a fraction

The impact of finite wordlength can be explained using an example of the *typical* operation involved in digital signal processing. Suppose

$$y = a x + w \tag{4.5.3}$$

In words, the signal sample corresponding to x (N bits) is multiplied by a coefficient a (M bits) and the sample w (N bits) is added to the product. The sum is restricted to N bits and stored as the resulting sample y (N bits). This operation can be achieved in the manner shown in Fig. 4.25, which is representative of what would happen when implemented in a **DSP** processor.

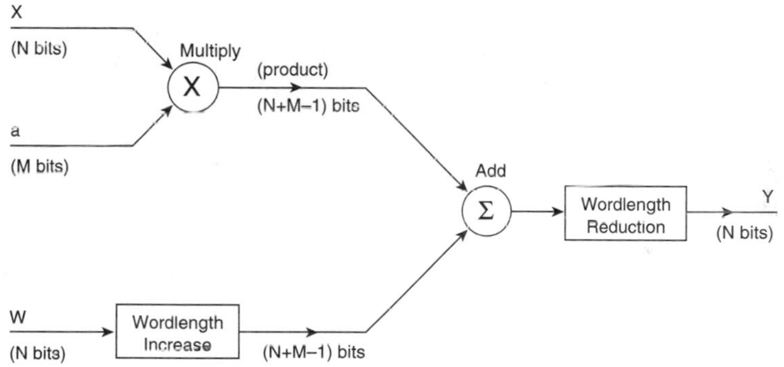

Fig. 4.25
Flow of operations in computing $y = a x + w$

If the true sum $(a x + w)$ has a value less than $2^{(N-1)}$ in magnitude, then the wordlength reduction involves the discarding of the (M–1) bits to the right of the binary point, while retaining the N bits to the immediate left of the binary point. This is the "normal" mode of operation. What if the true sum $(ax + w)$ has a value greater that $2^{(N-1)}$ in magnitude? Then, strictly speaking, we cannot represent the value in N bits and certain exception handling procedures are required. Most **DSP** processors provide for additional bits internal to the arithmetic unit that would allows signal values to exceed the limit of $2^{(N-1)}$. This allows us to handle overflow in a rational manner. The recommended approach, when overflow has occurred, is to assign the most positive or most negative number representable by N bits to the output (of the wordlength reduction block). This subsumes that we know that an overflow has occurred and for this we need additional bits in the number representation internal to the arithmetic unit. The DSP5600x signal processing chip from Motorola, for example, provides 8 additional bits of ***headroom*** to handle any interim values that might otherwise cause an overflow.

Assuming that overflow does not occur, the overall scheme can be modeled as shown in Fig. 4.26. The impact of wordlength reduction is treated as an additive noise signal ε. By our convention, ε corresponds to the fractional part of the true value of y. Thus the additive noise signal takes on values in the range

$$-1 + 2^{-M} \leq \varepsilon \leq 0 \tag{4.5.4}$$

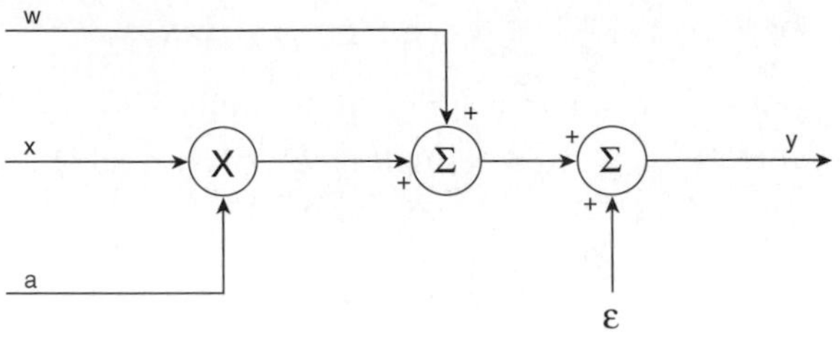

Fig. 4.26
Wordlength reduction modeled as an additive noise

As in the case of (analog) quantization noise, we model this roundoff noise as a uniform random variable with probability density function given by

$$p_{\varepsilon}(e) = \begin{cases} 1 & \text{for } -1 \leq e \leq 0 \\ 0 & \text{otherwise} \end{cases} \tag{4.5.5}$$

and is characterized by a mean value of 0.5 and a variance of (1/12). The nonzero mean arises because of the 2–s complement system. Clearly, the discarded quantity has a value given by

$$\varepsilon = \sum_{i=1}^{M-1} \alpha_i 2^{-i} \tag{4.5.6}$$

where the indices i = 1, 2, ... , M correspond to the M discarded bits to the right of the binary point. This value is always positive. This process of summarily discarding the least significant bits is called ***truncation***. Other methods of wordlength reduction include ***rounding*** whereby if $\alpha_1 = 1$ then 1 is added to the retained integer part. The error **pdf** for rounding is uniform over [0.5 , 0.5] and hence the mean of the additive noise is zero. The variance is still (1/12). The simplicity of truncation makes it the method of choice, the bias of the nonzero mean notwithstanding.

Analysis of roundoff noise relates to the study of the impact of each internal wordlength reduction operation on the final output of the digital filter. In the case of a direct-form **FIR** filter there is a single (logical) multiply-accumulate function. In the case of **IIR** filters implemented in cascade form, the error introduced by one stage is filtered by subsequent stages and thus the noise power estimate must incorporate this processing. We make the following assumptions:

i) The roundoff-error introduced is ***white***. In the case of truncation, which has a non-zero mean, we assume that the roundoff-error is the sum of a constant (zero-frequency) component and a white noise, zero-mean, process.

ii) Multiplication by a power of 2 does not introduce any roundoff error.

iii) The roundoff-error is ***uncorrelated*** with the signal(s) and coefficients.

4.5.2 Scaling in Digital Filters

When digital filters are implemented using finite wordlength, we try to optimize the ratio of signal power to the power of the roundoff noise. This involves a trade-off with the probability of arithmetic overflow. The optimization is achieved by introducing gain (or loss) in such a manner as to keep the signal level as high as possible without letting it get too large.

4.5.2.1 The Principle of Scaling

Consider the hypothetical example depicted in Fig. 4.27(a). An input signal, x(n), represented by B_1 bits, is processed by a arithmetic block corresponding to a fixed gain of K and the output is to be represented using B_2 bits. This can be modeled by the flowgraph of Fig. 4.27(b). In terms of signal processing the gain of K is not really accomplishing much other than changing the signal level. Denote by w(n) the intermediate product, Kx(n). The signal-to-noise ratio associated with {w(n)}, assuming we could represent it with great precision, is the same as that associated with {x(n)}. Other than absolute level, {w(n)} and {x(n)} are "identical." This is one of the basic precepts of scaling—a flat gain does not count as signal modification. The wordlength reduction to B_2 bits is considered an additive noise signal as shown in Fig. 4.27(c).

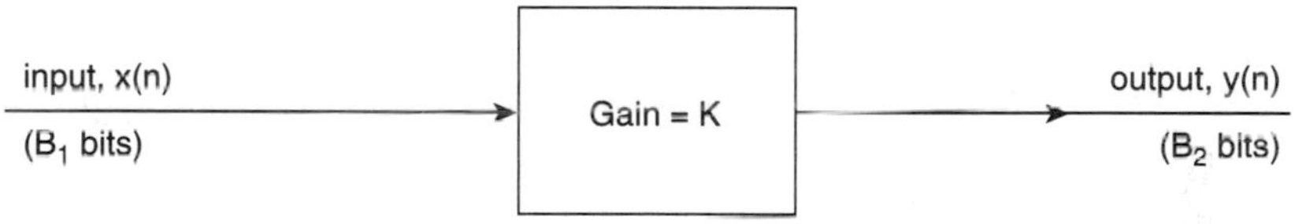

Fig. 4.27 (a)
Nominal view of a gain, K, in a discrete-time system

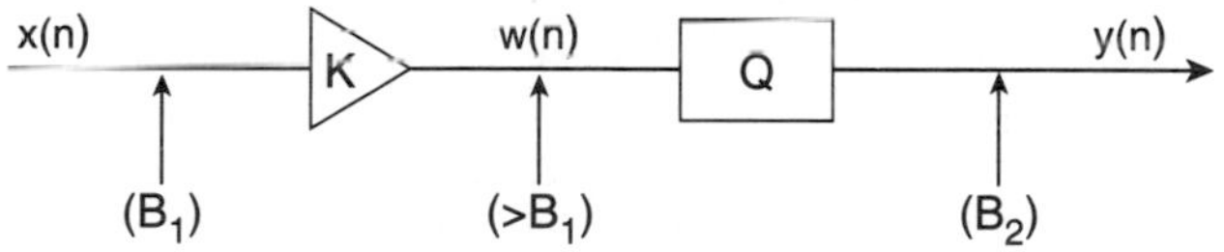

Fig. 4.27 (b)
Actual effect associated with a gain, K, in a digital system indicating the effect of finite wordlength

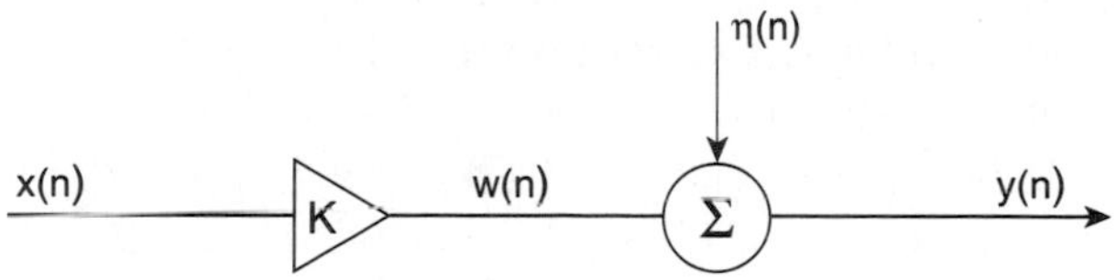

Fig. 4.27 (c)
Viewing the gain, K, as an ideal gain as in (a) with the finite-wordlength effect of (b) modeled as an additive noise

The **SNR** associated with $\{y(n)\}$ is clearly different because of the wordlength reduction to B_2 bits. Based on our convention of signal representation, the additive noise has a power of (1/12). We quantify the impact of this noise addition by the signal-to-noise ratio

$$(\mathrm{SNR})_y \cong \frac{K^2 \sigma_x^2}{\frac{1}{12} + K^2 \sigma_e^2} \tag{4.5.7}$$

which assumes $\{x(n)\}$ contains a noise component of power σ_e^2 that, after multiplication by K, adds to the roundoff noise. Eq. (4.5.7) predicts that there will be a degradation in **SNR**. The exception to this rule is when the gain K is a power of 2, 2^k, and implementing the multiplication involves a left shift of the bits. In this case there is a danger of overflow but there is no roundoff noise and the signal-to-noise ratio is unchanged. For a given input **SNR** it is clear that the degradation is worse when K is small. Since the noise power is fixed at (1/12), this **SNR** can be maximized by using as large a gain K as is feasible. This leads to the following rule of thumb. To achieve an overall gain of $K = K_1K_2$, it is advantageous to implement the larger of K_1 and K_2 first. Usually, but not always, one of the gains will be a (positive) power of 2 and clearly should be implemented first provided there is no danger of overflow.

The gain, K, cannot be unbounded since there is an inherent maximum value imposed by the finite wordlength. The wordlength reduction, obtaining y(n) from w(n), can be described by

$$\begin{aligned} w(n) \geq (2^{B_2-1}-1) &\Rightarrow y(n) = (2^{B_2-1}-1) \\ |w(n)| < (2^{B_2-1}-1) &\Rightarrow y(n) = \mathrm{Int}[w(n)] \\ w(n) < -(2^{B_2-1}-1) &\Rightarrow y(n) = -(2^{B_2-1}-1) \end{aligned} \tag{4.5.8}$$

In the discussion of quantizers it was seen that saturation causes a drastic reduction in **SNR**. To prevent this from happening, or at least keep the probability of this happening small, we limit the value of K. One way of determining the maximum value of K is to start from what the maximum value of the input can be. Since we have assumed that the input is represented by B_1 bits, the maximum value of $|x(n)|$ is $(2^{B_1} - 1)$, and thus a safe bound for K is

$$K \leq \frac{(2^{B_2-1}-1)}{(2^{B_1-1}-1)} \approx 2^{B_2-B_1} \tag{4.5.9}$$

If $B_2 > B_1$ then K is a power of 2 and there is no associated roundoff noise;

if $B_2 < B_1$ then K is a fractional value and roundoff noise cannot be avoided. As a "special" case, the situation where $B_1 = B_2$ seems trivial but in reality is fundamental to the understanding of scaling in digital filters. When the input wordlength and output wordlength are the same, then the *gain* should be unity or thereabouts.

For **FIR** filters the notion of scaling is somewhat easier than for **IIR** filters. In **IIR** filters, because of the recursive nature, the output is fed back and arithmetic overflow in computing an output value can thus be catastrophic; using saturation arithmetic, while overflow maybe not catastrophic, it is still very problematic. In nonrecursive filters the output is not fed back. Thus arithmetic overflow, or saturation, affects only a single output sample and thus, while not desirable, is not catastrophic (depending, of course, on the application).

4.5.2.2 Scaling a First-Order IIR Filter

For actual digital filters we have to extend the notion of the simple gain considered in Fig. 4.27(a) to the more general case depicted in Fig. 4.28. The notion of gain is necessarily different because the system is not a simple multiplier but represented by a difference equation. In order to bridge the gap between the two we shall consider a simple example and then generalize from there.

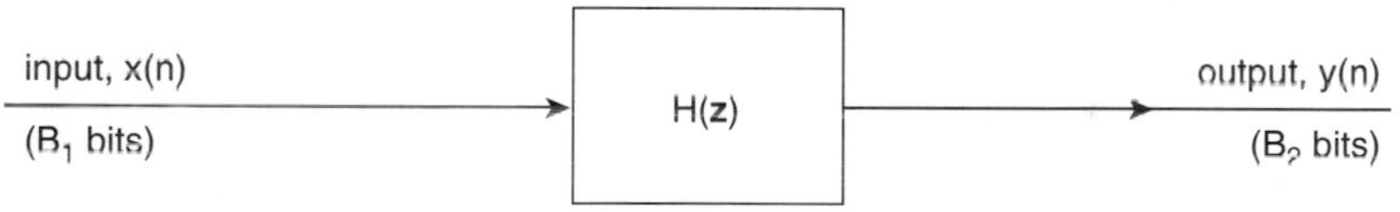

Fig. 4.28
Transfer function representation of a general digital filter

Suppose $H(\mathbf{z})$ is the first-order section described by

$$H(\mathbf{z}) = \frac{1}{1 - a\mathbf{z}^{-1}} \tag{4.5.10}$$

Stability requires that $|a| < 1$. What is actually implemented is the structure shown in Fig. 4.29, which includes a gain of K that is associated with scaling. The intention of including this gain is to minimize the probability of overflow; the gain is such as to ensure that the values of y(n) will not exceed $(2^{B_2} - 1)$ in magnitude. The behavior of the structure is therefore described by

$$G(\mathbf{z}) = KH(\mathbf{z}) = \frac{K}{1 - a\mathbf{z}^{-1}} \tag{4.5.11}$$

Recalling our earlier discussion, the *gain* of $G(\mathbf{z})$ must be (approximately) unity or less to avoid overflow and be as large as possible to preserve our signal-to-noise ratio. What is the gain of $G(\mathbf{z})$? The notion of gain as applied to scaling is a measure of how the signal "grows" in strength as a result of the filtering action. This notion of growth of signal level depends on the assumption made as to type of input $\{x(n)\}$ and based on this assumption we can define a suitable measure of signal growth.

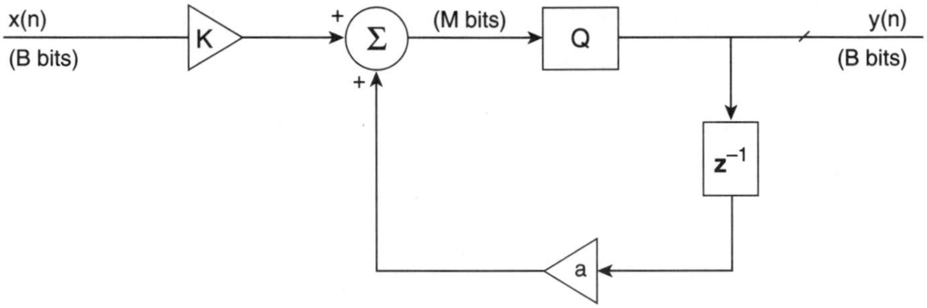

Fig. 4.29
Implementation of a first-order filter indicating the use of a scaling constant K

Suppose that $\{x(n)\}$ is a sinusoidal signal of frequency ω_0 (normalized radian frequency). Then the amplitude of the output is a factor of $|G(e^{j\omega_0})|$ greater than that of the input. For such signals an appropriate measure of *gain,* for scaling purposes, is

$$G_\infty = \max_{\omega} \left| G(e^{j\omega}) \right| \tag{4.5.12}$$

which, for the first-order filter under consideration, is

$$G_\infty = \frac{K}{1-|a|} \tag{4.5.13}$$

Thus if the signals being considered are principally sinusoidal in nature then a suitable scaling factor is given by

$$K_\infty \le (1-|a|) \tag{4.5.14}$$

If no specific knowledge of $\{x(n)\}$ is available, then we take the following approach. Since, in principle, we can obtain the impulse response of the filter (which is quite simple in the first-order case), the output can be written as

$$y(n) = \sum_{k=0}^{\infty} g(k)\,x(n-k) \tag{4.5.15}$$

and consequently bounded in value by

$$|y(n)| \le \Big[\sum_{k=0}^{\infty} |g(k)|\Big]\,[\max |x(n)|] \tag{4.5.16}$$

From this viewpoint we can define *gain* as

$$G_1 = \left[\sum_{k=0}^{\infty} |g(k)| \right] \tag{4.5.17}$$

indicating how the maximum value could "grow," in a worst-case scenario. For the first-order filter under consideration the value of G_1 is computed as

$$G_1 = \frac{K}{1-|a|} \tag{4.5.18}$$

Only in the first-order case is $G_1 = G$. In the general case $G_1 > G$. The scale factor using this definition of gain is K_1, where

$$K_1 \leq (1-|a|) \tag{4.5.19}$$

A third measure of growth is to treat $\{\mathbf{x}(n)\}$ as white noise of power σ_x^2 and in which case the power of the output, σ_y^2, is given by

$$\sigma_y^2 = \sigma_{\mathbf{x}}^2 \frac{1}{2\pi j} \oint_C G(\mathbf{z}) G(\mathbf{z}^{-1}) \mathbf{z}^{-1} d\mathbf{z} = \sigma_{\mathbf{x}}^2 \sum_{k=0}^{\infty} |g(k)|^2 \tag{4.5.20}$$

Viewing "growth" in terms of the increase in signal power, we can define *gain* as (the power gain)

$$G_2 = \sum_{k=0}^{\infty} |g(k)|^2 \tag{4.5.21}$$

For the first-order case this is simply

$$G_2 = \frac{K}{1-|a|^2} \tag{4.5.22}$$

corresponding to a scale factor of K_2 where

$$K_2 \leq \sqrt{(1-|a|^2)} \tag{4.5.23}$$

Unlike the simple multiplier, the quantizer error gets fed back in the case of a recursive filter. Modeling the quantizer error as additive noise, we can represent the filter output as coming from two sources. One is the actual input signal and the other is the error. This is depicted in Fig. 4.30. The negative sign associated with $\varepsilon(n)$ is to reinforce the notion that, with 2–s complement representations, the act of truncation always discards a positive value, corresponding to an error that is negative. Of special importance is the fact that the scaling factor K affects the input but not the quantizer error. This implies that we should keep K as large as is feasible.

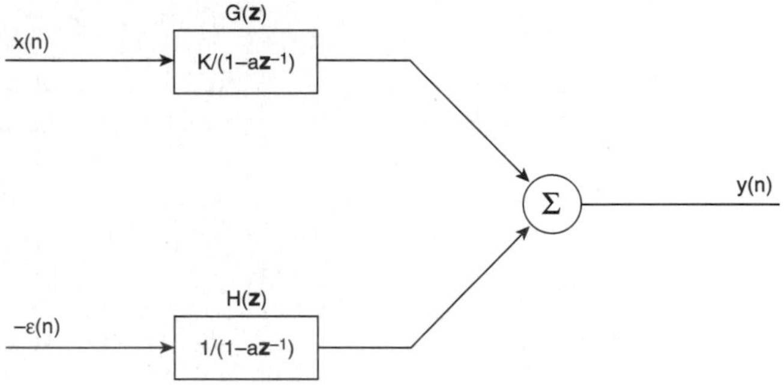

Fig. 4.30
Model of the roundoff noise for the scaled first-order filter shown in Fig. 4.29

Representing the output $\{y(n)\}$ as the sum of the desired signal $\{y_D(n)\}$ and a noise component, $\{\eta(n)\}$,

$$y(n) = y_D(n) + \eta(n) \tag{4.5.24}$$

the overall deleterious effect of quantization can be expressed in terms of the mean and variance of $\{\eta(n)\}$. If the quantizer is implemented as a straightforward truncation operation then $\{\mathbf{e}(n)\}$, and thus $\{\eta(n)\}$, has a nonzero mean. The mean and variance of $\{\eta(n)\}$ are evaluated as

$$\begin{aligned} m_\eta &= E\{\eta(n)\} = -\frac{1}{2}[H(\mathbf{z})]_{\mathbf{z}=1} \\ \sigma_\eta^2 &= \frac{1}{12}\frac{1}{2\pi j}\oint_C H(\mathbf{z})H(\mathbf{z}^{-1})\mathbf{z}^{-1}d\mathbf{z} \end{aligned} \tag{4.5.25}$$

In Eq. (4.5.25) we use the result that the mean and variance of $\{\mathbf{e}(n)\}$ are (1/2) and (1/12), respectively. Evaluating H(**z**) at **z**=1 provides the DC gain of H(**z**) and the contour integral expresses the noise gain or increase in variance of the input (white) noise. For the first-order case,

$$\begin{aligned} m_\eta &= -\frac{1}{2}\frac{1}{1-a} \\ \sigma_\eta^2 &= \frac{1}{12}\frac{1}{1-a^2} \end{aligned} \tag{4.5.26}$$

To maximize the signal-to-noise ratio we keep K, the scaling factor, as large as possible. Choosing K based on the *gain* expressed in Eq. (4.5.17) guarantees no saturation and is the *most conservative* of scaling factors. However it is the worst choice from the **SNR** viewpoint. Choosing K based on the *gain* expressed in Eq. (4.5.21) does not guarantee no saturation but is the best from the viewpoint of **SNR**. The third expression for gain, in Eq. (4.5.14), is somewhere in-between. In most cases $K_2 > K \quad > K_1$.

Architecting filter implementations based on the notion of *accumulation* introduces the possibility of reducing the impact of quantization error. The process of quantization involves keeping the most significant bits of the accumulant and discarding the fractional (least significant bits) part. This fractional part is actually available and can be extracted and stored and used in subsequent computations. For example, the normal sequence of events in implementing the first-order structure is expressed as:

Clear Accumulator:	ACC = 0;
Get first product term:	ACC = ACC + K*X;
Get second product term:	ACC = ACC + a*Y1;
Generate output (Quantize):	Y = **MSB**{ ACC };

After the last operation the least-significant bits are still available. Thus a variation of the operation is:

Initialize Accumulator:	ACC = E;
(product terms)	
Generate output (Quantize):	Y = **MSB**{ ACC };
Save LSBs of accumulator:	E = **LSB**{ ACC };

In this variation the quantization error in a sampling interval is saved and fed back to the computation in the subsequent sampling interval. What this achieves is depicted in Fig. 4.31.

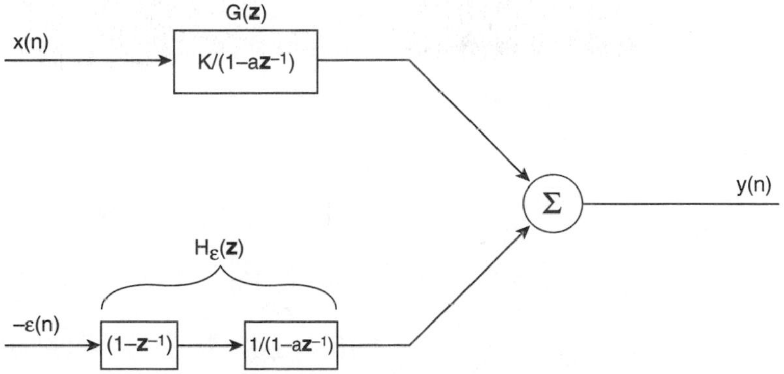

Fig. 4.31
Roundoff noise model with error feedback

The quantization error thus sees a transfer function with an additional $(1 - \mathbf{z}^{-1})$ compared to Fig. 4.30. The mean and variance of the quantization noise present in the output are computed as

$$m_\eta = -\frac{1}{2}\left[H_\varepsilon(\mathbf{z})\right]_{\mathbf{z}=1}$$

$$\sigma_\eta^2 = \frac{1}{12}\frac{1}{2\pi j}\oint_C H_\varepsilon(\mathbf{z})H_\varepsilon(\mathbf{z}^{-1})\mathbf{z}^{-1}d\mathbf{z} \tag{4.5.27}$$

First and foremost, the mean of the quantization error is forced to zero since $H_\varepsilon(\mathbf{z})$ has a transmission zero at DC. The impact on the variance depends on the transfer function being implemented. For the first-order filter, we can compute the variance of the noise power as

$$\sigma_\eta^2 = -\frac{1}{12}\frac{2}{1+a} \tag{4.5.28}$$

The ratio of roundoff noise power with feedback to the case without feedback is the improvement given by

$$\text{Improvement (ratio)} = \frac{2(1-a^2)}{1+a} = 2(1-a) \tag{4.5.29}$$

This technique is useful, corresponding to a reduction of noise power, only if the feedback shapes the noise spectrum appropriately. If the filter is lowpass the feedback should make the noise spectrum highpass and vice versa. This principle of feeding back the noise is applied in Delta Sigma Modulators studied in Chapter 8. Thus if a is a positive value (lowpass filter) then there is indeed an improvement is the sense that the noise power is reduced by the error feedback. If a is negative (highpass filter) then the error feedback of the form shown is contraindicated since the noise power increases because of the feedback.

4.5.2.3 Scaling Higher-Order IIR Filters

The application of scaling to higher-order filters is similar but has an additional twist. Suppose that we were trying to implement a fourth-order **IIR** filter. We will assume that the implementation will be of the cascade form and thus the desired transfer function can be written as

$$\begin{aligned} H(\mathbf{z}) &= \frac{(\alpha_0^1 + \alpha_1^1\mathbf{z}^{-1} + \alpha_2^1\mathbf{z}^{-2})(\alpha_0^2 + \alpha_1^2\mathbf{z}^{-1} + \alpha_2^2\mathbf{z}^{-2})}{(1 - \beta_1^1\mathbf{z}^{-1} - \beta_2^1\mathbf{z}^{-2})(1 - \beta_1^2\mathbf{z}^{-1} - \beta_2^2\mathbf{z}^{-2})} \\ &= \frac{N_1(\mathbf{z})}{D_1(\mathbf{z})}\frac{N_2(\mathbf{z})}{D_2(\mathbf{z})} = H_1(\mathbf{z})H_2(\mathbf{z}) \end{aligned} \tag{4.5.30}$$

and depicted by Fig. 4.32. The superscripts used in the notation are not "powers" but just indices to distinguish between the different second-order polynomials involved. The choice of negative sign in the denominator polynomials is for convenience only. $H_1(\mathbf{z})$ and $H_2(\mathbf{z})$ are not unique and depend on how the poles and zeros are associated with each other.

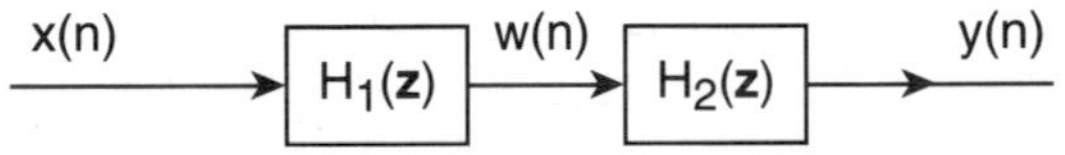

Fig. 4.32
Implementing a fourth-order **IIR** filter as the cascade of two second-order sections

Suppose we make the association

$$
\begin{aligned}
H_i(\mathbf{z}) &= K_i \frac{(\alpha_0^i + \alpha_1^i \mathbf{z}^{-1} + \alpha_2^i \mathbf{z}^{-2})}{(1 - \beta_1^i \mathbf{z}^{-1} - \beta_2^i \mathbf{z}^{-2})}\ ;\ i = 1, 2 \\
H(\mathbf{z}) &= K_0\, H_1(\mathbf{z})\, H_2(\mathbf{z})
\end{aligned}
\qquad (4.5.31)
$$

then the signal flow graph can take two forms shown in Fig. 4.33 and Fig. 4.34.

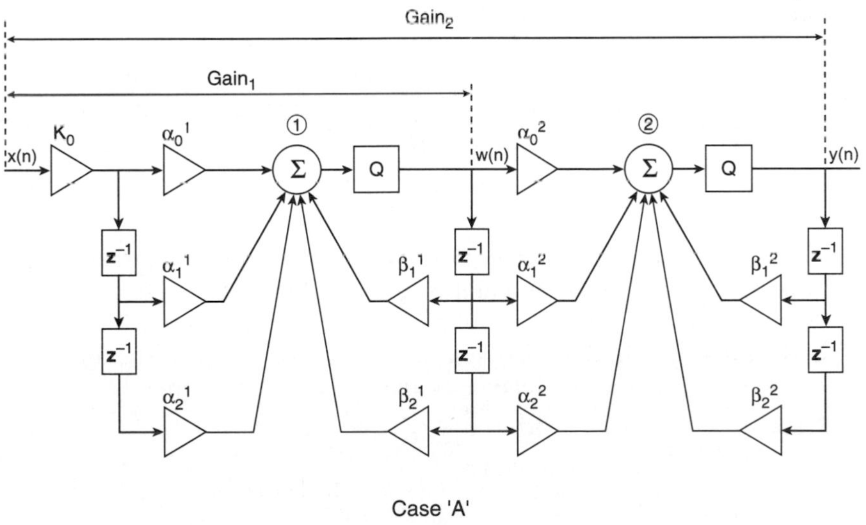

Fig. 4.33
Noncanonical implementation of two second-order sections. There are two points where roundoff noise gets injected

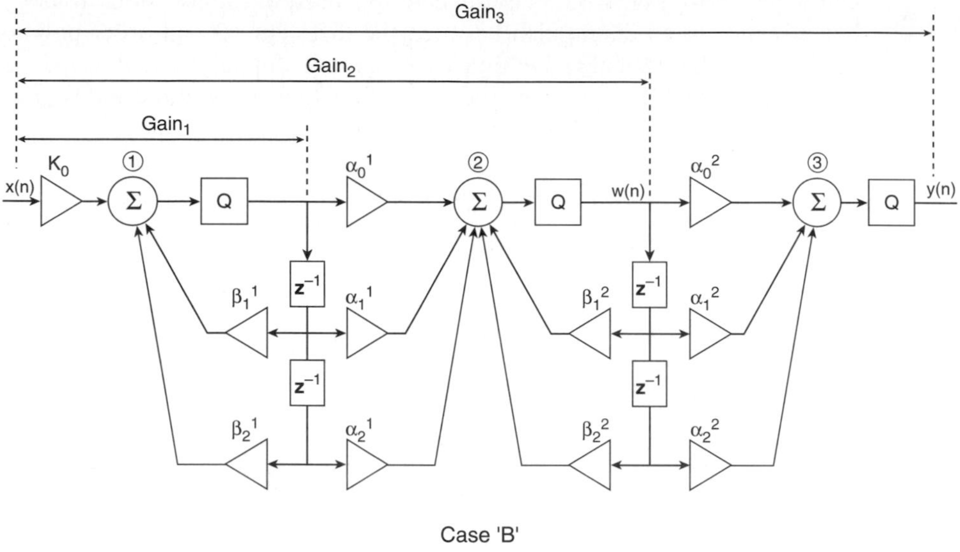

Fig. 4.34
Canonical implementation of two second-order sections. There are three points where roundoff noise gets injected

The *gain* terms represented by the K_i are the scaling factors that will be introduced. In the figures the actual coefficients of the numerator are modified by the scaling terms as

$$a_k^i = K_i \alpha_k^i \ ; \ i = 1, 2 \ ; \ k = 0, 1, 2 \tag{4.5.32}$$

The structure in Fig. 4.34 has the advantage of being canonical, using the least number of delay elements. The structure of Fig. 4.33 uses more delay elements but has fewer points where wordlength reduction takes place. Consequently, we have many permutations from which to choose the "best" implementation of $H(\mathbf{z})$.

The procedure of scaling involves choosing the scaling factors K_i such that the *gain* from the input to the point of quantization is unity. In case A (Fig. 4.33) we assume, without loss of generality, that $K_0 = 1$. As in the first-order case we can define this *gain* in three different ways. From a practical standpoint the gain based on frequency is the best choice in most cases. Consequently,

Case A:

$$K_1 \le \min_{\omega} \frac{|D_1(e^{j\omega})|}{|N_1(e^{j\omega})|} \tag{4.5.33a}$$

$$K_1 K_2 \le \min_{\omega} \frac{|D_1(e^{j\omega})||D_2(e^{j\omega})|}{|N_1(e^{j\omega})||N_2(e^{j\omega})|} \tag{4.5.33b}$$

and

Case B:

$$K_0 \le \min_{\omega} |D_1(e^{j\omega})| \tag{4.5.34a}$$

$$K_0 K_1 \le \min_{\omega} \frac{|D_1(e^{j\omega})||D_2(e^{j\omega})|}{|N_1(e^{j\omega})|} \tag{4.5.34b}$$

$$K_0 K_1 K_2 \le \min_{\omega} \frac{|D_1(e^{j\omega})||D_2(e^{j\omega})|}{|N_1(e^{j\omega})||N_2(e^{j\omega})|} \tag{4.5.34c}$$

In writing the above equations we assume that K_0 is chosen first, followed by K_1 and K_2.

The actual performance of the implementation is evaluated by modeling the quantization as an additive noise at each of the summation nodes labeled (1), (2), and (3) and calculating the resulting noise power at the output. To simplify the notation we define

$$\mathbf{I}\{G(\mathbf{z})\} = \frac{1}{2\pi j} \oint G(\mathbf{z}) G(\mathbf{z}^{-1}) \mathbf{z}^{-1} \, d\mathbf{z} \tag{4.5.35}$$

Then for the two cases the mean m_η and variance σ_η^2 of the roundoff noise present in the output are computed as:

Case A:

$$m_\eta = -\frac{1}{2}\left(\frac{K_2 N_2(\mathbf{z})}{D_1(\mathbf{z}) D_2(\mathbf{z})}\right)_{\mathbf{z}=1} - \frac{1}{2}\left(\frac{1}{D_2(\mathbf{z})}\right)_{\mathbf{z}=1} \tag{4.5.36a}$$

$$\sigma_\eta^2 = \frac{1}{12}\mathbf{I}\{\frac{K_2 N_2(\mathbf{z})}{D_1(\mathbf{z}) D_2(\mathbf{z})}\} + \frac{1}{12}\mathbf{I}\{\frac{1}{D_2(\mathbf{z})}\} \tag{4.5.36b}$$

and

Case B:

$$m_\eta = -\frac{1}{2}\left(\frac{K_1 K_2 N_1(\mathbf{z}) N_2(\mathbf{z})}{D_1(\mathbf{z}) D_2(\mathbf{z})}\right)_{\mathbf{z}=1} - \frac{1}{2}\left(\frac{K_2 N_2(\mathbf{z})}{D_2(\mathbf{z})}\right)_{\mathbf{z}=1} - \frac{1}{2} \quad (4.5.37a)$$

$$\sigma_\eta^2 = \frac{1}{12}\mathbf{I}\left\{\frac{K_1 K_2 N_1(\mathbf{z}) N_2(\mathbf{z})}{D_1(\mathbf{z}) D_2(\mathbf{z})}\right\} + \frac{1}{12}\mathbf{I}\left\{\frac{K_2 N_2(\mathbf{z})}{D_2(\mathbf{z})}\right\} + \frac{1}{12} \quad (4.5.37b)$$

These calculations are done for each possible pairing of zeros and poles and each possible association of pole-zero pair with $H_1(\mathbf{z})$ and $H_2(\mathbf{z})$. The chosen implementation will be the one deemed to have the least error measure that is some combination of the mean and variance of the overall roundoff noise in the output.

For higherorder filters greater than 4, there will be many permutations and deciding between different implementations requires a significant amount of computation. Further, the differences between the various choices may not be significant, especially when we consider the number of assumptions made to simplify the analysis. A "quick and dirty" way to establish a reasonable pairing and order is obtained using rules of thumb illustrated by the following example.

Consider the fourth-order filter with the poles and zeros depicted in Fig. 4.35. The poles are at, in polar notation, $r_1 \exp(j\theta_1)$ and $r_2 \exp(j\theta_2)$ along with the complex conjugates. The zeros shown are on the unit circle at angles ζ_1 and ζ_2. Such a pole-zero pattern is typical of a lowpass filter with its cutoff frequency between θ_1 and ζ_1 in normalized radian frequency units. A suitable pole-zero association to obtain second-order sections is to take the pole closest to the unit circle, p_1 in the figure, and associate with it the zero closest in frequency to the pole. The next closest pole to the unit circle is then associated with the closest remaining zero and so on. Thus we have second-order sections, $H_i(\mathbf{z})$, and the method described for associating poles and zeros tries to keep the associated G_i (see Eq. (4.5.14)) small.

Depending on whether we wish to use the implementation structure of Fig. 4.33 or that of Fig. 4.34, a slightly different philosophy is employed in deciding the order in which these second-order sections are placed.

i) The order in which these second-order sections are implemented, following the (non-canonical) structure of Fig. 4.33, is obtained by taking these gains G_i in ascending order. The section with the smallest G_i is made the first section (to the left in Fig. 4.33). Since the scaling factors are inversely proportional to these gains, we are implementing the larger multipliers first.

ii) For the structure in Fig. 4.34, the first section is chosen as the one that has the pole farthest away from the unit circle. Since the proximity to the unit circle is a measure of the Q of the pole (high Q is equivalent to high gain and narrow bandwidth), we arrange the second-order sections from left to right in ascending order of Q. Again, the intent is to implement larger multipliers first.

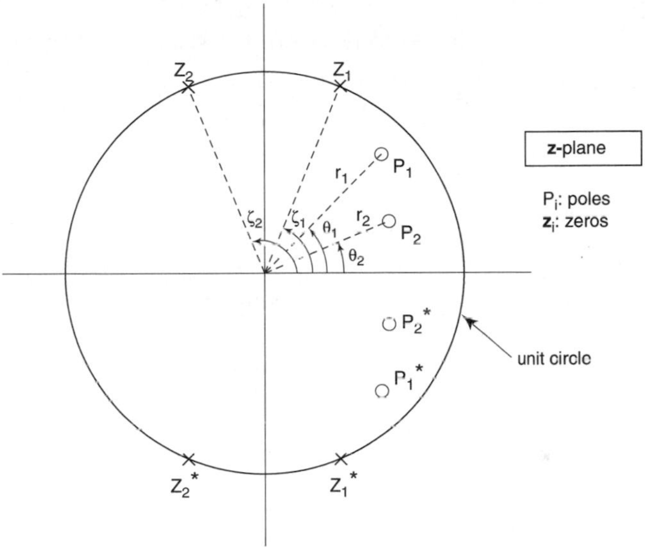

Fig. 4.35
Pole-zero plot for a typical fourth-order lowpass filter

4.5.2.4 Scaling FIR filters

For **FIR** filters the notion of scaling is somewhat easier to implement than for **IIR** filters. In **IIR** filters, because of the recursive nature, the output is fed back and arithmetic overflow in computing an output value can be catastrophic. In nonrecursive filters the output is not fed back. Arithmetic overflow, or saturation, will affect only a single output sample and thus, while not desirable, is not catastrophic (depending of course on the application).

Suppose the given **FIR** filter is

$$H(\mathbf{z}) = \sum_{i=0}^{N-1} \alpha_i \mathbf{z}^{-i} \tag{4.5.38}$$

The transfer function actually implemented will be KH(**z**) where K is a suitable scaling constant used to ensure that the overall *gain* of the filter is unity. This *gain* can be defined in one of the three ways we have been considering. Then

$$G(\mathbf{z}) = K_0 \sum_{i=0}^{N-1} K_1 \alpha_i \mathbf{z}^{-i} = K_0 \sum_{i=0}^{N-1} a_i \mathbf{z}^{-i} \tag{4.5.39}$$

where the overall scaling requirement of K has been split into $K = K_0K_1$, of which K_0 is implemented directly and K_1 is merged into the coefficients:

$$a_i = K_1 \alpha_i \text{ for } i = 1, 2, \ldots, (N-1) \tag{4.5.40}$$

as diagramed in Fig. 4.36.

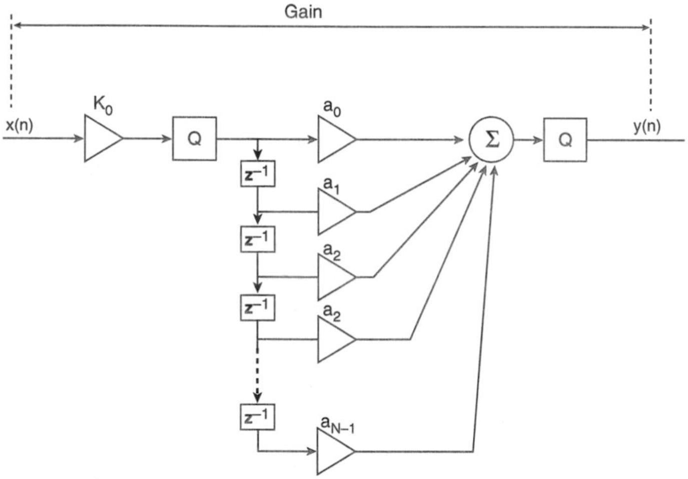

Fig. 4.36
Scaling an **FIR** filter

The reason for splitting the gain is that when the coefficients are quantized to finite precision, the quantizing error relative to the coefficient is worse for small values of the coefficients. Thus splitting the gain, which introduces an additional source of roundoff when K_0 is not unity (or a positive power of 2) is a compromise between roundoff noise and the degradation in accuracy of the transfer function (frequency response).

4.5.2.5 Concluding Remarks on Finite-Wordlength Effects

The effect of finite wordlength is a nonlinear effect. We have chosen to approximate this effect by "linearization," treating the roundoff noise as an additive phenomenon. There are other, more accurate means of analyzing the problem. When the wordlength is large then the simplified model we have presented is sufficient. When the wordlength is not large then the effects of finite wordlength have to be studied more carefully. **IIR** filters, in particular, are susceptible to certain deleterious effects. One such is the notion of a ***limit cycle***. Even when the input to an **IIR** digital filter is made identically zero, and we assume that the number representation used can treat the value zero exactly, a low level oscillation can be sustained. Furthermore, the impact of not being able to represent very large numbers can be catastrophic, with the appearance of very large amplitude oscillations if arithmetic overflow is not properly handled. References [1.3] and [1.4] provide several papers that deal with these nonlinear phenomena. With the advent of **DSP** chips, such as the Motorola DSP5600x series, which have a large wordlength and additional "headroom" to handle large numbers, the impact of these nonlinear phenomena, while not completely eliminated, can be mitigated by careful scaling.

4.6 EXERCISES

4.1. Develop a program to compute the signal-to-noise ratio for a quantizer. Consider the cases where the **pdf** of the input is uniform, exponential, and Gaussian. Consider M-level quantizers (where M is a parameter) where the decision levels and output levels are prescribed and thus consider as special cases, the uniform, A-law, and μ-law encoding laws. For a given set of output levels, convince yourself that the appropriate decision level is midway between the output levels.

4.2. Develop a set of routines to simulate the action of a quantizer and to measure the quantization noise power. In particular, develop the following:

a) Signal Generation. Generate {x(n)} where x(n) is a linear combination of white noise and a single frequency tone. The parameters are the frequency of the tone, the ratio of tone to noise (in power), the total signal power, and the **pdf** of the noise (uniform, exponential, Gaussian, other). Define the "peak" signal as +6 dBr. That is, a DC signal of peak amplitude has a power that is defined as +6dBr. A power of a sinewave of the same amplitude is +3dBr.

b) Quantization. Write routines to simulate the action of an N-bit uniform quantizer (N = 14, 13, 12, etc.), an A-law quantizer, and a μ-law quantizer. Denote the quantized version of x(n) by y(n). The quantization error is e(n) = y(n) – x(n).

c) Power measurement. Develop a routine to compute the power of the quantization error. One simple method is to evaluate the average of the square of e(n) over a large number, M, of samples.

d) Notch filter. Write a routine to implement a notch filter, the frequency of the notch being the frequency of the tone component of {x(n)}. If {y(n)} is filtered by the notch filter, the filter output will be principally the quantization error.

Then conduct the following simulations:

i) Compute and plot the **SNR** as a function of signal power when {x(n)} is a pure tone. Estimate the noise power either by computing the quantization error, e(n) = y(n) – x(n), or by applying a notch filter to {y(n)}.

ii) Increase the level of the noise component of {x(n)} and note the difference with the case when {x(n)} is a pure tone (the notch method of measuring quantization noise is not applicable).

4.7 REFERENCES AND BIBLIOGRAPHY

The literature that is directly alluded to in this chapter is listed in Section 4.7.1. There are several useful articles that contain material related to the subject matter. A sampling of these is provided in Section 4.7.2.

4.7.1 References

References [1.3] and [1.4] are collections of key papers in the development of digital signal processing as a body of knowledge and include papers from a variety of Journals including those published by the IEEE. A subset of these papers have been listed as being most pertinent vis-à-vis the material covered in this chapter.

[1.1] **CCITT** Recommendation **G.711**: Pulse code modulation (**PCM**) of voice frequencies.

[1.2] Jayant, N. S., Ed., *Waveform Quantization and Coding*, IEEE Press, New York, 1976.

[1.2.1] Max, J., "Quantization for minimum distortion," *IRE Transactions on Information Theory*, March 1960.

[1.2.2] Jayant, N. S., "Digital Coding of Speech Waveforms: PCM, DPCM, and ΔM quantizers," *Proc. IEEE*, May 1974.

[1.2.3] Paez, M. D., and Glisson, T. H., "Minimum mean-squared-error quantization in speech PCM and DPCM systems," *IEEE Trans. Comm.*, April 1972.

[1.3] Rabiner, L.R., and Rader, C.M., Editors, *Digital Signal Processing*, IEEE Press, New York, 1972.

[1.3.1] Liu, B., "Effect of finite word length on the accuracy of digital filters—a review," *IEEE Trans. Circuit Theory*, Nov. 1971.

[1.3.2] Rader, C.M., and Gold, B., "Effects of parameter quantization on the poles of a digital filter," *Proc. IEEE*, May 1967.

[1.3.3] Jackson, L.B., "On the interaction of roundoff noise and dynamic range in digital filters," *Bell System Tech. Journal*, Feb. 1970.

[1.3.4] Ebert, P.M., Mazo, J.E., and Taylor, M.G., "Overflow oscillations in digital filters," *Bell System Tech. Journal*, Nov. 1969.

[1.4] Digital Signal Processing Committee, IEEE Acoustics, Speech, and Signal Processing Society, Editors, *Digital Signal Processing, II*, IEEE Press, New York, 1976.

[1.4.1] Knowles, J. B., and Olcayto, E. M., "Coefficient accuracy and digital filter response," *IEEE Trans. Circuit Theory*, March 1968.

[1.4.2] Chan, D. S. K., and Rabiner, L. R., "Analysis of quantization errors in the direct form for finite impulse response digital filters," *IEEE Trans. on Audio and Electroacoustics*, Aug. 1973.

[1.4.3] Sandberg, I. W., and Kaiser, J. F., "A bound on limit cycles in fixed-point implementations of digital filters," *IEEE Trans. Audio and Electroacoustics*, June 1972.

[1.4.4] Long, J. L., and Trick, T. N., "An absolute bound on limit cycles due to roundoff errors in digital filters," *IEEE Trans. Audio and Electroacoustics*, Feb. 1973.

[1.4.5] Eckhardt, B., and Winkelnkemper, W., "Implementation of a second-order digital filter section with stable overflow behaviour," *Nachrichtentechnische Zeitschrift*, June 1973.

4.7.2 Bibliography

[2.1] Bellamy, J., *Digital Telephony*, John Wiley and Sons, New York, 1982.

[2.2] Jayant, N. S., and Noll, P., *Digital Coding of Waveforms*, Prentice Hall, Inc., Englewood Cliffs, NJ, 1984.

[2.3] Lane, J., and Hillman, G., "Implementing IIR/FIR filters with Motorola's DSP56000/DSP56001," *Motorola Application Note* APR7/D, 1990. [This is but one of several excellent Application Notes generated by the Digital Signal Processing Group at Motorola.]

[2.4] Shenoi, K., and Agrawal, B. P., "Selection of a PCM coder for digital switching," *IEEE Trans. on Acoustics, Speech, and Signal Processing*, Vol. ASSP-28, Oct. 1980.

[2.5] Digital Channel Bank. Requirements and Objectives. **Bell Pub. 43801**, *Bell System Technical Reference*, AT&T, Nov. 1982.

[2.6] *Notes on the BOC Intra-Lata Networks*, Technical Reference **TR-NPL-000275**, BELLCORE, 1986.

[2.7] Engineering Staff of Analog Devices, Inc., Sheinhold, D.H., Ed., *Analog-Digital Conversion Handbook*, Prentice Hall, Inc., Englewood Cliffs, NJ, 1986.

[2.8] The Applications Engineering Staff of Analog Devices, DSP Division, Mar, A., Ed., *Digital Signal Processing Applications using the ADSP-2100 Family*, Prentice-Hall Inc., Englewood Cliffs, NJ, 1990.

[2.9] *Digital Signal Processing Applications with the TMS320 Family*, Texas Instruments, 1986; Prentice Hall, Inc., Englewood Cliffs, NJ, 1987.

[2.10] *Programs for Digital Signal Processing*, Edited by the Digital Signal Processing Committee of the IEEE Acoustics, Speech, and Signal Processing Society, IEEE Press, New York, 1979.

[2.11] *Transmission Systems for Communications*, AT&T Bell Laboratories, Fifth Edition, Bell Telephone Laboratories, Inc., 1982.

DIGITAL COMPRESSION OF SPEECH SIGNALS IN TELECOMMUNICATIONS

5

5.1 INTRODUCTION

Speech signals are by far the most prevalent of signals transported over telecommunications networks. The conventional conversion from analog to digital provides a 64 kbps data rate for speech even though excellent subjective quality can be achieved at rates considerably lower. Methods of acheiving this compression are the subject of this chapter. The focus of compression is in the context of transmission, since in telecommunications, bandwidth or data carrying capacity is a scarce resource.

The compression methods discussed here can be qualified as ***predictive encoders***. The quantization, or wordlength reduction, is achieved at an implicit sampling rate commensurate with the bandwidth of the speech signal. In sharp contrast, noise-shaping techniques discussed in Chapter 8 use a sampling rate much higher than required by the sampling theorem. The quantization noise in a **ΔΣM** is shaped to occupy regions of frequency outside those occupied by the desired signal. Predictive encoders are based on the premise that the added noise is perceptually not annoying. Thus while **ΔΣM**s make no assumptions on the signal except its bandwidth, predictive encoders are efficient only if certain specific characteristics of the signal, or the recipient of the signal, can be exploited.

Research and development activities in digital processing applied to speech signals in telecommunications date back to the 1960s. Most of the results are available in publications such as the *Bell System Technical Journal* (now published as the *AT&T Technical Journal*), IEEE Transactions, Conference Proceedings, and so on. Classic texts on speech processing and coding are those by Rabiner and Schaefer [7] and Jayant and Noll [5]. Two IEEE Press publications [8] and [9] provide collections of significant articles published in a variety of journals. Specific details on standardized algorithms are available in published form through the **ANSI** standards that pertain to North American applications and through the **CCITT** Recommendations (**G** Series) that apply to Europe and most of the rest of the world. More recent publications of a tutorial nature include Dimolitsas [2], Jayant [4], and Steele [11]. Chapter 1 provides a description of the telephony channel with the intent of providing a flavor of the considerations that must be taken into account in the development of speech compression algorithms.

Section 5.2 introduces the principles of *predictive coding*. A simple case of first-order prediction is employed to explain the behavior. The extension to *adaptive predictive coding* is introduced in Section 5.3. One of the most widely deployed means for speech compression, and one that has been standardized, is the **ADPCM** algorithm. This is covered in Section 5.4. The most compression, for speech signals, is afforded by the techniques of *Linear Predictive Coding* (**LPC**). This is covered in Section 5.5. A brief description of an 8 kbps algorithm that is based on **LPC** and is one of two accepted standards for speech encoding in digital cellular telephony is discussed. The notion of *Digital Speech Interpolation* (**DSI**) is covered in Section 5.6. In that section, **DSI** is explained with respect to an implementation in an existing product and shows how the short-term variations in speech statistics, over an aggregate of several separate channels, can be "smoothed" over using *variable rate* **ADPCM**.

5.2 PREDICTIVE CODING

In the discussion of A/D methods in Chapters 3 and 4, a relationship was developed between the signal-to-quantization-noise ratio and signal level. If the maximum signal power expected is σ^2, then the crashpoint of the quantization characteristic is set at $K_1\sigma$ (K_1=4 is a common choice), where K_1 is a measure of the peak-to-rms ratio. Assuming N bits of quantizing accuracy, the additive quantization noise power is estimated as

$$\sigma_\varepsilon^2 = K_2(K_1\sigma)^2\, 2^{-2N} \qquad (5.2.1)$$

where K_2 is a constant depending on the shape of the **pdf** of the quantizing noise. Observe that the process of quantization can be forced to introduce a noise level that is proportional to the power of signal being quantized. Thus if N bits of quantization coarseness are applied to a signal of power σ^2, then (N-1) bits could be used to achieve the same quantization noise power if the signal power were reduced by a factor of four. This observation is the basis of predictive quantization.

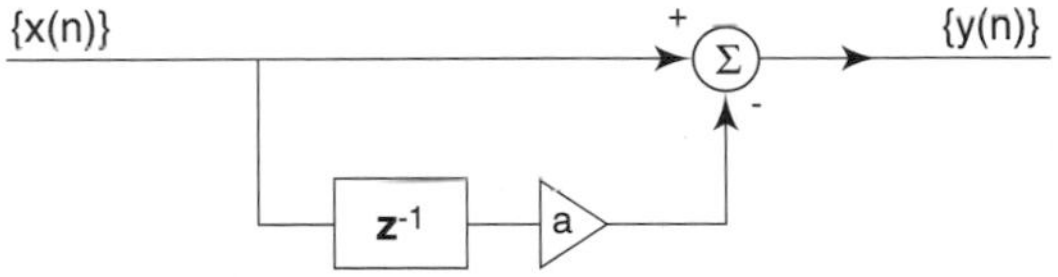

Fig. 5.1
First-order feed-forward prediction

Suppose $\{\mathbf{x}(n)\}$ is a discrete-time signal with associated autocorrelation function $\{R_{xx}(k)\}$. Then if one creates the difference signal as shown in Fig. 5.1

$$\mathbf{y}(n) = \mathbf{x}(n) - a\,\mathbf{x}(n-1) \qquad (5.2.2)$$

then the variance, i.e., the power, of $\{\mathbf{y}(n)\}$ can be expressed as (assuming $\{\mathbf{x}(n)\}$ and a are real values)

$$R_{yy}(0) = \mathbf{E}\{\mathbf{y}^2(n)\} = \mathbf{E}\{[\mathbf{x}(n) - a\,\mathbf{x}(n-1)]^2\}$$
$$= R_{xx}(0)[1 - 2a\,r_{xx}(1) + a^2] \tag{5.2.3}$$

where $r_{xx}(k)$ is the normalized autocorrelation of $\{\mathbf{x}(n)\}$. $R_{yy}(0)$, the variance of $\{\mathbf{y}(n)\}$, is minimized if

$$a = r_{xx}(1) \tag{5.2.4}$$

and with this choice

$$\sigma_y^2 = (1 - a^2)\sigma_x^2 \tag{5.2.5}$$

Clearly, if $r_{xx}(1)$ is close to unity, then the variance of $\{\mathbf{y}(n)\}$ is significantly less than that of $\{\mathbf{x}(n)\}$. Quantization of $\{\mathbf{y}(n)\}$, rather than $\{\mathbf{x}(n)\}$, requires N_1 bits, where

$$N_1 = N + \log_2(1 - a^2) \tag{5.2.6}$$

to keep the quantization noise power the same as if N bits were applied directly to quantize $\{\mathbf{x}(n)\}$. Therefore, in principle at least, the two systems depicted in Figure 5.2 are equivalent in terms of additive noise power. Since the wordlength reduction or quantization is performed on the difference signal, **predictive** coding is often called "**differential**" coding as well.

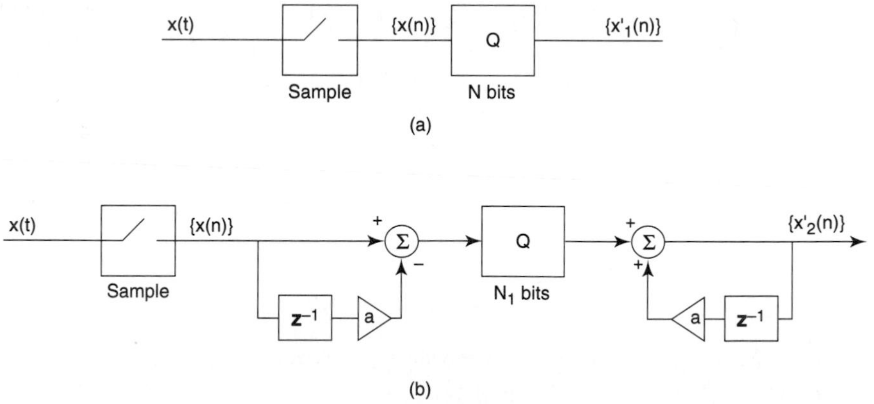

Fig. 5.2
First-order open-loop differential coding, $N_1 < N$

It should be noted that to ensure that the reconstruction filter is stable, it is necessary that $|a| < 1$. This is quite reasonable since the correlation coefficient $r_{xx}(1)$ is always less than one in magnitude.

Rather than view the conversion from analog to digital, it is more appropriate to view differential coding as a means for wordlength reduction. In Fig. 5.2 {x(n)} is viewed as a discrete-time signal, effectively, a level of quantization that corresponds to N= . It suffices that N >> N_1, for the system of differential encoding to be effective. That is, we consider the scheme depicted in Fig. 5.3 where the analog signal is first converted into an N-bit signal. Provided that the arithmetic used in implementing the filters utilizes a sufficiently large wordlength and thus introduces a negligible amount of roundoff noise, and if $N_1 << N$, then the two configurations shown in Figs. 5.2 and 5.3 are practically equivalent.

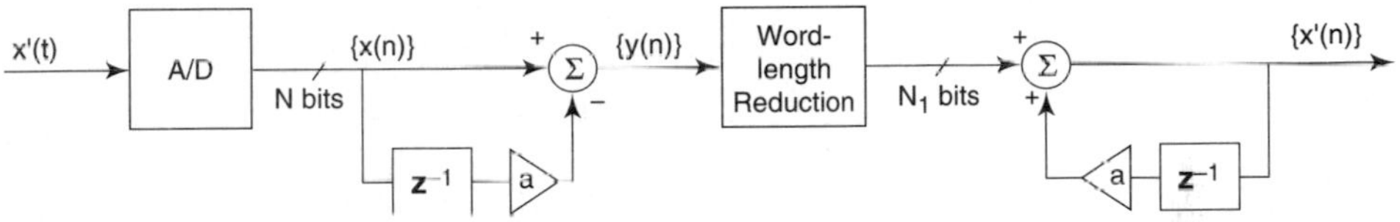

Fig. 5.3
First-order open-loop differential coding, N_1<N

The schemes depicted in Figs. 5.2 and 5.3 should never be implemented as shown. Rather, the "closed-loop" scheme depicted in Fig. 5.4 is preferred. The scheme for generating the lower wordlength is called the encoder and the scheme for reconstructing the signal is called the decoder.

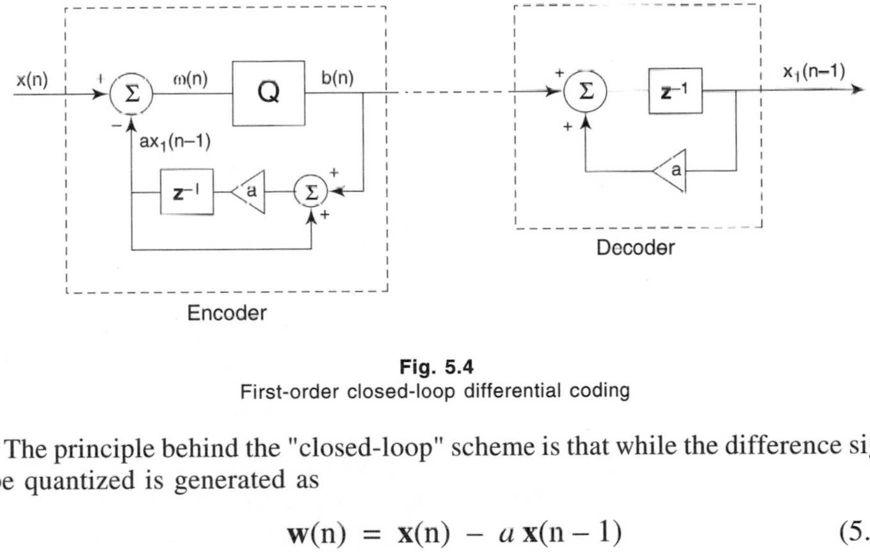

Fig. 5.4
First-order closed-loop differential coding

The principle behind the "closed-loop" scheme is that while the difference signal to be quantized is generated as

$$\mathbf{w}(n) = \mathbf{x}(n) - a\,\mathbf{x}(n-1) \tag{5.2.7}$$

in Fig. 5.3, the corresponding signal to be quantized is

$$\mathbf{w}(n) = \mathbf{x}(n) - a\,\mathbf{x}_1(n-1) \tag{5.2.8}$$

in Fig. 5.4. That is, since $\{\mathbf{x}_1(n)\}$, the reconstructed signal, is presumably an adequate representation of $\{\mathbf{x}(n)\}$, the difference signal that is quantized could well be generated in either manner. Note that the encoder includes a replica of the decoder.

The "closed-loop" scheme of Fig. 5.4 is preferred to the "open-loop" scheme of Fig. 5.3. To see why, consider the net quantization error in the two schemes. Denote by $\{\varepsilon(n)\}$ the intrinsic noise introduced by the quantizer. We usually assume that this signal is a white noise sequence, unrelated to the signal being quantized. The net quantization noise is a modified form of this "instantaneous" quantization noise because of the reconstruction filter. Writing

$$y(n) = x(n) - a\, x(n-1) \tag{5.2.9}$$

and

$$x'(n) = a\, x'(n-1) + y(n) + \varepsilon(n) \tag{5.2.10}$$

If we define

$$\eta(n) = x'(n) - x(n) \tag{5.2.11}$$

as the net quantization noise, then from Eq. (5.2.9) and Eq. (5.2.10) we get

$$\eta(n) = a\, \eta(n-1) + \varepsilon(n) \tag{5.2.12}$$

The white noise assumption for $\{e(n)\}$, and denoting the power of this signal as σ_ε^2, the variance of $\{\eta(n)\}$ is evaluated as

$$\sigma_\eta^2 = \frac{\sigma_\varepsilon^2}{(1 - a^2)} \tag{5.2.13}$$

Clearly, if a is close to unity than the variance of $\{\eta(n)\}$ can become quite large for the "open-loop" scheme.

For the "closed-loop" scheme, denoting by $\{\varepsilon(n)\}$ the error in quantizing $\{w(n)\}$, we get

$$w(n) = x(n) - \alpha x_1(n-1) \tag{5.2.14a}$$

$$x_1(n) = w(n) + \alpha x_1(n-1) + \varepsilon(n) = x(n) + \varepsilon(n) \tag{5.2.14b}$$

Clearly, the net quantization noise is just the error introduced by the quantizer itself and is not amplified by the encoder/decoder.

Another reason for choosing the "closed-loop" scheme is that in practice the predictor coefficient "a" is not fixed, but modified constantly. In order that the reconstruction filters in the decoder and encoder be the same, it is necessary that the algorithms for developing the predictor coefficient be applied to the same data. This is possible for the scheme depicted in Fig. 5.4 but not for the scheme shown in Fig. 5.3.

5.3 ADAPTIVE PREDICTIVE CODING

The principal drawback of the predictive coding scheme discussed in Section 5.2 is that it is applicable only if the signal {x(n)} is stationary and thus the correlation coefficient $r_{xx}(1)$ is well defined. Speech signals are approximately stationary over short time intervals. This short-term stationarity is borne out in practice. Over periods of about 100 msec speech signals have constant statistics. This requires that the predictor coefficient "a" be adaptive in nature in order to track the statistics of the signal.

Consider the purely feed-forward system depicted in Fig. 5.5 governed by

$$e(n) = x(n) - a^{(n)} x(n-1) \tag{5.3.1}$$

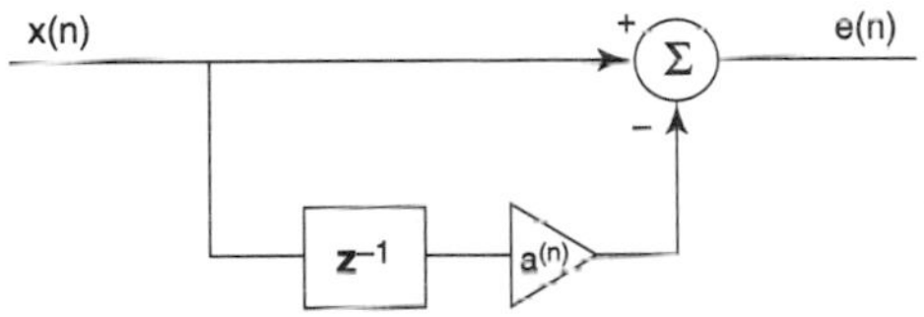

Fig. 5.5
First-order open-loop adaptive differential coding

The strategy for choosing a is to minimize the power of the "error" signal {e(n)}. This leads to the following approach for adapting the predictor coefficient:

$$a^{(n+1)} = a^{(n)} + G\, e(n)\, x(n-1) \tag{5.3.2}$$

where G is a suitable constant that determines the adaptation gain. The principles of adaptive filters are discussed in Chapter 6.

The closed-loop version is quite similar and is depicted in Fig. 5.6.

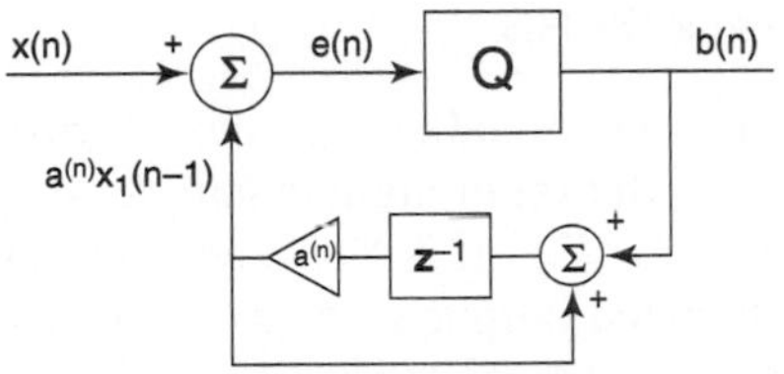

Fig. 5.6
First-order closed-loop adaptive differential coding

The governing equations defining the adaptation of the predictor coefficient

$$a^{(n+1)} = a^{(n)} + G\,e(n)\,x_1(n-1) \qquad (5.3.3)$$

The primary similarity between Eq. (5.3.2) and (5.3.3) is that in both it is necessary to constrain the coefficient $a^{(n+1)}$ to ensure that the recursive filter comprising the decoder is stable. This requires $|a^{(n+1)}| < 1$. The difference between the two schemes is the use of the input signal in one case and a reconstructed replica of the signal in the other.

In order to reinforce stability, Eq. (5.3.3) can be modified to

$$a^{(n+1)} = \boldsymbol{\alpha}^{(n)} + G\,e(n)\,x_1(n-1) \qquad (5.3.4)$$

where α is a constant close to, but less than, 1. The introduction of this feature, which drives $a^{(n+1)}$ towards zero (absolute stability), makes the adaptation very robust and, while not "optimal" in a mathematical sense, is the preferred method of adapting a (recursive) filter.

Note that the adaptation scheme in Eq. (5.3.3) that governs the structure of Fig. 5.6 utilizes $x_1(n-1)$, a quantity that is available at the decoder (receiver) as well and hence the decoder can track the encoder.

The quantity $e(n)\,x_1(n-1)$ in Eq. (5.3.4) is an estimate of the gradient of the error function, in our case the power of the difference signal, with respect to the coefficient a. It has been found through experimentation that this gradient can be estimated in a very crude manner and yet yield good results. In fact, the **ADPCM** algorithm specified later in this chapter uses an adaptation algorithm of the form

$$a^{(n+1)} = \boldsymbol{\alpha}^{(n)} + g\,\mathrm{sgn}(\mathbf{e}(n)\,\mathrm{sgn}(\mathbf{x}_1(n-1)) \qquad (5.3.5)$$

where the gradient is estimated from the *sign* of the quantities $e(n)$ and $x_1(n-1)$.

Since the speech signal $\{x(n)\}$ can, at best, be considered short-term stationary, there is no well-defined value to which $a^{(n)}$ can converge to. The optimal value will move with changes in the signal statistics. Consequently, the variance of the difference signal $\{e(n)\}$ is not necessarily constant. One approach used to accommodate this continual variation in signal power is to make the quantizer, represented by **Q** in Fig. 5.6, adaptive in nature. Adaptive quantizers are discussed in Jayant [3]. The principles underlying an adaptive quantizer are actually quite intuitive and discussed next.

5.4 ADAPTIVE QUANTIZERS

A quantizer can be visualized as shown in Fig. 5.7 for the case that has seven quantized levels. If the input signal, sample $x(n)$, is between two breakpoints, say between $(\Delta/2)$ and $(3\Delta/2)$, it is assigned a suitable codeword (3 bits for the 7-level case) and the associated output value, $b(n)$ is Δ. If the signal sample is small, with $|x(n)| < (\Delta/2)$, then the associated value is 0; if $|x(n)| > (5\Delta/2)$ then $|b(n)| = 3\Delta$. If several consecutive codewords correspond to an output level of 3Δ then it is likely that the quantizer is overloading. Conversely, if several consecutive codewords correspond to the output level 0, then it is likely that the quantizer is too coarse. If consecutive codewords indicate a reasonable mix of levels $\pm\Delta$ and $\pm 2\Delta$ (and 0 and $\pm 3\Delta$) then the quantizer is scaled correctly for the level of the signal being quantized. The strategy for adapting the quantizer is to increase the step size, Δ, if overload is likely, and to decrease the step size if it appears that the quantizer is too coarse. This intuitively satisfying algorithm is what is actually used.

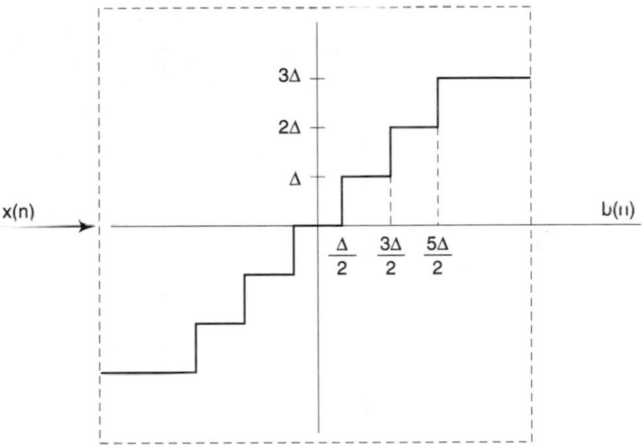

Figure 5.7
Seven-level uniform quantizer (wordlength reduction) characteristic

In the **ADPCM** algorithm, discussed in the next section, the quantizer used is not uniform but utilizes the principle of "optimal" spacing of decision levels. As long as these decision levels can be described as fractions of a (common) step size, the adaptive scheme described above is applicable.

5.5 ADAPTIVE DIFFERENTIAL PULSE CODE MODULATION

Conventional analog-to-digital converters used in telephony employ the μ-law (or A-law) coding structure, and an 8-kHz sampling rate, for a net bit rate of 64 kbps. Such an A/D technique is usually called **Pulse Code Modulation (PCM)**. Recognizing that 64 kbps is much more than really necessary for speech signals, a compression scheme based on adaptive differential means has been developed and standardized. This technique, **ADPCM** for **Adaptive Differential Pulse Code Modulation**, is the result of several thousand man-years of effort by researchers in several countries. The algorithm is specified to the extent that all the arithmetic operations are described at the bit-level; the standard is such that two units from different manufacturers will provide the same quantized signal and can thus be used interchangeably in standard configurations.

The signal processing involved with **ADPCM** is described in detail in the **American National Standards Institute** (**ANSI**) document **ANSI-T1.303** [15] and **CCITT** Rec. **G.721** [12] and **CCITT** Rec. **G.726** [13]. The standard for **ADPCM** describes a means for compressing, and subsequently expanding, the wordlength of a channel sample from 8 bits to either 3, 4, or 5 bits, corresponding to compression ratios of 2.67, 2, and 1.6, respectively. The compressed data rates are accordingly 24 kbps, 32 kbps, and 40 kbps respectively, since the underlying sampling rate is 8 kHz. This is achieved by using an **adaptive predictor** and an **adaptive quantizer**. Furthermore, the adaptive quantizer characteristic is not uniform but tailored to the type of signals the **ADPCM** device may encounter, primarily speech.

A considerable amount of research and experimentation went into the **optimization** of the algorithm. Consideration was given to the type of traffic as a mix of speech and voice-band data, the performance of the algorithm in the case of "tandems" (when the transmission chain involves multiple conversions from **PCM** to **ADPCM** and back to **PCM**), that the effect of transmission errors which would cause the encoder and decoder to lose tracking momentarily as well as the subjective quality of the reconstructed speech as compared to (uncompressed) 64-kbps μ-law (or A-law) coding. The core of the algorithm is identical in both the μ-law and A-law cases; the only difference is the format conversion between μ-law or A-law and the internal uniform format used for the arithmetic operations.

A channel employing **ADPCM** is not toll-quality in the strict sense. Compression to 40 kbps limits the use of the channel to modems operating at 9.6 kbps or less; compression to 32 kbps limits the modems to 4.8 kbps. However, from the viewpoint of perceptual speech quality, both 40-kbps and 32-kbps compression are virtually indistinguishable from the uncompressed (64 kbps, toll-quality) channel. Compression to 24 kbps is to be used only for speech though certain low speed modems may operate satisfactorily at this compression level. A trained listener can recognize that speech is being compressed to 24 kbps but the compressed signal does retain *all* intelligibility and tonal information.

With the advent of voice packetization, where the number of bits assigned to a channel may change dynamically, interest in what is known as the **embedded ADPCM** algorithm (see [16]) has grown. In 1991 **ANSI** standardized an algorithm that is a variable-rate algorithm. It is constructed in such a way that the decision levels of the quantizers at lower bit rates are subsets of the decision levels at higher bit rates. This permits bit reductions at any point without the need for coordination between transmitter and receiver. In contrast, the decision levels of the conventional **ADPCM** algorithm for the various rates are not subsets of each other and the transmitter and receiver must be in communication as to the coding rate. The principles of the embedded algorithms, in terms of adaptive prediction and adaptive quantization are not that different from the conventional (single-rate) algorithm to merit a separate explanation. The interested reader is referred to **ANSI**, **T1.303** [15] and **ANSI T1.310** [16] for the finer points of the differences. **ANSI T1.310** proposes a 16-kbps scheme (2 bits per sample) in addition to the 24-, 32-, and 40-kbps schemes. It has been observed that at 16 kbps the *quality degradation is noticeable*, even to an untrained ear, with a certain raspiness and accentuation of sibilant sounds.

In Section 5.5.1 we describe some of the features of the **ADPCM** algorithm. The intent is not to provide a detailed exposition of the standard but to point out some of the peculiarities. Based on our discussions of adaptive predictors, adaptive quantizers, and, in Chapter 4, nonuniform quantizers, the details of the algorithm as set forth in the standards literature ([12], [13], [15], [16]) should be readily understandable. As usual, the devil is in the details. These details can be better appreciated if the underlying rationale is known and it is this rationale that we shall try to provide.

5.5.1 Description of the ADPCM Algorithm

The prediction algorithm used in the **ADPCM** algorithm is more complex than that described earlier. Rather than a single predictor coefficient "a," corresponding to a single-pole reconstruction filter, the **ADPCM** algorithm specifies a reconstruction filter that has two poles and six zeros. It uses an adaptive quantizer to assign 3, 4, or 5 bits to the quantized (difference) signal. The quantizer itself is not uniform but utilizes a logarithmic type of characteristic reminiscent of the μ-law (or A-law) encoding characteristic. Furthermore, several "tricks" are employed to make the algorithm robust and ensure stability.

A simplified block diagram of the encoder is shown in Fig. 5.8 and a simplified block diagram of the decoder is shown in Fig. 5.9.

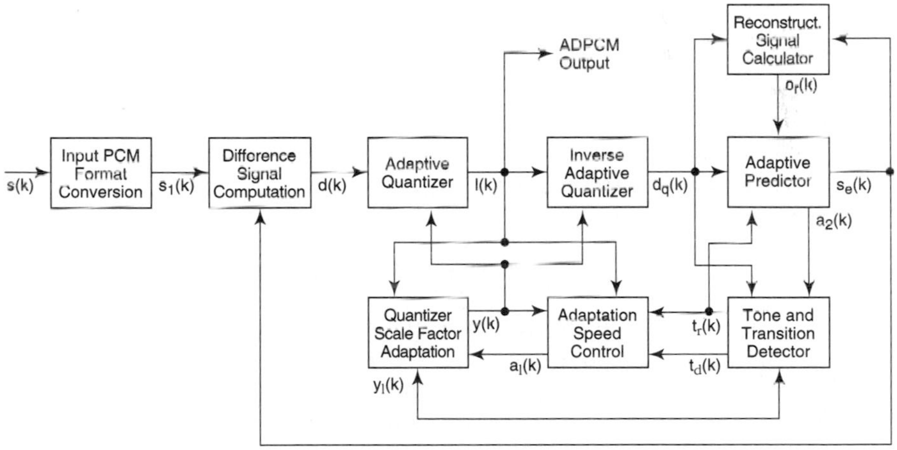

Fig. 5.8
Block diagram of ADPCM Encoder

In Fig. 5.8 the adaptive quantizer is split into an adaptive bit-encoding section and adaptive bit-decoding section. The decoder processing (Fig. 5.9) does not require both blocks. In our discussion in Section 5.3 the two blocks were lumped together as a single unit. Also, the apparent complexity associated with the adaptive predictor processing to create the signal estimate, relative to the simple structures shown in Sections 5.2 and 5.3, are caused by the **ADPCM** algorithm specifying both poles and zeros for the predictor (or reconstruction filter) rather than the simple single-pole structure assumed there.

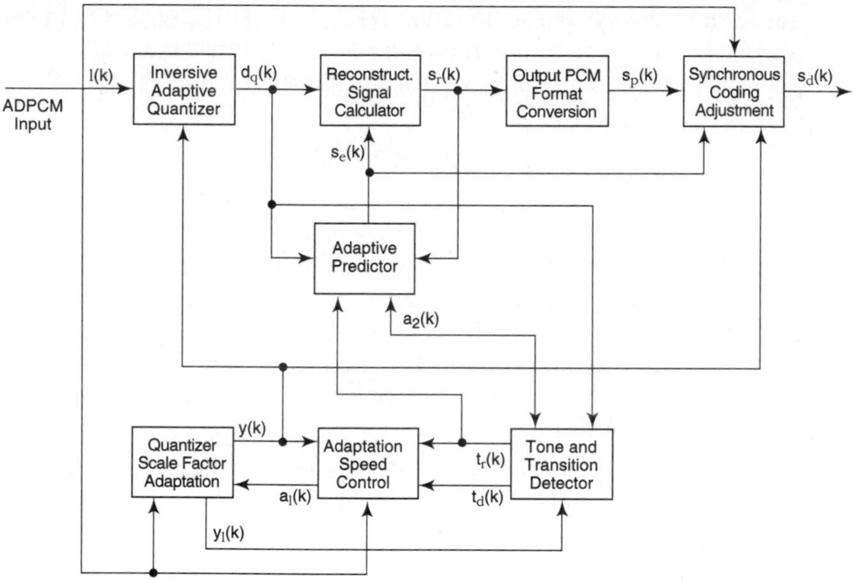

Fig. 5.9
Block diagram of ADPCM Decoder

The input signal is first converted into a uniform code to provide the (linear) signal $\{s_l(n)\}$. The difference signal $\{d(n)\}$ is computed from the uniform signal and the estimate $\{s_e(n)\}$.

$$d(n) = s_l(n) - s_e(n) \tag{5.5.1}$$

It is this difference signal that is quantized. Furthermore, all coefficient updates, changes to the adaptive quantizer, and so on, are performed based on the knowledge of the quantized difference signal. This is done because, in the absence of transmission errors, the receiver has a replica of this quantized difference signal.

5.5.1.1 The Adaptive Quantizer

A 7-, 15-, or 31-level nonuniform adaptive quantizer is used to quantize the difference signal $\{d(n)\}$, corresponding to 3, 4, or 5 bits per sample, respectively. The non-uniform nature of the quantizer is achieved by uniformly quantizing $\log_2(d(n))$. The adaptive nature is implemented by scaling the (logarithm of) the signal by a factor y(n), which is updated each sampling time. The details of the quantizer are available in **G.721** [12] or **ANSI T1.310** [16].

The updating of the scale factor y(n) is a unique feature of the standard. It utilizes a scheme that tracks the nature of the signal. Adaptation is fast for signals such as speech that produce difference signals with large fluctuations and is slow for signals, such as from voice-band modems that produce difference signals with small fluctuations. The speed is controlled by observing the rate of change of the difference signal values.

In the following discussion, the quantized version of the difference signal will be indicated by the subscript q, i.e., $\{d_q(n)\}$.

5.5.1.2 The Adaptive Predictor

The adaptive predictor computes the signal estimate $\{s_e(n)\}$ required for deriving the difference signal $\{d(n)\}$ in Eq. (5.5.1). This estimate is obtained from the quantized signal $\{d_q(n)\}$ that is available to both the receiver and transmitter (except, possibly in the case of transmission errors). In Section 5.3 the nature of the adaptive predictor was explained using a first-order reconstruction filter, i.e., a filter with one pole. In the **ADPCM,** algorithm the predictor uses two poles and six zeros. This is a reasonable compromise between complexity and performance and has proved effective in handling a wide variety of signals. The restriction to two poles is principally because the stability of a second-order **IIR** filter is easy to specify in terms of the two coefficients that constitute the pole pair (the coefficients are real and hence complex poles will occur in complex-conjugate pairs).

The signal estimate is computed by

$$s_e(n) = \sum_{i=1}^{2} a_i^{(n-1)} s_r(n-i) + s_{ez}(n) \tag{5.5.2}$$

where $\{s_{ez}(n)\}$ is the contribution of the zeros of the predictor filter and is given by

$$s_{ez}(n) = \sum_{i=1}^{6} b_i^{(n-1)} d_q(n-i) \tag{5.5.3}$$

The superscript (n-1) in the coefficients $\{a_i^{(n-1)};\ i=1,2\}$ and $\{b_i^{(n-1)};\ i=1, 2, \ldots, 6\}$ represents the time epoch. It is (n–1) for computing the current signal estimate $s_e(n)$ to reflect the fact that these coefficients were derived from information up to and including the previous input sample. The reconstructed signal $\{s_r(n)\}$ is computed as

$$s_r(n-i) = s_e(n-i) + d_q(n-i);\ i = 1, 2 \tag{5.5.4}$$

The description of a six-zero-two-pole filter by Eqs. (5.5.2), (5.5.3), and (5.5.4) may appear cumbersome. A direct-form implementation (see Chapter 3) is much simpler. Nevertheless the implementation scheme chosen for the **ADPCM** standard has an important feature. In order to adapt the coefficients, it is necessary to obtain estimates of the "gradient" and apply an update of the form of Eq. (5.3.4) or Eq. (5.3.5). This seemingly convoluted implementation provides the intermediate signals required to establish the necessary gradients.

Both sets of predictor coefficients are updated using a simplified gradient algorithm specified below

$$a_1^{(n)} = (1 - 2^{-8})\, a_1^{(n-1)} + 3\, \{ \mathrm{sgn}[p(n)]\, \mathrm{sgn}[p(n-1)] \} 2^{-8} \quad (5.5.5a)$$

$$a_2^{(n)} = (1 - 2^{-7})\, a_2^{(n-1)} + \{ \mathrm{sgn}[p(n)]\, \mathrm{sgn}[p(n-2)] - \mathbf{f}[a_1^{(n-1)}]\, \mathrm{sgn}[p(n)]\, \mathrm{sgn}[p(n-1)] \} 2^{-7} \quad (5.5.5b)$$

where

$$p(k) = d_q(k) + s_{ez}(k) \quad (5.5.6)$$

and

$$\mathbf{f}[a_1] = \begin{cases} 4a_1 & \text{for } |a_1| \le 0.5 \\ 2\,\mathrm{sgn}[a_1] & \text{for } |a_1| > 0.5 \end{cases} \quad (5.5.7)$$

The stability constraints applied to the $\{a_i\}$, that determine the **IIR** nature of the predictor, are

$$\left|a_2^{(n)}\right| \le 0.75 \text{ and } \left|a_1^{(n)}\right| \le 1 - 2^{-4} - a_2^{(n)} \quad (5.5.8)$$

The b_i are updated as

$$b_i^{(n)} = (1 - \alpha)\, b_i^{(n-1)} + \{ \mathrm{sgn}[d_q(n)]\, \mathrm{sgn}[d_q(n-i)] \} 2^{-7} \quad (5.5.9)$$

for i=1, 2, . . . , 6; where $\alpha = 2^{-8}$ for the 24- and 32- kbps modes and 2^{-9} for the 40-kbps mode of operation. The b_i are implicitly limited to ±2.

Implementations of an early version of the algorithm were put into service in 1985. It was observed that certain signals, primarily those originating from Frequency Shift Keying (**FSK**) modems, could cause the performance of the adaptation to degrade. That is, the adaptation algorithm, while properly "matched" to speech signals, could not operate satisfactorily on **FSK** signals. To alleviate this problem, and improve performance for these special cases, the following scheme was developed. Upon detection of an anomalous condition, the predictor coefficients are set to zero and the adaptive quantizer forced to adapt at its fastest rate. Through experimentation it was found that this detection of an untoward situation could be accomplished by observing $\{a_2^{(n)}\}$, the second pole coefficient, in conjunction with $\{d_q(n)\}$. In particular, if

$$a_2^{(n)} < -0.71875 \text{ and } |d_q(n)| > \text{threshold} \qquad (5.5.10)$$

then a "transition" is declared and the reset action taken.

5.5.1.3 The ADPCM Decoder

The **ADPCM** decoder looks very similar to part of the encoder. Notice that the decoder function is simply the process of reconstructing the signal from the bit-stream that defines the quantizing process and that the encoder has a decoder imbedded in a feedback loop. In the **ADPCM** standard an additional component called ***synchronous coding adjustment*** is added. The purpose of this is described here.

The conversion from **PCM** to **ADPCM** and back to **PCM** can be visualized as the addition of distortion. Now in a transmission system there may be situations that require multiple **synchronous tandems**, i.e., multiple conversions from **PCM** to **ADPCM** and back to **PCM** while remaining digital throughout. The synchronous coding adjustment block in a decoder is a technique used to ensure that after the **first** conversion to **ADPCM**, subsequent conversions do not accumulate distortion. Of course, this non-accumulation can be achieved only if the transmission is error free and if the intermediate bit-streams, **ADPCM** or **PCM**, are not disturbed by other processes such as "bit-robbing" for signaling, echo cancelers, or conversions between the A-law and μ-law.

Synchronous coding adjustment consists of modifying the output $\{s_p(n)\}$, the output of the uniform-to-**PCM** converter in Fig. 5.9. This modification is accomplished by first converting the μ-law signal $\{s_p(n)\}$ to a uniform **PCM** signal $\{s_{lz}(n)\}$ and computing the difference $d_z(n)$ between $s_{lz}(n)$ and the signal estimate $s_l(n)$ available in the adaptive predictor. This difference signal is then compared to the **ADPCM** quantizer decision intervals. If we denote by $s_d(n)$ the output of the synchronous decoder, then $s_d(n)$ is equal to $s_p(n)$ if the difference signal $d_z(n)$ is in the middle of the quantizer decision staircase. If $d_z(n)$ is less than the lower interval boundary, or greater than the upper interval boundary then $s_p(n)$ is "tweaked" to get $s_d(n)$. Specifically,

$$\begin{aligned} d_z(n) < \text{lower interval boundary} &\Rightarrow s_d(n) = s_p^+(n) \\ d_z(n) > \text{upper interval boundary} &\Rightarrow s_d(n) = s_p^-(n) \\ \text{otherwise } s_d(n) &= s_p(n) \end{aligned} \qquad (5.5.11)$$

The notation $s_p^+(n)$ implies that $s_d(n)$ is taken as the **PCM** code that represents the next more positive output level and $s_p^-(n)$implies that $s_d(n)$ is taken as the **PCM** codeword that represents the next more negative value than $s_p(n)$.

5.5.2 ADPCM Performance

There are several manufacturers that build equipment in conformance with the **ANSI** standards. The generic form of an **ADPCM** unit is shown in Fig. 5.10.

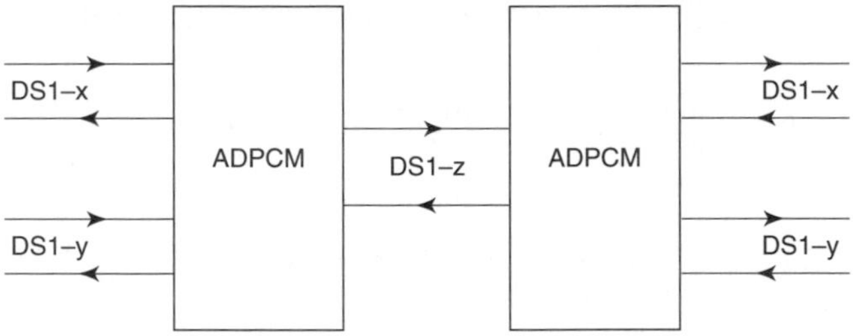

Fig. 5.10
Using **ADPCM** for 2:1 compression

Two full-rate **DS1** streams, each with 24 constituent 64-kbps speech channels are compressed into a single **DS1** stream comprising 48 constituent 32-kbps channels. A corresponding unit is necessary to convert the compressed **DS1** back into two full-rate **DS1** streams.

Such equipment has been in production and in use for several years and data is available on the impact of the encoding algorithm on actual traffic. Some of the significant observations are:

i. The added transmission delay is minimal. Introduction of **ADPCM** does not necessarily have to be accompanied by the addition of echo cancelers.

ii. Speech quality is indistinguishable from an uncompressed link.

iii. Facsimile machines that can operate at 4800 bps can transmit messages with no noticeable degradation in quality at the expense of a longer transmission interval than if the machine operated at 9600 bps.

iv. The usual (objective) measurement of quantizing distortion is only about 2 dB worse than if no **ADPCM** was used.

v. With as many as three synchronous tandems, perceived speech quality is virtually indistinguishable from the quality on a single 64-kbps **PCM** link.

vi. In synchronous tandem configurations, where no bits are robbed for signaling, up to six tandems have been employed with no noticeable loss in **SNR**.

vii. The idle channel noise performance of channel banks, which do the initial A/D conversion, often improves when the interconnection uses **ADPCM**.

viii. With **DS1** transmission errors introduced at a **BER** ("bit error rate") of 10^{-4}, the perceived quality of speech over an **ADPCM** link is **better** than the quality over a regular 64-kbps link at the **same BER**. However, most **DS1** links provide a performance much better than a **BER** of 10^{-6} and error rates better than 10^{-8} are common.

The performance data presented in this section are the results obtained by actual experimentation and reported in Skidanenko [10]. Since most units deployed in the field are configured in the so-called "48-channel robbed-bit-signalling " mode, this is the configuration that is emphasized. In the 48-channel RBS mode, the encoding method uses 4 bits per sample (32 kbps) except for every sixth sample from which one bit is "robbed" for signaling purposes. The term "unbundled " is often used to describe 48-channel operation.

Results are provided for multiple "**tandems**," both "synchronous" and "asynchronous." The definitions of these terms should be clear from Fig. 5.11. A tandem can be defined as an additional encoding. In the case of a synchronous tandem, the intermediate point between "encodings" is digital, **PCM**; for asynchronous tandems the intermediate stage is one where the signal has been converted back to analog.

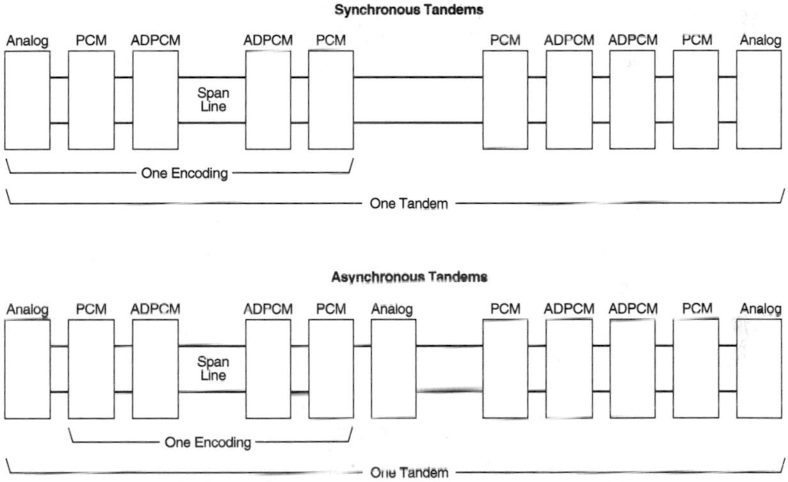

Fig. 5.11
Notion of synchronous and asynchronous tandems

5.5.2.1 Quantization Distortion (QD)

For testing the **QD** performance without the **ADPCM** transcoder, the channel bank output is looped as depicted in the test setup shown in Fig. 5.12. The **QD** performance measured for one, two, and three asynchronous tandems is shown in Fig. 5.13.

Synchronous tandems can be implemented using multiple transcoder units that are looped on the **Z** side ("Low Bit Rate Voice" or **LBRV**). The measured results for one, two, and three synchronous tandems are shown in Fig. 5.14. Note that there is a progressive degradation. This is because of the robbed-bit configuration chosen. Fig. 5.15 shows the results for up to six synchronous tandems when there are no bits robbed for signaling purposes (completely 32-kbps operation). Note that there is no accumulation of distortion.

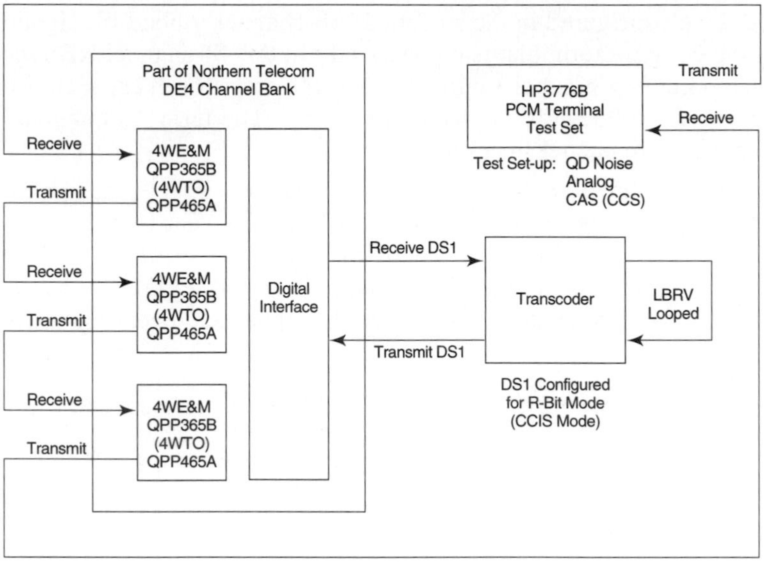

Fig. 5.12
Test setup for measuring **QD** introduced by **ADPCM**

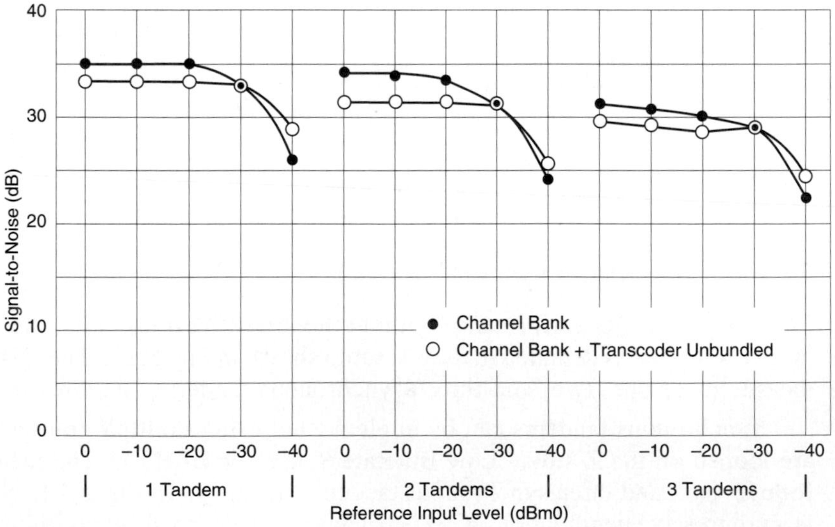

Fig. 5.13
SNR degradation introduced by **ADPCM**, asynchronous tandems

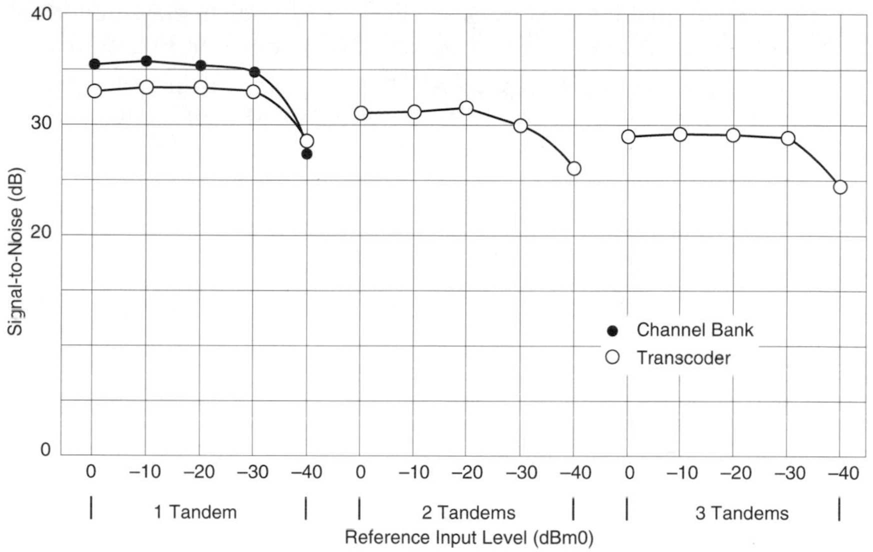

Fig. 5.14
SNR degradation introduced by **ADPCM**, synchronous tandems with robbed-bit signaling

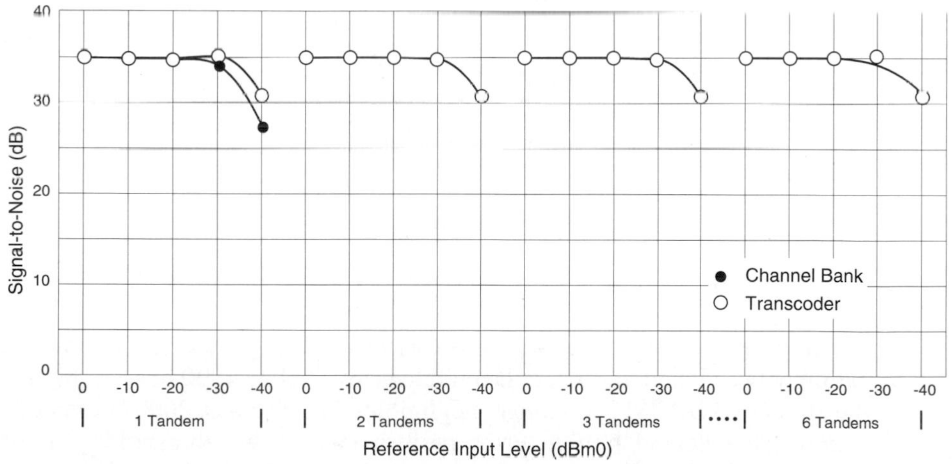

Fig. 5.15
SNR degradation introduced by synchronous **ADPCM** tandems without robbed bits

5.5.2.2 Idle Channel Noise

A test set up similar to that shown in Fig. 5.12 is used to measure idle channel noise. The idle channel noise for several tandems is shown in Fig. 5.16.

The apparent superiority of **ADPCM** in terms of idle channel noise is to be expected. The quantizer characteristic of **ADPCM** is of the mid-tread type and the coarser quantization of 4 bits, compared to 8 bits for the μ-law, puts a larger deadzone for low level signals.

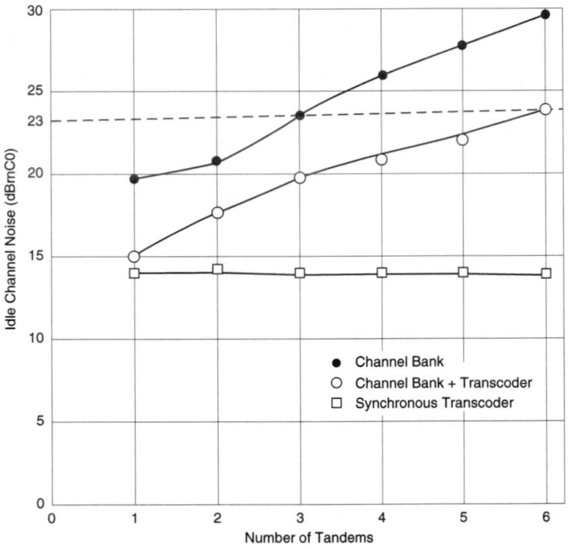

Fig. 5.16
Impact of tandems on idle channel noise

5.5.2.3 Performance with Modems

The test setup for measuring the impact of **ADPCM** on modems is shown in Fig. 5.17 and the results shown in Table 5.1.

It is immediately clear that **ADPCM** has a negligible effect on data communication between modems when the bit rate is low, below 4800 bps. At higher bit rates the effect of **ADPCM** is no longer negligible. In particular, 9600-bps modems suffer a noticeable degradation in error performance. This threshold is important to recognize because of the preponderance of "Group 3 Facsimile" machines that operate at a bit rate of 9600 bps in the "normal" mode. However, these machines can "downshift" their bit rate to 4800 (or 2400) bps and this increases the transmission time (but not quite by a factor of two in the case of multipage transmissions).

Table 5.1
Impact of **ADPCM** on the transmission between modems

Modem Type	Data Rate in B/S	1 Transcoding BER	2 Transcodings BER	3 Transcodings BER
103	300	<10E-8	<10E-8	<10E-8
202	1200	<10E-8	<10E-8	<10E-8
212	1200	<10E-8	<10E-8	<10E-8
208	4800	<10E-8	<10E-8	<10E-8
209	4800	<10E-8	<10E-8	<10E-8
2048	4800	<10E-8	<10E-8	<10E-8
V.22	2400	<10E-8	<10E-8	<10E-8
V.27	4800	<10E-8	<10E-8	<10E-8
V.28	4800	<10E-8	<10E-8	<10E-8
OMNI Multi-Port Modem	2400	<10E-8	<10E-8	<10E-8
	4800	<10E-8	<10E-8	<10E-8
	7200	7.2 X 10E-5	9.6 X 10E-5 Sync Loss	Unstable
	9600	2.1 X 10E-4	1.9 X 10E-3 Sync Loss	Unstable
V.29	4800	<10E-8	<10E-8	<10E-8
V.29**	4800	1.2 X 10E-6	<10E-8	1.6 X 10E-5
V.32	4800	<10E-8	<10E-8	<10E-8
V.32**	4800	2.0 X 10E-6	<10E-8	3.1 X 10E-5
TELEBIT Packetized High Speed Data Auto Baud	6700	3.3 X 10E-7	–	–
	4400	–	<10E-8	–
	3100	–	–	<10E-7

Notes:
**: Alternate data modem manufacturer tested. B/S: Bits per second. BER: Bit Error Rate.
Measurements taken with Fireberd 2000 Data Error Analyzer Test Set.
See reference test configuration.

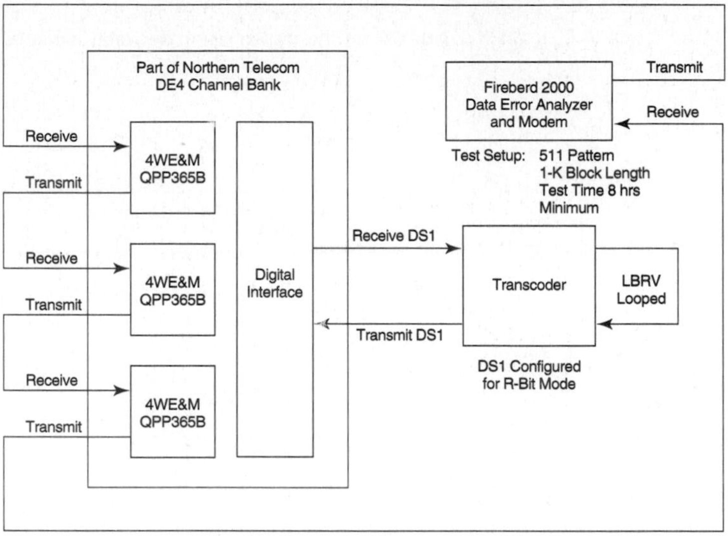

Fig. 5.17
Test setup for assessing impact of **ADPCM** on modem transmission

5.5.2.4 Performance with MF and DTMF Signals

The test setup used to measure the impact of **ADPCM** on signaling tones carried over the **PSTN** is depicted in Fig. 5.18.

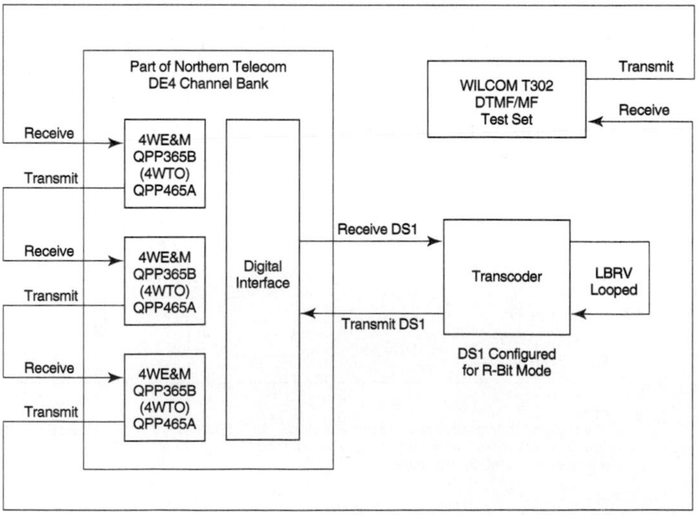

Fig. 5.18
Test setup for assessing impact of **ADPCM** on signaling tones

The results are presented in Table 5.2 for **DTMF** signals and in Table 5.3 for the **MF** case. **DTMF**, for "Dual Tone Multi-Frequency," corresponds to the tones used in pushbutton phones; **MF**, for "Multi-Frequency," is the scheme used between switches in a trunking configuration.

Table 5.2

Digit	XMT Composite Tone Pair Freqs. (Hz)	XMT Level (dBm)	RCV Composite Tone Pair Freqs. (Hz)	RCV Level (dBm)
1	697 + 1209	–10.0	696 + 1207	–10.3
2	697 + 1336	–10.0	697 + 1333	–10.3
3	697 + 1477	–10.0	697 + 1474	–10.3
4	770 + 1209	–10.0	771 + 1207	–10.3
5	770 + 1336	–10.0	771 + 1333	–10.3
6	770 + 1477	–10.0	771 + 1474	–10.3
7	852 + 1209	–10.0	854 + 1207	–10.3
8	852 + 1336	–10.0	854 + 1333	–10.3
9	852 + 1477	–10.0	854 + 1474	10.3
*	941 + 1200	–10.0	941 + 1207	–10.3
0	941 + 1336	–10.0	941 + 1333	–10.3
#	941 + 1477	–10.0	941 + 1474	–10.3

Notes:
1. Composite tone pait frequencies both sent at a –10.0-dBm level.
2. Composite tone pair frequencies both received at the level listed in the table.
3. Transmit time approximately 69 msec.
4. Receive on time approximately 65-67 msec.

Table 5.3

Digit	XMT Composite Tone Pair Freqs. (Hz)	XMT Level (dBm)	RCV Composite Tone Pair Freqs. (Hz)	RCV Level (dBm)
1	700 + 900	–10.0	700 + 900	–10.6
2	700 + 1100	–10.0	700 + 1102	–10.5
3	900 + 1100	–10.0	900 + 1101	–10.5
4	700 + 1300	–10.0	700 + 1300	–10.6
5	900 + 1300	–10.0	900 + 1301	–10.6
6	1100 + 1300	–10.0	1103 + 1301	–10.5
7	700 + 1500	–10.0	700 + 1500	–10.5
8	900 + 1500	–10.0	900 + 1500	–10.5
9	1100 + 1500	–10.0	1103 + 1500	–10.6
0	1300 + 1500	–10.0	1301 + 1500	–10.6

Notes:
1. Composite tone pait frequencies both sent at a –10.0-dBm level.
2. Composite tone pair frequencies both received at the level listed in the table.
3. Transmit time approximately 69 msec.
4. Receive on time approximately 65-67 msec.

5.5.2.5 Subjective Voice Tests with Random Line Errors

As part of the developmental effort that went into the design of the **ADPCM** algorithm, researchers at AT&T conducted several subjective tests and reported the results to **ANSI**. Transmission quality rating data were collected from several trials and numerous subjects. A numerical quality rating from 1 to 5 was assigned, with 5 representing "excellent." Tests were also conducted with random bit errors introduced into the span line. The test results reported involved one encoding/decoding in the case of **ADPCM**, and are summarized in Table 5.4.

Table 5.4

Coder	# of Encodings	Quality Rating
PCM 10^{-4}	1	≈ 3.80
PCM 10^{-3}	1	≈ 2.80
ADPCM 10^{-4}	1	3.93
ADPCM 10^{-3}	1	3.04

Note that at high error rates the subjective voice quality afforded by **ADPCM** is superior to **PCM**. However, it is quite unlikely that span lines with an error rate of 10^{-4} or 10^{-3} will be kept in service. At normal span-line-error rates, which range from 10^{-7} to 10^{-9}, the subjective quality difference between **ADPCM** and **PCM** is not noticeable except to possibly a very highly trained ear.

5.5.3 Concluding Remarks on ADPCM

Based on the experimental evidence, we can conclude that **ADPCM** is an excellent algorithm for compressing speech signals. It is robust enough to handle a wide variety of signals transported over a telephone network, such as speech, signaling tones, and signals from voice-band modems. Its sole drawback is that modems would have to fall back to rates of 4800 bps for acceptable performance over an **ADPCM** link. Thus while **ADPCM** is not "toll quality," it is certainly close to it.

The robustness, high quality, and low encoding delay of **ADPCM** have made it the method of choice in several applications where 64 kbps (regular encoding) is "too high" such as in cordless telephony. The low complexity of implementation, as evidenced by the availability of inexpensive chipsets, has made **ADPCM** a building block in **DCME** (Digital Circuit Multiplication Equipment) (see [18]) transport schemes. Even though **ADPCM** provides "only" a 2:1 compression ratio (i.e., at 32 kbps), it can be used in conjunction with **DSI** (Digital Speech Interpolation) techniques to obtain 4:1 and higher compression ratios. One such device, the **TC421** from DSC Communications Corp., which is based on the combination of **ADPCM** and **DSI**, and provides a robust 4:1 compression ratio on an aggregate of 96 channels, is described in Section 5.7.

5.6 LINEAR PREDICTIVE CODING OF SPEECH SIGNALS

Using **ADPCM** it is possible to get outstanding speech quality at 32 kbps and very good speech quality at 24 kbps. In order to get good quality at lower bit rates a different approach is used.

This technique, called ***Linear Predictive Coding*** (or **LPC**) was initially formulated in the mid-1960s and several variations have been proposed since then. Methods based on **LPC** are used to provide "understandable" quality at bit rates of 2.4 kbps and lower. Very good quality speech can be provided with bit rates of about 10 kbps.

LPC methods are a class of algorithms that are effective when the signal being compressed is speech or speech-like. By speech-like we mean that the signal can be modeled as shown in Fig. 5.19. The speech signal $\{s(n)\}$ is modeled as the output of an all-pole filter excited by a signal $\{t(n)\}$. The excitation $\{t(n)\}$ is a linear combination of a white noise component $\{q(n)\}$ and a periodic signal $\{p(n)\}$. That human speech can be approximated by the model in Fig. 5.19 is borne out by the success of compression algorithms that assume such a model.

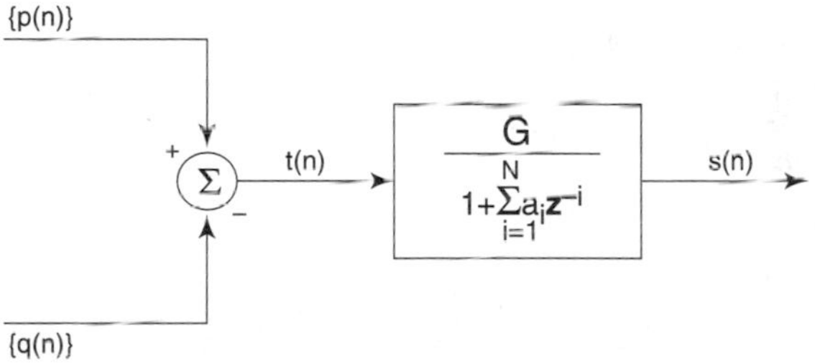

Fig. 5.19
Modeling speech as the output of an all-pole filter

Speech can be considered as the concatenation of "sounds" of duration approximately 64 msec. These sounds can be grossly classified as voiced, characterized by the dominance of the periodic component of the excitation, or unvoiced, in which case the noise component of the excitation is dominant. Consequently, the speech signal is analyzed in blocks, or frames, of samples corresponding to a duration of less than 64 msec. Typical frames are 20 to 32 msec in duration. The model parameters for these frames are computed and used for transmission rather than the speech waveform itself and in this manner we achieve data-rate compression.

In Section 5.6.1 we provide an overview of **LPC**. The intent is to explain the underlying principles of the algorithm. In Section 5.6.2 we introduce the normal equations. All **LPC** algorithms derive the coefficients of the model using the normal equations, or a simple variation thereof. To provide a flavor for the intricacies involved in implementing a speech coder using **LPC**, in Section 5.6.3 we describe, briefly, one implementation that has been standardized. The bibliography contains several references that describe other ways in which low-bit rate speech coders have been implemented.

5.6.1 LPC Algorithms

LPC coding techniques achieve data rate compression by transmitting information of the model rather than the speech waveform itself and the receiver reconstructs a representation, if not the actual waveform, of the information being transmitted. **PCM** and **ADPCM** techniques, on the other hand, transmit information by sending a code for each sample. By transmitting just a few parameters each analysis frame, considerable compression is achieved. For example, if 16 parameters, each represented by an 8-bit code, were transmitted every 32 msec, then the average bit rate would be 4 kbps, corresponding to a 16:1 reduction in bit rate compared to **PCM**. Since in **LPC** the intention is not to transmit the waveform but the informational content, **LPC** is a well suited for signals for which the model is applicable but may fail miserably on signals for which the model is not appropriate. **PCM** and **ADPCM**, on the other hand, are waveform encoding techniques and are thus better suited in situations where no assumptions on the nature of the signal are made.

Practical **LPC** schemes fall into three broad classes. In each class the information regarding the filter, i.e., the $\{a_i\}$, are transmitted for each block of samples. The classes differ in the manner by which information regarding the excitation is transmitted. What all **LPC** methods have in common is the notion of the model filter that defines the synthesis, and the notion of the residual $\{e(n)\}$ that characterizes the necessary excitation.

Given a set of coefficients $\{a_i\}$, an "inverse" filter can be applied to the observed speech signal $\{s(n)\}$. In particular, consider $\{e(n)\}$ given by (see Fig. 5.20(a))

$$e(n) = s(n) + \sum_{i=1}^{N} a_i s(n-i) \tag{5.6.1}$$

If the model is correct, and if the coefficients $\{a_i\}$ matched the model, then $\{e(n)\}$ is equal to the excitation $\{t(n)\}$, except possibly at the boundary of "model change." The signal $\{e(n)\}$ is called the ***prediction residual***. The three classes of algorithms differ in the manner by which information of the residual (excitation) is communicated.

The scheme that achieves maximum compression is, to coin an acronym, **MELP**, for "**M**odel **E**xcited **L**inear **P**rediction." In such a scheme we attempt to extract the pitch period, the voiced-unvoiced ratio, and a power normalization term from the prediction residual. The excitation is modeled as

$$t(n) = G\{\beta\varepsilon(n) + \alpha \sum_{k} \delta(n-kP)\} \tag{5.6.2}$$

where P is the pitch period, α and β determine the ratio of "voiced" power to "unvoiced" power, and G is a normalization constant to ensure that the reconstructed signal has the appropriate power. Assuming that the model is updated every M samples (256 samples at 8-kHz sampling corresponds to an observation, or analysis interval of 32 msec), then (N + 3) parameters need to be transmitted instead of M samples.

RELP, for "**R**esidual **E**xcited **L**inear **P**rediction," methods provide the least compression, albeit still more than **ADPCM**. The residual can be quantized quite severely while retaining adequate speech quality. Thus in **RELP**, every observation interval of M samples the N coefficients are transmitted in addition to kM bits, corresponding to quantizing each sample of the residual to k bits.

CELP, for "**C**ode **E**xcited **L**inear **P**rediction," methods are similar to **RELP** but use schemes to quantize the residual to, effectively, less than 1 bit per sample. This can be achieved by using a "code book" of possible excitation signals. Thus if the M samples of the excitation are "quantized" to one choice out of K possible M-point sequences, then only $\log_2(K)$ bits are required to describe the residual. The set of K choices is called the "code book" and hence the name **CELP**.

The techniques of linear predictive coding have been studied extensively and several implementations of LPC based encoders have been designed. They all share the same fundamental principles. Each implementation has a different "twist" to it, providing some advantage. These variations are based on extensive simulation and true ingenuity on the part of the developers. The availability of increased computational power in newer (faster, cheaper) digital signal processors has been a key technologically enabling factor. The underlying theoretical principles however, such as the pioneering works of Atal [1] and others (see Rabiner and Schaefer [7], for example) is still as appropriate today as it was twenty years ago and there have been few developments of that magnitude since.

5.6.2 The Normal Equations

Given a segment of the speech waveform, s(n) n = 0, 1, ... , (M–1), what are the appropriate coefficients $\{a_i\}$ to describe the model? One approach, and probably the simplest method, is to minimize the squared prediction residual averaged over the M samples. With {e(n)} given by Eq. (5.6.1), the error function E is

$$E = \sum_{n=0}^{M-1} [e(n)]^2 = \sum_{n=0}^{M-1} [s(n) + \sum_{i=1}^{N} a_i s(n-i)]^2 \qquad (5.6.3)$$

Differentiating E with respect to the coefficients and setting the derivatives to zero yields

$$\sum_{k=1}^{N} a_k R(i-k) = -R(i);\ i = 1, 2, \ldots, N \qquad (5.6.4)$$

where the R(i – k) are given by

$$R(i-k) = \frac{1}{M} \sum_{n=0}^{M-1} s(n-i)\, s(n-k) \qquad (5.6.5)$$

Eq. (5.6.4) describes a set of N linear equations called the ***normal equations***. Assuming that the speech signal is ergodic, then Eq. (5.6.5) corresponds to an estimate of the autocorrelation function of the speech signal. Thus we can assume

$$R(i-k) = R(k-i) = R(|i-k|) \qquad (5.6.6)$$

Writing Eq. 5.6.4 in matrix notation yields

$$\mathbf{R}\,\mathbf{a} = -\mathbf{r} \qquad (5.6.7)$$

The assumption in Eq. (5.6.6) implies that **R** has considerable structure. It is symmetric and, further, all diagonal elements are equal. Such a matrix is called **Toeplitz** and the solution of Eq. (5.6.7) can be sped up considerably by using all the implied symmetry. In particular, the following recursive procedure (see Makhoul [6] for an excellent discussion of the Normal Equations) is applied:

$$E_0 = R(0) \qquad (5.6.8a)$$

$$\mathbf{k}_i = -\frac{1}{E_{i-1}} \left[R(i) + \sum_{j=1}^{i-1} a_j^{(i-1)} R(i-j) \right] \qquad (5.6.8b)$$

$$a_i^{(i)} = \mathbf{k}_i \qquad (5.6.8c)$$

$$a_j^{(i)} = a_j^{(i-1)} + \mathbf{k}_i a_{i-j}^{(i-1)} \; ; \; 1 \le j \le (i-1) \qquad (5.6.8d)$$

Equations (5.6.8b) through (5.6.8d) are solved recursively for i=1,2,...,N. The final solution is given by

$$a_i = a_i^{(N)} \; ; \; i = 1, 2, \ldots, N \qquad (5.6.8e)$$

The recursive procedure described in Eq. (5.6.8) is such that all predictor coefficients for all model sizes of 1 through N are computed. Of special interest are the coefficients $\{\mathbf{k}_i \; ; i=1, \ldots, N\}$. The **k**-coefficients, also called ***reflection coefficients***, exhibit an invariance with respect to the model size. That is, $\mathbf{k}_i$ are the same value regardless of the size of the model N i. These coefficients are related to an equivalent lattice implementation of the filter, described by the direct form in Eq. (5.6.1) for the inverse filter, depicted in Fig. 5.20(b). Further, it can be shown that if, and only if, $\{a_i; i = 1, 2, \ldots, N\}$ describe the coefficients of a stable recursive filter described by

$$H(\mathbf{z}) = \frac{G}{1 + \sum_{i=1}^{N} a_i \mathbf{z}^{-i}} \qquad (5.6.9)$$

then the equivalent lattice filter will have

$$|\mathbf{k}_i| < 1 ; i = 1, 2, \ldots, N \qquad (5.6.10)$$

In other words, the recursive procedure described in Eq. 5.6.8 includes a built-in test for stability. The inverse filter, that is, the filter with the difference equation given in Eq. (5.6.1), will appear as shown in Fig. 5.20(a). The lattice implementation of the filter is shown in Fig. 5.20(b).

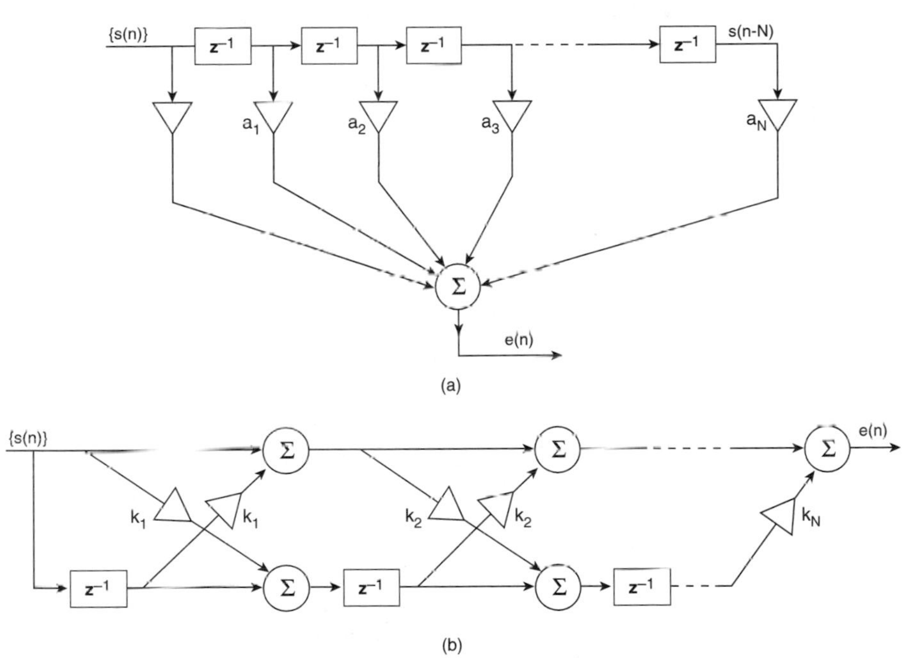

Fig. 5.20
Direct-form and lattice implementations of the *inverse* filter

In the late 1960s and early 1970s, a considerable amount of research effort went into characterizing the normal equations (Eq. (5.6.4) and Eq. (5.6.7)) and interpreting the nature of approximation to the model. An excellent summary is provided by Makhoul [6].

5.6.3 An 8-kbps LPC Algorithm

For application in digital cellular systems based on **TDMA** (Time Division Multiple Access), speech signals need to be encoded at approximately 8 kbps. In December 1989, a standard was published by the **EIA** (Electronic Industries Association), **IS-54** [19], which describes an algorithm that encodes speech at 8 kbps. The method is of the **CELP** variety and is called "**V**ector-**S**um **E**xcited **L**inear **P**rediction" or **VSELP**. A different version of an **LPC** encoder is specified in **IS-95** [20] for use in digital cellular systems employing spread spectrum, also called **CDMA** (Code Division Multiple Access). This latter method was developed at **Q**ualcomm Inc. and goes by the acronym **QCELP**. A brief description of **VSELP** follows.

5.6.3.1 Overview of VSELP

A block diagram of the decoder, or synthesizer, is shown in Fig. 5.21. The figure also indicates the various parameters that must be determined and encoded by the speech coder. The excitation is composed as the linear (vector) combination of two different code book excitations and hence the name **VSELP**.

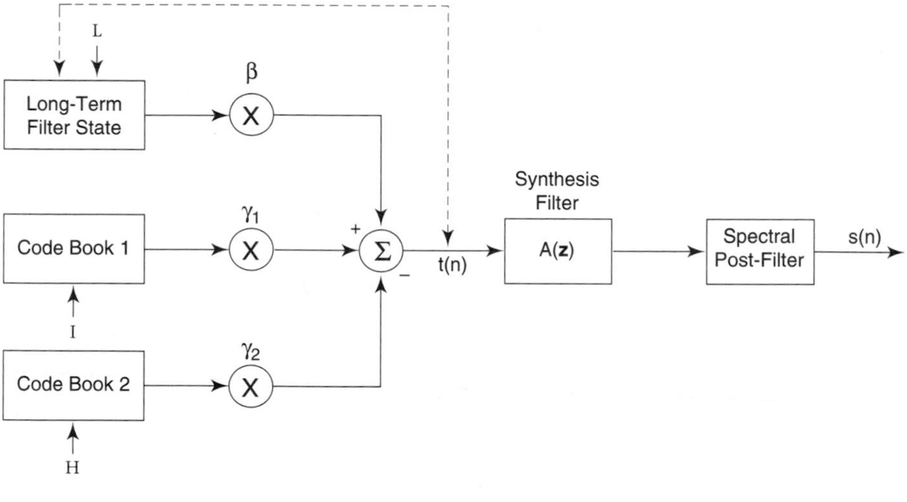

Fig. 5.21
VSELP model for synthesizing speech signals

The basic sampling rate used to convert the analog speech is 8 kHz. The frame size, i.e., the analysis interval, chosen is N_A=160 samples (20 msec). The parameters, encoded to a total of 159 bits, are transmitted every 20 msec, equivalent to a data rate of 7950 bps. The frame, 20 msec, is further subdivided into subframes of 40 samples (5 msec). Certain parameters are actually changed at the subframe rate to provide smooth transitions of parameters. The 159 bits per frame are divided between the parameters as shown in Table 5.5.

Table 5.5
Bit assignment for **VSELP** parameters

FILTER COEFFICIENTS $\{A_I\}$		38
FRAME ENERGY $R_{SS}(0)$		5
LAG L	7 BITS/SUBFRAME	28
CODEWORD I	7 BITS/SUBFRAME	28
CODEWORD H	7 BITS/SUBFRAME	28
GAINS $\beta, \gamma_1, \gamma_2$	8 BITS/SUBFRAME	32
TOTAL (PER FRAME)		159

The speech signal from the microphone is converted to digital using a μ-law codec and expanded to a uniform format using the same mapping specified in the **ADPCM** standard. Prior to analysis, a highpass filter is applied to the signal corresponding a fourth-order Chebyshev filter. This ensures that the speech signal has significant attenuation at 60 Hz, to minimize the impact of induced power line hum.

5.6.3.2 Processing Done Every Frame

The synthesis filter is a tenth-order all pole filter, i.e., N=10 (or **LPC-10**, as it is often referred to). The encoder computes the 10 coefficients $\{a_i\}$ using an algorithm similar to that described in Eq. (5.6.8). In **IS-54** care is taken to account for the finiteness of the analysis interval. The autocorrelation or "covariance" is computed as

$$\varphi(i, k) = \sum_{n=N}^{N_A - 1} s(n-i)\, s(n-k) \tag{5.6.11}$$

To further reduce the impact of the observation interval, the autocorrelation coefficients are windowed as in

$$\hat{\varphi}(i, k) = \varphi(i, k)\, w(|i-k|) \tag{5.6.12}$$

The window coefficients are specified in **IS-54** [19]. The purpose of windowing is to minimize the impact any discontinuity caused by the finite interval by weighting samples internal to the interval more than samples at the ends.

The reflection coefficients, $\{k_i\}$, that are computed (see Eq. (5.6.8)) are quantized as part of the algorithm using a bit allocation shown in Table 5.6. By quantizing the k_i as they are computed, the effect of quantization error is minimized. The quantization of the reflection coefficients is accomplished using a different code book for each of the 10 k_i. The quantized value is obtained by finding the code book value closest to the unquantized reflection coefficient. These quantized $\{k_i\}$ are converted into the direct form coefficients $\{a_i\}$.

Table 5.6
Bit assignment for the quantization of the reflection coefficients

COEFFICIENT	(BITS/FRAME)
K_1	6
K_2	5
K_3	5
K_4	4
K_5	4
K_6	3
K_7	3
K_8	3
K_9	3
K_{10}	2
TOTAL (BITS/FRAME)	38

To minimize the impact of a sudden change in the model, the speech coder linearly interpolates the $\{a_i\}$ for the first, second, and third subframes of each frame. This is accomplished by

$$\begin{aligned} a_i &= 0.75\, a_i(n-1) + 0.25\, a_i(n) \quad \text{(Subframe 1)} \\ a_i &= 0.5\, a_i(n-1) + 0.5\, a_i(n) \quad \text{(Subframe 2)} \\ a_i &= 0.25\, a_i(n-1) + 0.75\, a_i(n) \quad \text{(Subframe 3)} \\ a_i &= 0.0\, a_i(n-1) + 1.0\, a_i(n) \quad \text{(Subframe 4)} \end{aligned} \tag{5.6.13}$$

where $a_i(n)$ refer to the uninterpolated $\{a_i\}$ computed for the data in the current frame and $a_i(n\text{-}1)$ are the uninterpolated $\{a_i\}$ computed in the previous frame. The synthesis filter coefficients are thus modified every subframe (5 msec).

So that the synthesis procedure generates a signal of the same level as the input speech, an average power value is computed and encoded. This is done at the frame rate, once every 20 msec. This average power, $R_{ss}(0)$, is computed as

$$R_{ss}(0) = \frac{\varphi(0,0) + \varphi(N,N)}{2(N_A - N)} \tag{5.6.14}$$

and reflects an averaging method that is "biased" toward the latter half of the analysis window of N_A samples, again, to account for the finiteness of the frame. If we define R_{max} as the square of the maximum signal amplitude, the average power can be expressed in dB as

$$R_{dB} = 10 \log_{10} \left[\frac{R_{ss}(0)}{R_{max}} \right] \tag{5.6.15}$$

R_{dB} is quantized to 32 levels (5 bits). Signal powers of less than –72 dB are assumed to be null and the corresponding code is used to totally silence the decoder. The nominal power of the signal that the decoder, i.e., the synthesizer, must generate is $R_q(0)$, the power value corresponding to the uniform (linear) equivalent of the quantized form of R_{dB}. To smooth out changes from frame to frame, the actual power level, $R_q(0)$, utilized is modified every subframe using the following procedure.

$$R_q(0) = R_q(0)(n-1) \qquad \text{Subframe 1} \qquad (5.6.16a)$$

$$R_q(0) = [R_q(0)(n-1)R_q(0)(n)]^{0.5} \qquad \text{Subframe 2} \qquad (5.6.16b)$$

$$R_q(0) = R_q(0)(n) \qquad \text{Subframes 3 and 4} \qquad (5.6.16c)$$

where the (n) stands for current frame and (n-1) for the previous frame.

5.6.3.3 Processing Done Every Subframe

In each subframe the coder must determine the parameters that are updated on a subframe basis, namely the relative gains β, γ_1, and γ_2 (that determine the mix of the excitation signal), the two codewords I and H, and the long-term predictor lag, L.

The function of the long-term filter state block shown in Fig. 5.21 is to account for the (approximate) periodic behavior exhibited by voiced sounds. The quantity L can be viewed as the pitch period of voiced sounds. L is encoded using 7 bits for each subframe, where one of the 128 possible codes is reserved for the case when the pitch prediction is deactivated, as in the case of unvoiced sounds.

Conceptually, the quantities determining this make up of the excitation can be estimated using the model depicted in Fig. 5.22.

The prediction residual e(n) is computed using an inverse filter and the coefficients $\{a_i\}$ that are computed for the subframe being analyzed. The appropriate choice of the quantities L, I, H, β, γ_1, and γ_2 are those that minimize the strength of the difference between {t(n)}, the synthetic excitation, and {e(n)}, the prediction residual or "ideal" excitation. Experimentally it is observed that the subjective quality of the synthesized speech is superior if the minimization process is applied to a filtered version of $\{\eta'(n)\}$. The scheme specified in **IS-54** minimizes the energy, measured over the subframe, of the signal $\{\eta_1(n)\}$ obtained by filtering $\{\eta'(n)\}$ by a ***perceptual noise weighting filter*** given by

$$H_p(\mathbf{z}) = \frac{1}{1 + \sum_{i=1}^{N} a_i \lambda^i \mathbf{z}^{-i}} \qquad (5.6.17)$$

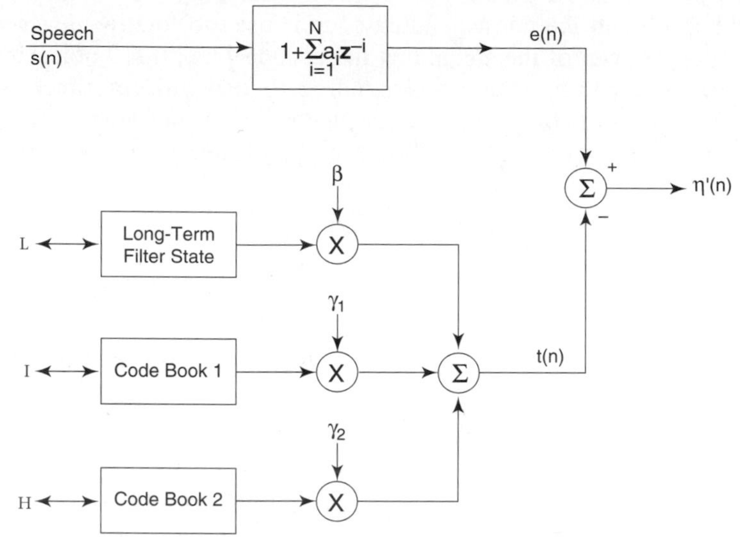

Fig. 5.22
Comparing code book excitation with the actual prediction residual

The perceptual noise weighting filter resembles the synthesis filter except that each coefficient a_i is weighted by λ^i where $0 < \lambda < 1$. **IS-54** specifies $\lambda = 0.8$. It is seen that if $re^{j\theta}$ is a pole of the synthesis filter, $(\lambda r)e^{j\theta}$ is a pole of $H_e(z)$. Consequently, in the frequency domain $|H_e(f)|$ will exhibit peaks where $|H(f)|$, the response of the synthesis filter, exhibits peaks. As a consequence, the matching of {t(n)} and {e(n)}, when viewed in terms of spectral content, will be better where $|H(f)|$ has peaks, albeit at the expense of spectral regions where $|H(f)|$ has valleys.

The **VSELP** algorithm specifies an "analysis-by-synthesis" approach to the minimization process and is depicted in Fig. 5.23. The intent is to choose the parameters so as to minimize the power of a signal {ε(n)} that incorporates the effect of the perceptual weighting. The notion of analysis by synthesis is nothing other than a search over all possible choices of the parameters in a systematic fashion.

The error signal {ε(n)}, whose strength must be minimized, is composed of

$$\varepsilon(n) = p(n) - p'(n) \tag{5.6.18}$$

where {p(n)} is obtained by filtering the speech {s(n)} by W(z) given by

$$W(\mathbf{z}) = \frac{1 + \sum_{i=1}^{N} a_i \mathbf{z}^{-i}}{1 + \sum_{i=1}^{N} a_i \lambda^i \mathbf{z}^{-i}} \tag{5.6.19}$$

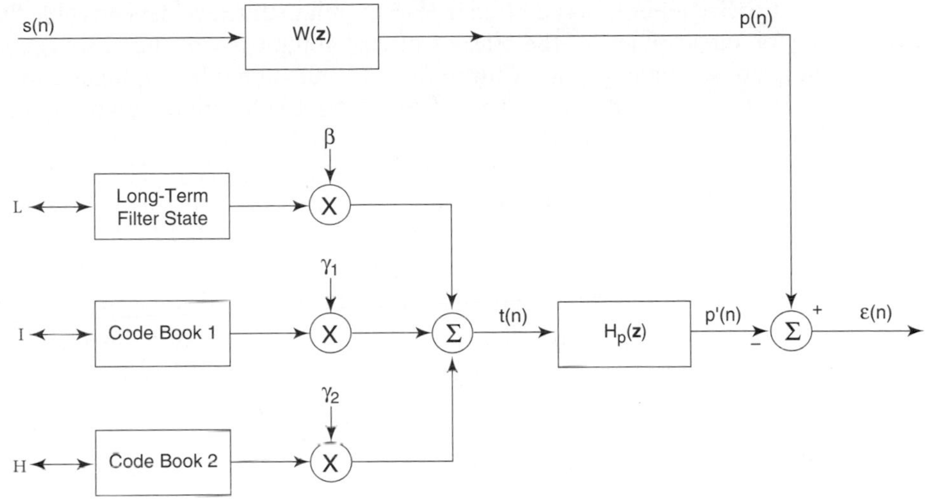

Fig. 5.23
Depicting the scheme for identifying optimal code book excitation

The parameters L, I, and H that control the signal generators, and their relative weighting β, γ_1, and γ_2 are chosen to minimize $\Sigma(e(n))^2$ over the subframe. The **VSELP** algorithm specifies a technique whereby L and β are determined first, using a correlation process with $\{p(n)\}$. With this choice of L and β a search process is outlined to obtained the best codeword I that describes the second component of $\{t(n)\}$, and then H and is obtained, also by a search procedure. Code books 1 and 2 each represent 128 normalized vectors. The "best" choice of a vector in the Code book is that which, when passed through the weighted synthesis filter represented by $H_e(z)$, provides the highest correlation with $\{p(n)\}$ over the length of the subframe. Finally the relative weights β, γ_1, and γ_2 are obtained to minimize $\Sigma(e(n))^2$ over the subframe.

In the process of obtaining β, γ_1, and γ_2, it is convenient to obtain three other related parameters P_0, P_1, and GS that uniquely determine $\{\beta, \gamma_1, \gamma_2\}$. A vector quantization scheme is used to encode the triplet $\{GS, P_0, P_1\}$ to 8 bits. In other words, the triplet $\{GS, P_0, P_1\}$ is approximated by one of a set of 256 triplets that are precomputed.

Thus on a subframe basis the bit allocation is

L	:	7 BITS
I	:	7 BITS
H	:	7 BITS
$\{GS, P_0, P_1\}$	:	8 BITS

for a total of 29 bits per subframe.

5.6.3.4 General Comments on VSELP

The **VSELP** algorithm specified in **IS-54** is relatively new. It is possible that revisions may be made either for the reasons of cost/complexity or the observation of certain anomalous situations that occur in practice but cannot be duplicated in a simulation or laboratory environment. The quality, from a subjective viewpoint, of the synthesized speech is extremely good, far superior to that obtained by the 16-kbps **ADPCM** algorithm and virtually on par with 32-kbps **ADPCM**. The trade-off is one of compression-ratio versus complexity—the **VSELP** implementation is much more expensive than the implementation of **ADPCM**. The **ADPCM** algorithm has the "advantage" that it is useful even when the signal is not speech-like; an advantage that is moot in the case where the nature of the signal is known.

The **VSELP** algorithm described also introduces **delay**. Considering that a block of speech samples must be collected and analyzed, there is an inherent minimum delay of one frame or 20 msec. This is in addition to other delays associated with other processing and transmission. While this is certainly an issue in a cellular telephone application, for a trunking application this delay may not be acceptable. With so much delay, which appears in the tail circuit, echo cancelers require "tail delays" in excess of 96 msec and such cancelers can be quite expensive (see Chapter 6, which discusses echo cancelation for the notion of tail delay, tail circuit, and the need for echo treatment). Further, long-tail-delay cancelers often degrade the signal, the degradation being roughly proportional to the tail delay. There are several algorithms being researched for low-bit rate encoding with low attendant delay for use in the general telephone network (see **CCITT** Recommendation **G.728** [14], for example).

5.7 DIGITAL SPEECH INTERPOLATION

Conversations between humans are inherently half-duplex, at least for most of the time. Normally, one party talks and the other party listens. Furthermore, normal speech has several pauses. Thus it is intuitively satisfying that from a statistical viewpoint, a speech signal is active only about 50% of the time. From the viewpoint of speech transmission therefore, "compression" can be achieved by allocating bandwidth only on demand. However, for such a compression scheme to be effective, bandwidth allocation has to be based on an aggregate of several separate channels, to take advantage of the statistical properties. In this section we discuss some of the issues related to this form of compression, also called Digital Speech Interpolation (**DSI**), by considering a specific example.

One such compression device, called the **TC421**, is available from DSC Communications Corp. The **TC421** aggregates the traffic from four "uncompressed" **DS1** lines, or 96 separate speech channels, into a single **DS1** transmission link, achieving a net compression of 4:1. It utilizes a proprietary scheme that combines speech interpolation and variable rate **ADPCM** to achieve its overall compression ratio. Being proprietary, a **TC421** unit is required on each end of the **DS1** span line, as depicted in Fig. 5.24.

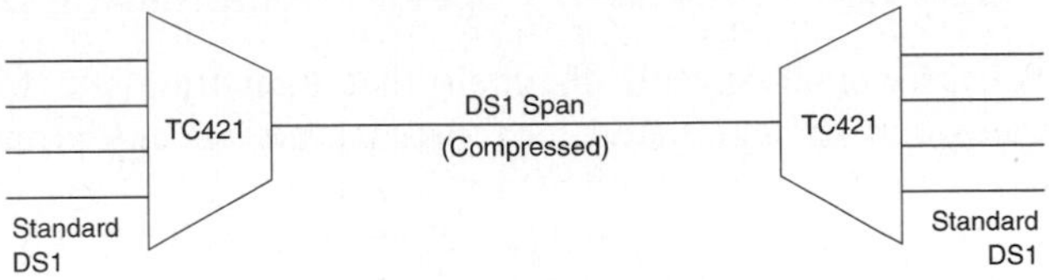

Fig. 5.24
Using the **TC421** for 4:1 compression

A simplified block diagram showing the various functions of a **TC421** is provided in Fig. 5.25. In the encoding direction, incoming speech channels are analyzed and

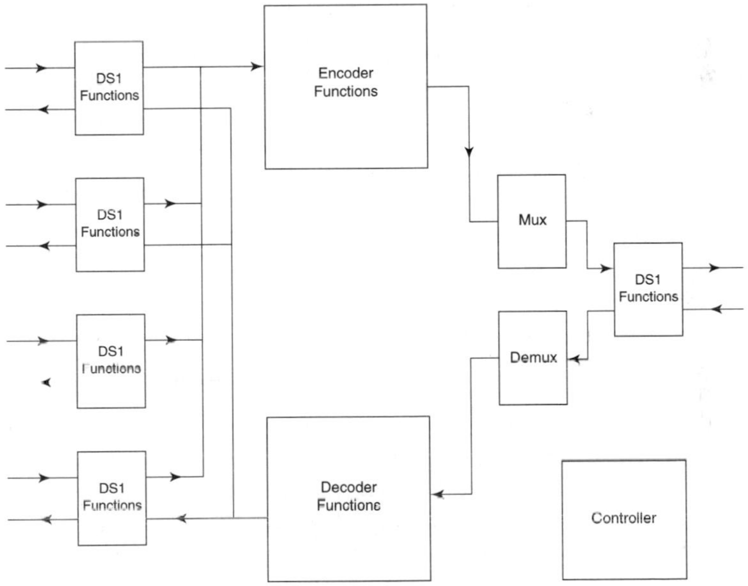

Fig. 5.25
Block diagram of the TC421

bandwidth suitably allocated. The multiplexed bit-stream is transported as a **DS1** signal to the companion **TC421**. In the reverse, or decoding direction, the incoming **DS1** signal from the distant **TC421** is demultiplexed and the speech signals for each channel synthesized. The external interfaces are all at the **DS1** level and thus **DS1** functions such as framing, synchronization, signaling bit extraction and insertion, line equalization, and so on, need to be performed. The ensuing description of the **TC421** concentrates on the key digital signal processing concepts utilized; specific details on the variety of features and functions can be obtained directly from the manufacturer.

5.7.1 Overview of the TC421 Speech Compression Device

The N (=96) input speech channels that constitute the four uncompressed **DS1** aggregates can be segregated into three classes at any point in time.

$$N = N_A + N_S + N_I \qquad (5.7.1)$$

Here N_I is the number of channels that are not involved in the transport of a conversation and can be considered idle. Determination of the idle/busy status of a channel is based on the channel associated signaling bits. If the channel is designated as "**CCIS**" (for Common Channel Inter-office Signaling where the state of the channel is communicated between switching machines using a separate channel), in which case the signaling bits are not available, the channel is deemed always busy, i.e., potentially carrying traffic. The busy channels, $N_A + N_S$, are divided into those that are carrying "silence" (N_S) and those that are carrying active speech (N_A). Bandwidth is allocated on the compressed **DS1** link only for the N_A active channels. The ratio (N_A/N) is the effective compression ratio achieved by the bandwidth allocation based on channel activity.

The **DS1** link has a payload capacity of 1.536 Mbps, less the bandwidth required for communication between the companion units at either end. Ignoring the latter component for the moment, the capacity of the link is equivalent to 24 64-kbps channels, 48 32-kbps channels, 64 24-kbps channels, or 96 16-kbps channels. Under normal conditions, only half the channels would be expected to be active. Thus 32 kbps **ADPCM** can be applied to the 48 active channels. Under light loading conditions, some or all of the active channels can be allocated more bandwidth by utilizing 40 kbps **ADPCM**. Under conditions of heavy loading, when several conversations are active in the same direction at the same time, the bandwidth allocated can be throttled down to 24- or even 16-kbps as necessary. Since 40-kbps and 32-kbps **ADPCM** provide essentially toll-quality (subjective) speech channels, the speech quality under normal conditions is indistinguishable from an uncompressed link. The bandwidth throttling scheme provides a graceful degradation under conditions of peak loading.

Because some of the **DS1** payload is utilized for overhead, there may be instances where the available bandwidth may be insufficient to transport all the N_A channels requiring transmission bandwidth. This results in ***blocking***, which is a very undesirable phenomenon. In the **TC421** the intermachine link uses 80 kbps to transmit the various parameters describing each of the 96 channels to the distant end. This means that if all 96 channels are active simultaneously, 91 channels can be transported at 16 kbps and 5 channels are blocked. Viewed another way, with an average channel activity factor of less than 95%, blocking is extremely rare. Since in practice the average speech channel is active only about 40% to 50% of the time in one direction, the phenomenon of blocking is virtually nonexistent in the **TC421**. The allocation of bandwidth to the overhead channel involves an interesting trade-off. If the control channel uses a smaller data rate, the probability of blocking is reduced. However, the rapidity of parameter updates is also reduced. The assignment of the overhead channel is thus a compromise between blocking, the ability to update the state of the constituent channels, and, needless to say, complexity of implementation.

One of the primary drawbacks of **ADPCM**, operating at 32 kbps, is the inability of the compression scheme to support signals from 9.6-kbps modems. While not a major issue in the 1980s, the proliferation of facsimile (FAX) machines that operate at 9.6 kbps have made this shortcoming of **ADPCM** the reason why it is not as widely deployed as it could be. 40-kbps **ADPCM** algorithms, however, do support 9.6-kbps modems. This observation is made use of in the **TC421**. A scheme for detecting the compressibility of the channel signal is employed and the signal in any particular channel is categorized as "speech," i.e, compressible, or "data" implying that 40 kbps is required. Data channels are automatically assigned 40 kbps. By opening up bandwidth for such data calls, the available bandwidth for channels carrying speech is reduced, implying an increased probability of blocking. For example, with 10 data calls in progress, with 80 kbps for the control channel, the link can support 66 16-kbps pipes. Blocking occurs if the (average) channel occupancy is greater than 66/86 or 76.7%. Clearly, the scheme can support several data channels before the probability of blocking becomes an issue.

5.7.2 Speech Interpolation

The underlying principle of speech interpolation is that periods of silence can be filled in an artificial manner. From a bandwidth perspective, the transmission medium is utilized by a speech channel only when it is active. From a signal processing perspective, **DSI** involves two processes—first determining whether the channel is active or silent, and second, generating a signal to fill in the gaps.

If the short-term power of a speech signal is plotted against time, it would take the form exemplified in Fig. 5.26. When the channel is silent, the power is low, less than some threshold. Active speech is characterized by the short-term power being above the threshold. If the threshold is large then the detector will have the propensity for classifying active speech as silent, whereas a low threshold may characterize a period of silence as active. The effective compression ratio is thus related to the threshold. Since the average speech power may vary with time, from call to call and sometimes during a call, the threshold needs to be adaptive. In telephony, the speech levels observed are such that the threshold level is in the range of –55 dBm0 and –40 dBm0. That is, signals below –55 dBm0 (35 dBrN0) are most probably background noise and signals above –40 dBm0 are most probably active speech. A simple scheme for adapting the threshold is discussed later.

However, a simple comparison of power to threshold is not sufficient for accurately classifying speech signals as active or silent. The normal cadences in speech make the graph of power with time appear as numerous hills and valleys during active speech. It is possible for these valleys to drop below the threshold for short durations. Categorizing these dips as silence and inserting a synthesized signal in their place gives the resultant speech an unnatural quality that is quite unpleasant. The active/silent detection must therefore bridge such gaps, which could be tens of milliseconds

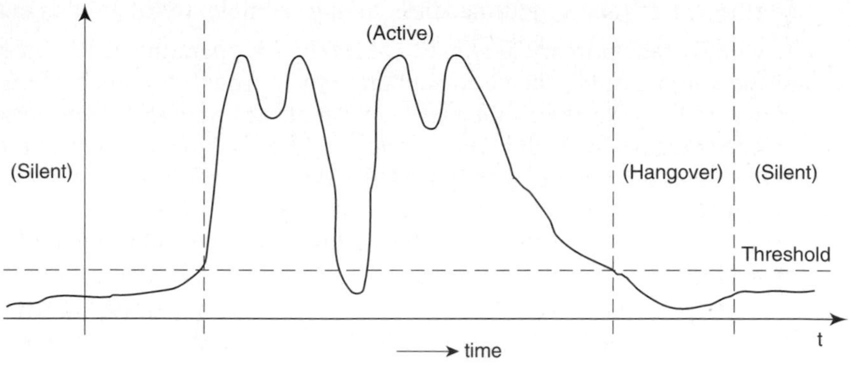

Fig. 5.26
Classification of silence via short-term power

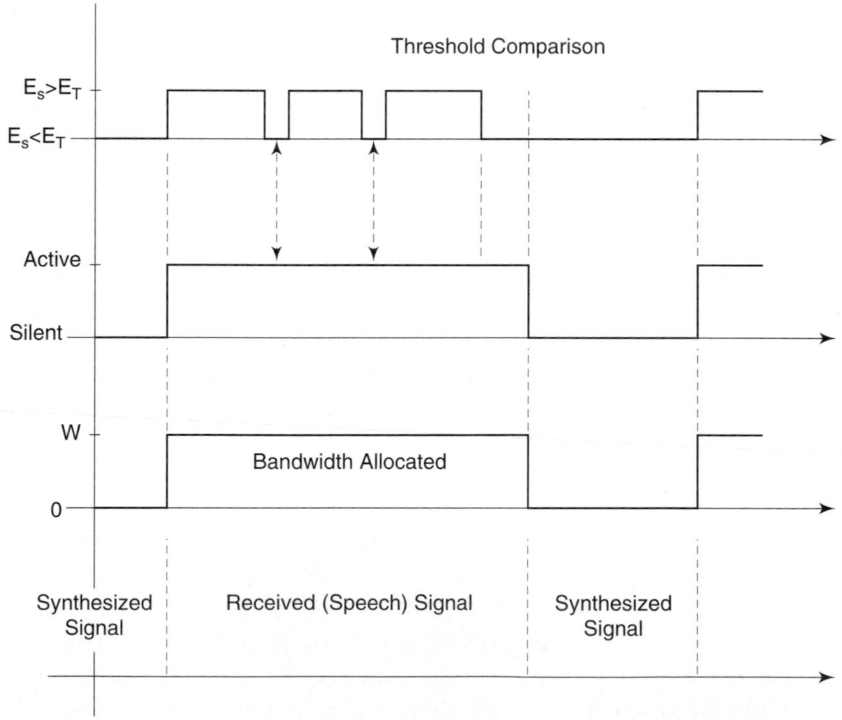

Fig. 5.27
DSI: Bandwidth is allocated only when signal classified *active*

long. A second property observed is that at the end of an active speech interval, the signal power may drop below the threshold even though the segment of the signal is important from a perceptual standpoint. This overhang, or transition time, is of the order of 50 msec and for this period the signal, even though the power is below the threshold, must be considered active. The attack time of an utterance, that is the onset of an active interval, can be quite rapid. It is important that the signal be considered active with minimal delay. The **TC421** achieves these operations of bridging, hangover, and fast attack in a manner described next.

Denoting by E_s the short-term power of the signal, by E_T the threshold level, and W the bandwidth utilized to transport the active speech segments, the operation of digital speech interpolation can be depicted in Fig. 5.27. The decoder thus outputs the received speech during active segments and a synthesized signal during silent segments.

The synthesized signal during periods of silence is quite important. Absolute silence between utterances is quite unnerving and is quite contrary to what the human ear is expecting. Random (or pseudo-random) noise is by far the most pleasing to the ear and is the logical choice of signal with which to replace silent segments. For best subjective results the power of the synthesized noise should be commensurate with the power of the background noise the ear can perceive during active speech.

The adaptive nature of threshold setting can be combined with background noise computation in the following simple, but effective, algorithm. Let E_N denote the estimate of the noise power, E_s the signal power, and E_T the threshold, at the current decision instant. Then E_N and E_T are updated as follows:

$$\text{If } E_s \geq E_T^{(old)} \text{ Then } E_T^{(new)} = E_T^{(old)} + \Delta E_+ \quad \text{And } E_N^{(new)} = E_N^{(old)} \tag{5.7.2a}$$

$$\text{If } E_s < E_T^{(old)} \text{ Then } E_T^{(new)} = E_T^{(old)} - \Delta E_- \quad \text{And } E_N^{(new)} = \alpha E_N^{(old)} + (1-\alpha) E_s \tag{5.7.2b}$$

$$\text{If } E_T^{(new)} < E_{MIN} \text{ Then } E_T^{(new)} = E_{MIN} \tag{5.7.2c}$$

$$\text{If } E_T^{(new)} > E_{MAX} \text{ Then } E_T^{(new)} = E_{MAX} \tag{5.7.2d}$$

The intent of the algorithm is to increase E_T during periods of active speech and decrease E_T during periods of silence. The increment and decrement need not be the same. Also, the threshold is constrained to be between two extremes E_{max}, and E_{min}, which are approximately –40 dBm0 and –55 dBm0, respectively. The averaging process for estimating the background noise power is applied only when the signal level is below the threshold for active speech.

The **TC421** utilizes the so-called **R**udimentary **D**igital **S**peech **I**nterpolation (**RDSI**) scheme. Since the burden of compression is shared between speech interpolation and variable rate **ADPCM**, a simplified form of **DSI** can be applied. The underlying premise of **RDSI** is to minimize the probability of incorrectly categorizing active speech as silent, albeit at the expense of increasing the probability of categorizing silent periods as active. This may reduce the compression ratio associated with the **DSI** but results in very high quality recovered speech without the clipping and other annoyances associated with other implementations of speech interpolation.

The speech signal being analyzed is broken into analysis intervals, termed "masterframes," of duration 12 msec. The term masterframe stems from the nomenclature of frame, corresponding to an interval of 125 μsec, associated with **DS1** signals. A masterframe is thus 96 **DS1** frames. The distinction of active or silent is applied to the entire masterframe. Since speech signals exhibit a high peak-to-rms ratio, the criterion to determine whether a segment is active utilizes both the short-term-average power E_S over the interval as well as a measure of the peak signal sample E_P. To ensure that the reconstructed speech sounds as natural as possible, the **RDSI** scheme tends to err, if at all, in classifying silent periods as active. In order to do this, we use a two-step procedure for classifying whether the masterframe is active or silent. Thus an interval is classified as "potentially active" if

$$E_S > E_T \ \ \textbf{OR} \ \ E_p > \text{Threshold} \tag{5.7.3}$$

and "potentially silent" otherwise. A masterframe is classified as silent if it is potentially silent **and** (M–1) previous masterframes were also potentially silent. With a 12-msec masterframe, a suitable value for M is 6.

The impact of this scheme is to classify as many as M (=6) masterframes as active in excess of the true period of active speech, resulting in a reduction of the compression ratio. However, the method ensures that valleys in the short-term power are bridged and that there is adequate hangover time. The use of the peak value as part of the decision ensures that the attack time is rapid. The net result is reconstructed speech that is of very high quality, without any discrepancies at the start, end, or middle, of utterances.

5.7.3 Signal Processing in the Encoder

A simplified block diagram of the encoding functions is shown in Fig. 5.28.

The input speech signal {s(n)} comprising a 64-kbps μ-law encoded **PCM** signal is organized into blocks of 12 msec, i.e., in masterframes and fed to the **ADPCM** encoder.

The **ADPCM** encoder is of variable bit rate. The bit rate assignment comes from the bandwidth allocation module and is changed on masterframe boundaries. In so doing, it is possible to synchronize the changes in bit rate at the encoder and decoder. Commercial chip sets are available, for example from Base-2 Systems, Inc., that implement the **ADPCM** algorithm on a multiplicity of channels. Base-2 provides a 2-chip chip set that, with some additional memory, can implement the **ADPCM** algorithm on 48 channels simultaneously. The algorithm implemented can be changed, on an individual channel basis between 16/24/32/40 kbps. Furthermore, this change can be implemented in a "hit-less" manner if the encoder and decoder are suitably synchronized.

The computation of short-term power and peak detection functions are straightforward and are implemented as

$$E_s(n) = \alpha E_s(n-1) + (1-\alpha)[s(n)]^2 \tag{5.7.4}$$

and

$$E_p(n) = \begin{cases} |s(n)| \text{ if } |s(n)| \geq E_p(n-1) \\ \alpha E_p(n-1) \text{ if } |s(n)| < E_p(n-1) \end{cases} \tag{5.7.5}$$

where in both cases α is a constant close to unity, approximately 0.95 to 0.99. The threshold computation and determination of "potentially active" follow the schemes laid out in Eq. (5.7.2) and Eq. (5.7.3), respectively, and are done once per masterframe, utilizing the values of $E_S(n)$ and $E_p(n)$ corresponding to the last sample (of the masterframe).

The background noise power of the channel is communicated to the decoder using the overhead (control) channel in the compressed **DS1** stream. Consequently, it must be quantized to a small number of bits, about 2. Recognizing that the noise power is nominally a constant and is not expected to change rapidly, the quantization to 2 bits can be accomplished using a **ΔΣM** scheme of the type described in Chapter 8.

The signal classification block refers to the processing associated with determining the compressibility of the signal. A speech signal can be compressed, if necessary, to as low as 16 kbps whereas a *data* signal can be compressed to 40 kbps and not lower (if the signal classification could further classify the data into high speed and low speech it is possible to use 32 kbps, or lower, for the lower speed data signals. However, this classification requires an implementation of significant complexity and cost). There are several coarse methods for this discrimination. The **TC421** utilizes two such methods and combines the results, declaring the signal to be data if **either** scheme determines the signal as such. Thus the classification is biased, minimizing the probability of falsely classifying a data signal as compressible at the expense of the probability of incorrectly identifying a speech signal as incompressible.

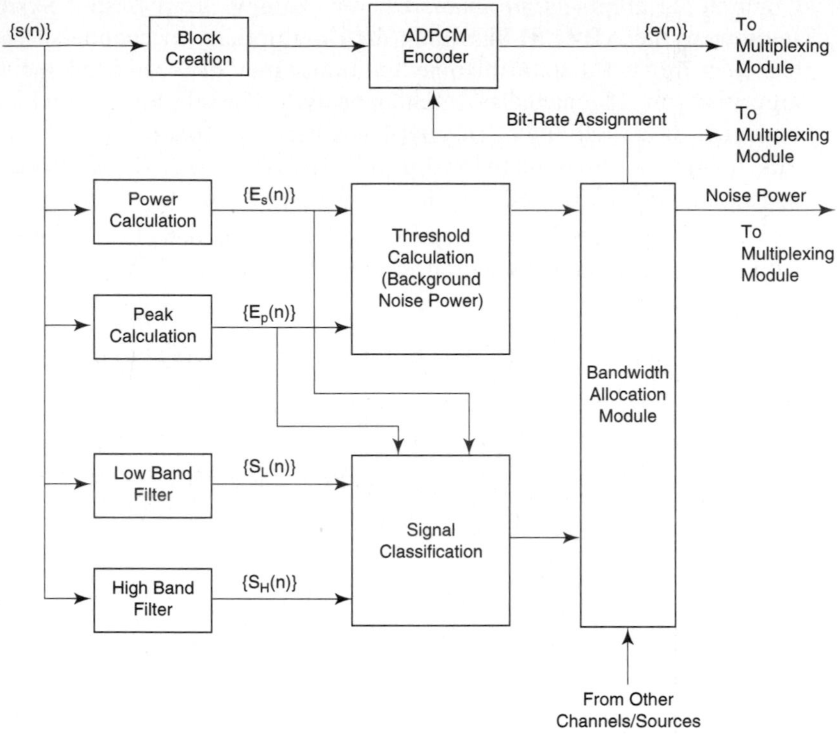

Fig. 5.28
Signal processing in the **TC421**, a speech compression device employing digital speech interpolation together with variable-rate **ADPCM**

The simplest form of discrimination uses the ratio of peak-to-rms value. Let

$$R = 10 \log_{10} \{ \frac{E_s}{E_p^2} \} \tag{5.7.6}$$

If the signal is a pure sinusoid, then R is about 3 dB. Modem-generated signals, even those that use multiple carriers, exhibit a more Gaussian-like **pdf** and thus the value for R is around 9 to 11 dB for such traffic. Speech signals however, are characterized by strong peaks and will have R in the range of 13 to 17 dB. Thus the classification algorithm considers the signal to be data if R < 12 dB (or $E_P^2 < 4\,E_s$).

The second discriminator utilizes a frequency domain criterion. Speech signals are characterized by having most of their power in lower frequencies, typically below 1 kHz. Modem-generated signals tend to have almost flat spectra with the propensity of having more power in the higher frequency region, above 1 kHz. Thus by applying two bandpass filters, one extracting the signal in the range [300, 1000] Hz and the other in the range [1000, 3400] Hz, the nominal bandwidth of [300, 3400] Hz is classified via the power in the low band and high band. If there is more power in the high frequency band then the signal is classified as data; greater low frequency power implies the signal is probably speech.

There are other, effective, discriminators that can be applied. For instance, if the signal {s(n)} is analyzed using **LPC** techniques and the power of the residual signal (prediction error) is small, then the conclusion is that the **LPC** model is indeed effective, implying that the signal is, in all probability, speech. A large residual signal power indicates a nonspeech-like nature. A signal with a low prediction residual power is compressible; a large prediction residual power indicates that the signal is not compressible.

The **ADPCM** algorithm also can provide a clue as to the nature of the signal. The adaptation speed control parameter of the adaptive quantizer sheds light on the nature of the signal. However, this internal parameter may not be readily available outside the chip set implementing the **ADPCM** algorithm. The particular choice of peak-to-rms ratio and spectral division of power was chosen for the **TC421** based on considerations of implementation complexity.

The bandwidth allocation module utilizes information from all 96 channels, as well as from the other portions of the device, such as the **DS1** functions and the controller, and is described in the next section dealing with the compressed **DS1** link.

5.7.4 The Compressed DS1 Link

The compressed **DS1** bit-stream follows the standard **D4/ESF** framing format. Thus the payload comprises of 192 bits per **DS1** frame. Of these, 10 are allocated for overhead and 182 for transporting the data from the 96 channels. These frames are organized as masterframes, where each masterframe is 96 **DS1** frames or 12 msec. This assembly is depicted in Fig. 5.29. The "F" bits correspond to the framing bits of the **DS1** signal. Of the 10 overhead bits, one is used to establish a masterframe alignment pattern. This permits the numbering of frames modulo-96, from frame 0 through frame 95. Since this master-framing pattern is carried within the payload, there is no connection with the **D4**-superframe. This ensures that the compressed **DS1** can be transported and cross-connected at the 1.536-Mbps (=24*64 kbps) level by **DACS** (<u>D</u>igital <u>A</u>ccess and <u>C</u>ross-connect <u>S</u>ystem) machines that conform to accepted industry standards.

The information that must be transmitted on a per-channel basis comprises bit rate, signaling bits, and background noise power. In addition, information on a per-**DS1** basis, such as ***alarms***, must be communicated to the distant end. There are two basic approaches to transmitting this overall information. One approach is to provide updates; the other is to provide a *complete snapshot every* masterframe. The first approach is more efficient from the point of view of bit rate required for transmission since the state of the machine is not likely to change drastically from masterframe to masterframe. The second approach utilizes more control channel bandwidth. However, by providing a complete picture of the state of the machine every masterframe, the link is more robust and can recover rapidly in the event of transmission errors. The **TC421** control channel philosophy is geared toward the latter approach, opting for robustness over minimum bandwidth.

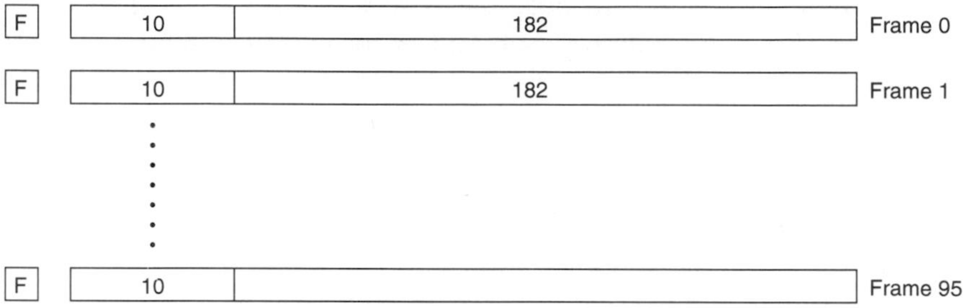

Fig. 5.29
Notion of a *masterframe* as 96 **DS1** frames

The bandwidth allocation algorithm is quite simple. The N=96 channels are subdivided, on a per-masterframe basis, into

$$N = N_A + N_S + N_F + N_I + N_D \tag{5.7.7}$$

The **TC421** allows the operator to predetermine the bandwidth allocation for certain prescribed channels. For example, some of the constituent **DS0**s of the uncompressed **DS1**s might be carrying 64-kbps digital data. **ADPCM** or **DSI** on these **DS0**s is meaningless and these (sub)channels must be transported as intact 64-kbps streams. Eq. (5.7.7) indicates that there are N_F such "fixed-rate" (the bit rate prescribed for these channels is under operator control and can be 0, 16, 24, 32, 40, or 64 kbps) channels. N_I refers to the number of channels that are idle as implied by their signaling state, because of an existing alarm condition, or because the **DS1** may be *out-of-service*. N_D are those channels that have been designated as data and must be assigned 40 kbps, or 5 bits per frame (125 msec). N_A is the number of channels that are active and N_S those that are silent.

Thus 182 bits per frame must be assigned to the $(N_F + N_D + N_A)$ channels. The $(N_F + N_D)$ channels are assigned bandwidth first and the remaining bits per frame divided "equally" over the remaining N_A channels. Denoting by B_F the bit assignment per frame for the N_F channels, then B_S bits, where

$$B_S = 182 - B_F - 5N_D \tag{5.7.8}$$

remain for assignment to the N_A active speech channels. Blocking occurs if

$$B_S < 2\,N_A \tag{5.7.9}$$

since the minimum bit rate for an active speech channel is 16 kbps or 2 bits per frame.

The bit assignment algorithm is:

if B_S $5N_A$, then all N_A channels are assigned 5 bits (40 kbps);
if $5N_A > B_S$ $4N_A$, then $(B_S - 4N_A)$ channels are assigned 5 bits and the rest 4 bits;
if $4N_A > B_S$ $3N_A$, then $(B_S - 3N_A)$ channels are assigned 4 bits and the rest 3 bits;
if $3N_A > B_S$ $2N_A$, then $(B_S - 2N_A)$ channels are assigned 3 bits and the rest 2 bits;
if B_S $2N_A$, then $(2N_A - B_S)$ channels are blocked, and the rest assigned 2 bits.
This bit allocation is assessed every masterframe.

5.7.5 Signal Processing in the Decoder

The principal functions of the decoder are shown in Fig. 5.30.

The noise power estimate from the encoder is quite coarse, < 2 bits wide. This is smoothed out using a lowpass filter and converted to an equivalent ***gain***. The noise generation process is such that, in conjunction with the ***gain***, the output is a μ-law encoded noise signal of appropriate power level. Thus this locally generated noise at the receiver has a power level that is approximately the same as the power of the background noise measured at the encoder.

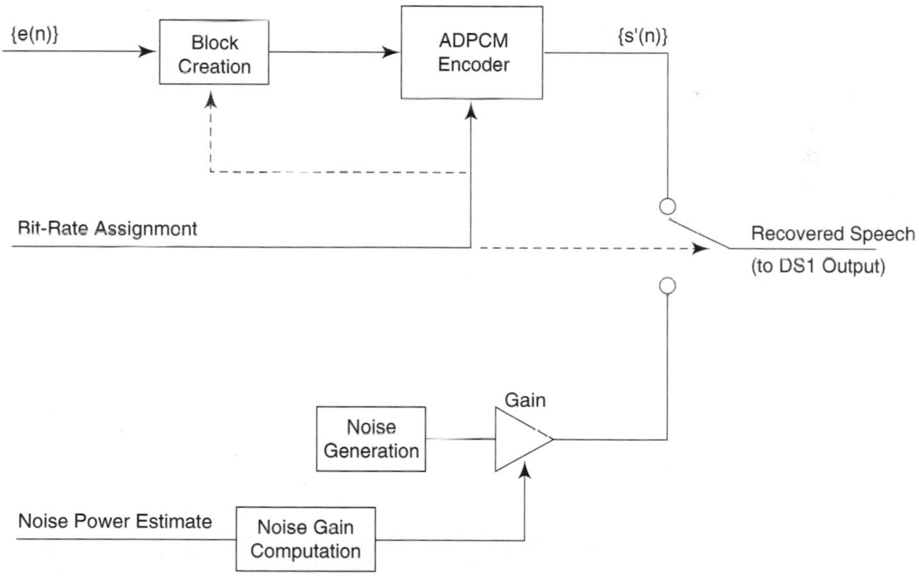

Fig. 5.30
Received speech with noise injection during *silence*

The bit rate assignment controls the action of the **ADPCM** decoder. By ensuring that changes occur on the boundaries of masterframes, the encoder and decoder are in synchrony. In the event that the bit assignment for the particular channel is 0, indicating silence, then there is no signal available from the demultiplexing module for the duration of the masterframe in question. A synthetic zero-valued signal is applied to the decoder and the bit-value of the previous frame retained. The bit rate assignment also controls which signal is considered as the output. If 0, then the recovered speech signal is the output of the noise generator; for all other bit rates the output of the **ADPCM** decoder is used.

5.7.6 Concluding Remarks on DSI

DSI is the enabling technology that makes voice packetization efficient. If these packets are allowed to be of variable length, then full advantage can be taken of the (variable) duration of active speech. Compression is achieved because, statistically at least, speech is active less than 50% of the time. There is equipment available, based on what is known as "Fast Packet" or "Frame Relay" technology which achieves compression of greater than 2:1 by using **DSI**. Coupled with **ADPCM**, compression ratios of better than 4:1 are achieved. Such equipment is, however, quite expensive and introduces a significant amount of transmission delay.

The Rudimentary Digital Speech Interpolation (**RDSI**) scheme described above, takes a conservative approach, utilizing in effect, packets of fixed duration, equivalent to a masterframe, albeit of variable length in bits. By intentionally being conservative on the classification of active versus silent, the quality of the interpolated speech is excellent and, further, the implementation is less expensive compared to a traditional **DSI** approach. To account for short intervals of "peak" activity, variable bit rate **ADPCM** is used to reduce the bit rate to prevent overflow of the capacity of the compressed link. The alternative to variable bit rate **ADPCM** is blocking. In other words, the *a priori* model assumes that continuous speech, without any "breaks," but encoded at a lower bit rate is subjectively better in quality than a "choppy" signal encoded at a higher bit rate. Furthermore, the **RDSI** scheme introduces much less transmission delay than conventional **DSI** implementations.

5.8 REFERENCES

[1] Atal, B. S., and Hanauer, S. L., "Speech Aanalysis and synthesis by linear prediction of the speech wave," *J. Acoustic Soc. of America*, Vol. 50, No. 2, 1971.

[2] Dimolitsas, S., "Standardizing speech-coding technology for network applications," *IEEE Communications Magazine*, Vol. 31, No. 11, Nov. 1993.

[3] Jayant, N. S., *Waveform Quantization and Coding*, IEEE Press, New York, 1976.

[4] Jayant, N. S., "High-quality coding of telephone speech and wideband audio," *IEEE Communications Magazine*, Vol. 28, No. 1, Jan. 90. (This issue focuses on Speech Processing and Applications.)

[5] Jayant, N. S., and Noll, P., *Digital Coding of Waveforms*, Prentice Hall Inc., Englewood Cliffs, NJ, 1984.

[6] Makhoul, J., "Linear prediction: A tutorial review," *Proc IEEE*, Apr. 1975.

[7] Rabiner, L. R., and Schafer, R. W., *Digital Processing of Speech Signals*, Prentice Hall, Inc., Englewood Cliffs, NJ, 1978.

[8] Schafer, R. W., and Markel, J. D., Eds., *Speech Analysis*, IEEE Press, New York, 1979.

[9] Sibul, L. H., (Ed), *Adaptive Signal Processing*, IEEE Press, New York, 1987.

[10] Skidanenko, C., "ADPCM Performance," presented at Network 90, San Francisco, June, 1990.

[11] Steele, R., "Speech codecs for personal communications," *IEEE Communications Magazine*, Vol. 31, No. 11, Nov. 1993.

[12] Recommendation **G.721**: 32 kbit/s adaptive differential pulse code modulation (**ADPCM**).

[13] Recommendation **G.726**: 40, 32, 24, 16 kbit/s adaptive differential pulse code modulation (**ADPCM**).

[14] Recommendation **G.728**: Coding of speech at 16 kbit/s using low delay-code excited linear prediction (**LD-CELP**).

[15] Digital Processing of Voice-Band Signals—Algorithm for 24, 32, and 40 kbit/s **ADPCM, ANSI T1.303**—1989.

[16] Digital Processing of Voice-Band Signals—Algorithms for 5-, 4-, 3-, and 2-bit/sample Embedded **ADPCM, ANSI T1.310**—1991.

[17] Network Performance Standards—32 kbit/s **ADPCM** Tandem Encoding Limits, **ANSI T1.501**—1988.

[18] **DCME**—Interface Functional and Performance Specification, **ANSI T1.309**—1991. (**DCME**: **D**igital **C**ircuit **M**ultiplication **E**quipment)

[19] Cellular System: Dual-Mode Mobile Station—Base Station Compatibility Standard, **IS-54** , Electronic Industries Association (**EIA**), Washington, DC, Dec. 1989.

[20] Mobile Station—Base Station Compatibility Standard for Dual-Mode Spread Spectrum Cellular System, **IS-95**, Electronic Industries Association (**EIA**) and Telecommunications Industry Association (**TIA**), Washington, DC, July 1993.

ECHO SUPPRESSORS, ECHO CANCELERS, AND ADAPTIVE FILTERS

6

6.1 INTRODUCTION

In Chapter 1 we modeled the inability to match the impedance of the subscriber loop as an echo path, $H_e(f)$. This echo has significant impact on the perceived quality of a telephone call. Echo is probably the most devastating degradation for long distance telecommunications, especially when the two parties are separated by great distances, since in this case we have a large time delay in the transmission path.

In this chapter we first discuss the need for echo control. There are two devices employed in the Public Switched Telephone Network (**PSTN**) for addressing echo. The echo suppressor, described in Section 6.2 is the "older" method, one that is rapidly being phased out. However, the underlying principles of the echo suppressor are still employed in such "modern" devices as hands-free telephones and even echo cancelers. The treatment of echo suppressors here tries to embellish these fundamental operating principles as well as provide a discussion of the relevant specification, **CCITT** Recommendation **G.164** [8].

The modern method for providing echo control is the echo canceler, discussed in Section 6.4. The treatment, as in the case of suppressors, tries to emphasize the fundamental operating principles as well as discuss the relevant specification, **CCITT** Recommendation **G.165** [9]. Echo cancelers include an internal component that is essentially an echo suppressor and the principles developed in Section 6.2 are very relevant. cancelers, to be applicable in the modern telephone network, need to be adaptive in nature and understanding their operation requires some knowledge of adaptive filters. In Section 6.3 we provide a discussion of adaptive digital filters that is quite general but is slanted toward the application of echo cancelation.

6.1.1 The Need for Echo Control

For purposes of studying the problem of echo, we can model the transmission path associated with the telephone call as shown in Fig. 6.1. The connection path between the two endpoints is referred to, in telecommunications terminology, as a ***circuit***.

At the extreme ends are the telephone sets. Within a standard set there is a signal path directly between the mouthpiece and the earphone to provide what is called ***sidetone***. Sidetone is a very important feedback mechanism for maintaining automatic volume control. Without sidetone a talker does not have any indication as to his or her loudness level. We mention this to emphasize the fact that an echo is not always bad—provided it is not delayed significantly, as in the case of sidetone. In fact sidetone, as a device for controlling volume, is so effective that it allows network planners and designers to assume that there is a "typical" level for speech signals. This level is approximately -13 dBm0. Telephone calls that are problematic tend to be those involving a telephone set that does not have adequate sidetone, such as "hands-free" telephones.

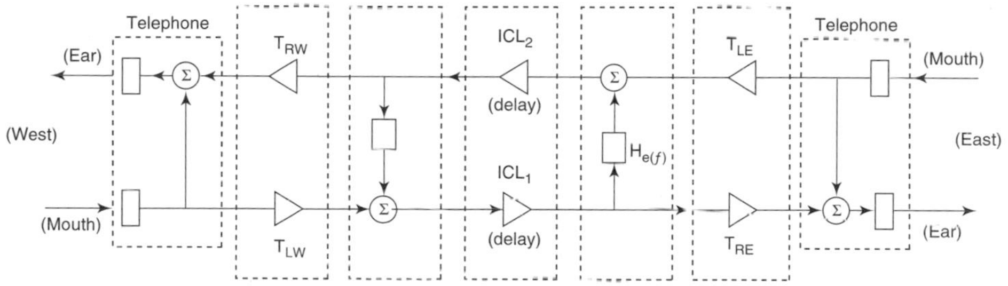

Fig. 6.1
Transmission path associated with a telephone call indicating gains and delays encountered by the signal and echo

The speech from the West talker traverses the local access subscriber loop and experiences a loss, $\mathbf{T_{LW}}$, which is a measure of the reduction in power level because of the frequency response of the cable. If the cable is very short then $\mathbf{T_{LW}}$ 0 dB. This speech then goes over the transmission network, encountering a one-way delay of T. The network may introduce a loss, called the ***Inserted Connection Loss*** (**ICL**). At the hybrid in the East line circuit, impedance mismatch causes an echo governed by the transfer function $H_e(f)$, which returns to the talker after experiencing an additional T units of delay and a loss of ($\mathbf{ICL} + \mathbf{T_{RW}}$) dB. Defining the **ERL** (echo return loss) as the power "gain" (loss), assuming the upper cutoff frequency of the telephone channel is f_0,

$$\mathbf{ERL} \;=\; -\log_{10}\{\frac{1}{f_0}\int_0^{f_0}\left|H_e(f)\right|^2 df \qquad (6.1.1)$$

we can express the power of the returned echo in terms of a transmission loss called ***Talker Echo Path Loss*** (**TEPL**) where

$$\mathbf{TEPL_W} = \mathbf{T_{LW}} + \mathbf{R_{LW}} + \mathbf{ICL_1} + \mathbf{ICL_2} + \mathbf{ERL_E} \text{ (dB)} \qquad (6.1.2)$$

Thus the talker hears his own speech attenuated by an amount $\mathbf{TEPL_W}$ (dB) and delayed by 2T units of time. Some of the returned echo signal may be reflected again by the hybrid in the West line circuit, by an amount $\mathbf{ERL_W}$ and be returned to the East (listener). Thus the East listener hears the West talker first directly and then again, delayed by the round-trip transmission time of 2T. This is called listener echo. Methods used to control talker echo generally control listener echo as well and thus listener echo is a secondary problem.

ERL and **TEPL** are measures of power gain (or loss). There is an additional consideration, as is clear from Fig. 6.1, namely the presence of a closed loop. From control systems theory we derive a notion of the stability of this closed loop. A closed loop is unstable, or oscillates, if the loop gain and loop phase response at any frequency satisfy the condition of unity gain and 180-degree phase shift (π radians). The ***gain margin*** is the worst-case loop gain (actually the least loss) at any frequency at which the loop phase response is 180 degrees or greater. Since even for small values of transmission delay the loop phase response exceeds the 180-degree limit, at some frequency, the stability of the loop can be guaranteed only by ensuring that the loop gain is strictly less than unity at all frequencies. With this in mind we define the ***Singing Return Loss*** (**SRL**) associated with a hybrid as

$$\mathbf{SRL} \;=\; -\log_{10}[\max|H_e(f)|] \text{ dB} \;;\; 0 \le f \le 4 \text{ (KHz)} \qquad (6.1.3)$$

With this definition, the ***Singing Margin*** (**SM**) for the circuit is expressed as

$$\mathbf{SM} = \mathbf{SRL}_E + \mathbf{SRL}_W + \mathbf{ICL}_1 + \mathbf{ICL}_2 \qquad (6.1.4)$$

Circuits with inadequate singing margin are quite easy to identify—talking over a circuit with inadequate **SM** is reminiscent of talking into a barrel and hence the phrase "rain-barrel effect" to describe the "hollow" sensation. **SRL** and **ERL** are both measures of the adequacy of the hybrid. One is a worst case, the **SRL**, and the other is an average, the **ERL**. It is easy to show that **SRL** < **ERL** and thus even if the hybrid provides a good echo cancelation on the average, it does not necessarily imply that it provides a good **SRL**. In practice we normally assume that **SRL** is no more than 3 dB worse than the **ERL**, i.e., **SRL** **ERL**-3 and further that any echo control technique that provides sufficient **TEPL** also provides adequate **SM**.

The perceived quality of a telephone call can be related to three quantities. Any additive noise is a degradation and the greater the noise the worse the circuit (connection) is perceived to be. The second aspect is loss. The greater the (one-way) transmission loss, the lower will be the talker level as perceived by the listener and this is clearly a degradation. The greater the loss, the worse the circuit is perceived to be. The third aspect is the echo path loss. The echo path loss by and of itself is not so meaningful, considering that sidetone can be viewed as a echo path with little to no loss. The echo path loss in conjunction with the transmission delay is quite meaningful. From a subjective viewpoint the talker (listener) can tolerate a smaller **TEPL** if the time delay of echo return is short and, conversely, will demand a large **TEPL** if the time delay of the echo return is large.

The subjective effects of echo have been studied extensively, primarily by researchers at Bell Labs in the United States and by researchers at various laboratories associated with the telephone administration in various countries. The following table, Table 6.1, is derived from the results published by AT&T (see [4]). The table provides an indication of how much the **TEPL** should be in order to satisfy 90%, 70%, and 50% of the general public.

What the table indicates is that for an echo path delay (roundtrip) of, for example, 80 msec, the talker echo path loss must be greater than 42 dB to satisfy the general public at the 90% level. That is, 90% or greater of the public would <u>not</u> find the circuit objectionable if the **TEPL** is greater than 42 dB; if the **TEPL** for the same delay is 32 dB then up to 50% of the public would find the circuit objectionable, again from an echo standpoint.

Table 6.1

Relationship between round-trip echo delay and **TEPL** required

Echo Path Delay (msec)	TEPL (dB)	TEPL (dB)	TEPL (dB)
1.5	5	0	-
5	15	9	5
10	21	15	11
20	27	21	18
30	32	26	22
40	35	20	25
60	39	33	28
80	42	36	32
100	44	38	34
200	50	44	40
300	53	47	43
600	56	51	47
1200	58	53	48
	90% level	70% level	50% level

From the viewpoint of the network provider, the loss in the local loop is a variable and cannot be counted upon to provide any contribution to the **TEPL**. Thus the overall echo performance must be satisfied by the hybrid and some form of ***echo control***. The introduction of inserted connection loss (**ICL**) is one technique used to mitigate the problem of echo. The **ICL**, usually about 6 dB (of loss), is introduced into both directions of transmission. This degrades the perceived circuit quality based on the signal being attenuated by **ICL** (dB) but would attenuate the echo by an amount of 2***ICL** (dB). Modern transmission networks follow the **0-dB Loss Plan**, whereby the network does not (is not supposed to) introduce any (intentional) loss except for special cases, implying that the **ICL** is zero, putting the entire burden of providing circuit quality on the hybrid and devices whose sole purpose is to reduce the impact of echo.

Providing adequate hybrid balance to meet the echo attenuation needs for arbitrary delays is not economically feasible. In a recent publication from **BELLCORE** [10], a model is described for the echo return loss performance of hybrids. Two cases are developed. In one it is assumed that a single balance network is used for all loops. In the second case it is assumed that loops can be segregated into a few (typically four) categories and a suitable balance network designed for each category. Thus, depending on the loop classification, an appropriate balance network is used. Fig. 6.2 depicts the distribution of the **ERL** over the population of subscriber loops that is achieved. The model for the distributions follows the shape of a normal **pdf** with $(\mu, \sigma) = (11, 3)$ for the case of the single compromise balance network and $(\mu, \sigma) = (13.6, 2.8)$ for the segregated loop-balancing scheme. The minimum **ERL**, computed as μ-3σ, is 2 dB and 5.2 dB for the two cases. From Fig. 6.2 and Table 6.1 it is clear that echo control methods are required except for the limited cases where the round-trip delay is very short.

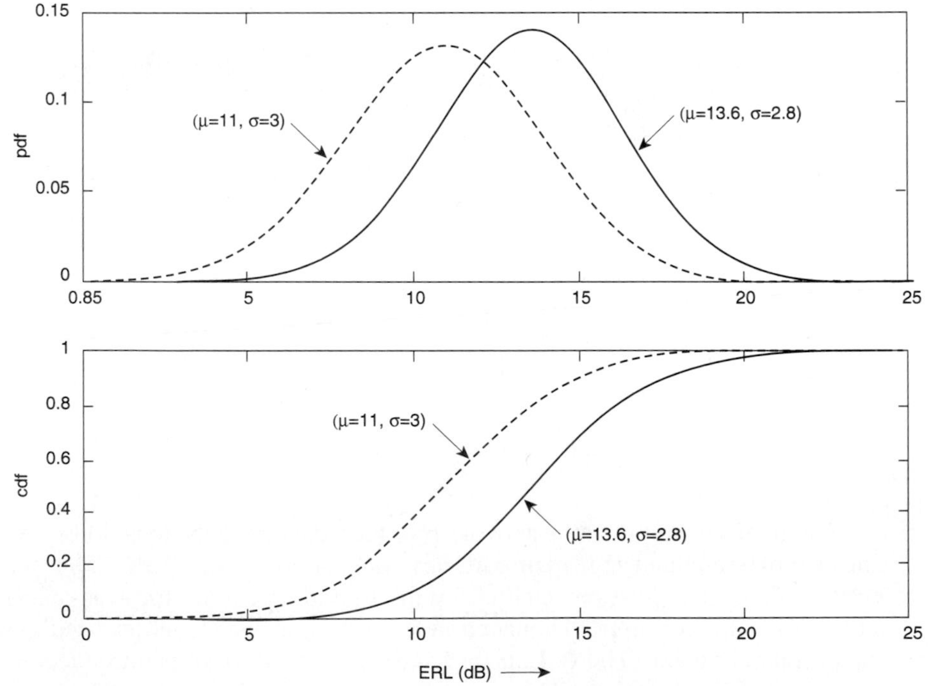

Fig. 6.2
Probability density and cumulative distribution functions of **ERL** assuming a single compromise balance network or segregated loop balance

If the sole network element involved in the circuit is a local switch, the round-trip delay is small, especially if the switch is analog, and no special echo treatment is necessary. Echo control is then required only for long-distance calls involving two or more switches geographically separated by a reasonable distance. In the modern network with the advent of digital switches, digital transmission, and a plethora of digital equipments such as cross-connects and multiplexers, all adding transmission delay, the notion of "long distance" is rapidly becoming quite short indeed. In fact, echo treatment is required for **all** calls that traverse two or more central offices, be they cross-town, cross-country, or even international.

6.2 ECHO SUPPRESSORS

One technique for providing echo control, other than introducing **ICL**, and which is used primarily for international calls and calls completed over satellite trunks, is the method of ***echo suppression***. Echo suppressors are still in use but all modern echo control equipment utilizes the principle of (digital) echo cancelation. As will be seen in the Section 6.4, echo cancelers do have some of the characteristics of echo suppressors, so a brief explanation of the considerations that apply to suppression is provided here. For reference, the specifications that apply to echo suppression are provided in **CCITT** Recommendation **G.164** [8].

6.2.1 The Principle of Echo Suppression

The operation of an echo suppressor can be explained with reference to Fig. 6.3. The underlying premise of echo suppressors is that conversation, i.e., a telephone call, is half-duplex. That is, at any point in time, only one talker is active and the other party is a listener. The echo suppressor introduces a loss in each direction of transmission that is not a fixed loss but can be changed depending on a decision as to which direction is "dominant." Thus if the West talker is active, or louder than the East talker, the loss G_1 will be nominally 0 dB whereas the loss G_2 will be large, usually greater than 35 dB. Thus all signal in the East-to-West direction will be blocked, or at least attenuated greatly. This signal contains the echo that West may have heard and consequently this attenuation ***suppresses*** the echo. Conversely if East is active and West is silent, then the signal in the West-to-East direction is principally the echo of East's speech and by making G_1 a large loss, this echo is suppressed.

A fundamental problem in determining whether the loss in any direction should be zero or large relates to the problem of detecting silence. The gain control mechanism depicted in Fig. 6.3 has this responsibility. The procedure requires that the device compute the short-term power of the signal and compare this with a threshold. If the power, **P**, is greater than a threshold, say T_B, we can say that we have active speech. Detection of silence involves the determination that the power **P** has dropped below a threshold T_A and a timer to verify that the power is below this threshold for a sufficient length of time. This is depicted in Fig. 6.4.

Consider the case when the West talker is active and the East talker is silent. Then the power level $\mathbf{P}_W$ is greater than $\mathbf{P}_E$, the latter being principally the echo of West's speech arising at the East hybrid. By setting $G_1 = 0$ dB and $G_1 =$ dB, the echo returned to West is suppressed. Now suppose that West "stops" talking. The power level $\mathbf{P}_W$ then diminishes and we enter a hangover period. Depending on the round-trip delay between the suppressor and the East hybrid, the level $\mathbf{P}_E$ might remain substantial for a while. It is necessary to hold the gain conditions as is until we know that this round-trip delay has elapsed.

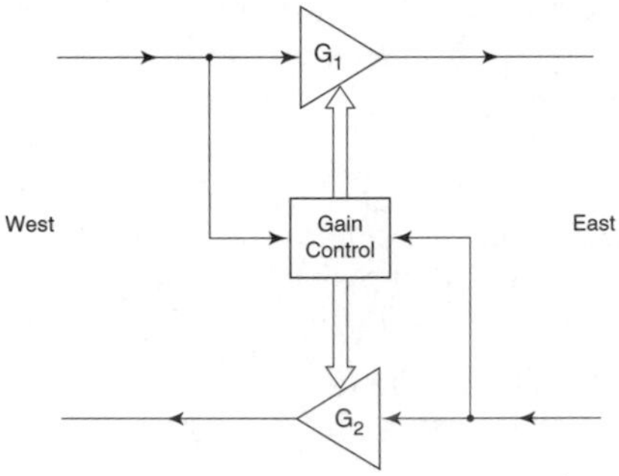

Fig. 6.3
Echo suppression achieved by insertion of a variable gain in each direction

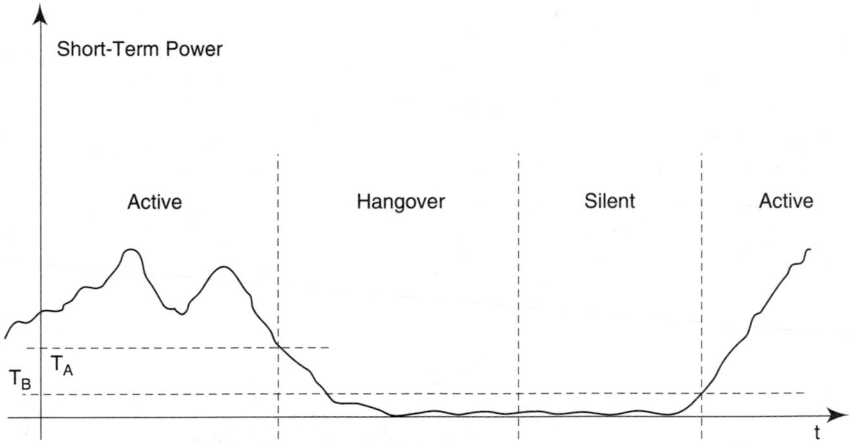

Fig. 6.4
Using short-term power to establish silence or presence of speech signal

Unfortunately, we cannot really distinguish whether $\mathbf{P}_E$ is comprised of echo power or whether the East talker is indeed active. The implication is that we will almost certainly clip either the trailing segment of West's speech by having too little hold time, or clip the initial segment of East's speech if the hold time is too long. Further, by forcing G to be either 0 dB (no loss) or dB (very large loss), we are guaranteeing that only one direction of transmission can be achieved. The half-duplex nature and the clipping are annoying at best for human voice communication but are disastrous for data communication.

6.2.2 Improving the Subjective Quality of Echo Suppressors

One method of alleviating this half-duplex behavior and allowing "break-in," whereby a loud talker is not suppressed regardless of the power level in the reverse direction, is to control the loss in a "soft" manner. Rather than have just two values of loss, 0 dB and dB (corresponding to linear units of 1 and 0, respectively), the gain takes on a variety of values in between. This method of setting gain, say G_1, is depicted in Fig. 6.5.

If the ratio $\mathbf{P}_W/\mathbf{P}_E$ is substantial, then G_1 is approximately 0 dB and the West speech (along with any echo) is passed on in East's direction with little or no attenuation. This is shown as region (C) in the figure. If the ratio $\mathbf{P}_W/\mathbf{P}_E$ is very small then G_1 is a large loss, provided $\mathbf{P}_E$ is substantial. If $\mathbf{P}_E$ is very small then the actual value used for G_1 is moot in this region, labeled (A) in the figure. The middle region, (B), is where certain subjective optimization is required. The subjective quality of the circuit will depend on an appropriate choice of the thresholds $\mathbf{T}_1$ and $\mathbf{T}_2$.

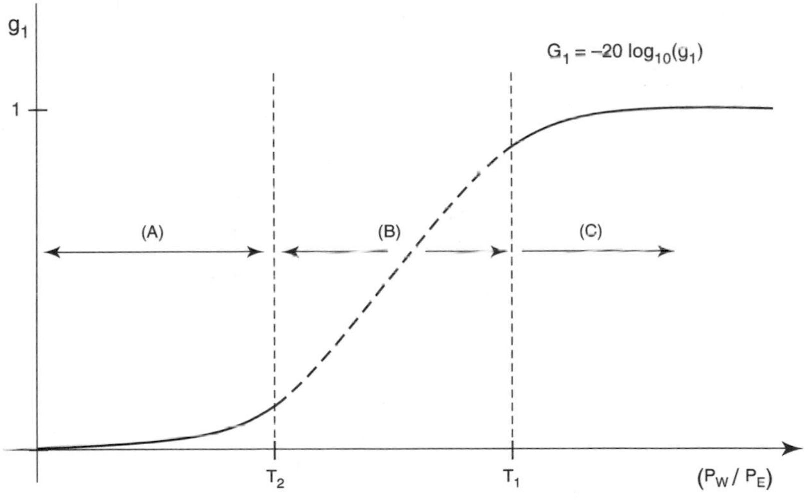

Fig. 6.5
Varying gain as a function of the ratio of signal powers

A heuristic explanation of how these thresholds can be established is based on the minimum expected **ERL** and an estimate of the actual **ERL** for the circuit. If the minimum **ERL** is 6 dB, then if $\mathbf{P}_W/\mathbf{P}_E$ is greater than 6 dB we are reasonably sure that the signal coming from the West direction is primarily speech generated by West. We can thus set $\mathbf{T}_1 = \min\{\mathbf{ERL}\}$. If the estimate for the **ERL** is **X** dB then a suitable value for the lower threshold $\mathbf{T}_2$ is approximately **X**. Considering that the estimate, **X**, may not be very accurate, we can include a safety factor by setting $\mathbf{T}_2 = \mathbf{X+Y}$. A negative value for **Y** means that we are willing to trade-off increased echo transmission for reduced clipping. A positive value for **Y** is the reverse situation where we believe that echo is worse than clipping.

It turns out that a positive value for **Y**, approximately 10 dB, is suitable provided we take into account the notion of a hangover time and provided that the estimate of the **ERL** is substantial, 20 dB or better. That is, the loss G_1 is never changed from 0 to abruptly, but gradually over a period of about 200 msec. This is the reason that the gain is shown as a dashed line for region (B). If the **ERL** estimate is less than 20 dB then **Y** is reduced accordingly but the notion of hangover time is still applicable.

The terminology used for the three regions is quite expressive of the situations depicted. Region (C) is called ***double-talk*** or ***hard double-talk***, implying, from the viewpoint of controlling G_1, that there is a substantial amount of West's speech traversing the West-to-East direction. Region (A) is called ***single-talk***, implying that the West-to-East signal does not have much content attributable to the West talker. Region (B) is called ***soft double-talk*** indicating the possibility of having measurable quantities of West's speech mixed with (East's) echo.

One method for achieving this **level dependent loss** is the use of a nonlinear device whose input-output characteristic is shown in Fig. 6.6.

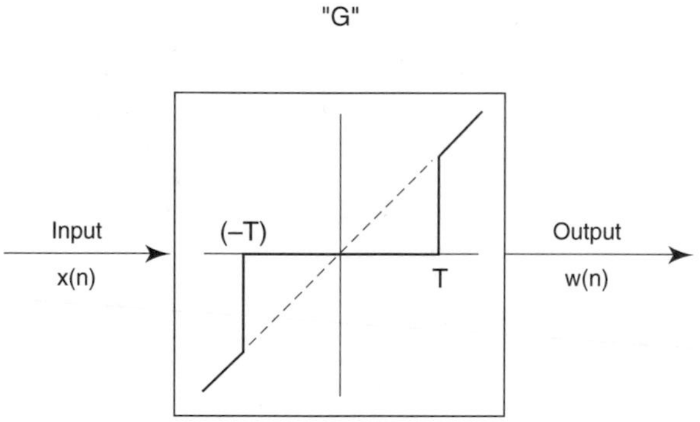

Fig. 6.6
Center-clipper method for achieving a level-dependent loss

The operation of this memoryless nonlinearity can be written as

$$w(n) = \begin{cases} 0 & \text{for } |x(n)| \leq T \\ x(n) & \text{for } |x(n)| > T \end{cases} \tag{6.2.1}$$

Thus large sample values go through unchanged but small sample values are suppressed. The overall "loss" achieved depends on the threshold **T**. If **T** is 0 (linear units) then the device corresponds to no loss at all. A large value of **T** corresponds to infinite loss for small signal levels. Since small signal levels are consistent with echo, the device achieves the function of echo suppression. Since large signal levels correspond to West's speech, these will be transmitted, albeit with some distortion, providing the "break-in" behavior.

The complete suppression of signal has an adverse subjective effect. When the West-to-East signal is completely suppressed, the East listener hears absolute silence (provided the transmission medium between the suppressor and the East hybrid does not add much noise by itself) and may believe that the circuit is dead. To alleviate this problem, we add a low level noise, referred to as ***comfort noise***, when the suppression is total. This is depicted in Fig. 6.7. When a sample is suppressed, it is replaced by a sample of a locally generated noise, $\{\mathbf{e}(n)\}$, as

$$w(n) = \begin{cases} e(n) \text{ for } |x(n)| \le T \\ x(n) \text{ for } |x(n)| > T \end{cases} \tag{6.2.2}$$

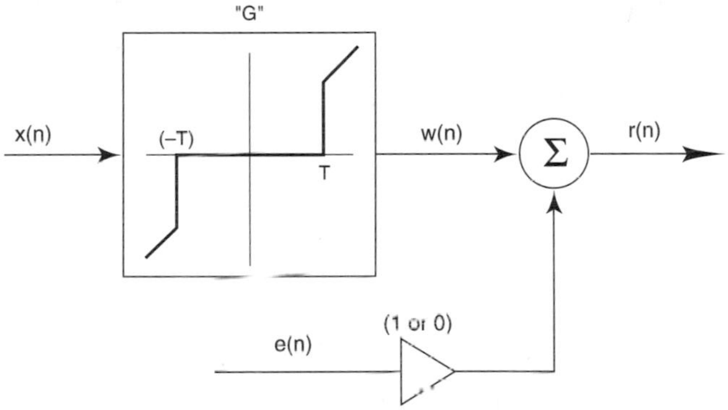

Fig. 6.7
Injection of comfort noise when center clipper squelches signal for improving the subjective quality of the circuit

The power of the noise is important. It should match the background noise, in power if not form. The background noise is that which is present when neither talker is active. Or, more correctly, the background noise perceived by the East listener is the noise present when East is silent and West is active and is discerned in the inter-syllabic or interword silences that are present in normal conversational speech. The term ***noise matching*** refers to the process of estimating the power of the background noise and generating a local noise signal of the same power to replace samples suppressed by the threshold device. The choice of the threshold is equally if not more important than the power of the comfort noise. A suitable approach for estimating this threshold is discussed in the Section 6.4 which discusses echo cancelers. It may be argued that the form of this comfort noise should match the background noise as well. That is, the spectrum of the comfort noise should match the spectrum of the background noise. A limited number of experiments conducted by the author indicate that this is not true. In fact, having a comfort noise that is white noise, with a flat spectrum, actually **improves** the subjective quality of the circuit. While not conclusive, the experiments did indicate that the manner with which the suppression and noise-power matching is accomplished is far more important than spectral matching of the background noise. The concept of comfort noise is exactly the same as the "fill-in" signal used in digital speech interpolation schemes as discussed in Chapter 5 (Section 5.7).

As mentioned before, echo suppressors are being replaced by echo cancelers as the preferred mode of echo control. It is also true that all echo cancelers include provision for a modicum of echo suppression and thus the principles discussed above are quite applicable to echo cancelers as well.

6.2.3 Disabling Echo Suppressors

While echo suppressor operation can be optimized for subjective quality, they cannot be used for calls where the communication is between data transmission devices, namely modems. To allow data communications devices to share a common transmission and switching fabric with voice communications, some technique is required to ***disable*** echo suppressors for the duration of a data call, i.e., a call between modems. The notion of ***disabling*** an echo suppressor is equivalent to freezing the loss in both directions at 0 dB (clear); in a sense the suppressor removes itself from the transmission path. Conversely, the notion of ***re-enabling*** is when the suppressor decides that the call for which it was disabled is over and it must "unfreeze" its loss values and revert to normal operation.

The method employed by echo suppressors for detecting whether they should go into the disabled state utilizes the answer tone employed by modems. When a modem is alerted to an incoming call, it first goes off-hook and then transmits a tone of (nominally) 2100 Hz. Therefore the suppressor is continually examining both directions of transmission for this tone. When the tone is detected it will set both G_1 and G_2 to 0 dB (the threshold **T** to 0 if the nonlinear device is used) and wait for the end of the call.

The requirements on the tone disabling function are depicted in Fig. 6.8 (not drawn to scale). What the figure shows is that if the tone-disabling module of the suppressor sees a single-frequency tone of power greater than –31 dBm0 within the band (f_L, f_H) it must operate, i.e., disable echo suppression function. If the tone frequency is less than 1900 Hz or greater than 2350 Hz then the tone should not be recognized as a disabling tone. The region in between is a transition band where the suppressor may or may not recognize the tone as a disabling tone. The operate region, as specified by **CCITT** Recommendation **G.164** [8], is the frequency band 2100 ± 21 Hz, i.e., $f_L = 2079$ Hz and $f_H = 2121$ Hz. In North America a slightly different band is specified by some long distance carriers. This is to accommodate the existence of modems of older vintage that may not provide a suitable 2100-Hz tone. These older modems are serviced by extending the "must disable" range to f_L = 2010 Hz and f_H = 2240 Hz, keeping the other parameters the same.

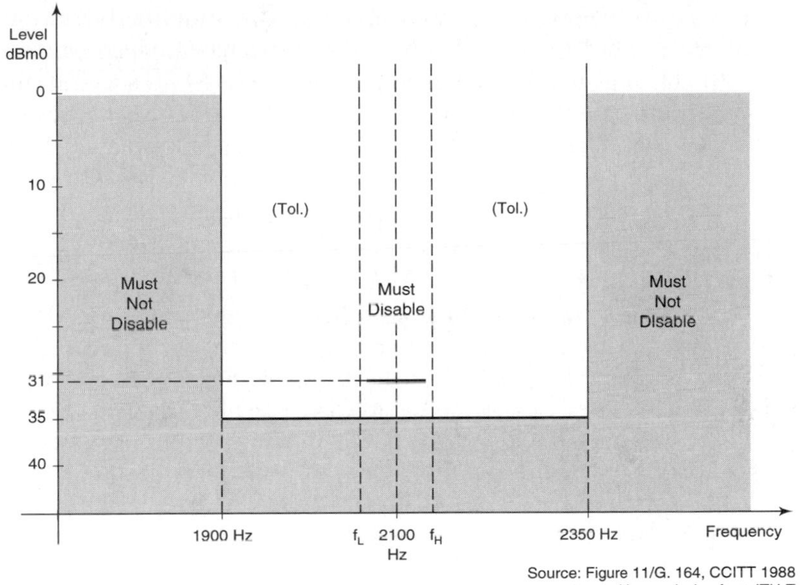

Source: Figure 11/G. 164, CCITT 1988
with permission from ITU-T

Fig. 6.8
Frequency domain description of requirements of the tone disabling function

One problem faced in designing a tone disabler is that normal speech should not ***talk-off*** the suppressor; the tone disabler must distinguish between a tone at 2100 Hz and short-term power around 2100 Hz from a (loud) talker. In pre-modern days the speech was filtered first by a telephone set that had a lowpass characteristic that started rolling off (gradually) at about 700 Hz. Modern electronic handsets do not have this rolloff characteristic and the possibility of talk-off is increased. The way the tone disabler protects against talk-off is to require that the power in a narrow band about 2100 Hz be sustained for a period of time and that, further, the power outside the frequency band [1900 Hz, 2350 Hz] be sufficiently low so as to distinguish between speech and tone. Since actual data transmission cannot proceed until the suppressor has been disabled, the period of time, also referred to as ***operate time***, cannot be made arbitrarily long. The **CCITT** recommends a limit of 400 msec. Thus the requirement is that the tone disabler operate within 300 ± 100 msec after receipt of the sustained disabling tone having a level in the range (–28 dBm0, 0 dBm0). For disabling tones of low level, between –28 dBm0 and –31 dBm0, the tone disabler must operate but does not have to do so with rapidity.

Having disabled the suppressor function upon receipt of a 2100-Hz disabling tone, the tone disabler must hold this mode even after the 2100-Hz tone has gone away and actual (modem) data transmission has commenced and continue in this mode for the duration of the call. How then does the tone disabler know to release? That is, under what conditions should the suppressor re-enable? The suppressor in and of itself has no notion of the actual ***end-of-call*** and must detect this condition based on the signals transmitted in the two directions. A simplified description of the end-of-call state is that the power in both directions has dropped below some threshold for a suitable length of time. Recommendation **G.164** [8] specifies this holding level as –36 dBm0 and the release time as greater than 100 msec. Thus the tone-disabler does not release for signal dropouts, short durations of low power, less than 100 msec. Since it is still possible to falsely operate on speech, it should release within 250 ± 150 msec after the signals in **both** directions fall at least 3 dB below the holding threshold of –36 dBm0. Thus a minimal impairment is caused by such accidental speech disabling, since re-enabling will occur at the first pause in conversation.

The signal processing functions associated with the tone disabler are depicted in Fig. 6.9. This must be implemented for the two directions of transmission separately. Some designs have been proposed that "cheat" and use, as $\{x(n)\}$, the sum of the signals in the two directions. The power within and without a band around 2100 Hz is computed where the nominal bandwidth (pass and stop) is about 21 Hz. The operations associated with these measurements are provided in Table 6.2.

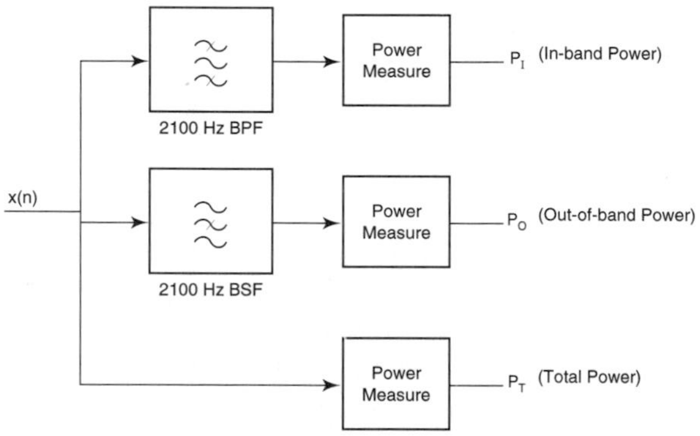

Fig 6.9
Signal processing implied for the proper operation of the tone-disabler.

Table 6.2
Operational requirements of tone-disabler module in an echo suppressor

Condition	Duration	Action
1. $P_I > -31$dBmO	long	operate
2. $P_I > -28$dBmO $P_I > P_O + 5$	300 ± 100 msec	operate
3. $P_I < P_O$	-	do not operate
4. $P_T < -36$dBmO	reasonably long	release
5. $P_T < -39$dBmO	250 ± 150 msec	release

6.2.4 Concluding Remarks on Echo Suppressors

The advent of communication satellites heralded a new era in telecommunications. Connectivity could be established between distant locations using a satellite for "bouncing off" signals. Unfortunately the round-trip delay of such a circuit is of the order of seconds rather than milliseconds. Inadequate **ERL** renders circuits with such delays useless for human conversation. Echo suppressors were introduced to defeat this large delay. The first units were bulky, expensive, power-hungry analog devices, but were the only choice available. Suppressors have since been phased out and replaced with digital echo cancelers.

Nevertheless, the concept of echo suppression is still alive and well and every echo canceler contains an echo suppressor to remove the "last vestiges of echo." The notion of tone disabling is applicable to echo cancelers as well and the requirements of the tone disabling function for cancelers is quite similar to that of suppressors.

6.3 ADAPTIVE FILTERS

As the term implies, adaptive filters are filters that are changed or adapted while in operation. The theory of adaptive filters deals with methods for describing how to change these filters in order to achieve a predefined performance criterion. The theory also provides tools for analysing how well the adaptation scheme will, in a statistical sense, perform in terms of this performance criterion. Adaptive filter theory is quite vast and complex and we will just touch upon those aspects that apply to telecommunications, specifically to the operation of devices called **echo cancelers.** An excellent treatment of adaptive filters, that is eminently readable, is available in the book by Treichler et al. [5]. Other references include Sibul [2] and Widrow and Stearns [6]. Echo cancelers usually employ **FIR** filters, primarily because such filters are never unstable, and also because their behavior is more easily analyzed than that of adaptive **IIR** filters.

Echo cancelation is applied in many distinct applications. From a network standpoint an echo canceler is a device that removes echos in (long-distance) telephone calls. In data communications, high speed modems that utilize the Public Switched Telephone Network (**PSTN**) for transmitting data employ echo cancelation techniques as well. In Integrated Services Digital Network (**ISDN**) applications echo canceling techniques are used in providing the digital data stream between the customer premise and serving central office. These three applications are different in scope and have different constraints. In Section 6.4 we will discuss network echo cancelers and thus our discussion of adaptive filters will be slanted in that direction.

6.3.1 Introduction to Echo cancelation

The principle of adaptive filters is introduced by considering the situation depicted in Fig. 6.10. A ***reference*** signal, $\{\mathbf{x}(n)\}$, is modified by some system, **G**, and returns as the signal $\{\mathbf{y}(n)\}$, which is comprised of the output of system **G** together with an additive signal $\{\mathbf{s}(n)\}$. In the context of network echo cancelers, the signal $\{\mathbf{x}(n)\}$ is the speech signal from the ***far-end*** talker, $\{\mathbf{s}(n)\}$ represents the speech signal of the ***near-end*** talker, and $\{\mathbf{e}(n)\}$ is the ***echo*** signal that is returned toward the distant end. An echo canceler is a filter $H(\mathbf{z})$ that operates on the same reference signal $\{\mathbf{x}(n)\}$ to produce a local signal, $\{\zeta(n)\}$, which, in an ideal situation, is identically equal to the echo. By subtracting this replica of the echo signal from $\{\mathbf{y}(n)\}$ the signal actually returned to the distant end is substantially free of echo and consists of only the near-end speech.

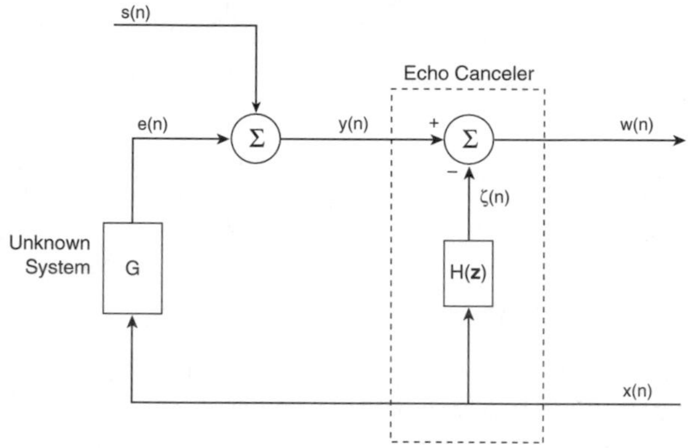

Fig. 6.10
Principle of echo cancelation—subtracting a replica of the echo. Since only one echo is removed, the structure is called a "split-type canceler." An equivalent device at the other end of the circuit will remove any echo generated there

We make the assumption that this unknown system is linear and time invariant and thus can be described by a transfer function $G(\mathbf{z})$. Clearly, echo cancelation is achieved if $H(\mathbf{z})$ $G(\mathbf{z})$. Thus from an operational viewpoint it suffices that we design a filter that mimics this unknown system. For example, we could choose a suitable input $\{x(n)\}$, apply this input to the system, record the output and then, in principle, obtain the impulse response $\{g(n)\}$ and/or the transfer function $G(\mathbf{z})$ of the unknown system. Since this $\{g(n)\}$ is, in general, representative of an infinite impulse response, an **FIR** filter $H(\mathbf{z})$ can only approximate $G(\mathbf{z})$ but by making the length of $H(\mathbf{z})$ sufficiently long this approximation error can be made small.

This approach is not feasible for two main reasons. First, in a telecommunication network we cannot arbitrarily choose to inject a specified input except in certain controlled test situations. Therefore we cannot really choose what $\{x(n)\}$ is but have to rely on live traffic that is usually, but not always, speech. Second, an echo canceler is a shared resource. At any point in time the signals traversing the canceler will be from a particular call between two end stations. When the telephone call is over, the same channel is made available to a different call between two completely different end stations. This implies that the unknown system is quite variable and changes from call to call. Even during the progress of a call there could be changes if, for example, an additional telephone goes off-hook or an additional party is conferenced in. For these reasons it is necessary to make $H(\mathbf{z})$ ***adaptive***.

The way we measure the efficacy of cancelation is in terms of the power of the echo component of $\{\mathbf{w}(n)\}$ which, ideally, is zero. Thus the *modus operandi* of echo cancelers is to modify $H(\mathbf{z})$ to minimize the power of the residual echo. The problem can be formulated in the following way. First, we make the assumption that the unknown system is fixed over the short-term and can be described by the input-output relation

$$e(n) = \sum_{k=0}^{N-1} g(k)x(n-k) \tag{6.3.1}$$

The locally generated echo replica is

$$\xi(n) = \sum_{k=0}^{N-1} h_n(k)x(n-k) \tag{6.3.2}$$

where we have indexed the coefficients of the **FIR** filter $H(\mathbf{z})$ to indicate that they may be altered in a time-dependent manner. Hence the transmitted signal, which is ideally echo free, is given by

$$\mathbf{w}(n) = \mathbf{y}(n) - \xi(n) = \mathbf{s}(n) + [\mathbf{e}(n) - \xi(n)] = \mathbf{s}(n) + \mathbf{r}(n) \tag{6.3.3}$$

where $\{\mathbf{r}(n)\}$ is the residual, or uncanceled, echo signal. Our intent is to minimize σ_r^2, the power of $\{\mathbf{r}(n)\}$. We shall first assume that the coefficients are also "fixed" and try to evaluate what is the *best* values for these "fixed" coefficients. We also assume that $\{\mathbf{x}(n)\}$ is a zero-mean random process and so σ_r^2 is given by

$$\sigma_r^2 = \mathbf{E}\{\mathbf{r}(n)^2\} = \mathbf{E}\{[\mathbf{e}(n) - \xi(n)]^2\} =$$

$$\mathbf{E}\{\mathbf{e}(n)^2\} - 2\sum_{k=0}^{N-1} h_n(k)R_{\mathbf{ex}}(k) + \sum_{k=0}^{N-1}\sum_{j=0}^{N-1} h_n(k)h_n(j)R_{\mathbf{xx}}(j-k) \tag{6.3.4}$$

where the expected value of the square of the residual echo is expressed in terms of the filter coefficients, the correlation between the initial echo signal and the reference process $\{\mathbf{x}(n)\}$, and the autocorrelation of the reference process. The optimal coefficients, $\{h^*(k)\}$, are obtained by setting the first derivative of σ_r^2 with respect to each of the coefficients to zero. This yields the following system of linear algebraic equations, called the **normal equations**:

$$\sum_{k=0}^{N-1} h^*(k)R_{\mathbf{xx}}(k-j) = R_{\mathbf{ex}}(j);\ j=0, 1, \ldots, (N-1) \tag{6.3.5}$$

If the coefficients and the correlations are organized as vectors and matrices then the normal equations take the form

$$\mathbf{h}^{*} = \mathbf{R}_{\mathbf{xx}}^{-1}\,\mathbf{R}_{\mathbf{ex}} \tag{6.3.6}$$

As a special case, when $\{\mathbf{x}(n)\}$ is **white noise**, then $R_{xx}(k) = \sigma_x^{\,2}\,\delta(k)$ and the normal equations reduce to

$$\mathbf{h}^{*} = \frac{1}{\sigma_{x}^{2}}\,\mathbf{R}_{\mathbf{ex}} \tag{6.3.7}$$

We next make the assumption that the locally generated signal $\{\mathbf{s}(n)\}$ is uncorrelated with $\{\mathbf{x}(n)\}$. With this assumption we could write

$$\begin{aligned} R_{\mathbf{yx}}(k) &= \mathbf{E}\{\mathbf{y}(n)\,\mathbf{x}(n-k)\} = \mathbf{E}\{\mathbf{s}(n)\,\mathbf{x}(n-k)\} + \mathbf{E}\{\mathbf{e}(n)\,\mathbf{x}(n-k)\} \\ &= R_{\mathbf{sx}}(k) + R_{\mathbf{ex}}(k) = R_{\mathbf{ex}}(k) \end{aligned} \tag{6.3.8}$$

which indicates that we could also perform the correlation between the observed signal, $\{\mathbf{y}(n)\}$, and $\{\mathbf{x}(n)\}$ in order to ascertain $R_{ex}(k)$ for the normal equations. In one sense the normal equations represent the way we would have computed the optimal coefficients if we were able to specify the signal and measure the output. Solving the normal equations requires the inversion of a matrix which in turn is usually too computationally intensive for any cost-effective real-time implementation. Nevertheless, examination of the normal equations provides some insight into the pitfalls we may encounter, regardless of the method used to obtain the coefficients.

Consider the autocorrelation matrix $\mathbf{R}_{xx}$ which is described by

$$\mathbf{R}_{\mathbf{xx}} = \begin{bmatrix} R_{\mathbf{xx}}(0) & R_{\mathbf{xx}}(1) & \cdots & R_{\mathbf{xx}}(N-1) \\ R_{\mathbf{xx}}(1) & R_{\mathbf{xx}}(0) & \cdots & \cdots \\ \cdots & \cdots & \cdots & \cdots \\ R_{\mathbf{xx}}(N-1) & \cdots & \cdots & R_{\mathbf{xx}}(0) \end{bmatrix} \tag{6.3.9}$$

where we have exploited the symmetry of the autocorrelation sequence. The autocorrelation matrix is seen to have considerable structure. It is symmetric and further has the same terms along "diagonals." Such a matrix is called ***Toeplitz*** and has been studied extensively. Inversion of this matrix is feasible only if the matrix is "well conditioned." The notion of well conditioned relates to how close the matrix is to being singular. If the matrix is close to singular then calculating the inverse is difficult, often impossible, since ordinary arithmetic rounding errors may make the matrix "look" singular. In the case of white noise, the autocorrelation matrix is

$$\mathbf{R_{xx}} = \sigma_{\mathbf{x}}^2 \begin{bmatrix} 1 & 0 & \ldots & 0 \\ 0 & 1 & \ldots\ldots & \\ \ldots & \ldots & \ldots & \ldots \\ 0 & \ldots\ldots & & 0 \end{bmatrix} \tag{6.3.10}$$

and is very well conditioned and its inverse is obvious. When a matrix is singular, its rows are not linearly independent. For example, if {x(n)} is a sinusoidal signal with base period of **m** samples, then rows seperated by **m** are identical! It can be shown that the conditioning of the autocorrelation matrix can be related to the flatness of the power spectrum, $S_{\mathbf{xx}}(f)$. If $S_{\mathbf{xx}}(f)$ is reasonably flat (like white noise) then the autocorrelation matrix will be well conditioned. If $S_{\mathbf{xx}}(f)$ has very distinct peaks and valleys (the sinusoid has delta functions that correspond to very, very high peaks) then the autocorrelation matrix is ill conditioned. If $S_{\mathbf{xx}}(f)$ has nulls then the autocorrelation matrix will be very ill conditioned and may even be singular (no inverse exists).

We shall assume that {x(n)} is white. By doing so we will keep the analysis quite simple. In situations where we know the signal is not white we can suitably change our expectations as to the performance of the filter. Considering the number of other assumptions we make anyway, we know that the analysis and the derived results must be taken in context. The trick, as it were, is to complete an analysis using a variety of assumptions and, mostly from experience, ascertain a level of confidence in the derived results. In the case of adaptive filters we know that if {**x**(n)} is not white noise then our confidence in the results will diminish in proportion to the peakedness of the power spectrum $S_{\mathbf{xx}}(f)$; the greater the peaks or lower the valleys in $S_{\mathbf{xx}}(f)$ the less confidence we have in our results (based on the white noise assumption).

To indicate the impact of the spectrum of {**x**(n)} on the design problem, we write the error, that is the residual echo, power as

$$\begin{aligned} \sigma_{\mathbf{r}}^2 &= \mathbf{E}\{\mathbf{r}(n)^2\} = \mathbf{E}\{[\sum_{k}(g(k)-h(k))\mathbf{x}(n-k)]^2\} \\ &= \sum_{k}\sum_{n} \Delta h(k)\,\Delta h(n)\,R_{\mathbf{xx}}(n-k) \end{aligned} \tag{6.3.11}$$

where {Δh(n)} represents the impulse response of the *filter error* arising because we have not matched H(**z**) and G(**z**) exactly. After some algebraic manipulation the residual error power is given by

$$\sigma_{\mathbf{r}}^2 = \frac{1}{f_s}\int_{-\frac{1}{2}f_s}^{-\frac{1}{2}f_s} |\Delta H(f)|^2 S_{\mathbf{xx}}(f)\,df \tag{6.3.12}$$

Thus the spectrum of the reference signal $\{\mathbf{x}(n)\}$ is a weighting applied to the frequency response error. This implies that if $S_{xx}(f)$ (or $S_{xx}(e^{j\omega})$) is zero (or very small) at any frequency (or range of frequencies), we cannot control the frequency response error at that frequency!

When $\{\mathbf{x}(n)\}$ is **white noise** the correlation between $\{\mathbf{e}(n)\}$ and $\{\mathbf{x}(n)\}$ can be derived as

$$R_{\mathbf{ex}}(k) = \mathbf{E}\{\,\mathbf{e}(n)\mathbf{x}(n-k)\,\} = \mathbf{E}\{\sum_{j=0}^{\infty} g(j)\mathbf{x}(n-j)\mathbf{x}(n-k)\,\}$$

$$= \sum_{j=0}^{\infty} g(j)R_{\mathbf{xx}}(k-j) = \sigma_{\mathbf{x}}^2 g(k) \tag{6.3.13}$$

since $R_{xx}(k) = \sigma_x^2\,\delta(k)$. This yields the intuitively satisfying result that the optimal coefficients for the filter are given by

$$h^*(k) = g(k)\,;\ k = 0, 1, 2, \ldots, (N-1) \tag{6.3.14}$$

That is, the optimal set of N coefficients is just the first N coefficients of the impulse response of the unknown system, **G**. What is not so intuitive is why this result should not be true in general, regardless of the spectrum of $\{\mathbf{x}(n)\}$. The reason for this is that we are minimizing the power of the residual signal and not trying specifically to make $H(\mathbf{z})$ exactly the same as $G(\mathbf{z})$.

With this choice of coefficients the power of the residual echo is given by

$$\mathbf{E}\{\,\mathbf{r}(n)^2\,\} = \mathbf{E}\{[\sum_{k=N}^{\infty} g(k)\mathbf{x}(n-k)]^2\,\}$$

$$= \sigma_{\mathbf{x}}^2 \sum_{k=N}^{\infty} g(k)^2 = \sigma_{\mathbf{x}}^2 G_{MIN} \tag{6.3.15}$$

where we have again used the fact that the autocorrelation is a delta function. This represents the minimum value of residual echo power, which cannot be reduced further, and is a consequence of our choosing to eliminate echo using a finite-length (N samples) filter. In practice the unknown is system is indeed unknown and we do not have the *a priori* values of $\{R_{ex}(k); k=0, 1, 2, \ldots, (N-1)\}$ to obtain the coefficients of $H(\mathbf{z})$. What we can do is estimate this correlation by observing M samples, say $\{y(m);\ m = 0, 1, \ldots, (M-1)\}$, and using our knowledge of $\{\mathbf{x}(n)\}$, which has been observed for (all) time up to and including sample epoch (M–1), approximate the correlation by a sum of the form

$$R_{ex}(k) \approx \frac{1}{M} \sum_{n=0}^{M-1} y(n)\,x(n-k) \tag{6.3.16}$$

The quality of this estimate will improve as we increase M, the observation period. We are implicitly assuming that the process is ergodic.

One approach to adaptive behavior would be to make this estimate every M samples and change H(**z**) accordingly. That is, we use Eq. (6.3.16) to estimate $R_{ex}(k)$ and estimate the variance of {**x**(n)} by a similar summation and compute, that is estimate, $h_M(k)$ by

$$h_M(k) \approx \frac{\frac{1}{M} \sum_{n=0}^{M-1} y(n)\,x(n-k)}{\frac{1}{J} \sum_{n=0}^{J-1} x(M-1-n)^2} \tag{6.3.17}$$

where we have assumed that J terms will be used in the estimate of σ_x^2 (J can be larger than M since we have knowledge of past values of the reference signal). While this approach provides a new estimate of H(**z**) every M samples, it is not really "adaptive." It is not adaptive because it does not take into account the fact that having estimated {$h_M(k)$} we do not use our knowledge of {w(n)}, which contains the residual echo, to decide whether our estimate of the filter coefficients was any good at all. Intrinsic to adaptive filters is the notion that, having made a guess of what the filter coefficients should be, we use the resulting (error) signal {w(n)} to refine our estimate. This notion is explained next.

6.3.2 Adaptive Echo cancelation

The notion of refining our estimate can be formulated as shown in Fig. 6.11. The overall unknown system is the combination of G(**z**) and our most recent estimate of what H(**z**) should be, which is labeled H_0(**z**). Refinement of the coefficients is then equivalent to establishing what the (correction) filter ΔH(**z**) is to minimize the residual error in {$w_1(n)$}. The same mathematical principles apply whether we are obtaining ΔH(**z**) or H(**z**). That is, calling the residual echo in {w(n)} as {**r**(n)}, the normal equations for Δ**h** would be

$$\Delta\mathbf{h}^* = \mathbf{R}_{xx}^{-1} R_{rx} \tag{6.3.18}$$

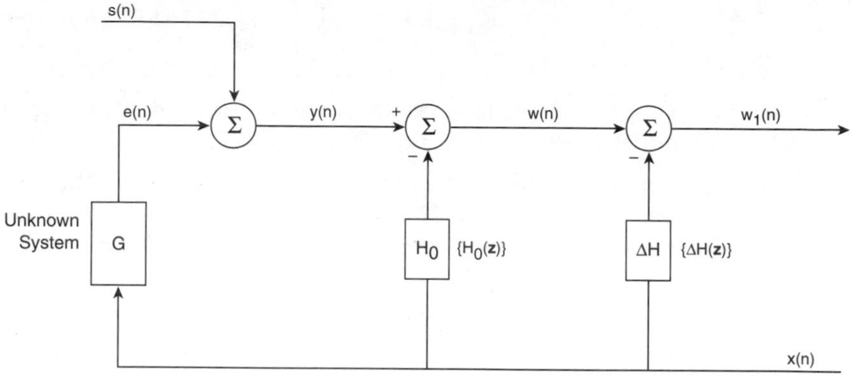

Fig. 6.11
Reconfiguring the echo-replicating filter of Fig. 6.10 with the notion of a correction

With the same set of assumptions as before, we can estimate $R_{\mathbf{rx}}(k)$ by

$$R_{\mathbf{rx}}(k) \approx \frac{1}{M} \sum_{n=0}^{M-1} \mathbf{w}(n)\,\mathbf{x}(n-k) \tag{6.3.19}$$

and estimate $\Delta h_M(k)$ by

$$\Delta h_M(k) \approx \frac{\dfrac{1}{M} \displaystyle\sum_{n=0}^{M-1} \mathbf{w}(n)\,\mathbf{x}(n-k)}{\dfrac{1}{J} \displaystyle\sum_{n=0}^{J-1} \mathbf{x}(M-1-n)^2} \tag{6.3.20}$$

The adaptive behavior of the algorithm is expressed in words in the following way. If we have a current estimate, $H_{rM}(k)$, we observe the output for M samples and compute a correction term $\Delta H_{rM}(k)$ and set

$$h_{(r+1)M}(k) = h_{rM}(k) + \Delta h_{rM}(k) \tag{6.3.21}$$

The expression for the update given in Eq. (6.3.20) is correct only if the reference signal is actually white noise. In practice we cannot guarantee this since we have no control over $\{\mathbf{x}(n)\}$; it is what it is. However, since we are computing corrections to the filter it is reasonable to expect that the white noise assumption is not quite so critical as the case when we are computing the filter coefficients themselves (by Eq. (6.3.17) for example). Second, to reduce the computational burden, we do not compute an estimate of $\sigma_{\mathbf{x}}^2$ (the denominator of Eq.(6.3.17) or, (6.3.20). Instead we use what is known as the ***Least Mean Square*** (**LMS**) algorithm, which uses for the correction term

$$\Delta h_{rM}(k) = \mu \frac{1}{M} \sum_{m=0}^{M-1} \mathbf{w}(rM+m)\,\mathbf{x}(rM+m-k) \qquad (6.3.22)$$

where the constant μ is called the adaptation gain. In the classical form of the **LMS** algorithm the correlation estimate is exceedingly coarse, using just one term ($M = 1$). Thus the classical **LMS** algorithm is described by

$$h_{n+1}(k) = h_n(k) + \mu\, \mathbf{w}(n)\,\mathbf{x}(n-k) \qquad (6.3.23)$$

The expected behavior of the adaptation algorithm is that the coefficients $\{h_n(k)\}$ converge to the optimal values described by $\{h^*(k)\}$. It can be shown that larger values of gain provide for more rapid convergence. However, if the gain is too large then the algorithm will not converge! We refer the interested reader to Widrow et. al. [3] for a detailed analysis of the stability of the **LMS** algorithm. The summary result is that the largest allowable value of gain, μ, relates to how well the the autocorrelation matrix is conditioned. In particular, if the eigenvalues of $\mathbf{R}_{xx}$ are λ_1, λ_2, λ_3, ... , λ_N, with the N eigenvalues listed in descending order of magnitude, then for the adaptation scheme of Eq. (6.3.23) (the classical **LMS** algorithm), the stability constraint is

$$0 < \mu < \frac{2}{\lambda_1} \qquad (6.3.24)$$

That is, the gain must be less than twice the reciprocal of the largest eigenvalue (magnitude). If $\{\mathbf{x}(n)\}$ is white noise then all N eigenvalues are the same value, σ_x^2, and then

$$0 < \mu < \frac{2}{\sigma_x^2} \qquad (6.3.25)$$

Following Messerschmitt [1], the rate of convergence of the algorithm can be described in terms of "modes," where each eigenvalue of the autocorrelation matrix defines a mode. The rate of convergence of a mode follows the trajectory

$$(1-\mu\lambda_i)^n \rightarrow 0 \qquad (6.3.26)$$

which depicts the convergence to zero of the residual echo power. This convergence to zero requires that

$$|1-\mu\lambda_i| < 1 \qquad (6.3.27)$$

which is true provided the adaptation gain satisfies Eq. (6.3.24). Further, it is seen that the rate of convergence to zero is governed by the smallest (in magnitude) eigenvalue. Combining the stability requirement and the rate of convergence, we can state that the behavior of the adaptation is governed by the ratio of the least to greatest eigenvalue. Because of the specific structure of the autocorrelation matrix (Toeplitz) it can be shown that this ratio can be translated into the behavior of the power spectrum. That is,

$$\frac{\lambda_{\min}}{\lambda_{\max}} \approx \frac{\min\limits_{f} S_{xx}(f)}{\max\limits_{f} S_{xx}(f)} \tag{6.3.28}$$

Thus the flatter the spectrum, the less is the eigenvalue spread and consequently the better we can expect the adaptive algorithm to perform (from the viewpoint of convergence). If $S_{xx}(e^{j\omega})$ has "holes" or zeros then we cannot expect the algorithm to converge at all!

The choice of adaptation gain, μ, is influenced by an additional consideration. To keep the analysis reasonable we shall assume that the reference signal is white Gaussian noise. This means that the optimal coefficients are given by Eq. (6.3.14). After we have converged any further corrections should be zero. For the classical **LMS** scheme, the corrections are

$$\Delta h_n(k) = \mu\, \mathbf{w}(n)\, \mathbf{x}(n-k) \tag{6.3.29}$$

Since we have one observation of what is actually a random process, we cannot guarantee that $\Delta h_n(k)$ will be zero even if we have converged. This can be illustrated by first writing the expression for the output $\{\mathbf{w}(n)\}$:

$$\mathbf{w}(n) = \mathbf{s}(n) + \sum_{k=0}^{\infty} g(k)\,\mathbf{x}(n-k) - \sum_{k=0}^{N-1} h(k)\,\mathbf{x}(n-k) \tag{6.3.30}$$

Defining the coefficient error by

$$\chi_n(k) = h^*(k) - h(k) \tag{6.3.31}$$

the output $\{\mathbf{w}(n)\}$ can be written as

$$\begin{aligned} \mathbf{w}(n) &= \mathbf{s}(n) + \sum_{k=N}^{\infty} g(k)\,\mathbf{x}(n-k) - \sum_{k=0}^{N-1} \chi_n(k)\,\mathbf{x}(n-k) \\ &= \mathbf{s}(n) + \mathbf{r}(n) \end{aligned} \tag{6.3.32}$$

Thus $\{\mathbf{w}(n)\}$ is comprised of three components. $\{\mathbf{s}(n)\}$ represents the locally generated signal (near-end speech plus any additive interference); the second term represents the echo that cannot be canceled because of the finite length of our **FIR** filter; the third term represents the remnant of the echo because the filter coefficients are not optimal. Because we continually modify (adapt) the coefficients, and our correction is not guaranteed to be zero, our notion of convergence must be defined appropriately. We treat convergence as the condition whereby all the systematic errors between our coefficients and the optimal values has been removed and the actual coefficients are fluctuating about the mean (optimal) values. That is, we treat $\{\chi_n(k)\}$, the coefficient error vector, as a random process that, after convergence has been achieved, will be zero-mean:

$$\text{Convergence} \Rightarrow \mathbf{E}\{\chi_n(k)\} = 0;\ k = 0, 1, 2, \ldots, (N-1) \qquad (6.3.33)$$

The variance or power associated with $\{\chi_n(k)\}$ then represents how good this convergence actually is. That is, the goodness of convergence is represented by

$$\sigma^2_{\chi(k)} = \mathbf{E}\{\chi_n(k)^2\};\ k = 0, 1, 2, \ldots, (N-1) \qquad (6.3.34)$$

or, to take into account all N coefficients simultaneously, we define

$$\sigma^2_{\chi} = \mathbf{E}\{\sum_{k=0}^{N-1} \chi_n(k)^2\} \qquad (6.3.35)$$

The efficacy of the adaptive filter from a system viewpoint is established in terms of the Echo Return Loss (**ERL**). If there is no filter ($H(\mathbf{z})$ 0), then the power of the echo component in $\{\mathbf{w}(n)\}$ is given by

$$\sigma^2_{\mathbf{e}} = \sigma^2_{\mathbf{x}} \sum_{k=0}^{\infty} |g(k)|^2 = \sigma^2_{\mathbf{x}} G_E \qquad (6.3.36)$$

and we define **ERL** in terms of the ratio of echo power to the power of the reference signal. That is,

$$\mathbf{ERL} = \frac{\sigma^2_{\mathbf{e}}}{\sigma^2_{\mathbf{x}}} = \sum_{k=0}^{\infty} |g(k)|^2 = G_E \quad (\text{linear scale})$$

$$\mathbf{ERL} = -10 \log_{10}(G_E)\ \text{dB} \quad (\text{log units for } \textbf{loss}) \qquad (6.3.37)$$

By applying the filter we reduce the power of the echo. This is measured as an **Echo Return Loss Enhancement** (**ERLE**). At time epoch of n (in samples), the power of the residual echo is given by

$$\sigma^2_{\mathbf{r}} = \sigma^2_{\mathbf{x}} \sum_{k=N}^{\infty} |g(k)|^2 + \sigma^2_{\mathbf{x}} \mathbf{E}\{\sum_{k=0}^{N-1} \chi_n(k)^2\} \qquad (6.3.38)$$

(note that this is a statistical expectation of the square of $\mathbf{r}(n)$), which means that the **ERLE** can be expressed as

$$\mathbf{ERLE} = \frac{\sigma_r^2}{\sigma_e^2} = \frac{\sum_{k=N}^{\infty} |g(k)|^2 + \mathbf{E}\{\sum_{k=0}^{N-1} \chi_n(k)^2\}}{\sum_{k=0}^{\infty} |g(k)|^2}$$

$$= \frac{G_{MIN} + \sigma_\chi^2}{G_E} \quad (\text{linear scale}) \tag{6.3.39a}$$

$$\mathbf{ERLE} = -\log_{10}\left[\frac{G_{MIN} + \sigma_\chi^2}{G_E}\right] \text{ dB} \quad (\text{log scale}) \tag{6.3.39b}$$

In taking the expectation we make the simplifying assumptions that the reference signal is uncorrelated with the coefficient error and that the coefficient error sequence itself is "white," $\chi_n(j)$ uncorrelated with $\chi_n(k)$ for $j \neq k$. While these assumptions are not really applicable, we make them anyway since without them the analysis would be intractable. We can see there is a lower bound to the **ERLE** we can achieve because of the "model error" represented by G_{MIN}. By making N long enough we can probably make G_{MIN} small enough to be ignored. The next item that prevents our enhancement from being large (infinite) is the uncertainity in the coefficients, encapsulated by the variance term σ_χ^2. This variance can be analyzed in the following manner.

We write the coefficient error as

$$[h^*(k) - h_{n+1}(k)] = [h^*(k) - h_n(k)] - [h_{n+1}(k) - h_n(k)] \tag{6.3.40}$$

and thus express the coefficient error as it relates to our coefficient updates, $\Delta h_n(k)$, as

$$\chi_{n+1}(k) = \chi_n(k) + \Delta h_n \tag{6.3.41}$$

The coefficient uncertainity is thus tied directly into the estimate of the correlation coefficient. If our estimate is good (exact), then when we reach the state of convergence this update is zero and our coefficient error remains zero. Unfortunately, this is not the case. In the classical **LMS** algorithm our coefficient update is given by

$$\Delta h_n(k) = \mu\, \mathbf{w}(n)\, \mathbf{x}(n-k) \tag{6.3.42}$$

and even if, by some chance, $\chi_n(k)$ is zero, $\chi_{n+1}(k)$ is, in general, nonzero. Thus the variability expressed by σ_χ^2 is related to how good the estimate (Eq. (6.3.41)) is. We can get a feel for this variability by assuming that, at time epoch n, $\chi_n(k)$ was zero. Then

$$\Delta h_n(k) = \mu\, \mathbf{s}(n)\mathbf{x}(n-k) + \mu \sum_{j=N}^{\infty} g(j)\mathbf{x}(n-j)\mathbf{x}(n-k) \qquad (6.3.43)$$

and we can see that the expected value of the coefficient update is

$$\mathbf{E}\{\Delta h_n(k)\} = 0 \qquad (6.3.44)$$

In statistical nomenclature $\Delta h_n(k)$ is an ***unbiased estimator*** since, on the average, it provides us with the correct value. The expected value of the square of $\Delta h_n(k)$ is given by the rather formidable expression

$$\mathbf{E}\{\Delta h_n(k)^2\} = \mathbf{E}\Big\{\mu^2 \mathbf{s}(n)^2 \mathbf{x}(n-k)^2 + 2\mu^2 \mathbf{s}(n)\mathbf{x}(n-k)^2 \sum_{j=N}^{\infty} g(j)\mathbf{x}(n-j) + \mu^2 \sum_{j=N}^{\infty}\sum_{i=N}^{\infty} g(j)g(i)\mathbf{x}(n-j)\mathbf{x}(n-i)\mathbf{x}(n-k)^2\Big\} \qquad (6.3.45)$$

The assumptions that we have made, namely that $\{\mathbf{s}(n)\}$ and $\{\mathbf{x}(n)\}$ are uncorrelated, and that $\{\mathbf{x}(n)\}$ is white Gaussian noise, allow us to evaluate this expression as

$$\mathbf{E}\{\Delta h_n(k)^2\} = \mu^2 \sigma_s^2 \sigma_{\mathbf{x}}^2 + \mu^2 \sigma_s^2 G_{MIN} \qquad (6.3.46)$$

More important than the actual expression is the information we can derive from it. If we make the filter long enough, then G_{MIN} is small. If we ignore it, the variability σ_χ^2 is approximately equal to

$$\sigma_\chi^2 = N\mu^2 \sigma_{\mathbf{s}}^2 \sigma_{\mathbf{x}}^2 = N\left(\mu\sigma_{\mathbf{x}}^2\right)^2 \left(\frac{\sigma_{\mathbf{s}}^2}{\sigma_{\mathbf{x}}^2}\right) \qquad (6.3.47)$$

This tells us that the best echo return loss enhancement, in a statistical sense, is limited by the presence of local signal. This effect is exacerbated by the length of the filter. The impact of the local signal is mitigated, to some degree, by using a small value for the adaptive gain. For the generalized **LMS** algorithm, whereby the coefficients are updated only every M samples, the coefficient update is governed by Eq. (6.3.22). It can be shown that the variability σ_χ^2 is reduced by a factor of M, that is,

$$\sigma_\chi^2 = \frac{N}{M}\mu^2 \sigma_s^2 \sigma_x^2 = \frac{N}{M}\left(\mu\sigma_x^2\right)^2\left(\frac{\sigma_s^2}{\sigma_x^2}\right) \tag{6.3.48}$$

at the expense of the rate of adaptation (we update the coefficients only every M samples as opposed to every sample for the classical **LMS** algorithm). The best achievable **ERLE** is thus directly proportional to the signal-to-noise ratio, treating **all** locally generated signal as noise.

To summarize, an adaptive (**FIR**) filter is described by the following operations in the context of an echo canceler:

a) **Filter**. A local replica of the echo is generated by an **FIR** filter operating on the most recent samples of the reference input:

$$\xi(n) = \sum_{k=0}^{N-1} h_n(k)\,\mathbf{x}(n-k) \tag{6.3.49}$$

b) **Cancelation.** This replica is subtracted from the input signal to *cancel* the echo component:

$$\mathbf{w}(n) = \mathbf{y}(n) - \xi(n) \tag{6.3.50}$$

c) **Adaptation.** Based on the result of the cancelation, the coefficients are updated to get ready for the next sample:

$$h_{n+1}(k) = h_n(k) + \mu\,\mathbf{w}(n)\,\mathbf{x}(n-k) \tag{6.3.51}$$

The algorithm is governed by a **stability constraint**, whereby the adaptation gain cannot be made arbitrarily large. A large value of adaptation gain provides a rapid convergence at the expense of achievable echo return loss enhancement. The presence of local signal, either interference (noise) or near-end speech has the deleterious effect of reducing the achievable **ERLE**. The overall behavior of the adaptation algorithm is depicted in Fig. 6.12, which shows how the echo power diminishes with time (provided the adaptation gain is small enough to ensure stability). A small value of adaptation gain yields a gradual slope that levels off at some value of minimum power (maximum echo return loss including enhancement). A larger value of adaptation gain provides a steeper initial slope but the final value is larger (worse).

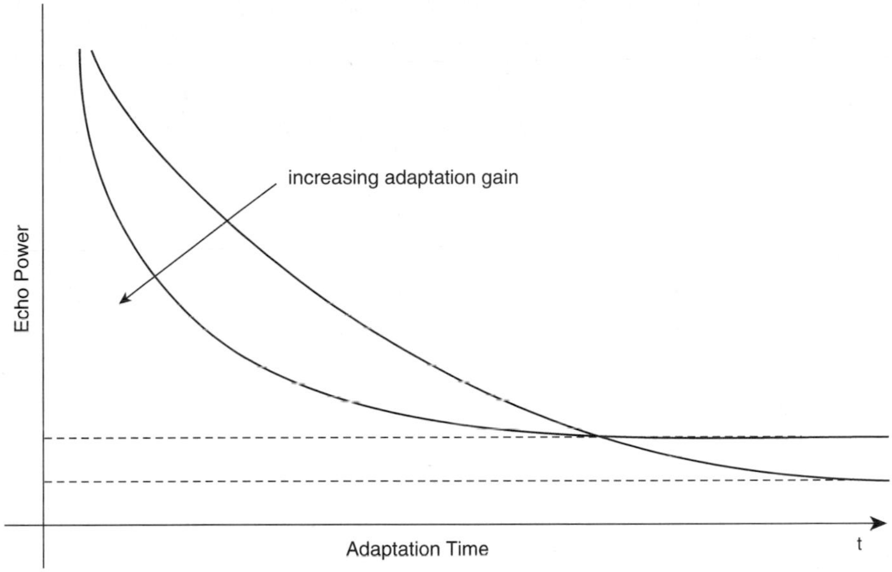

Fig. 6.12
Indicating trade-off between adaptation rate and achievable **ERL** as a function of adaptation gain

d) Performance. The performance of an adaptive filter depends not just on the adaptation gains, μ, but also on the filter length, N, and the signal-to-noise ratio. Provided the length is large enough to correct for the (generally infinite) length of the echo path impulse response, increasing N further is detrimental; doubling N will reduce the **ERLE** by 3 dB. The **ERLE** is also affected by the signal-to-noise ratio. An x dB decrease in **SNR** translates directly to an x dB reduction in the (expected) **ERLE**. While we did not analyze the impact of nonlinearities in the echo path, it is obvious that such behavior cannot be modeled by a linear time-invariant system and thus an adaptive **FIR** may not provide any echo-cancelation if the nonlinear behavior is pronounced. If the nonlinear effects are slight then they can be modeled as an additive noise that degrades the **ERLE** much the same way as locally generated signal.

6.4 NETWORK ECHO CANCELERS

Echo cancelers are the preferred method of echo control in the modern network. The principal advantage is that echo cancelers over suppressors is that the former do not necessarily restrict the communication to a half-duplex mode. All echo cancelers used in the network are digital in nature and operate on digital signals. A detailed description of the requirements of an echo canceler are provided in **CCITT** Recommendation **G.165** [9].

The location of an echo canceler is on the ***trunk side*** of a switch since echo control is required only for interswitch calls. This is diagramed in Fig. 6.13.

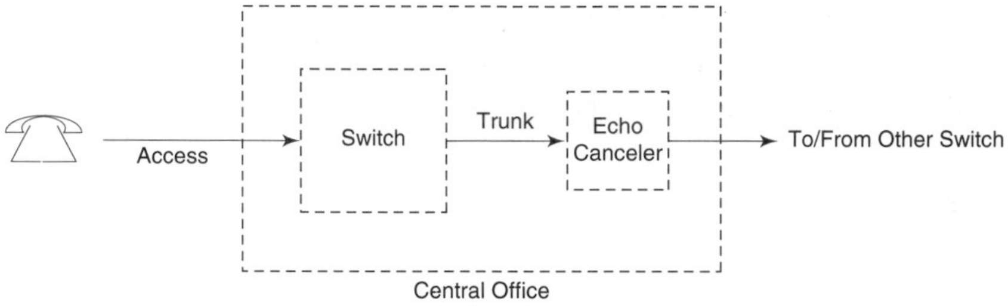

Fig. 6.13
Echo cancelers are usually deployed on the trunk side of a switch

The trunk-side location implies that the interconnections to and from the echo canceler are **DS1** or **E1**. **DS1** and **E1** represent Time Division Multiplexed (**TDM**) assemblies. As we described in Chapter 1, **DS1** is comprised of 24 voice channels, each corresponding to an 8-kHz sampling rate, μ-law encoded signal, plus some framing overhead to make up a serial bit-stream of (24*8*8 + 8) = 1544 kbits/sec. **E1** is comprised of 30 voice channels with overhead, each channel representing a signal sampled at 8 kHz and quantized according to the A-law for a net serial bit rate of (30*8*8 + 128)= 2048 kbits/sec. It is important to recognize that since echo cancelers operate on an assembly of 24 channels it is possible to share computational power and amortize the cost of hardware over 24/30 channels.

The reason for showing a telephone and access line is to indicate which side of the echo canceler the echo originates. This side is called, variously, the ***near-end,*** or ***tailside***. Conversely, the other side of the canceler is referred to as the ***long-haul***, ***far-end***, or ***network*** side. Other terminologies are used in the literature but the meaning is usually obvious from the context. The speech signal originating from the near-end side is always considered *local*.

6.4.1 Echo Canceler Functional Summary

A simplified block diagram indicating the functions of an echo canceler is shown in Fig. 6.14. The structure is called a "split-type echo canceler" since it removes echo only in one direction. All echo cancelers manufactured today are of the split type since it is expected that there will be an echo canceler at each side of the circuit, with the tails pointing in opposite directions, to remove echo generated from both "local" ends.

The signal {x(n)} from the far end, which can be observed by the canceler, is not modified in any way but passed on toward the near-end. This signal reaches the tail side hybrid and some fraction of the power is returned as echo. The block diagram also shows the injection of near-end speech and noise. These are important. The noise injected is a primarily quantization noise (and aliasing) introduced by the A/D converter but may have other sources as well. The signal from the tail side, {y(n)}, is thus a combination of near-end speech, noise, and echo. Since the physical location of the canceler may be far removed from the hybrid, the source of the echo, we have shown a transmission path with dotted lines. This transmission path may include various equipment including switches, cross-connects, and multiplexers, in addition to transmission media such as copper cable, fiber optic cable, radio, and so on. We shall assume that no signal processing, as far as the particular speech channel is concerned, is performed in the transmission path and thus it appears just as a delay.

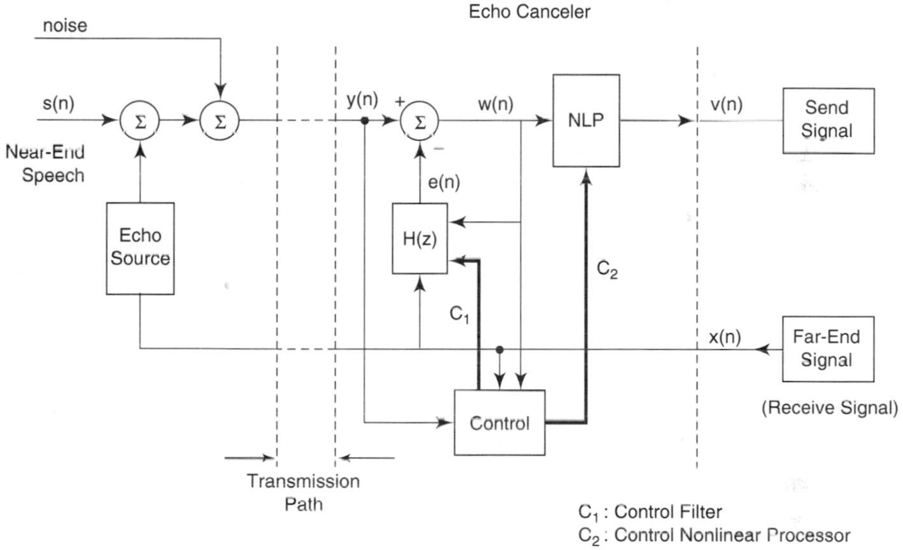

Fig. 6.14
Principal components of an echo canceler are the control mechanism, adaptive filter, and nonlinear Processor

The canceler implements an adaptive filter, denoted by H(**z**), that generates a local replica of the echo and subtracts this replica from {y(n)} to generate {w(n)} which, ideally, comprises just the near-end speech and, unfortunately, any locally generated noise in the tail circuit. The operation of the adaptive filter is described in Section 6.3, where we saw that w(n) is not completely echo-free. Consequently all echo cancelers include what is known as a ***Nonlinear Processor*** (**NLP**), which is actually an echo suppressor, as depicted in Fig. 6.15, to remove the last vestiges of remaining echo. Thus the *Send* signal, {v(n)}, which is transmitted in the long-haul direction is, for thc most part, echo free.

Also shown in the diagram is a block labeled **Control**. The functions of this block are to control the adaptive filter ($\mathbf{C}_1$) and the non-linear processor ($\mathbf{C}_2$). This is achieved by processing the various signals that can be observed by the block, namely {x(n)}, {y(n)}, and {w(n)}. An additional function of the control block is to determine whether the echo canceler should be disabled for the duration of the call and to determine when, if the canceler has been disabled, to re-enable the canceler. The notions of disabling and re-enabling are discussed in more detail later.

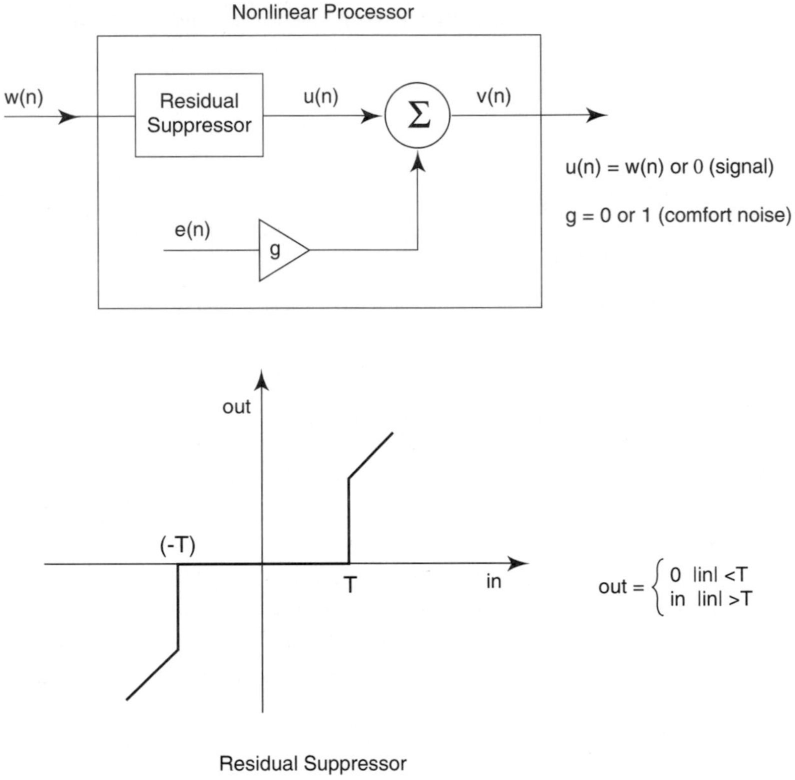

Fig. 6.15
Nonlinear processor comprises a residual suppressor (center clipper) and means to inject comfort noise.

A simplified block diagram of an echo canceler in a **DS1** environment is shown in Fig. 6.16. It comprises circuitry to perform the multiplexing and demultiplexing of the 24 constituent channels contained in the **DS1** assembly. The adaptive filters are done on a per-channel basis, implying that there are logically 24 such units within the equipment. The block labeled **Control** includes the functions associated with the **DS1** nature of the input-output signals and also includes the control functions associated with each of the 24 separate per-channel cancelers on a shared basis. A similar diagram can be generated for the **E1** case where instead of 24 we have 30 channels.

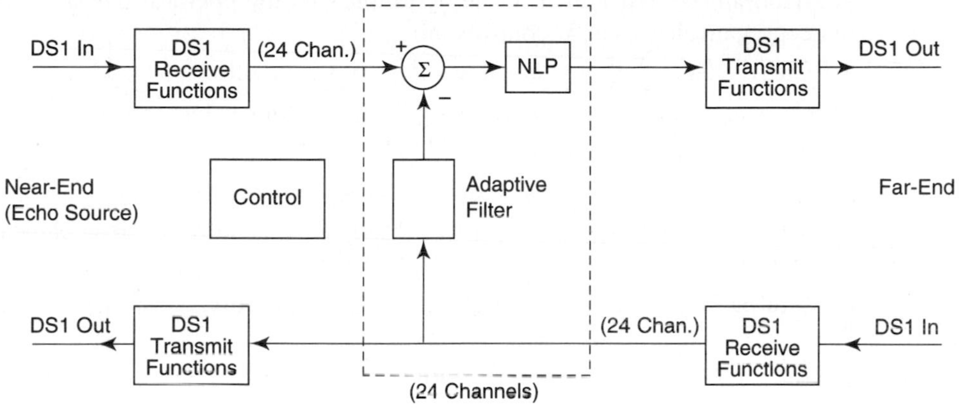

Fig. 6.16
Block diagram of a **DS1** level echo canceler

In discussing the operation of an echo canceler we introduce the term "state" of an echo canceler. Just as we saw for echo suppressors, the notion of the "state" of an echo canceler refers to the relative strengths of the far-end signal vis-à-vis the near-end signal. "Single-talk" is the condition in which the power of the reference signal, P_x, is much greater than the power of the near-end signal, P_s. "Hard double-talk" is the condition where P_s is much greater than P_x and is usually indicative that the far-end talker is silent and the near-end talker is active. "Double-talk" is the condition where P_s and P_x are approximately equal. "Soft double-talk" is the condition between single-talk and double-talk and reflects our concern of hangover times. The "idle" state is when P_x and P_s are small and have been so for an extended period of time. The idle state corresponds to both the near end and far-end talkers being silent.

6.4.2 Summary of Requirements on the Echo Canceler

Being a **DS1** or **E1** device, an echo canceler has several requirements to satisfy that have little or nothing to do with echo cancelation. These are not addressed here. We shall consider only those that pertain to the canceler on a per-channel basis. Thus for all intent and purposes we could consider a single channel only, the requirements being the same for all 24/30 channels. These channels are completely independent and care must be taken in an actual implementation that sharing resources does not compromise the operation of any individual channel.

A complete set of requirements is provided in **CCITT** Recommendation **G.165** [9]. Several network providers have developed their own requirements but these are based on **G.165** with variations, if necessary, to reflect the provider's networking philosophy. We shall discuss some of the requirements in detail, those that provide some indication of the signal processing needs and constraints, and not dwell much on the others.

The requirements are described in the form of experiments that can be performed in a laboratory environment. They presume that the operator has some control on the echo canceler. These controls are:

a) **Clear.** The coefficients of the filter are set to zero.

b) **Freeze.** The coefficients are held at the present value and not updated.

c) **Adapt.** Allow the adaptation to resume.

d) **Disable NLP.** Remove the **NLP** from the circuit path. Equivalent to disabling the echo suppressor part of the canceler. The echo return loss enhancement is then just that provided by the adaptive filter.

e) **Enable NLP.** Allow the **NLP** to perform normally.

6.4.2.1 Echo Return Loss Enhancement Requirement and Compliance Testing

The echo return loss enhancement that can be achieved with an adaptive filter depends on various factors such as, for example, the nature of the signal {x(n)} (which we called the reference signal in Section 6.3) in terms of power and spectral content. The enhancement is also influenced by the **ERL** itself, that which the hybrid provides, power of the near-end signal, impulse response, and so on. To set an appropriate baseline, the requirements are provided using a hypothetical case, one that might never occur in practice, but that can be constructed in a laboratory environment.

Consider the scheme shown in Fig. 6.17, which represents a single channel. The equipment that performs the **DS1/E1** multiplexing/demultiplexing is not indicated in the figure. A signal generator is used to provide {x(n)}, which is a white noise signal (8 kHz sampling rate, μ-law/A-law encoded) of power that can be varied over a range from –10 dBm0 to –30 dBm0. An echo path is created that can provide 6+ dB of **ERL** and care is taken to ensure that the effective impulse response of the echo path introduced is not longer than the length of the adaptive filter.

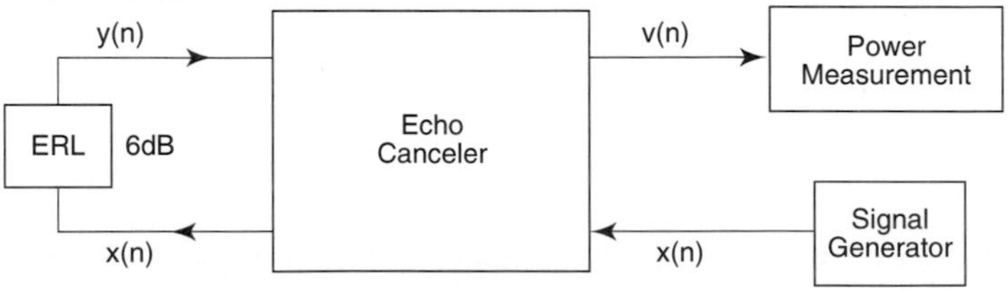

Fig. 6.17
Test setup for quantifying echo canceler performance

With the **NLP** disabled and the adaptive filter cleared, we expect the power reading to be exactly **X** dB below the power of the reference signal {x(n)}, where **X** is the **ERL** provided in the tail circuit. With the **NLP** held disabled, the transmit power, the power of {v(n)} is measured for different choices of reference power and the **ERLE** thus computed. Denoting the powers of reference and transmit signals by P_x and P_v (in dBm0), respectively, the **ERLE** is given by

$$\textbf{ERLE} = P_x - P_v - \textbf{X} \tag{6.4.1}$$

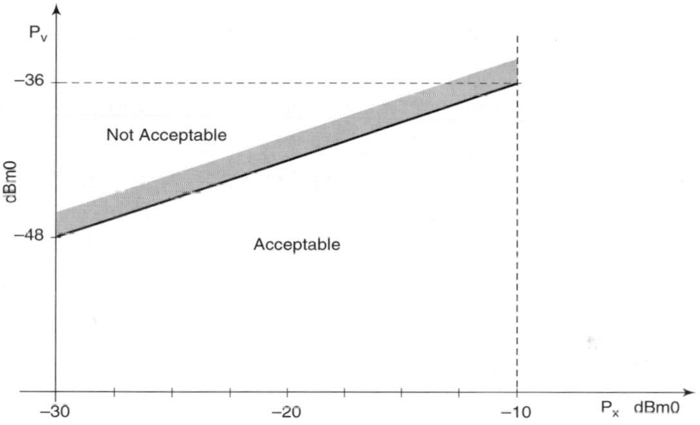

Fig. 6.18 (a)
Echo cancelation performance requirement in terms of send signal power

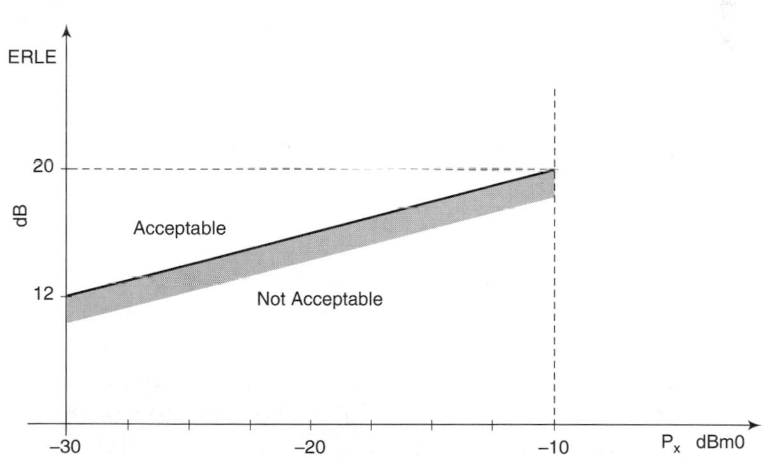

Fig. 6.18 (b)
Echo cancelation performance requirement in terms of **ERLE**

The requirement on **ERLE** is provided in an indirect fashion. Since the primary function is to reduce the echo to an "acceptable" level, it suffices that the returned power, P_v, be sufficiently small. Thus **G.165** specifies that for all values of **ERL**

6 dB, the returned power, P_v, as a function of P_x, shall be less than the value indicated in Fig. 6.18(a). This can be translated to an **ERLE** requirement by assuming that the base **ERL** is 6 dB and is indicated in Fig. 6.18(b). From the figure we see that the minimum **ERLE**, at –10 dBm0 signal and 6 dB **ERL** should be 20 dB or greater. This is quite lax. Most echo cancelers in the network achieve an **ERLE** of 30 dB or more under these conditions. The intent of this requirement is to include all the deleterious effects that might be present in an actual scenario that reduce the **ERLE** from its "best" value. That is, the 20-dB **ERLE** is supposed to reflect actual operational conditions. A commonly accepted value for the minimum **ERLE** in this situation is 28 dB.

With the nonlinear processor enabled, the **ERLE** is to be "infinite." The power measured at the transmit point, P_v, should be less than –65 dBm0 and there should be no echo component in {v(n)}, which should be composed solely of comfort noise. No technique is specified to verify that this is indeed so. Consequently, all echo canceler manufacturers include a control mechanism that shuts off the comfort noise (does not add it in). With the comfort noise out of the picture, the power of {v(n)} should be null, the lowest possible reading on the power measurement device, indicating that the residual suppressor has squelched the echo remnants completely.

6.4.2.2 Convergence Time

How quickly does the canceler react to a different environment? Since the canceler is in the trunking network, the echo path changes from call to call. Consequently, it is necessary for the adaptive filter to converge quite rapidly, preferably within the first few syllables of speech at the start of a new call. This is embodied in the graph shown in Fig. 6.19. Prior to time T the canceler is held in a clear and frozen state while a power of –10 dBm0 is applied at the reference input. At time T the canceler is enabled and adaptation starts. At time (T+500 msec) the canceler is frozen, and the filter coefficients are held in that state. The drop in power level of the transmit section is observed and must be greater than 21 dB at a nominal **ERL** of 6 dB. Another way of describing the requirement is that for all values of **ERL** greater than 6 dB and for reference power between –30dBm0 and –10dBm0, the power P_v, should be 27+ dB less than P_x. The **CCITT** requirement allows the nonlinear processor to be enabled in this test.

Again, the intention in **G.165** is that this convergence be demonstrated in an actual network application. In a controlled environment the convergence time is considerably more rapid. The graph in Fig. 6.20 is indicative of the manner in which the convergence time is usually indicated by the manufacturer. The reference signal is kept at –10 dBm0 and the **ERL** set to 6 dB and the canceler placed in a clear state with the **NLP** disabled. At time T the canceler is allowed to start adapting and the power, P_v, monitored. The time taken for the power to drop below –41 dBmo is indicative of the time taken for the echo return loss enhancement to go from 0 (clear) to 25 dB. This time interval, ΔT, is a measure of the convergence time of the canceler. Most cancelers available today specify this time as "less than 250 msec" though it has been observed in practice that the actual time is much less than 250 msec. In interpreting this requirement, the measurement of power should be taken in the spirit in which it is intended. Since power measurement requires some finite time (see Chapter 2), it is not possible to track the power P_v in an instantaneous manner. What is actually done is to initiate adaptation and freeze the canceler after a time ΔT and see what the power level is then. The experiment is repeated as often as necessary, increasing ΔT each time and each time providing a sufficient time interval for the power measurement to "stabilize," until a power value of –41 dBm0 (approximately) is achieved.

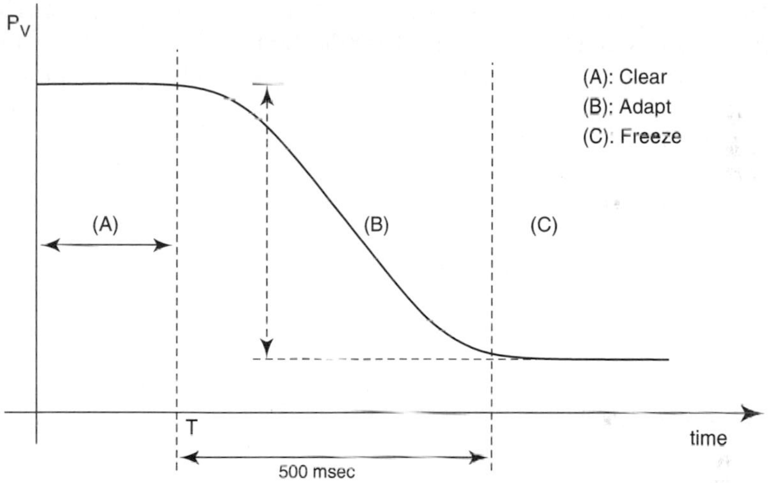

Fig. 6.19
Convergence time requirement for an echo canceler

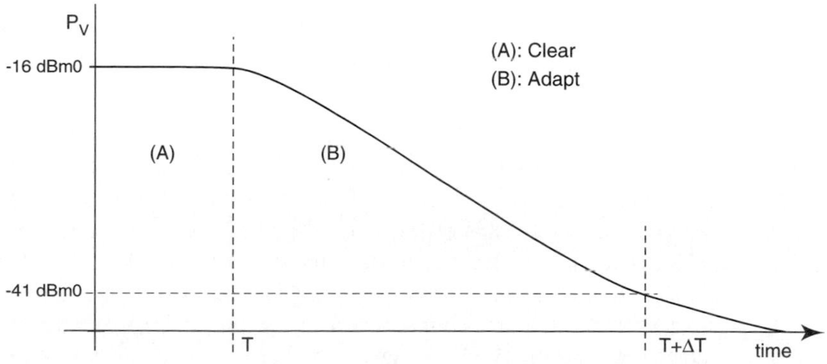

Fig. 6.20
Convergence time is, roughly, the interval for send signal power to drop by 25 dB

Applying a freeze command manually is not feasible, especially when we are talking of applying a freeze command about 200 msec after applying an adapt command. This is achieved in practice by computer-controlled operation of the echo canceler and the ancillary equipment such as the signal generators and level measurement devices. Very often the freeze command is "faked" as will be clear from the next section.

6.4.2.3 Performance under Double-Talk Conditions.

The behavior of an adaptive filter is degraded considerably when the power of the near-end speech is significantly high. Consequently, an echo canceler has to detect this double-talk condition and freeze the coefficients autonomously. The setup used to determine proper operation is shown in Fig. 6.21. The sequence of events is depicted in Fig. 6.22.

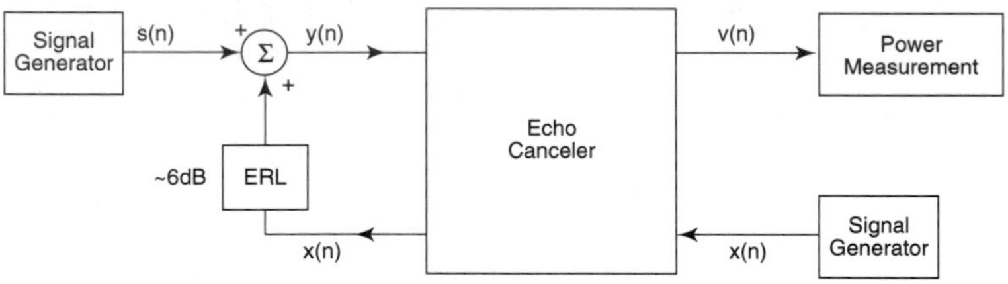

Fig. 6.21
Setup for investigating impact of near-end signals on echo canceler performance

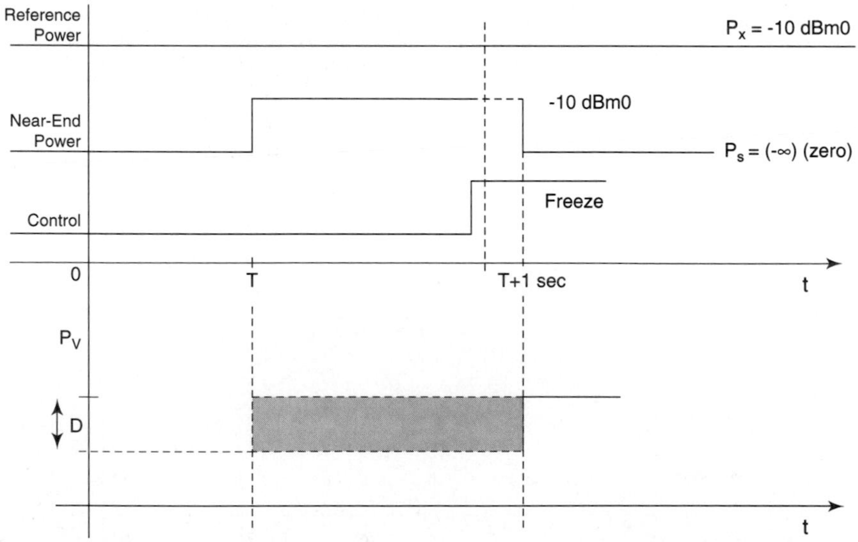

Fig. 6.22
Methodology for measuring divergence caused by near-end signal

At first the near-end signal is kept small (– dBm0), the **ERL** set and maintained at 6 dB, the reference signal power set at –10 dBm0, the **NLP** is held disabled, and the canceler is allowed to adapt. Sufficient time is allowed to elapse for convergence and the transmit power, P_v, recorded. At some time T, the near-end signal level is raised abruptly to –10 dBm0 (comparable to the reference level). One second later, at T+1 sec, the canceler is put into a "freeze" state. Subsequently, the near-end signal is removed and the transmit power measured. The difference, D dB, is a measure of how much the canceler diverged as a result of the double-talk. This deviation is required to be less than 10 dB. In practice it will be a lot less and application of a strong near-end signal is one way to "freeze" the canceler and thus allow a human operator enough time to issue a freeze command.

6.4.2.4 Infinite Echo Return Loss Behavior

There is the danger that an echo canceler might actually introduce echo! For example, if we were in a 4-wire end-to-end situation, with no hybrids in the circuit, there would be no echo to cancel. Any attempt by the canceler to "remove" echo would actually create echo. The infinite return loss requirement is meant to ensure that the canceler has some means to prevent this undesirable echo generation. The test procedure is depicted in Fig. 6.23.

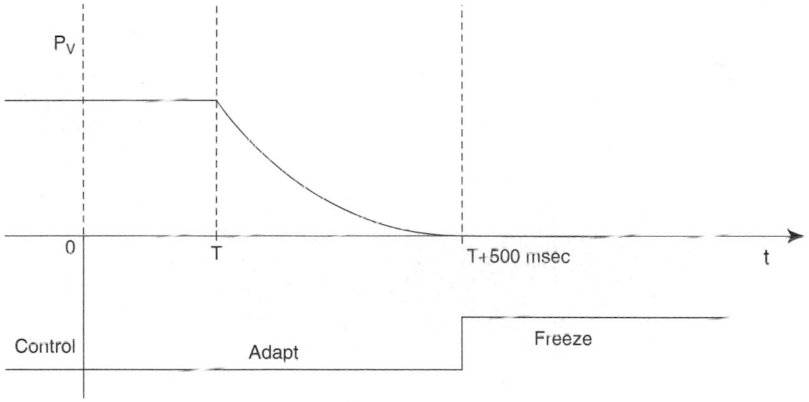

Fig. 6.23
Methodology for verifying that an echo canceler will not be the source of echo

The canceler is allowed to fully converge for a given echo path while applying a suitable reference signal power, P_x. This power is maintained throughout the experiment. At some time T the echo path is opened (infinite echo return loss) and 500 msec following that the canceler is "frozen." The power of the transmit signal is an indication of the echo introduced by the canceler and should be small. The requirement as stated in **G.165** is that for all initial **ERL** greater than 6 dB and for all power levels P_x in the range –30 dBm0 to –10 dBm0, the power P_v should drop below –37 dBm0 within 500 msec after the echo path is interrupted.

6.4.2.5 Tone Disabling Requirements

Unlike echo suppressors, which cause the transmission path to appear half-duplex, echo cancelers provide full duplex operation. For most modems the echo canceler can remain in operation throughout the call. For certain modems, however, it is advisable to disable the echo canceler.

Modern high speed modems include signal processing that counteracts the impact of a bad channel, to some extent. This is achieved by using a training sequence at the start of the call to determine the condition of the channel and the modems at the endpoints store this information appropriately. The assumption is that the channel will not change during the transmission of data, the main body of the call. Echo cancelers, being adaptive devices, do change the channel charcteristics and thus could, by "improving" the channel, degrade the performance of the modems.

The method prescribed for disabling echo cancelers is similar to that of disabling echo suppressors with the following twist. The echo canceler tone disabler requires the detection of a 2100-Hz tone with phase reversals of that tone. The answering modem emits a 2100-Hz tone and periodically bumps the phase by 180 degrees. The waveform of a sinusoid with phase reversals is shown in Fig. 6.24. Other than the additional requirement of detecting phase reversals, the requirements for the tone disabler for an echo canceler are the same as for an echo suppressor and are discussed in Section 6.2.

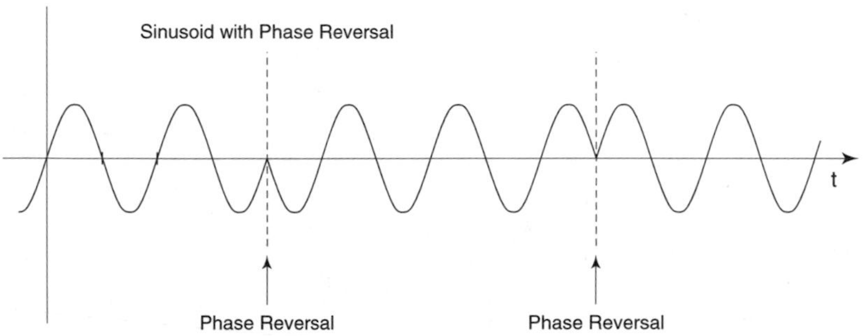

Fig. 6.24
Waveform depicting a (2100-Hz) tone with periodic phase reversals

When and how to disable an echo canceler has been the source of considerable debate. Certain network providers believe that the echo canceler must always be disabled for a call between modems, regardless of the modem type. Other network providers believe that, other than for some highspeed (4800 and above bits/sec), the echo canceler should be enabled for a call between modems. Examples have been given using specific modems to support both points of view. The correct answer is probably somewhere in between and, regardless of the philosophy of disabling, there will be modems that "fail."

One method, which has considerable merit, is based on the fact that an echo canceler has two parts, the filter for "canceling" the echo and and a nonlinear processor for suppressing the residual echo. Consequently we could define the operation of an echo canceler tone disabler as follows:

if (2100-Hz tone is detected) then disable **NLP**;
if (2100-Hz tone has phase reversals) then disable filter (also);
if (re-enable conditions are met) re-enable **NLP** and adaptation.

The notion of disabling the filter is to clear the coefficients to zero and inhibit adaptation.

6.4.2.6 Other Requirements

There are a few, slightly more esoteric, requirements imposed on cancelers.

For example, it is common to provide a "leak" in the adaptation process whereby each coefficient is adapted according to the **LMS** algorithm and in addition to the normal coefficient update so generated, a small "correction" is added that reduces the magnitude of the coefficient. In the absence of any reference signal, i.e., if P_x is extremely small, these corrections will eventually drive the coefficient to zero. The intention is to keep the canceler robust. When P_x is small, the returned echo is small as well so the actual values of the coefficients are not critical. When the signal power increases at a later time, having a zero coefficient means we are not canceling any echo, but we are surely not adding echo.

Another requirement has to do with stability. When the reference signal $\{x(n)\}$ is narrowband, akin to a sinusoid, we know that the adaptive filter response cannot be controlled at frequencies other than that of the sinusoid. If the reference signal later changes to a broadband signal then the canceler may actually become an echo generator. The intent of the stability requirement is that this does not happen.

6.4.3 Performance Limitations of the Adaptive Filter

The behavior of the adaptive filter can be explained using Fig. 6.25 on the next page. The echo path is modeled as a linear time-invariant filter with impulse response $\{g(k)\}$ or transfer function $G(\mathbf{z})$ where

$$\text{Echo Path:}\quad G(\mathbf{z}) = \sum_{k=0}^{\infty} g(k)\,\mathbf{z}^{-1} \tag{6.4.2}$$

The signal $\{y(n)\}$ received by the canceler is a combination of echo, near-end speech $\{s(n)\}$, and noise $\{\eta(n)\}$, where the principal contribution to the noise is the A/D converter in the line circuit. In practice the echo path is only approximately linear and the expected echo return loss enhancement achieved should be derated to account for the inability of a linear time-invariant system, $H(\mathbf{z})$, to account for such behavior. The base **ERL**, that provided by the hybrid, can be expressed as

$$(\text{Baseline})\ \mathbf{ERL} = -\log_{10}\left\{\sum_{k=0}^{\infty} |g(k)|^2\right\}\ \text{dB} \tag{6.4.3}$$

We will assume that this **ERL** is greater than 6 dB, which means that

$$\text{ERL} \;\geq 6\,\text{dB} \;\Rightarrow\; \{ \sum_{k=0}^{\infty} |g(k)|^2 \} \;\leq\; \frac{1}{4} \;\Rightarrow\; |g(k)| \;\leq\; \frac{1}{2}\; \forall k \qquad (6.4.4)$$

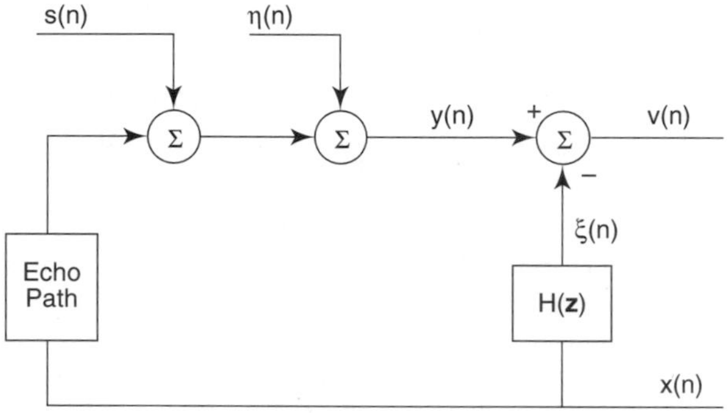

Fig. 6.25
Presence of signals other than echo can limit the canceling capacity of the adaptive filter

A plot of {g(k)} takes the form shown in Fig. 6.26.

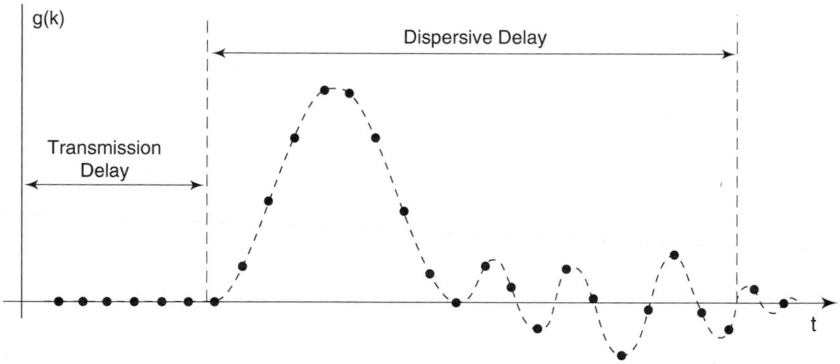

Fig. 6.26
Notion of tail delay as a combination of transmission, or flat, delay and dispersive delay caused by hybrid imbalance

The initial part of {g(k)} is all zeros, representing a flat transmission delay between the canceler and the hybrid. Following that is the nonzero portion that represents the effective impedance mismatch in the hybrid. This portion is called the *dispersive* delay since it has frequency response implications. The total of flat and dispersive delays is called the ***tail delay***. The flat delay is a variable, since we do not know the location of canceler vis-à-vis the hybrid. The dispersive delay is usually less than 8 msec for each 2-to-4-wire conversion in the tail circuit (there are some cases that involve multiple conversions). cancelers are usually specified in terms of tail delay. A 32-msec canceler, for example, is able to accommodate up to 32 msec of tail delay. The common choices are 16, 32, 48, 64, and 96 msec. Since the underlying sampling rate is 8 kHz, these tail delays can be expressed in samples as 128, 256, 384, 512, and 768 samples, respectively. Thus a 32-msec tail delay would have an **FIR** filter, H(**z**), of length 256.

The overall **ERL**, including H(**z**), is

$$\mathbf{ERL} = -\log_{10}\left\{ \sum_{k=N}^{\infty} |g(k)|^2 + \sum_{k=0}^{N-1} |g(k) - h(k)|^2 \right\} \text{dB} \qquad (6.4.5)$$

and we recognize the **ERLE** as

$$\mathbf{ERLE} = -\log_{10}\left\{ \frac{\sum_{k=N}^{\infty} |g(k)|^2 + \sum_{k=0}^{N-1} |g(k) - h(k)|^2}{\sum_{k=0}^{\infty} |g(k)|^2} \right\} \text{dB} \qquad (6.4.6)$$

By choosing the cancelable tail delay of the canceler to be large enough, we can ignore the impact of the echo path impulse response beyond N samples. The limitation on the **ERLE** achievable by the filter is thus governed solely by how closely the H(**z**) can match G(**z**).

One principal contributor to this limitation is the finite-wordlength constraint on the coefficients. Assume we use (B+1) bits to represent the coefficients. Then since we know that |g(k)| 1/2, we can bound the quantization error involved with quantizing any coefficient by

$$|\Delta h| \le \frac{1}{2} 2^{-B} \qquad (6.4.7)$$

If we assume that this quantization is uniformly distributed, then

$$\mathbf{E}\{ |\Delta h|^2 \} \le \frac{1}{12} 2^{-2B} \qquad (6.4.8)$$

Recognizing that this quantization is responsible for degrading the total **ERL** we could achieve, even if the {g(k)} were known, we can write

$$\mathbf{E}\{ \text{best case } \mathbf{ERL} \} \le N\,\mathbf{E}\{ |\Delta h|^2 \} \le \frac{N}{12} 2^{-2B} \quad (\text{linear scale}) \qquad (6.4.9)$$

which, in dB, yields

$$\mathbf{ERL}_{TOT} \approx -10\log_{10}(N) + 6B + 10.79 \text{ (dB)} \qquad (6.4.10)$$

In deriving this result we have assumed that the coefficient errors are independent and identically distributed.

This bound should be interpreted carefully. Eq. (6.4.9) provides an expression for the expected value for the "best case **ERL**" assuming, quite logically, that the best case corresponds to *a priori* knowledge of the echo path impulse response. If the coefficients {g(k)} just happened to have values that could be represented exactly by the available wordlength then the (total) **ERL** would be infinite. The expression in Eq. (6.4.10), being the expected value, should be considered "our best guess of the best case **ERL**." Any conclusions made from this result would be guidelines rather than absolutes.

For the case of a 32 msec echo canceler, this implies that the wordlength used to represent the coefficients must be greater than 9 bits for the overall **ERL**, including the filter, to be in excess of 30 dB. That is

$$(B+1) \geq 1 + \frac{1}{6}[30 + 10\log_{10}(256) - 10.79] \text{ bits} \qquad (6.4.11)$$

On the one hand this number is conservative since several coefficients, corresponding to the flat transmission delay, are zero and thus have no associated quantizing error, provided our number representation can express zero as a value. On the other hand, this is not that conservative since there are several other factors that prevent us from achieving the ideal h(k) = g(k). The expected error in coefficient estimation may be large enough such that the adaptation algorithm may assign nonzero values for these coefficients (that are supposed to be zero) and the notion of having no quantization error because the value is zero is not relevant.

Note that the length of the filter appears directly. Consequently, an echo canceler of minimum required tail delay should be used. If we know that the overall tail delay is less than 16 msec, then using a 16-msec canceler would provide 3 dB of improvement over a 32-msec canceler. If the tail delay were greater than 16 msec then a 16-msec canceler would not be correcting for the echo path impulse response beyond 128 samples and the deterioration could be large. The rule of thumb is to use a 16-msec version if the canceler is "co-located" with the originating switch (where the 2-to-4-wire conversion takes place). This accounts for all cases of dispersive delay and intraoffice delay caused by intervening equipment between the canceler and the switch. In the long-distance network, where there may be an intervening Switch between the canceler and the originating switch, a 32-msec canceler is advised. In extreme cases where there may be much transmission delay between the canceler and the hybrid, a 48-msec canceler could be used. If the flat transmission delay is so great that 48-msec tail delay is not sufficient, it is recommended that the network design be re-examined.

Another reason the **ERLE** would be limited is the inability of the **LMS** (or any adaptive algorithm) to recognize convergence and stop adaptation. In the case of the **LMS** algorithm, we derived the following expression in Section 6.3:

$$E\{\text{converged } \mathbf{ERL}\} = -10\log_{10}\{N\mu^2\sigma_s^2\sigma_x^2\} \quad (6.4.12)$$

which provides the expected value of the final combined **ERL** in dB. A small value for the adaptation gain, μ, a short filter (small N), and a good signal-to-noise ratio are required for the converged filter to provide good echo cancelation. The signal-to-noise ratio implicit in Eq. (6.4.12) is the ratio of reference signal power to the total additive signal power comprising the near-end signal. Even if the near-end talker is silent, the A/D quantization noise will be present. Assuming that we choose μ to be

$$\mu \approx \frac{1}{N\sigma_x^2} \quad (6.4.13)$$

to guarantee a stable adaptation, the expected **ERLE** is limited only by the signal-to-noise ratio. If the near-end talker is silent, then the "noise" power, σ_s^2, is principally that of the A/D converter, assuming that the transmission path between the canceler and the hybrid is very good and does not add any appreciable noise component. Since the A/D converter is of the companded variety, there is an upper limit to the signal-to-noise ratio of 37 dB, as shown in Chapter 4, and thus

$$E\{\text{converged } \mathbf{ERL}\} = 10\log_{10}\{\frac{\sigma_x^2}{\sigma_s^2}\} \le 37 \text{ dB} \quad (6.4.14)$$

The expected final **ERL** is limited by the quantization noise of the converter in the line circuit. Again this "bound" should be interpreted carefully since it is an expected value rather than an absolute limit.

It turns out that the limit of 37 dB in Eq. (6.4.14), even though it is derived statistically, is an extremely good bound. Achieving this level of final **ERL** is possible **only** for short filters and when there are **no significant nonlinearities** in the tail circuit, and in **very controlled** environments where the spectrum of the reference signal is indeed **flat**. Laboratory experiments that have indicated higher values of final **ERL** have created an **ideal** echo path in which there are only a **few nonzero coefficients** in {g(k)}. While in operation, the reference signal will be, most likely, speech and it is well known that the spectrum of speech is not flat. The number of nonzero echo path impulse response coefficients is about 60 to 100 corresponding to a dispersive delay of about 8 msec. In such a situation, achieving 37 dB of final **ERL** with the filter alone is possible, but not likely.

A more realistic expectation on the (maximum) **ERL** afforded by the filter alone is about 30 dB, implying that for most long-distance calls requiring echo treatment, the nonlinear processor is required to suppress the last vestiges of the echo.

6.4.4 Algorithms for Controlling an Echo Canceler

The operation of the echo canceler involves an adaptive filter and a nonlinear processor. Controlling the operation of these entities to achieve a level of performance that can be considered "optimal" is more an art than a science and most echo canceler manufacturers guard the algorithms that manipulate the filter and **NLP** most carefully. The **CCITT**, in Recommendation **G.165** provides some guidelines as to the proper operation but in the final analysis what matters is how the speech traversing the canceler sounds. Standards bodies may decide in their infinite wisdom to provide specifications, limits, and even implementation details; all for naught if the subjective quality is perceived to be low. A canceler that is rated as high quality, on a subjective basis, may not meet these standards; a canceler that meets these standards may not sound very good. This is a very contentious area regarding echo cancelers and it is doubtful that any sensible resolution will ever be reached.

In this section we will not describe any particular algorithm. Rather, we will discuss some of the heuristics that go into the design of algorithms for controlling the adaptation and the nonlinear processor.

The adaptation of the filter follows the following equations:

$$\mathbf{w}(n) = \mathbf{y}(n) - \sum_{k=0}^{N-1} h_n(k)\,\mathbf{x}(n-k) \qquad (6.4.15a)$$

$$h_{n+1}(k) = h_n(k) + \mu\,\mathbf{w}(n)\,\mathbf{x}(n-k) \qquad (6.4.15b)$$

which describe the echo cancelation and the update of the coefficients. The **NLP** implements

$$\mathbf{r}(n) = \begin{cases} \mathbf{w}(n) & \text{if } |\mathbf{w}(n)| > T \\ \mathbf{e}(n) & \text{if } |\mathbf{w}(n)| \le T \end{cases} \qquad (6.4.16)$$

where $\{\mathbf{e}(n)\}$ is a locally generated noise.

Control of the canceler includes the determination of the following important variables:

a) adaptation gain, μ

b) **NLP** threshold, T

c) Comfort noise level, σ_e^2

Recall that the adaptation gain controls the rate of adaptation, in that a large value of μ implies rapid convergence; too large a value of adaptation gain gives rise to instability. The adaptation gain also determines the "final" **ERL** in that the larger the value of μ, the worse is the expected deviation from the "norm"; this deviation is directly related to both the adaptation gain and the power of the near-end speech.

The nonlinear threshold parameter, T, determines how "choppy" the speech will sound with respect to the echo level. A large value of T suppresses all the echo but also deteriorates the quality of the near-end speech. In principle, during hard double-talk, the threshold must be zero; during single-talk the threshold must be at its "maximum value," during soft double-talk the threshold must be "in between."

The comfort noise level is not quite as important as the other two factors. Nonetheless if the estimate of the comfort noise is quite wrong, it will be registered by the user as a "low quality" connection. The level of sophistication in echo canceler control algorithms has reached the point that differentiating between vendors of equipment depends on how well this comfort noise level is controlled. Separating the wheat from the chaff based on published specifications is not easy; distinguishing good from very good can only be accomplished by actual demonstration of equipment carrying **live traffic**.

Proper control requires that we be able to extract a good "picture" of the situation based on the four signals that can be observed. These signals are the reference, $\{x(n)\}$, the composite signal $\{y(n)\}$, the echo replica $\{\xi(n)\}$, and the "echo canceled" signal $\{w(n)\}$. The power, or some appropriate measure of strength, of these signals is computed on an ongoing basis. Let us denote these strengths by P_x, P_y, P_ξ, and P_w, respectively. Based on the relative strengths of these signals we determine the adaptation gain, the **NLP** threshold, the level of comfort noise, and an estimate of the echo cancelation achieved. The adaptation gain is also influenced by the type of reference signal in the sense of spectral composition. To determine whether the reference signal is principally noise or is essentially a sinusoid requires additional information; we have seen (in Chapter 2 and Chapter 5) that the peak value, in conjunction with the power measurement, can be used to classify the signal appropriately.

Now the near-end signal, $\{s(n)\}$, is not available except indirectly via $\{y(n)\}$, which includes the echo, so P_s has to be estimated in some way. To a first approximation $P_s = P_w$, the power of the "canceled signal," which contains $\{s(n)\}$ along with residual echo. In the same vein, P_ξ is an estimate of the echo power and $(P_y - P_w)$ is an estimate of the reduction of the echo power. In the idle state we would expect P_y to be a measure of the background power, σ_η^2, which we need to establish in order to generate a suitable comfort noise.

6.4.4.1 Controlling the Adaptation Gain, μ

Controlling the adaptation gain is a compromise between rapid convergence, which is achieved using a high gain, and "converged" **ERL,** which is better with smaller values of μ. In choosing the adaptation gain we must also keep in mind that too large a value causes instability and that the stability is also influenced by the spectral content of the reference signal, $\{x(n)\}$.

From the viewpoint of stability, we must ensure that the adaptation gain satisfies

$$\mu < \frac{2}{\lambda_{max}} \tag{6.4.17}$$

where λ_{max} is the largest eigenvalue of the autocorrelation matrix of the reference signal. However, in practice, the individual eigenvalues are rarely known and thus Eq. (6.4.17) is not easy to apply. It can be shown that

$$\frac{1}{N\sigma_x^2} \le \frac{1}{\lambda_{max}} \tag{6.4.18}$$

from which we can derive a (weak) bound for μ. Including a safety factor of 2, a "typical" maximum value for the adaptation gain is

$$\mu \le \frac{2}{N\sigma_x^2} \;;\; \mu \approx \frac{1}{N\sigma_x^2} \tag{6.4.19}$$

This is a proper setting for the adaptation gain provided that the reference signal has a "flat" spectrum and provided that there is no near-end speech.

We can make a guess as to the flatness of the reference signal spectrum from the peak-to-rms ratio. Denoting by M_x the peak value of $\{x(n)\}$,

$$\Gamma_x = \text{peak-to-rms ratio} = 20 \log_{10}(\frac{M_x}{\sigma_x}) \text{ dB} \tag{6.4.20}$$

If the signal is very narrowband, such as a sinusoid, the peak-to-rms ratio is "small." For a sinusoid, for example, the peak-to-rms value is 1.41... or 3 dB. For a random signal, say with a Gaussian **pdf**, defining the peak value is done in terms of a probability, but with any "reasonable" definition of peak value, the peak-to-rms ratio is about 10 dB and for the exponential **pdf** about 16 dB. The peak-to-rms ratio by itself does not tell us anything about the flatness of the spectrum. However, we "know" that the reference signal will be either sinusoidal-like, noise-like, or speech-like and the peak-to-rms ratio can differentiate between these three cases. Of these the noise case has the flattest spectrum and the sinusoidal case the least flat with the speech case somewhere in between. This ratio allows us to modify our choice of μ. Having computed an initial value of μ as in Eq. (6.4.19) we reduce the magnitude of μ if we believe that the reference signal is narrowband. For values of Γ_x close to 3 dB we make the adaptation gain zero. Thus, based on the signal power and peak-to-rms ratio we can arrive at a suitable value for adaptation gain that is suitable for adaptation in the case of single-talk.

The converged **ERL** is dependent on the total power of the near-end signal. Thus if the near-end signal strength is great then we should not really be adapting at all. This is the notion of "freezing" the adaptation upon detection of hard double-talk. For hard double-talk and double-talk we reduce the adaptation gain to zero. For soft double-talk the adaptation gain is chosen to somewhere between the optimal value for single-talk and zero.

The converged **ERL** is also dependent on the adaptation gain. A smaller value of gain provides better final **ERL**. The value of μ for the single-talk case is normally large to allow for rapid convergence. If we can detect the "converged" state, with the coefficients fluctuating about their optimal values, we reduce the adaptation gain.

Thus if

$$P_y \approx P_\xi \text{ and } P_w \text{ is small then converged} \tag{6.4.21}$$

so then we reduce the value of μ since Eq. (6.4.21) indicates that we are in a single-talk situation and should be reducing the "error" of the **LMS** algorithm itself.

In the idle state the value of adaptation gain is moot since all the signal levels are small and thus coefficient corrections, if any, are also small. From the viewpoint of "safety," reflecting our inability to detect rapid changes of state from an "averaging" measurement such as power, it is advisable to keep the adaptation gain small (approximately zero).

6.4.4.2 Controlling the Nonlinear Processor Threshold, T

Based on the various power levels mentioned, and the state in which we believe the canceler to be, we estimate the echo return loss and the echo return loss enhancement. In particular, if we are in a *single-talk* state then

$$\textbf{ERL} \text{ (echo path)} \approx -10 \log_{10}\left(\frac{P_\xi}{P_x}\right) \text{ dB} \tag{6.4.22a}$$

$$\textbf{ERLE} \approx -10 \log_{10}\left(\frac{P_w}{P_\xi}\right) \text{ dB} \tag{6.4.22b}$$

$$\textbf{ERL}_{\text{TOT}} = \textbf{ERL} + \textbf{ERLE} \text{ (dB)} \tag{6.4.22c}$$

In practice we average these values over several readings to reduce the effect of measurement noise. The estimate of **ERLE**, in particular, is affected greatly by the presence of near-end speech; the estimate of **ERL** is incorrect if we have not converged. Consequently, we do the averaging over those readings that satisfy Eq. (6.4.21) and P_x is sufficiently high (reference signal is not silent).

Based on our estimate of **ERL** and **ERLE** we estimate the total **ERL** as shown. The power of the echo residual is then be given by

$$\sigma_{er} = K_1 \sigma_x \text{ where } 20 \log_{10}(K_1) \approx \textbf{ERL}_{\text{TOT}} \tag{6.4.23}$$

Since the intent of the residual suppressor is to eliminate this residual echo, we want to keep the "cut-in" threshold large enough to suppress all samples that are potentially echo. The threshold is chosen as

$$T = K_2 \sigma_{er} \text{ where } 20 \log_{10}(K_2) \approx 16 \tag{6.4.24}$$

which reflects our estimate of the peak-to-rms ratio of speech as about 16 dB. This is the "optimal" value for T in the single-talk case.

Having decided on the value of the threshold for the single-talk case we can modify this value according to the actual state of the canceler. In particular, for hard double-talk and double-talk we reduce the threshold to zero. For soft double-talk we choose a value somewhere between the optimal value for single-talk and zero.

In the idle state the value of the threshold is moot since all the signal levels are small and thus echo, if any, is also small. From the viewpoint of "safety," reflecting our inability to detect rapid changes of state from an "averaging" measurement such as power, we are advised to keep the threshold small (approximately zero) for the idle state.

6.4.5 Concluding Remarks on Echo Cancelers

As is clear from the prior discussion, designing an echo canceler for an environment as "hostile" as the public switched telephone network is as much of an art as a science. It has been the author's experience that mathematical analyzes and computer simulations based, as such studies must be, on any reasonable set of assumptions, do not match the "real world." Such analyses and simulations are useful in that they build intuition, the ability to translate between an observed phenomenon and a possible cause. There is no substitute for actually designing, building, and deploying equipment, along with the trials and tribulations of laboratory testing and field trials with actual live traffic. In describing the control algorithms, for example, we have shied away from hard numbers and the myriad special cases that need consideration, the principal reason being that such details may be relevant in a discussion of echo cancelers but shed little extra light on the applicability of digital signal processing.

There are presently five major suppliers of echo cancelers in North America. Alphabetically, these are AT&T Network Systems, Coherent Communications, Ditech Inc., DSC Communications Corp., and Tellabs Inc. Each has its own specific strengths. The unit from Ditech, developed by Dr. Charles Davis, is the only one of the five that is based solely on **DSP** processors and implemented for the most part in firmware. The other four manufacturers have designed ASICs to perform at least the adaptive filter part of the canceler, with the control algorithms implemented in firmware. It is not uncommon to have a combination of ASIC, **DSP** processor, and a general-purpose microprocessor operating in concert. The units from DSC Communications and Tellabs, for example, use ASICs for the adaptive filter, Motorola 56000 **DSP** processors to implement the tone-disable and similar measurements, and a general-purpose processor such as the Motorola 68000 for control, operator interface, timing, and so on.

6.5 EXERCISES

6.1. Consider the following "derivation" of an adaptive algorithm for an **FIR** filter. The reference signal is $\{x(n)\}$, the returned signal $\{y(n)\}$ is comprised of the echo and, possibly, near-end signal plus noise. The **FIR** filter is an N-tap filter $H(\mathbf{z}) = A(\mathbf{z})$ whose coefficients are $\{a_k, k = 0, 1, 2, \dots, (N-1)\}$. We can formally write the error signal $\{e(n)\}$ as

$$E(\mathbf{z}) = Y(\mathbf{z}) - A(\mathbf{z})\,X(\mathbf{z}) \tag{6.P.1}$$

The intent is to find coefficients to minimize the strength of $\{e(n)\}$. Since

$$e(n) = y(n) - \sum_{k=0}^{N-1} a_k x(n-k) \tag{6.P.2}$$

we can write the partial derivative of the squared error signal with respect to the coefficient a_k as

$$\frac{\partial e(n)^2}{\partial a_k} = -2e(n)x(n-k) \tag{6.P.3}$$

This is the "slope" (or "gradient") of the instantaneous error power $e(n)^2$ relative to a_k. Minimization is achieved when this slope is zero. When this slope is not zero, it tells us how much the a_k should have differed in order to minimize the error power, all other signals and coefficients being constant. Since the coefficients are not constant and the signals are not deterministic, this slope is not deterministic either. If nothing else, it at least tells us the direction in which to modify the coefficient. Hence the algorithm

$$a_k^{(n+1)} = a_k^{(n)} + \mu \operatorname{sgn}(e(n)) \operatorname{sgn}(x(n-k)) \tag{6.P.4}$$

Now suppose that the filter was **IIR**, $H(\mathbf{z}) = A(\mathbf{z})/B(\mathbf{z})$, with coefficients $\{a_k\}$ and $\{b_k\}$ for the numerator and denominator, respectively. Then we can "derive" an adaptive algorithm in the following manner. Note that

$$E(\mathbf{z}) = Y(\mathbf{z}) - \frac{A(\mathbf{z})}{B(\mathbf{z})} X(\mathbf{z}) \tag{6.P.5}$$

a) Write an expression for the error, $\{e(n)\}$, in the time domain similar to Eq. (6.P.2).

b) Show that the slope relative to a_k has the same form as Eq. (6.P.3). Convince yourself that the slope with respect to the b_k cannot be expressed in the same fashion as Eq. (6.P.3). Hence the simple algorithm of Eq. (6.P.4) is not applicable for adapting the b_k.

In order to derive a simple algorithm similar to Eq. (6.P.4) for an **IIR** filter, (the $\{b_k\}$), consider the following modified error criterion. Suppose we define the error signal as $\{e_1(n)\}$, where

$$E_1(\mathbf{z}) = B(\mathbf{z})Y(\mathbf{z}) - A(\mathbf{z})X(\mathbf{z}) \tag{6.P.6}$$

c) Write a time domain expression for $\{e_1(n)\}$ similar to Eq. (6.P.2).

d) Show that the gradient of $e_1(n)^2$ with respect to the b_k is similar in form to that shown in Eq. (6.P.3). Comment on the differences, if any.

e) Devise an adaptation scheme suitable for the denominator coefficients, $\{b_k\}$.

f) Draw a block diagram showing the various signals involved and the associated modifications by A(**z**) and B(**z**).

g) For further study on adaptive **IIR** filters, the following are helpful (in addition to the references in Section 6.6):

Crespo, P. M., and Honig, M. L., "Pole-zero decision feedback equalization with a rapidly converging adaptive **IIR** algorithm," *IEEE Journal on Selected Areas in Communications*, Vol. 9, No. 6, Aug. 1991.

Johnson, C. R., "Adaptive IIR filtering: current results and open issues," *IEEE Trans. Information Theory*, Vol. IT-30, No. 2, Mar. 1984.

Lynch-Aird, N. J., "Review and analytical comparison of recursive and nonrecursive equalization techniques for **PAM** transmission systems," *IEEE Journal on Selected Areas in Communications*, Vol. 9, No. 6, Aug. 1991.

Honig, M. L., "Convergence models for adaptive gradient and least squares algorithms," *IEEE Trans. Signal Processing*, Vol. 31, No. 2, Apr. 1983.

Shynk, J. J., "Adaptive IIR filtering," *IEEE ASSP Magazine*, Vol. 6, No. 2, Apr. 1989.

6.2. Write a program to simulate the action of an adaptive **FIR** filter. The principal routines required are:

a) Signal generation: A suitable signal is the combination of noise and a tone. The power of the signal, the frequency of the tone, and the ratio of noise power to tone power must be parameters that can be specified. The spectrum of the signal can be shaped by applying a simple **IIR** first-order **IIR** filter defined by the coefficient a (which is less than 1); the sign of a determines whether the spectrum of the signal is "lowpass" or "highpass."

b) Echo path: A suitable model is an **FIR** filter of M taps, $\{g(n)\}$, with a simple choice being $g(n) = \mathbf{k}\,\alpha^n$, where **k** is chosen such that $\Sigma g(n)^2 = 0.5$ (6 dB **ERL**) (or any other preferred choice of **ERL**).

c) Adaptive filter: Is an N-point **FIR** filter $\{a(n)\}$. The adaptation follows Eq. (6.P.4). The length of the adaptive filter is not necessarily equal to the length of the **FIR** filter that simulates the echo path.

d) Power Measurement: Especially of the error signal, $\{e(n)\}$. Achieved using a "running average" by applying a first-order **IIR** filter to the square of the error signal. The signal $\{v(n)\}$ represents the "short-term error power," an indication of the power of $\{e(n)\}$ as computed over a finite window up to and including sample number "n." The coefficient a of the first order filter needs to be of the order of 0.99.

e) **Error Measurement**: The computation of the "ideal filter error" which is $\Sigma(g(k) - h(k))^2$ = **ERL**(n) at each iteration (sample).

f) **Output Format**: A plot, or table, or equivalent, which shows the variation of the filter error and the (short-term) power of the error signal as time progresses is especially useful.

The intent of the simulation program is to obtain a good understanding of the impact of the following:

i) signal power;

ii) ratio of noise to tone in the reference signal;

iii) ratio of reference signal power to additive noise power in the echo path;

iv) impact of the magnitude of the adaptation gain on the rapidity of convergence and the minimum filter error;

v) impact of using a length N that is too small, too large, and just right;

6.6 REFERENCES

[1] Messerschmitt, D. G., "Echo cancelation in speech and data transmission," *IEEE Journal on Selected Areas in Communication*, Mar. 1984.

[2] Sibul, L. H., (Ed), *Adaptive Signal Processing*, IEEE Press, New York, 1987.

[3] Widrow, B., McCool, J. M., Larimore, M. G., and Johnson, C. R., "Stationary and nonstationary learning characteristics of the LMS adaptive filter," *Proc. IEEE*, August, 1976.

[4] Staff of AT&T Bell Laboratories, *Transmission Systems for Communications*, Fifth Edition, Bell Telephone Laboratories, Inc., 1982.

[5] Treichler, J. R., Johnson, C. R., and Larimore, M. G., *Theory and Design of Adaptive Filters*, John Wiley and Sons, New York, 1987.

[6] Widrow, B., and Stearns, S.D., *Adaptive Signal Processing*, Prentice Hall, Inc., Englewood Cliffs, NJ, 1985.

[7] T1A1.6 Working Group on Specialized Signal Processing, *A Technical Report on Echo canceling*, Report No. 27, Alliance for Telecommunications Industry Solutions (formerly Exchange Carriers Standards Association), 1993.

[8] CCITT Recommendation **G.164**: Echo Suppressors.

[9] CCITT Recommendation **G.165**: Echo cancelers.

[10] Region Digital Switched Network Transmission Plan. Science and Technology Series **ST-NPL-000060, BELLCORE**, 1988.

BANDPASS FILTERS, TRANSMULTIPLEXERS, AND THE DISCRETE FOURIER TRANSFORM (DFT)

7

7.1 INTRODUCTION

Bandpass filter banks have several applications. The most recognizable application pertains to the spectral analysis of signals. In this context, the signal component centered at a prescribed frequency is extracted using a (narrow) bandpass filter. The strength of the filter output is representative of the signal content around the center frequency of the bandpass filter. In traditional spectrum analyzers a single filter is employed and frequency scanning, or sweeping, is achieved using amplitude modulation or heterodyning. The local oscillator frequency is chosen so that the frequency band of interest is shifted and positioned within the passband of the fixed filter. Frequency sweeping is thus a sequential procedure. In modern spectrum analyzers a bank of filters is implemented and, provided the passbands of the bank of filters span the frequency range of interest, the frequency "sweep" is achieved in a parallel fashion. Traditional spectrum analyzers use analog filters and, while implementing a bank of filters is possible in principle, achieving a tight "match" between separate filters is difficult. Using digital signal processing it is possible to have a perfect match between the various filters of the bank and hence the "parallel" method is quite feasible. The **DSP** implementation of a bank of filters is based on the principle of heterodyning and uses the Discrete Fourier Transform (**DFT**) as a basic building block to provide the several filters simultaneously. In fact, the **DFT** can be viewed as a bank of bandpass filters, the viewpoint expounded in Section 7.3. Using Fast Fourier Transform (**FFT**) techniques, the computational burden is reduced significantly when filter banks are built around the **DFT**.

Banks of bandpass filters apply quite naturally in the multiplexing and demultiplexing of Frequency Division Multiplexed (**FDM**) signals. In **FDM**, several narrowband channels are stacked in frequency and separation of the composite signal into its separate constituent channels requires a bank of filters, appropriately centered, to pick off the desired channel(s). In telephony, digital signal processing techniques are used to convert between a 12-channel group, which is an analog signal comprised of 12 separate 4-kHz-wide voice channels, and a digital transmission scheme, namely **DS1**, which is a digital bit-stream comprised of 24 separate **PCM** channels. In Chapter 1 we provide a description of a **DS1** signal. The conversion is embodied in the **Transmultiplexer**, which provides a bidirectional conversion between a 24-channel **DS1** and two 12-channel Groups. The transmultiplexer is the subject of Section 7.4, where the treatment follows Narasimha and Peterson [1.3].

In the transmultiplexer, and in most implementations of filter banks, we are faced with the situation of having multiple sampling rates. On the **TDM**, or **DS1** side, each voice channel is comprised of a stream of samples at 8 kHz. On the **FDM** side, the combination of several channels in the frequency domain implies that the composite signal has a wider bandwidth and can be supported only with a sampling rate much higher than 8 kHz. Thus in the **TDM**-to-**FDM** direction there is an increase in sampling rate. Conversely, in the **FDM**-to-**TDM** direction there is a reduction in sampling rate. The filter banks in the transmultiplexer provide the frequency selectivity and bandlimiting required to combat aliasing effects (**FDM-TDM**) and replicate-rejection (**TDM-FDM**). In order to better understand the impact of sampling rate changes, an introduction to interpolation (sampling rate increase) and decimation (sampling rate decrease) is provided in Section 7.2. More details on multirate filters can be found in Crochiere and Rabiner [1.2] and Vaidyanathan [1.5].

DSP techniques based on the Discrete Fourier Transform (**DFT**) or its cousin the Discrete Cosine Transform (**DCT**) provide filter banks that have a very unique characteristic. Each filter is "identical" to the other filters of the bank, other than, of course, the center frequency. Each filter is a shifted version of a "prototype" filter. What this means is that the relative magnitude and phase characteristics of the filters comprising the bank can be very tightly controlled. Use of **FIR** filters with linear phase characteristics renders the group-delay distortion moot. The amplitude characteristics can be chosen such that, even if the adjacent filters "overlap," implying that the bandlimiting is not perfect, the impact of aliasing and/or incomplete replicate rejection can be combated. Such filter banks have received much attention and an excellent tutorial article is provided by Vaidyanathan [1.4] and a detailed treatise on the subject available in Vaidyanathan [1.5].

Somewhat similar to the notion of the transmultiplexer is a scheme for data transmission that is called, variously, Discrete Multi-Tone (**DMT**) or Multi-Carrier Modulation (**MCM**). Such modulation schemes are eminently suitable for transporting data over channels that require a lot of equalization. That is, the channel is far from an ideal bandpass filter. The **DMT/MCM** schemes transport by breaking up the channel into several narrower bandwidth subchannels, the premise being that these subchannels have a more ideal frequency response (over the narrow bandwidth) and thus equalization can be dispensed with. Again there is the notion of a bank of bandpass filters. In Section 7.5 we provide an introduction to the **DMT** technique that, it so happens, is an ingenious application of some of the properties of the Discrete Fourier Transform. The principles of the **DMT** are explained in Cioffi [1.1].

7.2 INTERPOLATION AND DECIMATION

In our discussions up to now the sampling rate was always a **fixed** quantity. In this section we consider cases where the rate is changed. When we implement banks of filters, the change in signal bandwidth permits a change in sampling rate, a situation that is quite important when the filters are **FIR** in nature.

The notion of ***interpolation*** is to increase the sampling rate, implying that we are introducing new samples ("oversampling") between sampling instants. ***Decimation*** is the converse, whereby the sampling rate is decreased implying that we are ignoring certain sampling instants ("downsampling"). The implications of such changes are related to replication and aliasing, concepts introduced in our discussions on the sampling theorem. We shall assume that all sample rate changes are done in an integer fashion, with the ratio of the two sampling rates in question an integer. Extending this to cases where the ratio is a rational number is easy; cases where the ratio is not a rational number are beyond the scope of this book.

In Sections 7.2.1 and 7.2.2 we discuss the principles of interpolation and decimation, respectively. In Section 7.2.3 we describe a technique called the "polyphase representation," which permits us to describe systems that involve interpolation and decimation using block diagrams, similar to discrete-time system block diagrams for fixed sampling rates. The polyphase represented is introduced by way of the examples of time division multiplexing and frequency division multiplexing. Simple though these examples are, they illustrate a useful property, namely that the "inverse" of a system can be generated quite easily if the system is described by its polyphase representation.

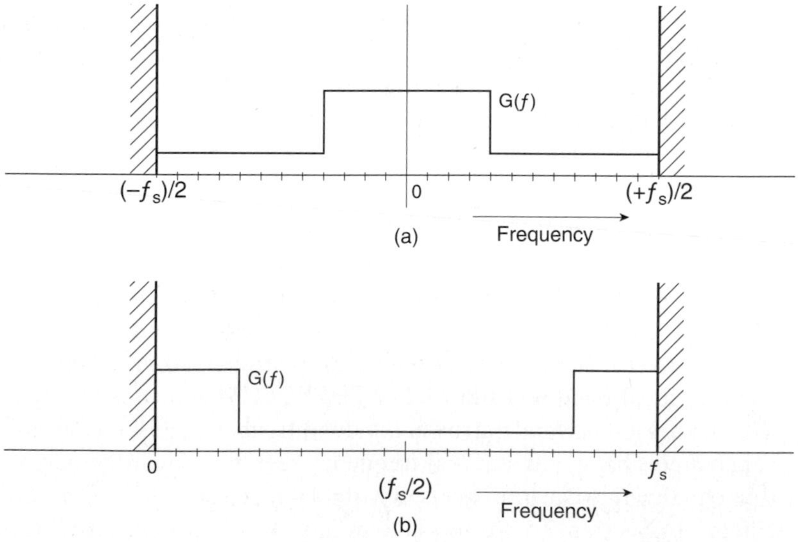

Fig. 7.1
Equivalent frequency domain ranges to describe discrete-time signals and systems

In studying sample rate changes it is convenient to introduce the notion of a frequency window. When we are using a sampling rate of f_s, the only frequencies we need to consider are those in the range $[-f_s/2, f_s/2]$ or, equivalently, $[0, f_s]$. All functions of frequency specified in this range can be extended by periodic continuation. For example, a frequency response $G(f)$ could be depicted in either of the forms shown in Fig. 7.1. Considering $G(f)$ as a lowpass filter is somewhat more palatable in Fig. 7.1(a) but nonetheless both pictures represent the same $G(f)$. The functional description within this window is the ***base*** function and by periodic continuation outside this window we obtain the replicates.

7.2.1 Interpolation

Suppose we have a discrete-time signal $\{x(nT_{s1})\}$ corresponding to the samples of some analog signal, w(t), taken at a sampling rate of $f_{s1} = 1/T_{s1}$. Now suppose we wish to generate a new signal, called $\{y(nT_{s2})\}$, corresponding to the samples of the same analog signal w(t) taken at a sampling rate of $f_{s2} = 1/T_{s2}$. The procedure followed in achieving this goal is called interpolation. We shall assume that the ratio $f_{s2}/f_{s1} = N$ is an integer. The rationale for the terminology is that we have to fill in (N-1) sample values between two successive samples $x(nT_{s1})$ and $x((n+1)T_{s1})$.

It is convenient to visualize the process as two steps as depicted in Fig. 7.2. The first step is to increase the sampling rate by a factor of N. This is achieved by inserting (N-1) zero-valued samples between every two successive samples of the input, $\{x(nT_{s1})\}$. The second step is to implement a digital filter (discrete-time filter) operating at the higher sampling rate, f_{s2}, to obtain the desired output $\{y(nT_{s2})\}$. The two steps can be visualized as interpolation of the sampling rate followed by interpolation of the sample values. In describing the digital filter by a transfer function $H(\mathbf{z})$, keep in mind that $\mathbf{z}^{-1}$ corresponds to a unit delay at the higher sampling rate.

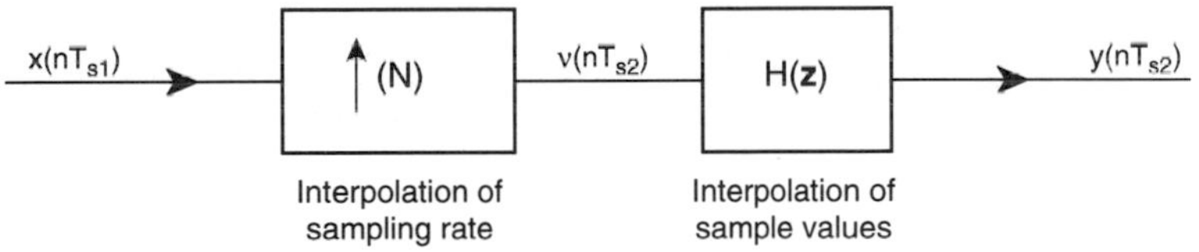

Fig. 7.2
Interpolation as the two-step process of increasing the sampling rate and filling in samples

The action of the interpolation of sampling rate is expressed mathematically as

$$v(nT_{s2}) = \begin{cases} x(mT_{s1}) & \text{for } n = mN \\ 0 & \text{for } n \neq mN \end{cases} \tag{7.2.1}$$

which is depicted in Fig. 7.3. The sample values of the signal $\{v(nT_{s2})\}$ are equal to the values of $\{x(nT_{s1})\}$ where the sampling instants coincide and are zero for the (N–1) intervening sampling instants.

It is reasonable to assume that the two signals are the "same." In fact, an "equivalence" can be established if the two signals are converted into a continuous-time (analog) signal using ideal converters. With reference to Fig. 7.4, the signals $y_1(t)$ and $y_2(t)$ are derived from $\{x(nT_{s1})\}$ and $\{v(nT_{s2})\}$, respectively. That is,

$$y_1(t) = \sum_{n} x(nT_{s1})\,\delta(t-nT_{s1}) \tag{7.2.2a}$$

$$y_2(t) = \sum_{m} v(mT_{s2})\,\delta(t-mT_{s2}) \tag{7.2.2b}$$

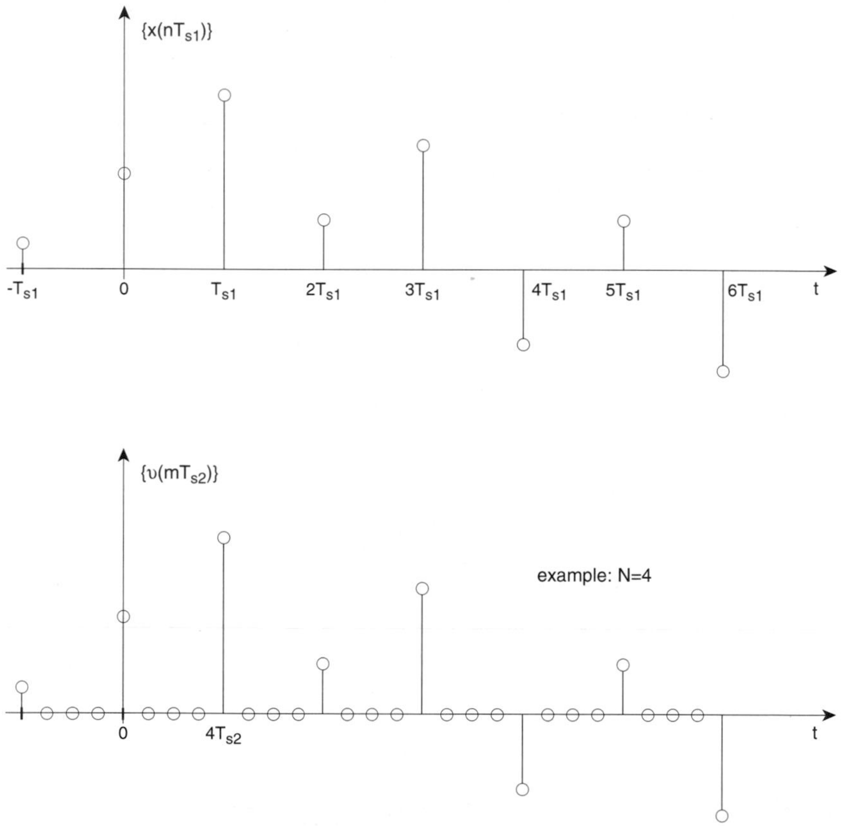

Fig. 7.3
Sampling rate increase by insertion of zero-valued samples

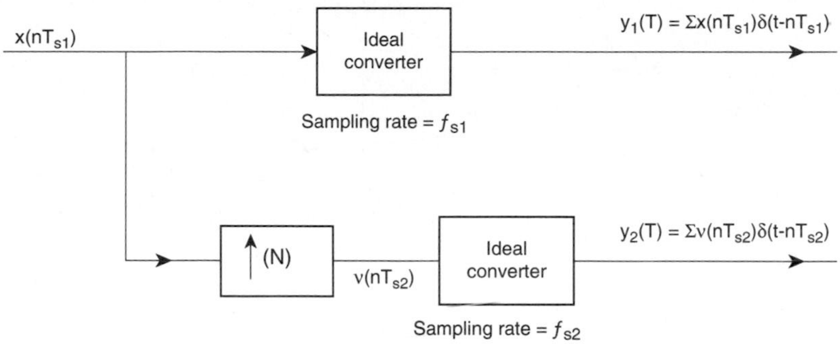

Fig. 7.4
Illustrating the equivalence of $\{x(nT_{s1})\}$ and $\{v(nT_{s2})\}$

Clearly, $y_1(t)$ and $y_2(t)$ are the same, indicating the equivalence of $\{x(nT_{s1})\}$ and $\{v(mT_{s2})\}$. When viewed in the frequency domain, a similar relationship can be derived. If $X(f)$ and $V(f)$ are the discrete-time Fourier transforms (**DTFT**s) of $\{x(nT_{s1})\}$ and $\{v(mT_{s2})\}$, then we can write

$$Y_1(f) = f_{s1}X(f) \text{ and } Y_2(f) = f_{s2}V(f) \qquad (7.2.3a)$$

$$\Rightarrow V(f) = \frac{f_{s1}}{f_{s2}}X(f) = \frac{1}{N}X(f) \qquad (7.2.3b)$$

The factor of N, the ratio of sampling rates, arises because of the sample rate change. Recall that the **DTFT** expresses the strength of the signal in terms of units that are construed as "per-sample." The inclusion of (N–1) zero-valued samples thus reduces the strength of the signal, in terms of "per-sample" units, by a factor of N. That is, since the **DTFT** is actually a function of the normalized sampling frequency, the rate change, with inclusion of zeros, reduces the frequency domain components by a factor of N. The translation from $X(f)$ to $V(f)$ is depicted in Fig. 7.5. Since the sampling rate associated with $\{x(nT_{s1})\}$ is f_{s1}, we normally consider $X(f)$ in a window of width f_{s1} and obtain the values of $X(f)$ for any frequency outside this window by periodic continuation. This is indicated in the figure by representing the replicates outside $[0,f_{s1}]$ by dashed lines. For $V(f)$, on the other hand, the window extends from $[0,f_{s2}]$ and what is considered a replicate from the viewpoint of $X(f)$ is part of the *base* function for $V(f)$. The figure uses a sampling rate increase factor of four as an example and it is clear that the *base* $V(f)$ is comprised of the *base* $X(f)$ along with three replicates of $X(f)$. In general, the base $V(f)$ is comprised of the base $X(f)$ together with (N–1) replicates.

If the spectral content of $X(f)$ and $V(f)$ is so similar then why bother to increase the sampling rate at all? The reason is that the "sphere of influence" of a digital filter is limited to a window of one period (equal to the sampling frequency) in the frequency domain. Increasing the sampling frequency thus widens the scope of applicability of the filtering process. This is illustrated by the following hypothetical example.

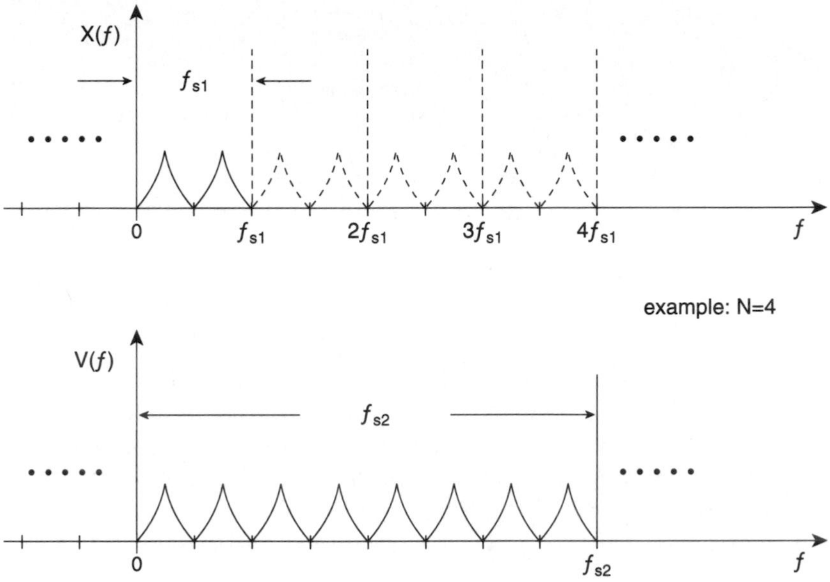

Fig. 7.5
Frequency domain equivalence of $X(f)$ and $V(f)$; i.e., $\{x(nT_{s1})\}$ and $\{v(nT_{s2})\}$

Suppose we have an analog signal with two sinusoidal components at frequencies 9 kHz and 10 kHz and we wish to remove the 10-kHz component. This can be achieved by the scheme shown in Fig. 7.6. The **ADC** block comprises the anti-aliasing (analog) filter and A/D converter to provide a digital signal $\{x(nT_{s2})\}$ with a sampling rate of f_{s2} = 32 kHz. A digital filter is used to remove the 10-kHz component and the resultant signal converted to analog by the block **DAC,** which includes the replicate-rejection (analog) filter. The frequency domain representation of the operation is shown in Fig. 7.7, which is self-explanatory.

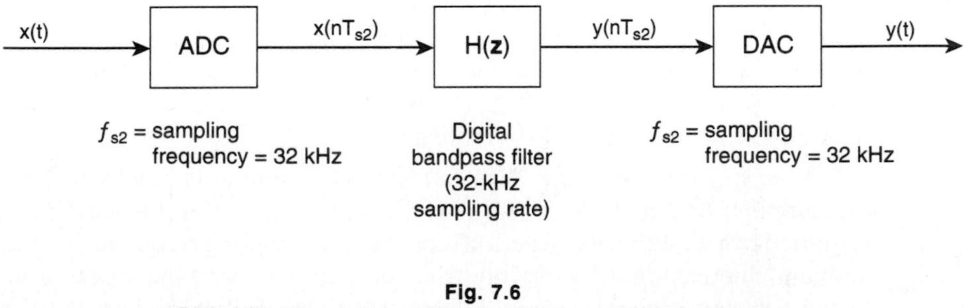

Fig. 7.6
Digital filter operating at 32 kHz with no sample rate changes

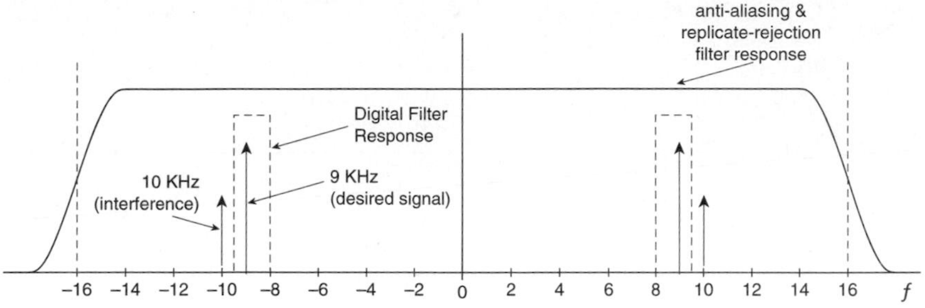

Fig. 7.7
Frequency domain view of filtering operation for removing the 10-kHz component

A second scheme that achieves the same overall effect is shown in Fig. 7.8. The **ADC** converts the analog signal to digital using a sampling rate of 8 kHz but the anti-aliasing filter is the same as that used in Fig. 7.6 (normally the filter would have a cutoff frequency of 4 kHz to accommodate the 8 kHz sampling rate) in order that the 9 kHz component not be attenuated. Because of aliasing the 9-kHz (10-kHz) components appears in the digital signal $\{x(nT_{s1})\}$ as a 1-kHz (2-kHz) component.

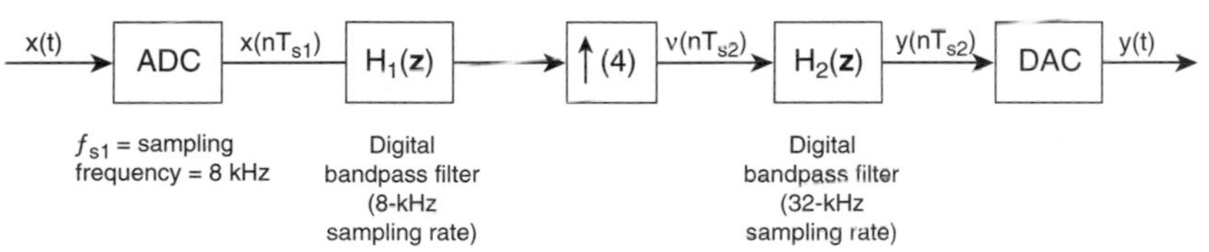

Fig. 7.8
Alternate method for removing the 10 kHz component while retaining the 9-kHz component but operating on aliased signals

The digital filter H_1 operates at a sampling rate of 8 kHz and introduces a frequency shaping that eliminates the 2-kHz component as shown in Fig. 7.9. The process of interpolation creates three additional replicates. At the higher sampling rate, 32 kHz, the digital filter chooses the appropriate replicate as depicted in Fig. 7.10. In essence the digital filter $H_2(\mathbf{z})$ operating at 32 kHz together with the associated analog replicate-rejection filter constitute a bandpass replicate-rejection (or replicate selection) filter as though the output of $H_1(\mathbf{z})$ had been converted to analog at the 8-kHz sampling rate.

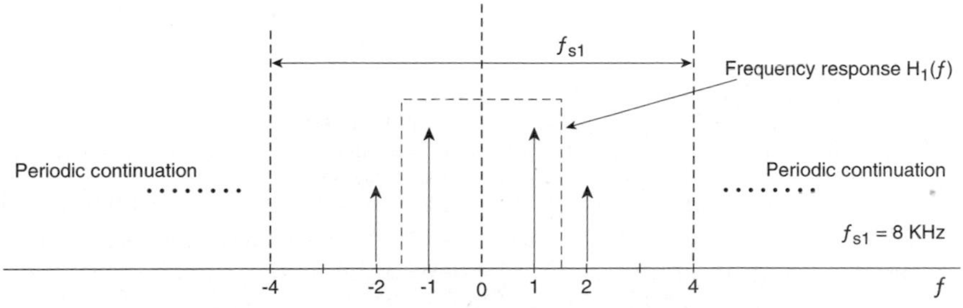

Fig. 7.9

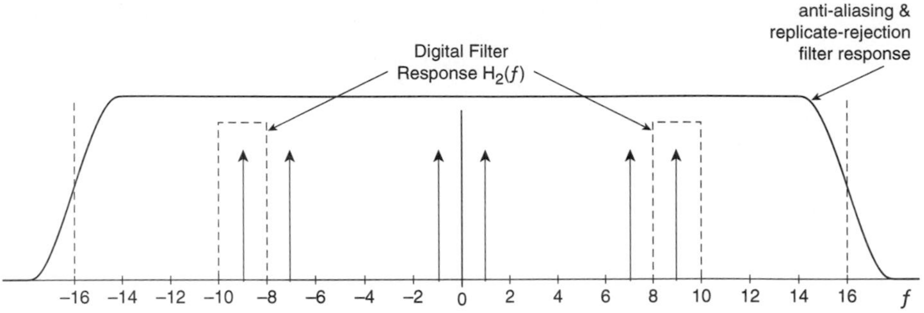

Fig. 7.10

To summarize, the process of interpolation is the means by which we increase the sampling rate and in so doing we expand the range of frequency over which we can provide digital filters. Interpolation by a factor of N creates N replicates of which we can choose one as the primary and reject the others using (digital) bandpass filters operating at the higher rate.

7.2.2 Decimation

Decimation is the logical reverse of interpolation. It is the process by which the sampling rate is reduced, with the concomitant effect of aliasing. Decimation by a factor of N involves reducing the sampling rate from a high value of f_{s2} to f_{s1}, which is N times slower. Decimation is depicted in Fig. 7.11. The signal, $\{y(nT_{s2})\}$, is filtered by H(**z**) (which is optional) at the high sampling rate f_{s2} to yield $\{x(nT_{s2})\}$, also at the high sampling rate. The function of H(**z**) is akin to the anti-aliasing filter associated with A/D conversion; all unwanted signals are removed. Decimation of the sampling rate is achieved by choosing every Nth sample of $\{x(nT_{s2})\}$ to yield the digital signal $\{v(nT_{s1})\}$, which has the underlying sampling rate of $f_{s1} = (1/N)f_{s2}$.

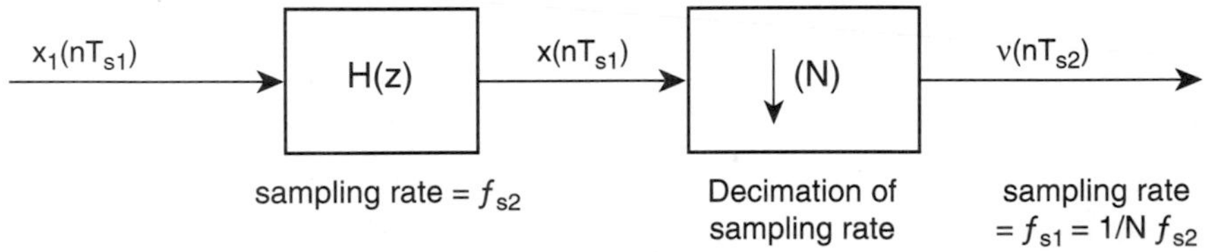

Fig. 7.11
Decimation as the two-step process of anti-aliasing filtering and sampling rate reduction

The operation of decimation, that is undersampling, can be visualized as shown in Fig. 7.12. The samples $\{x(nT_{s2})\}$ can be viewed as samples of an analog signal, x(t), taken at intervals of T_{s2}. Undersampling by a factor of N (a factor of four in the figure) would involve retaining every Nth sample.

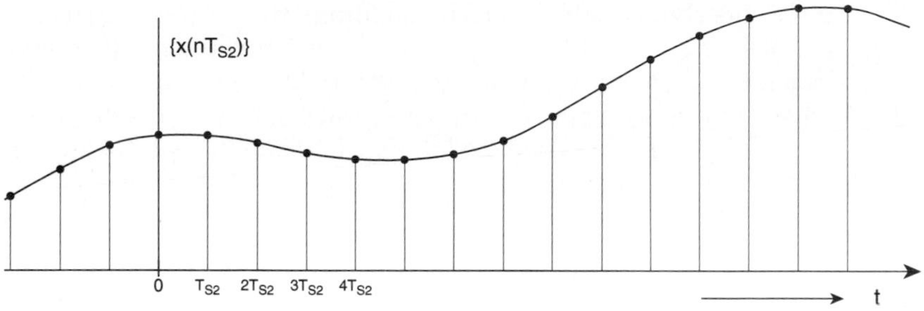

Fig. 7.12 (a)
Discrete-time signal at the higher sampling rate along with the notion of the underlying (lowpass) analog signal

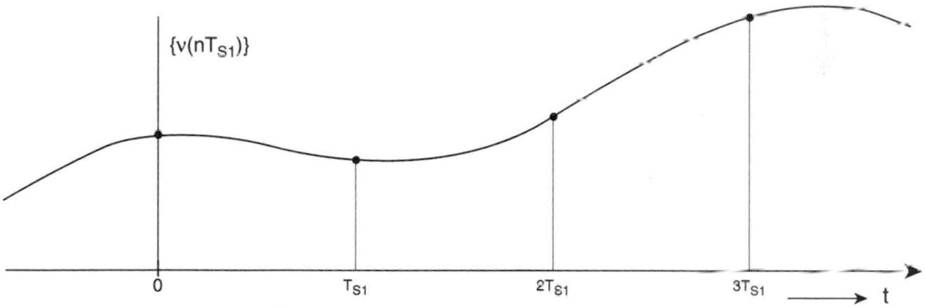

Fig. 7.12 (b)
Illustration of the process of undersampling—choosing every Nth sample

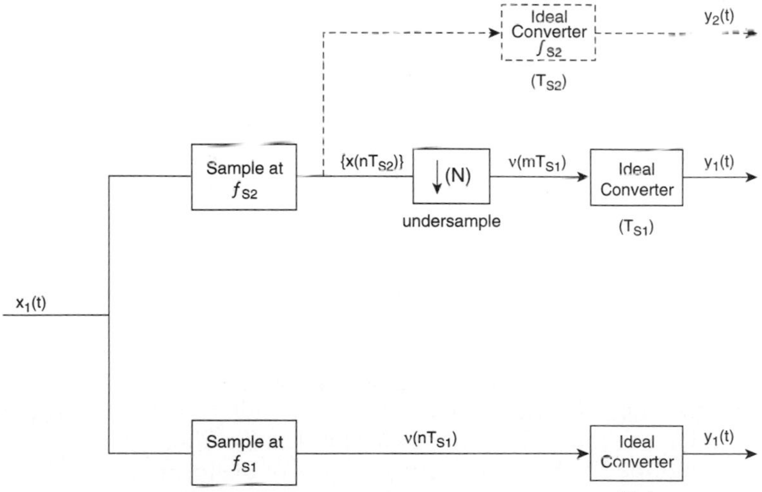

Fig. 7.13
Scheme to illustrate the equivalence of $\{x(nT_{s2})\}$ and $\{v(nT_{s1})\}$

If the underlying analog signal is bandlimited to $[-f_{s1}/2, f_{s1}/2]$ then the information content of $\{v(nT_{s1})\}$ and $\{x(nT_{s2})\}$ is the "same" in the sense that there is no overlap between the spectral replicates and thus we could, in principle, extract x(t) from either $y_1(t)$ or $y_2(t)$ by applying an ideal lowpass filter. The conversion back to analog from $\{x(nT_{s2})\}$ and $\{v(nT_{s1})\}$ is depicted in Fig. 7.13 and the spectra are illustrated in Fig. 7.14 which shows what the spectra might appear if there was no overlap.

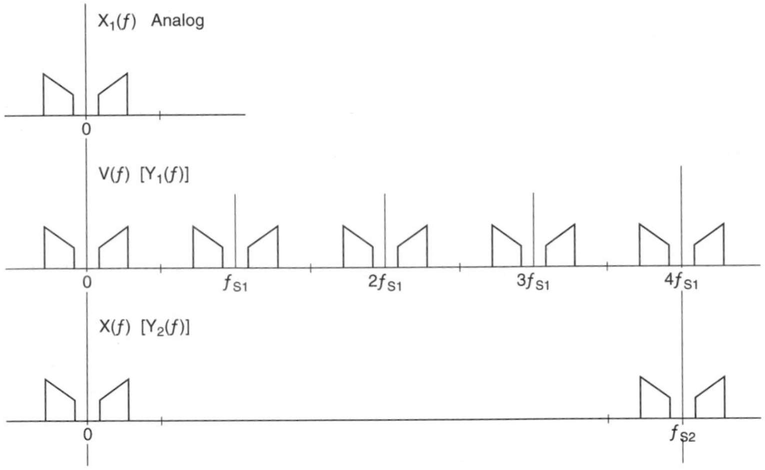

Fig 7.14
Frequency domain view of sample rate reduction illustrating the creation of replicates (aliasing)

If x(t) is not bandlimited to $[-f_{s1}/2, f_{s1}/2]$ then we expect the informational content of $\{v(nT_{s1})\}$ and $\{x(nT_{s2})\}$ to be different, the difference arising from aliasing. For example, x(t) may have been bandlimited to $[-f_{s2}/2, f_{s2}/2]$ and sampling at the high rate would not induce aliasing. The undersampling process does, however, introduce aliasing. This is depicted in Fig. 7.15, which shows that all frequency components of $X(f)$ fold into the band of interest to form $V(f)$. This is the same effect observed if x(t) is sampled at f_{s1} directly. When we sample at f_{s2} we have the opportunity of introducing the digital filter $H(\mathbf{z})$ (see Fig. 7.11) prior to undersampling. If this filter is an ideal lowpass filter depicted in Fig. 7.15, then $V(f)$ is of the form shown in part (d) of the figure. Note that this is the same as we obtain if we apply an (ideal) analog lowpass filter to x(t) prior to sampling at f_{s1}.

In drawing, Fig. 7.15 it is assumed that x(t) is real and thus $X(f)$ demonstrates (conjugate) symmetry about $f = 0$. In the case of the digital signals, real-valued samples imply (conjugate) symmetry as well and if our "window" is $[0, f_s]$ the symmetry is about $f_s/2$. In the digital domain the filters, if need be, can be real-valued, as exemplified by the symmetric response of the ideal lowpass filter, or, be complex-valued. An illustrative response of a complex-valued (ideal) bandpass filter is depicted in Fig. 7.15 (b). Applying this filter implies that $\{v(nT_{s1})\}$ is complex-valued and thus need not have a (conjugate) symmetric spectrum. Thus if $H(\mathbf{z})$ (see Fig. 7.11) is such a complex-valued filter, the spectrum of the undersampled signal could take the form shown as $V_2(f)$ in Fig. 7.15 (e). Complex-valued analog signals and complex-valued analog filters (filters with responses that do not exhibit conjugate symmetry about $f = 0$) are difficult to implement. Digital systems on the other hand, by providing for complex-valued arithmetic, can implement complex-valued filters quite naturally.

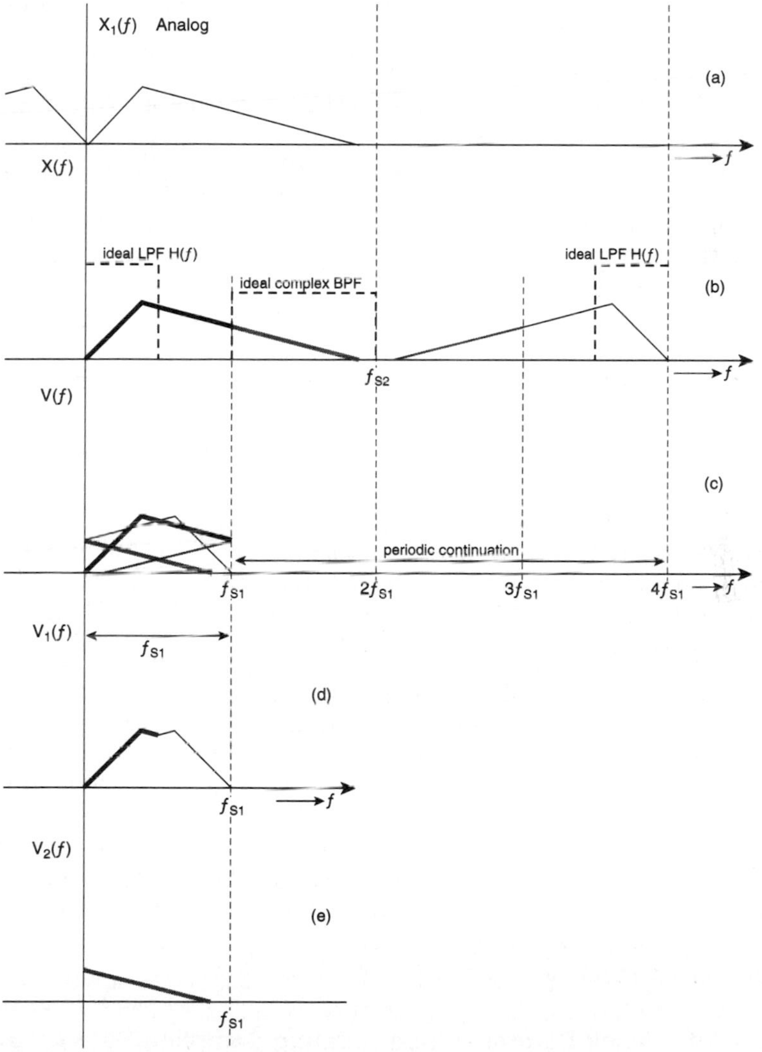

Fig. 7.15
Illustrating the aliasing associated with undersampling

Inherent in the notion of decimation is the implication of the sampling rate being higher than that specified by the sampling theorem. By doing so some of the burden of anti-aliasing can be shouldered by digital filters. Further, by allowing for complex-valued digital filters, we have great flexibility in choosing the frequency window that will correspond to our primary, or "base," spectrum of the digital signal. Examination of the aliasing process associated with the undersampling indicates that choosing every Nth sample can be described in the time domain in the following way. The time index of the signal $\{x(nT_{s2})\}$ is converted to a "double index" (m, r) by setting

$$n = mN + r \text{ where } 0 \quad r \quad (N-1) \tag{7.2.4}$$

Choosing every Nth sample is achieved by associating, for some r,

$$v(mT_{s1}) = x(mNT_{s2} - rT_{s2}) \tag{7.2.5}$$

Recognize that there are N possible ways in which this undersampling can be achieved. The ***normal*** way is for r = 0, implying that

$$v(mT_{s1}) = x(mNT_{s2}) = x(nT_{s2}) \quad (r = 0) \tag{7.2.6}$$

The case r = 0 has a simple description in the frequency domain, namely

$$V(f) = \frac{1}{N} \mathop{}_{k=0}^{N-1} X(f - kf_{s1}) \tag{7.2.7}$$

The factor of N is often ignored. The reason for retaining this factor is that our frequency domain functions have the dimension of per-unit sample and thus need to be suitably normalized when a sampling rate change is done. The equivalent of Eq. (7.2.7) for other values of r is covered in the next subsection.

7.2.3 The Polyphase Decomposition

In dealing with digital filters with a fixed underlying sampling rate, the Z-transform is a useful tool. The notion of a unit delay is encapsulated by a multiplicative term $\mathbf{z}^{-1}$ in the Z domain. Various filters can be described as transfer functions and manipulated as polynomials and the ratios of polynomials. Filters can be decomposed into smaller units and manipulated as blocks. The polyphase decomposition is an extension of Z-transform theory that allows us to do similar manipulations when there are sampling rate alterations.

7.2.3.1 Block Diagrams Incorporating Sampling Rate Changes

Consider the case when a signal $\{x(nT_{s2})\}$ is decimated by choosing every Nth sample to yield a lower sampling rate f_{s1}. By writing

$$n = mN + r \text{ where } 0 \quad r \quad (N-1) \tag{7.2.8}$$

we can choose one of N phases of undersampling corresponding to $\{v_r(mT_{s1})\}$ as

$$v_r(mT_{s1}) = x(mT_{s1} + rT_{s2}) \, ; 0 \le r \le (N-1) \qquad (7.2.9)$$

By defining a ***unit delay*** as a delay of one sample at the high rate, f_{s2}, the operation of decimation and choosing the appropriate sampling phase can be diagramed as shown in Fig. 7.16.

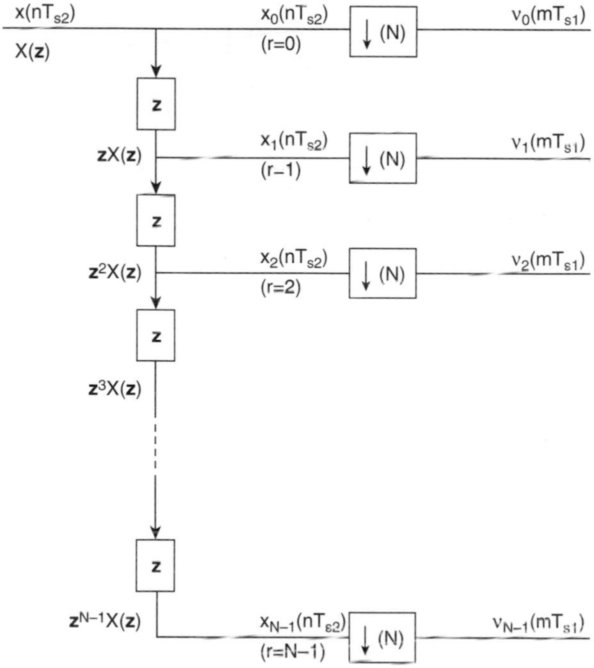

Fig. 7.16
Illustrating the relation between the N possible choices inherent in sampling rate reduction by a factor of N

Thus each choice of undersampled signal is equivalent to delaying/advancing the signal $\{x(nT_{s2})\}$ by r samples and undersampling according to

$$v_r(mT_{s1}) = x_r(mT_{s1}) \, ; 0 \le r \le (N-1) \qquad (7.2.10)$$

Fig. 7.16 shows the input signal being advanced, its Z-transform multiplied by **z** at each stage. This could be altered to show delays, whereby each block indicating an advance is replaced by a block indicating a delay, if we chose to make our association of single index to double index by

$$n = mN - r \text{ where } 0 \quad r \quad (N-1) \tag{7.2.11}$$

A similar situation arises in interpolation. If $\{v(mT_{s1})\}$ is interpolated to give $\{x(nT_{s2})\}$, the ***normal*** process of interpolation is described by

$$x(nT_{s2}) = \begin{cases} v(mT_{s1}) & \text{for } n = mN \\ 0 & \text{for } n \neq mN \end{cases} \tag{7.2.12}$$

Strictly speaking we can choose which (N–1) samples are made zero. If we decompose the time index at the high sampling rate as shown in Eq.(7.2.8) then we can choose some value of r and make

$$x_r(nT_{s2}) = x(mNT_{s2} + rT_{s1}) = \begin{cases} v(mT_{s1}) & \text{for some } r \\ 0 & \text{otherwise} \end{cases} \tag{7.2.13}$$

It can be seen that $\{x_r(nT_{s2})\}$ for r = 1, 2, ... , (N–1), are delayed version of the normal case, $\{x_0(nT_{s2})\}$. This is diagramed in Fig. 7.17.

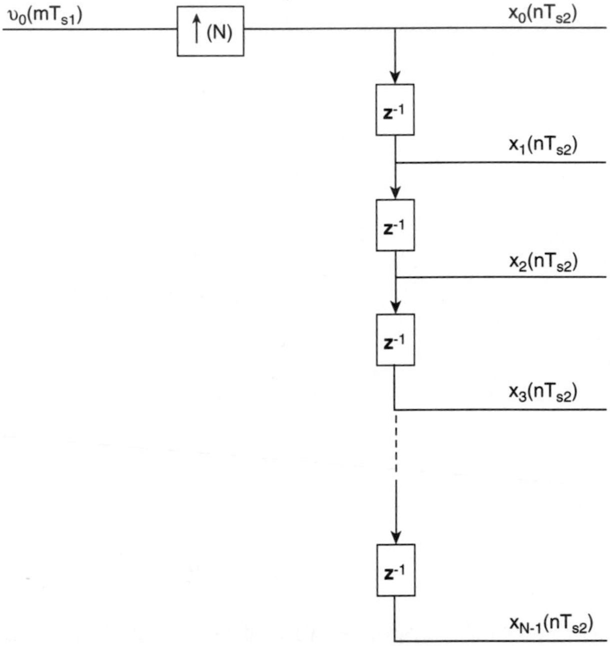

Fig. 7.17
Illustrating the relation between the N ways in which a sampling rate increase by zero-fill of (N–1) samples can be achieved

7.2.3.2 Sampling Rate Changes and their Z-Transform Implications

It is instructive to relate the Z-transform of $\{x_0(nT_{s2})\}$ to the sequence $\{v(mT_{s1})\}$. Since

$$X_0(\mathbf{z}) = \sum_n x_0(nT_{s2})\,\mathbf{z}^{-n} \tag{7.2.14}$$

where $\mathbf{z}^{-1}$ corresponds to a unit sample delay at the sampling rate f_{s2}, we can use the index mapping of Eq. (7.2.8) to get

$$X_0(\mathbf{z}) = \sum_n \sum_{r=0}^{N-1} x_0(mNT_{s2} + rT_{s1})\,\mathbf{z}^{-r}\mathbf{z}^{-mN}$$

$$= \sum_m x_0(mNT_{s2})\mathbf{z}^{-mN} = \sum_m v(mT_{s1})\mathbf{z}^{-mN} = V(\mathbf{z}^N) \tag{7.2.15}$$

where

$$V(\eta) = \sum_m v(mT_{s1})\eta^{-m} \tag{7.2.16}$$

Recognize that Eq. (7.2.16) describes the Z-transform of the sequence $\{v(mT_{s1})\}$ in terms of a transform domain variable η to keep the notions of unit delay separate. The quantity η^{-1} is representative of a unit delay at the lower sampling rate f_{s1}. Thus the operation of interpolation described by Eq. (7.2.12) can be described in the Z-transform domain by replacing the transform domain variable by $\mathbf{z}^N$. We know that the relationship between the Z domain and frequency is such that as the Z domain variable traverses one complete revolution on the unit circle the frequency variable traverses one complete period, i.e., an interval of f_s. Thus as η completes one revolution, frequency traverses an interval of f_{s1}; the variable η^N completes N revolutions, with frequency traversing the interval of f_{s1} N times. Replacing η by η^N is thus the Z domain equivalent of ***replication***.

Deriving a Z domain description of decimation is a little involved since the impact of aliasing needs to be addressed. First we define

$$\mathbf{W} = e^{-j\frac{2\pi}{N}} \tag{7.2.17}$$

as an aid to simplifying the subsequent notation. Note that $\mathbf{W}$ is related to the **Discrete Fourier Transform (DFT)** and satisfies the following property:

$$\sum_{k=0}^{N-1} W^{nk} = N\,\delta(n-k) \tag{7.2.18}$$

The Z-transform of the sequence $\{x(nT_{s2})\}$ can be written as

$$X(\mathbf{z}) = \sum_{n} x(nT_{s2})\mathbf{z}^{-n} = \sum_{m}\sum_{r=0}^{N-1} x_0(mNT_{s2} + rT_{s1})\,\mathbf{z}^{-r}\,\mathbf{z}^{-mN} \tag{7.2.19}$$

where the single-index to double-index mapping of Eq. (7.2.8) has been used. Substituting our notion of decimation from Eq. (7.2.9) we can write

$$X(\mathbf{z}) = \sum_{m}\sum_{r=0}^{N-1} v_r(mT_{s1})\,\mathbf{z}^{-r}\,\mathbf{z}^{-mN} = \sum_{r=0}^{N-1} V_r(\mathbf{z}^{N})\,\mathbf{z}^{-r} \tag{7.2.20}$$

If we replace $\mathbf{z}$ by $\mathbf{z}\mathbf{W}^k$ we get

$$X(\mathbf{z}\mathbf{W}^{k}) = \sum_{r=0}^{N-1} V_r(\mathbf{z}^{N})\,\mathbf{z}^{-r}\,\mathbf{W}^{-rk} \tag{7.2.21}$$

and summing over k =0, 1, ... , (N–1) yields

$$\sum_{k=0}^{N-1} X(\mathbf{z}\mathbf{W}^{k}) = \sum_{r=0}^{N-1} V_r(\mathbf{z}^{N})\,\mathbf{z}^{-r}\,\{\sum_{k=0}^{N-1} \mathbf{W}^{-rk}\} = NV_0(\mathbf{z}^{N}) \tag{7.2.22}$$

where the property of $\mathbf{W}$ mentioned in Eq. (7.2.18) has been used to cull the value of $r = 0$. Therefore,

$$V_0(\mathbf{z}^{N}) = \frac{1}{N}\sum_{k=0}^{N-1} X(\mathbf{z}\mathbf{W}^{k}) \tag{7.2.23}$$

provides a way for translating the Z-transform of the higher-rate signal to that of the signal at the lower sampling rate. It is useful to consider the implication of Eq. (7.2.23) in the frequency domain. Setting $\mathbf{z} = \exp(j2\pi f T_{s2})$ (remember that $\mathbf{z}^{-1}$ is a delay at the high sampling rate f_{s2}) we get

$$V_0(f) = \frac{1}{N}\sum_{k=0}^{N-1} X(f - k\frac{f_{s2}}{N}) = \frac{1}{N}\sum_{k=0}^{N-1} X(f - kf_{s1}) \tag{7.2.24}$$

which is exactly the same result we obtained, from the viewpoint of aliasing, in Eq. (7.2.7).

7.2.3.3 Examples of the Polyphase Decomposition

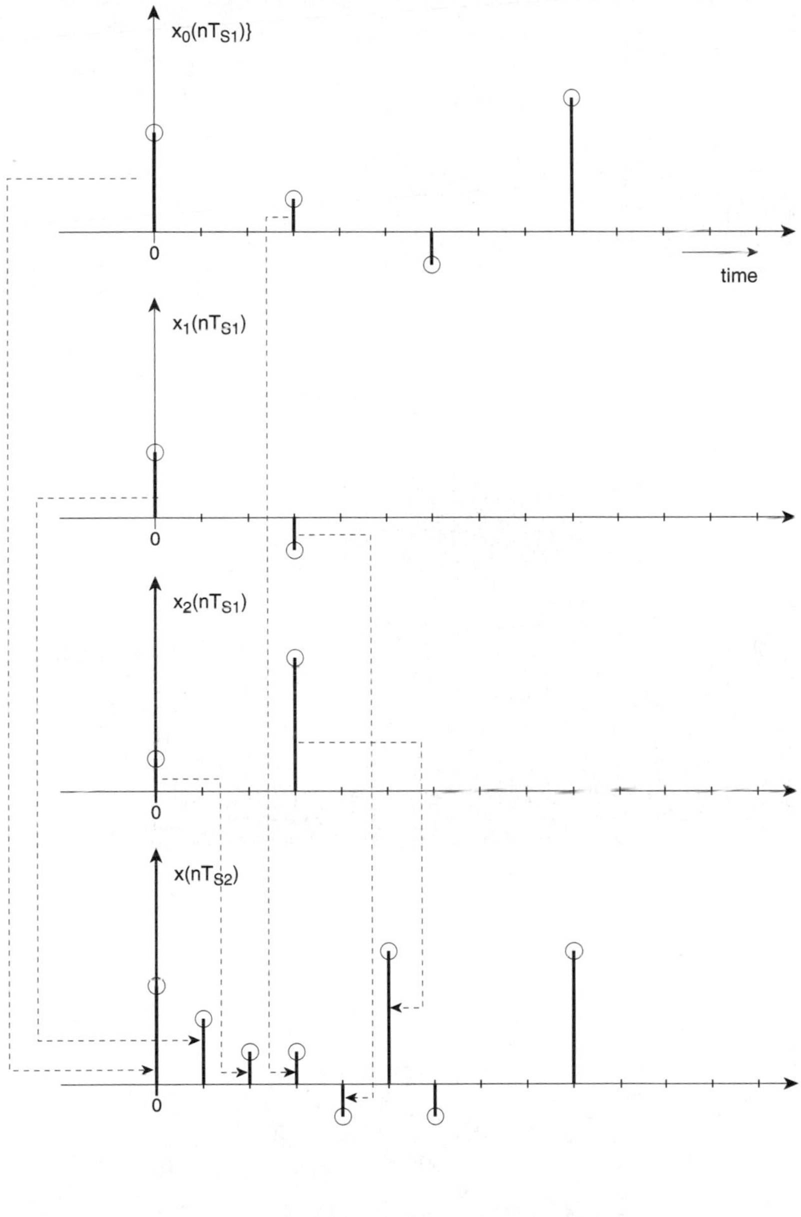

Fig. 7.18
Time Division Multiplexing as the interleaving of samples from different signals

One example of the polyphase representation is in **Time Division Multiplexing** (**TDM**). The notion of **TDM** is very simple. Consider N separate digital signals $\{x_k(nT_{s1})\}$, k=0, 1, ... , (N–1), which have a common sampling rate f_{s1}. A new signal, with the underlying sampling rate of $f_{s2} = Nf_{s1}$, can be created by ***interleaving*** samples from each of the N separate signals. This illustrated in Fig. 7.18 for the case N=3.

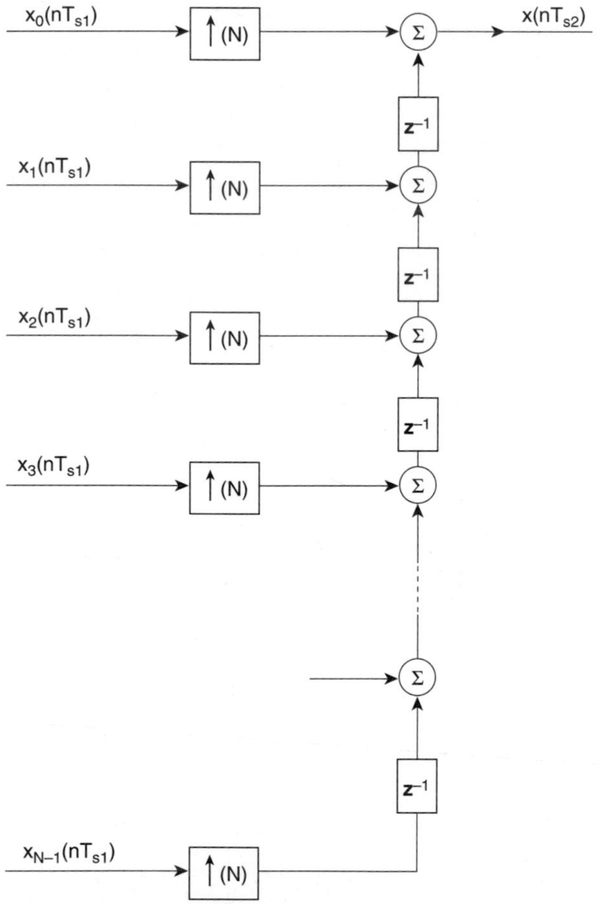

Fig. 7.19
Polyphase representation of **TDM**

The operation of **TDM** is described by

$$x(mNT_{s2} + rT_{s2}) = x_r(mT_{s1}) \tag{7.2.25}$$

where the association of n with m, N, and r is given by Eq. (7.2.8). The block diagram of the operation is shown in Fig. 7.19. In terms of the polyphase representation, the Z-transform, X(**z**), is given by

$$X(\mathbf{z}) = \sum_{r=0}^{N-1} X_r(\mathbf{z}^N)\,\mathbf{z}^{-r} \tag{7.2.26}$$

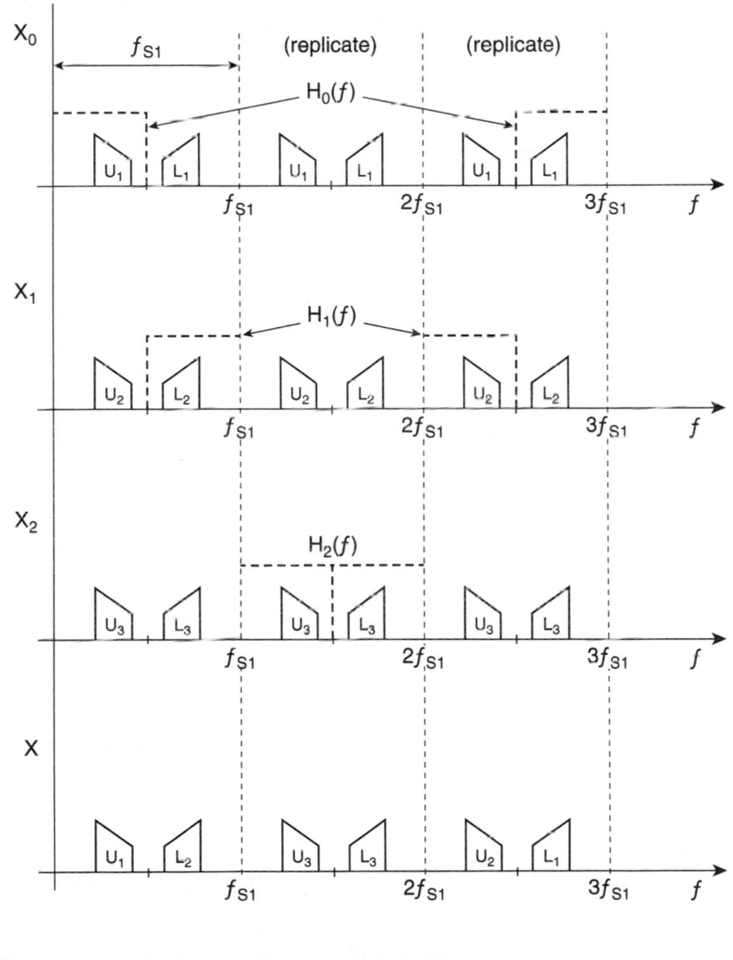

Fig. 7.20
Frequency Division Multiplexing achieved by selection of the appropriate replicate created by zero-fill oversampling

A second example where the polyphase representation can be used is in **Frequency Division Multiplexing (FDM)**. The notion of **FDM** can be explained with respect to Fig. 7.20, which depicts the case of what is known as **SSB-FDM** for **"Single Side Band Frequency Division Multiplexing**."

Suppose we have N separate digital signals $\{x_k(nT_{s1})\}$ as before but rather that interleave these signals in time we wish to interleave these signals in the frequency domain. Consider the example of N = 3 shown in the figure.

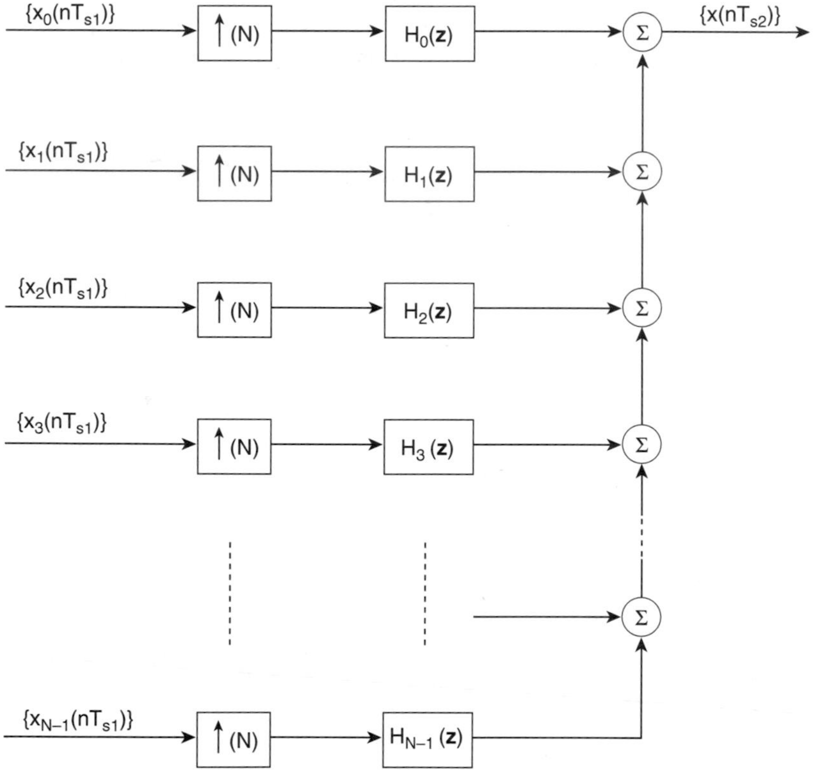

Fig. 7.21
Filter-bank representation of Frequency Division Multiplexing

Interpolating $\{x_i(nT_{s1})\}$ by inserting zero samples expands our view to a window of $f_{s2} = Nf_{s1}$ and each signal has N replicates within this window. We use digital filters $H_i(f)$ to choose which replicate of each signal we want to combine into $\{x(nT_{s2})\}$. We assume that we are dealing with real-valued signals, implying a certain symmetry in the frequency domain, as is evident in Fig. 7.20. The polyphase representation of the operation is shown in block diagram form in Fig. 7.21 and can be expressed as

$$X(\mathbf{z}) = \sum_{r=0}^{N-1} X_r(\mathbf{z}^N)\, H_r(\mathbf{z}) \tag{7.2.27}$$

The inverse processes, namely **demultiplexing**, can be obtained very simply from the *forward* process of multiplexing. In particular, the demultiplexing of the **TDM** signal $\{x(nT_{s2})\}$ into its components is laid out in Fig. 7.22. **Notice that it can be derived from Fig. 7.19 by reversing the sense of the arrows, replacing summing nodes by branch points, and replacing the interpolation block by decimation.** Because of our particular index mapping convention, we replace blocks of $\mathbf{z}^{-1}$ with blocks of $\mathbf{z}$ (replace delay by advance).

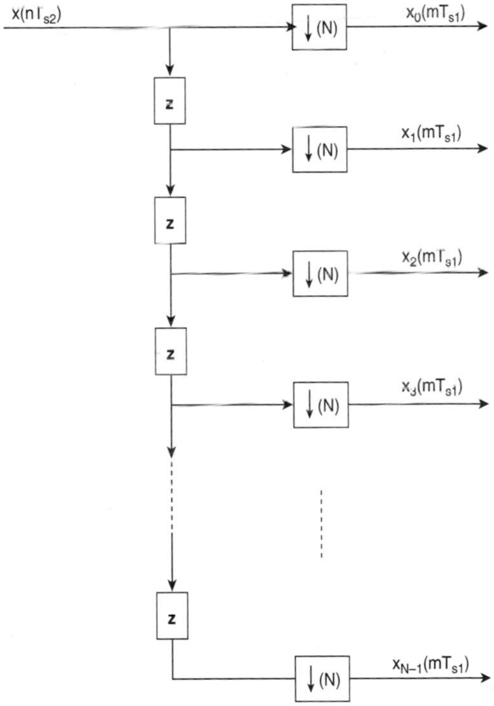

Fig. 7.22
Polyphase representation of **TDM** demultiplexing

The demultiplexing of an **FDM** signal $\{x(nT_{s2})\}$ into its constituent parts follows Fig. 7.23. The filters $H_i(\mathbf{z})$ used are the **same** as those for the multiplexing process.

Again note that Fig. 7.23 can be derived from Fig. 7.21 by reversing the sense of the arrows, replacing summing nodes with branch points, and replacing interpolation by decimation.

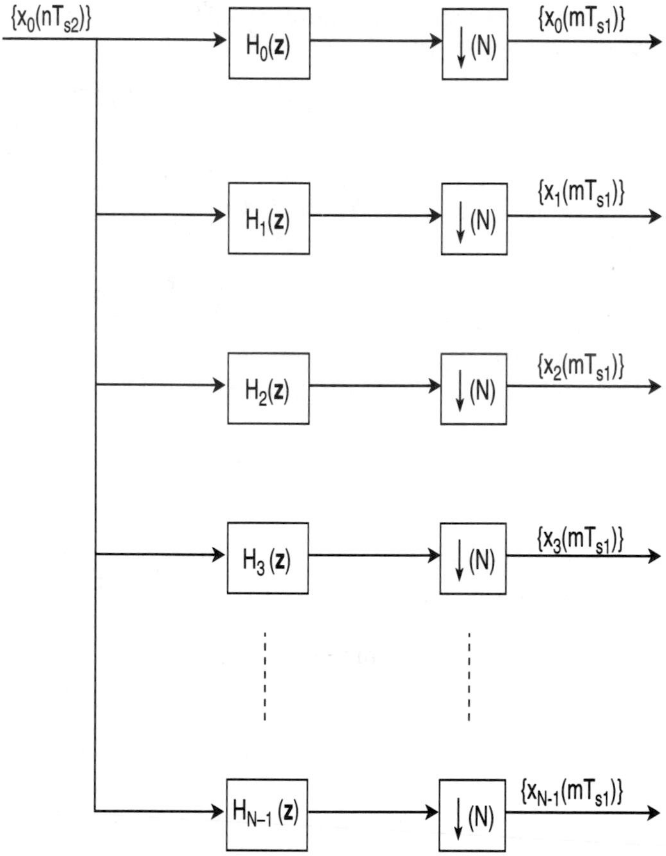

Fig. 7.23
Filter-bank representation of **FDM** demultiplexing

A superficial similarity is discernible between the **FDM** and **TDM** structures. In fact, if $H_k(\mathbf{z}) = \mathbf{z}^{-k}$ in Fig. 7.21 (or $\mathbf{z}^k$ in Fig. 7.22) then the two structures are indeed the same. In this special case the filters provide no frequency selectivity, though. Provided there is no (unknown) delay between and no filtering action on the composite signal between the multiplexer and the demultiplexer, then both schemes reproduce the constituent signals $\{x_k(nT_{s1})\}$ faithfully. If there is a flat delay of an integer number of samples between the multiplexer and demultiplexer, then the constituent signals are reproduced but with an ambiguity of channel numbering! Consequently, in actual **TDM** transmission schemes additional information in the form of "framing" is appended to the composite signal in order to establish channel identity. If there is any filtering action on the composite stream, such as, for example, conversion to analog and then back to digital (with the associated replicate-rejection and anti-aliasing filters) then the identity of the channels is destroyed. Therefore, in practice, digital transmission channels are required to provide "bit-transparency." The composite signal at the demultiplexer is, other than differences arising from transmission errors, the same, albeit delayed, as the composite signal at the multiplexer. The embedded framing establishes the correct channel numbering.

By separating the channels in frequency, using nontrivial $H_k(\mathbf{z})$, it is possible to "recover" the appropriate channel signals even if there is some filtering in the path between the multiplexer and the demultiplexer. The recovered (sub)channel signal may not be identical to the transmitted (sub)channel signal on a sample-by-sample basis but will be the same on a spectral basis (frequency domain equivalence). Of course the nonideal behavior of the channel separation filters will have some impact on the frequency content of the (sub)channel signals but this impact can be kept within pre-specified limits by suitably increasing the complexity of the channel separation filters. In fact, the composite signal may be converted to analog and then back to digital while retaining the spectral content of the constituent channels.

If the channel separation filters are not ideal, then they impact the (sub)channel signals in the following way. In the multiplexer, the passband distortion of the filter (passband ripple, nonlinear phase) will impact the base of the channel signal directly; noninfinite stopband attenuation means that the replicate of the channel is not suppressed completely and will overlap the base of a different channel. This latter effect is called crosstalk. In the demultiplexer we have a similar situation for the passband distortion. The noninfinite stopband attenuation implies that certain channels are not completely suppressed prior to sampling rate reduction and could alias into other (sub)channels leading, again, to crosstalk.

Equipment that provides the conversion to and from the two forms of multiplexing is called a **transmultiplexer** and is discussed in Section 7.4. It is shown there that the N filters used for the **FDM** frequency separation are related and this relationship can be exploited.

7.3 THE DISCRETE FOURIER TRANSFORM (DFT)

This section deals with the Discrete Fourier Transform (**DFT**) and its companion, the Discrete Cosine Transform (**DCT**). In particular, the **DFT** and **DCT** are portrayed in terms of banks of bandpass filters and the applications described in Sections 7.4 and 7.5 utilize the **DCT** and **DFT**, respectively, for implementing filter banks. The popularity of the **DFT** and **DCT** stems from the availability of algorithms to compute these matrix operations in an efficient manner. Such algorithms are given the generic name of Fast Fourier Transforms (**FFT**s) or Fast Cosine Transforms (**FCT**s).

In Sections 7.3.1 and 7.3.2 we provide a rationale for considering the **DFT** and **DCT**, respectively, a bank of filters where the underlying frequency response follows that of a filter whose response follows a sin(Nf)/sin(f) rule. The method for providing "better" filters is the subject of Sections 7.3.7 and 7.3.8. In Section 7.3.3 we provide a formal definition of the **DFT** and **DCT**. The terminology "transform" is arguably a misnomer. A more appropriate view is that of representing vectors using basis vectors that are not the rows/columns of an identity matrix but a set of basis vectors based on the complex exponential or real cosine for the **DFT** and **DCT**, respectively, and in Section 7.3.4 a relationship between the **DFT** and the discrete-time Fourier transform (the **DTFT**) is provided. Some important properties of the **DFT**, related to convolution and symmetry, are treated in Sections 7.3.5 and 7.3.6. These properties are used later when we talk about the Discrete Multitone (**DMT**) scheme in Section 7.5.

7.3.1 The DFT as a Bank of Bandpass Filters

Consider the situation depicted in Fig. 7.24, where a signal $\{x(nT_s)\}$ is filtered by $H_0(\mathbf{z})$ to provide $\{y_0(nT_s)\}$. The filter is the **FIR** rectangular window defined by

$$h_0(n) = \begin{cases} 1 \text{ for } 0 \le n \le (N-1) \\ 0 \quad \text{otherwise} \end{cases} \tag{7.3.1}$$

and the output is given by

$$y_0(nT_s) = \sum_{k=0}^{N-1} x((n-k)T_s)\,h_0(n) = \sum_{k=0}^{N-1} x((n-k)T_s) \tag{7.3.2}$$

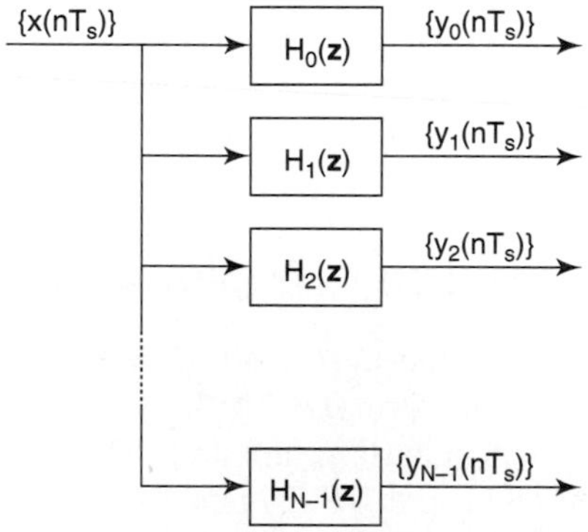

Fig. 7.24
Multichannel bandpass filtering

Clearly, $\{y_0(nT_s)\}$ is a *moving average*, over N consecutive samples, of $\{x(nT_s)\}$. The frequency response achieved is shown in Fig. 7.25. The filter is lowpass in nature and approximates, albeit not very well, the ideal lowpass response shown.

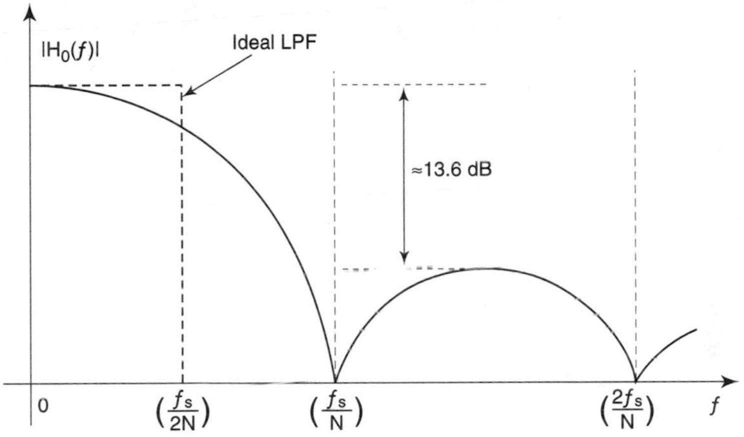

Fig. 7.25
Rectangular window frequency response as an approximation to an ideal lowpass filter

Now consider the output of $H_1(\mathbf{z})$, given that $H_1(\mathbf{z})$ is a complex-valued **FIR** filter with impulse response given by

$$h_1(n) = h_0(n)\, e^{j2\pi n f_1 T_s} \tag{7.3.3}$$

$H_1(\mathbf{z})$ can be considered a bandpass filter centered at the frequency f_1. To see this, we observe that

$$H_1(f) = H_0(f)\ (*)\ \delta(f - f_1) = H_0(f - f_1) \tag{7.3.4}$$

and $H_1(f)$ is the shifted version of $H_0(f)$. Now $\{h_1(n)\}$ is not real-valued and thus $H_1(f)$ does not have to demonstrate the conjugate symmetry displayed by $H_0(f)$. If, further, we make the assumption that

$$f_1 = 1\left(\frac{f_s}{N}\right) \tag{7.3.5}$$

then the frequency response of $H_1(\mathbf{z})$ is depicted as in Fig. 7.26. The response is that of a bandpass filter of *nominal* bandwidth f_s/N and centered at f_s/N.

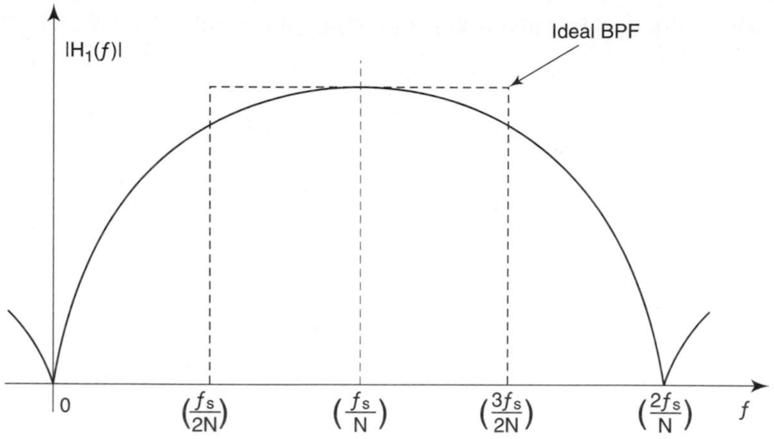

Fig. 7.26
Bandpass filter derived from the rectangular window *prototype*

In the same vein we define $H_k(\mathbf{z})$ by the impulse response

$$h_k(n) = h_0(n)\, e^{j2\pi n f_k T_s} \Rightarrow H_k(f) = H_0(f - f_k) \tag{7.3.6}$$

and where

$$f_k = k\left(\frac{f_s}{N}\right) \Rightarrow h_k(n) = h_0(n)\, e^{j2\pi\left(\frac{nk}{N}\right)} \tag{7.3.7}$$

The collection $H_k(\mathbf{z})$, k=0, 1, ... , (N–1) can be viewed as a bank of N bandpass filters, each with a nominal bandwidth of f_s/N and centered on a grid of frequencies kf_s/N. This is illustrated in Fig. 7.27, which uses the ideal bandpass characteristic solely for illustrative purposes. Since $H_0(\mathbf{z})$ is a real-valued filter, it exhibits the conjugate symmetry which in turn implies that "half" its bandwidth is from 0 through f_s/2N and "half" its bandwidth is between ($f_s - f_s$/2N) and f_s. The other filters, being complex valued can be depicted as contiguous passbands, nominally between kf_s/N ± f_s/2N. The assembly of bandpass filters in Fig. 7.24 thus comprises a "covering" of all frequencies between 0 and f_s.

The output of the kth bandpass filter can be written as

$$y_k(mT_s) = \sum_{n=0}^{N-1} x((m-n)T_s)\, h_k(n) = \sum_{n=0}^{N-1} x((m-n)T_s)\, e^{j2\pi\left(\frac{nk}{N}\right)} \tag{7.3.8}$$

Defining, with some abuse of notation,

$$x_{N-1-n} = x((m-n)T_s) \text{ and } y_k = y_k(mT_s) \tag{7.3.9}$$

we can derive the outputs of the bank of filters as

$$y_k = \sum_{n=0}^{N-1} x_n \, e^{j2\pi(\frac{nk}{N})} \; ; \; k = 0, 1, 2, \ldots, (N-1) \tag{7.3.10}$$

This is the defining relation for the (inverse) Discrete Fourier Transform. It can be expressed in matrix notation by

$$\mathbf{y} = \mathbf{W}_N \mathbf{x} \tag{7.3.11}$$

where the matrix $\mathbf{W}_N$ is given by

$$\mathbf{W}_N = \left[W_{nk}\right] \text{ where } W_{nk} = e^{j2\pi(\frac{nk}{N})} \tag{7.3.12}$$

and the vectors **x** and **y** are $\{x((m-N+n)T_s);\ n=0, 1, \ldots, (N-1)\}$ and $\{y_k(mT_s);\ k=0, 1, \ldots, (N-1)\}$, respectively.

Pictorially the **DFT** computation can be shown as a "black box" that implements a matrix multiplication. That is, a block with N inputs and N outputs as shown in Fig. 7.28 where the input-output relation is specified by Eq. (7.3.10).

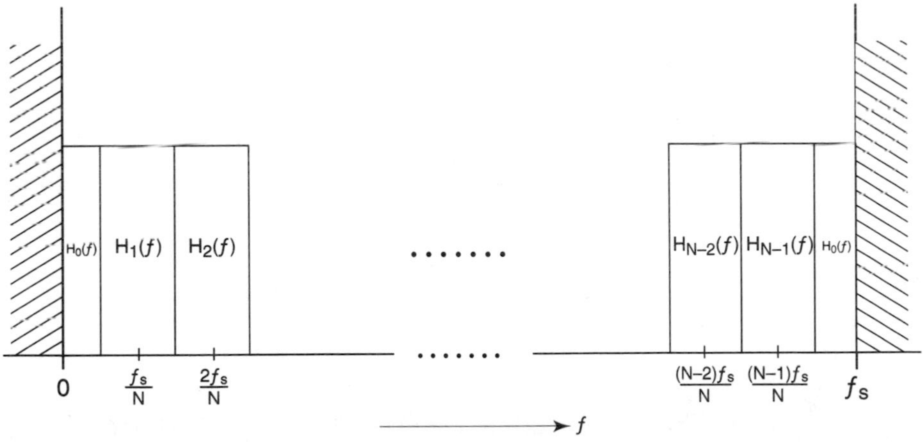

Fig. 7.27
Covering the frequency range $[0, f_s)$ with N bandpass filters

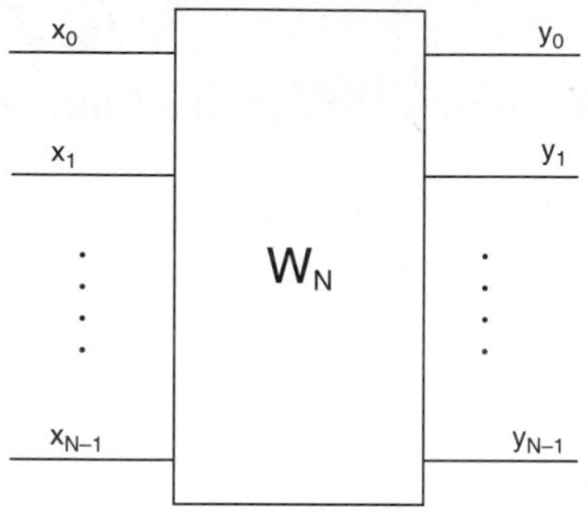

Fig. 7.28
Block representation of **DFT** computation

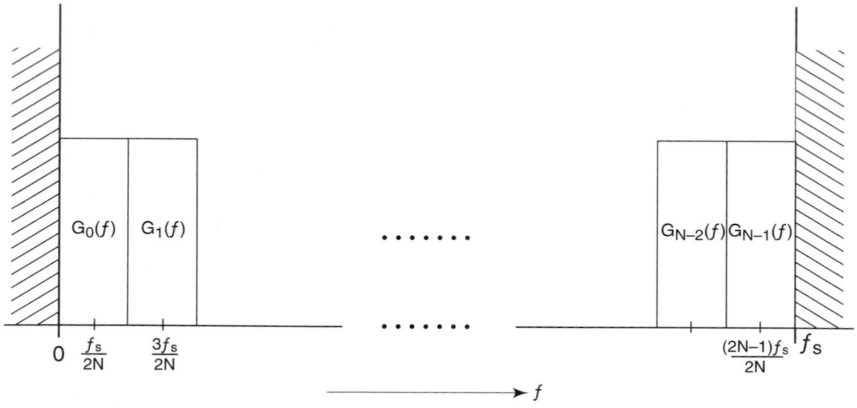

Fig. 7.29
Symmetric covering of [0, f_s] with N bandpass filters

Fig. 7.27 shows one way to cover the frequency window [0, f_s] using N bandpass characteristics, each of width f_s/N. The center frequencies of these characteristics follow the rule kf_s/N. An alternative is to use the covering depicted in Fig. 7.29, where each $G_k(f)$ is related to the corresponding $H_k(f)$ of Fig. 7.27 by

$$G_k(f) = H_k(f - \frac{f_s}{2N}) \tag{7.3.13}$$

That is, the center frequencies have been shifted by f_s/2N. The corresponding impulse responses are given by

$$g_k(n) = h_0(n) \exp(-j \frac{2k+1}{2N} 2\pi n) \tag{7.3.14}$$

Examination of Fig. 7.29 indicates that $G_k(f)$ and $G_{N-1-k}(f)$ occupy symmetric regions of frequency. Thus if we consider real-valued filters, we can construct a bank of N/2 filters (if N is even; (N+1)/2 filters if N is odd) by associating these symmetric pairs. Specifically, if we defined the filters $\{p_k(n)\}$ by

$$p_k(n) = \frac{1}{2}[g_k(n) + g_{N-1-k}(n)] \tag{7.3.15}$$

from which

$$p_k(n) = h_0(n)\cos\left(\frac{2k+1}{2N} 2\pi n\right) \tag{7.3.16}$$

then $\{p_k(n)\}$ are real-valued and the frequency responses $\{P_k(f)\}$ exhibit the complex conjugate symmetry as expected. Such an approach to a bank of filters is especially pertinent when the signal $\{x(nT_s)\}$ is real-valued and thus $X(f)$ exhibits conjugate symmetry in the frequency domain. The corresponding bank of filters is depicted in Fig. 7.30.

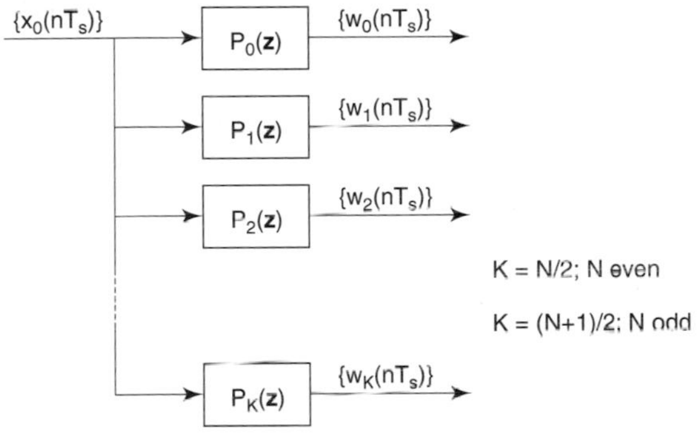

Fig. 7.30
Bank of real-valued bandpass filters derived from Fig. 7.29

The outputs of this bank of filters, now characterized by real-valued impulse responses for each constituent filter, can be written as

$$w_k(mT_s) = \sum_{n=0}^{N-1} x((m-n)T_s)\cos\left(\frac{2k+1}{2N} 2\pi n\right) \tag{7.3.17}$$

Defining, with some abuse of notation,

$$x_n = x((m-n)T_s) \text{ and } w_k = w_k(mT_s) \tag{7.3.18}$$

we can write the outputs of the filter bank as

$$w_k = \sum_{n=0}^{N-1} x_n \cos\left(\pi \frac{2k+1}{N} n\right) \tag{7.3.19}$$

The expression in Eq. (7.3.19) is the defining relation for the Real Discrete Fourier Transform (**RDFT**), which is just a special case of the regular **DFT**, and can be expressed in matrix notation as

$$\mathbf{w} = \mathbf{R}_N \mathbf{x} \tag{7.3.20}$$

where the matrix $\mathbf{R}_N$ is given by

$$\mathbf{R}_N = \left[R_{nk}\right] \text{ where } R_{nk} = \cos\left(\pi \frac{2k+1}{N} n\right) \tag{7.3.21}$$

and the vectors **x** and **w** are $\{x((m-n)T_s);\ n=0, 1, \ldots, (N-1)\}$ and $\{w_k(mT_s); k=0, 1, \ldots, (N-1)\}$, respectively. Strictly speaking the number of distinct outputs $\{w_k(mT_s)\}$ is not N, but N/2 for even N and (N+1)/2 for odd N. Nevertheless it is convenient to write Eq. (7.3.20) as an N*N matrix operation with the knowledge that $w_k(mT_s) = w_{N-1-k}(mT_s)$. Recognize also that the total bandwidth of each bandpass filter is now $2(f_s/N)$, of which (f_s/N) stems from the "positive" frequencies between 0 and $f_s/2$ and (f_s/N) arises from the "negative" frequencies between $f_s/2$ and f_s.

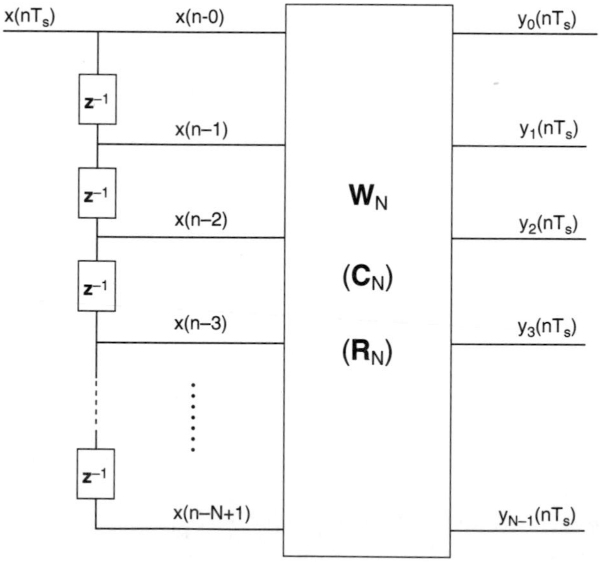

Fig. 7.31
Using the **DFT** to implement a bank of bandpass filters derived from the N-point rectangular window

The implementation of a bank of bandpass filters employing the **DFT** as the principal computational element is shown in Fig. 7.31. N consecutive input samples are processed by the matrix multiplication depicted by $\mathbf{W}_N$ to provide one output sample for each of N channels. In the case of the **DFT** the output signals are, in general, complex-valued, even if the input is real-valued. In this case, the filter centered at zero frequency and, if N is even, the filter centered at $f_s/2$, have real-valued outputs. The coverage of frequency is shown in Fig. 7.27. The implementation of a bank of real-valued bandpass filters that covered the frequency window in the fashion shown in Fig. 7.29 can be obtained by replacing $\mathbf{W}_N$ in Fig. 7.31 by $\mathbf{R}_N$ and keeping in mind that we do not have N independent outputs but, rather, N/2 for even N and (N+1) /2 for odd N.

7.3.2 The DCT as a Bank of Bandpass Filters

Obtaining N independent outputs from a bank of real-valued bandpass filters that covers the frequency window in the fashion of Fig. 7.29 can be achieved in the following manner. If we have N bandpass filters that, with real-valued impulse responses, exhibit conjugate symmetry, then we can place them in the frequency domain as shown in Fig. 7.32. Each filter has a (nominal) total passband width of f_s/N but this is comprised of a width of $f_s/2N$ for positive frequency and $f_s/2N$ for negative frequency. The periodic nature of the frequency response of a digital filter allows us to depict the negative frequencies between $[-f_s/2, 0]$ as the frequency range $[f_s/2, f_s]$ and stick to the window in the frequency domain of $[0, f_s]$ as used before.

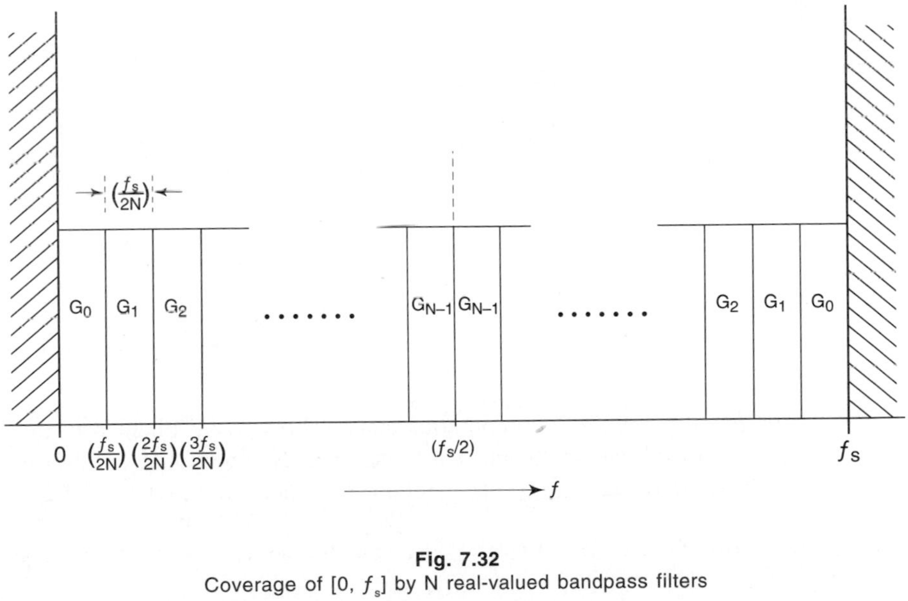

Fig. 7.32
Coverage of $[0, f_s]$ by N real-valued bandpass filters

The center frequencies, f_k, k=0, 1, ... , (N–1), of the N filters are given by

$$f_k = \frac{(2k+1)}{4N} f_s \tag{7.3.22}$$

and the reflection at $(f_s - f_k)$. The impulse response of the kth bandpass filter is

$$g_k(n) = h_0(n) \cos\left(\frac{2k+1}{2N} \pi n\right) \tag{7.3.23}$$

and the output of the kth filter, $\{v_k(nT_s)\}$ is therefore obtained as

$$v_k(mT_s) = \sum_{n=0}^{N-1} x((m-n)T_s) \cos\left(\frac{2k+1}{2N} \pi n\right) \tag{7.3.24}$$

where the explicit dependence on T_s has been suppressed for convenience. Defining, with some abuse of notation,

$$x_n = x((m-n)T_s) \text{ and } v_k = v_k(mT_s) \tag{7.3.25}$$

we can write the outputs of the filter bank as

$$v_k = \sum_{n=0}^{N-1} x_n \cos\left(\pi \frac{2k+1}{2N} n\right) \tag{7.3.26}$$

The expression in Eq. (7.3.26) is the defining relation for the Discrete Cosine Transform (**DCT**). It can be expressed in matrix notation as

$$\mathbf{v} = \mathbf{C}_N \mathbf{x} \tag{7.3.27}$$

where the matrix $\mathbf{C}_N$ is given by

$$\mathbf{C}_N = \left[C_{nk}\right] \text{ where } C_{nk} = \cos\left(\pi \frac{2k+1}{2N} n\right) \tag{7.3.28}$$

and the vectors $\mathbf{x}$ and $\mathbf{v}$ are $\{x((m-n)T_s);\ n=0,1,...,(N-1)\}$ and $\{v_k(mT_s);\ k=0, 1, ... , (N-1)\}$, respectively.

Pictorially, the **DCT** computation can be shown as a "black box" that implements a matrix multiplication. That is, a block with N inputs and N outputs as shown in Fig. 7.33, where the input-output relation is specified by Eq. (7.3.26).

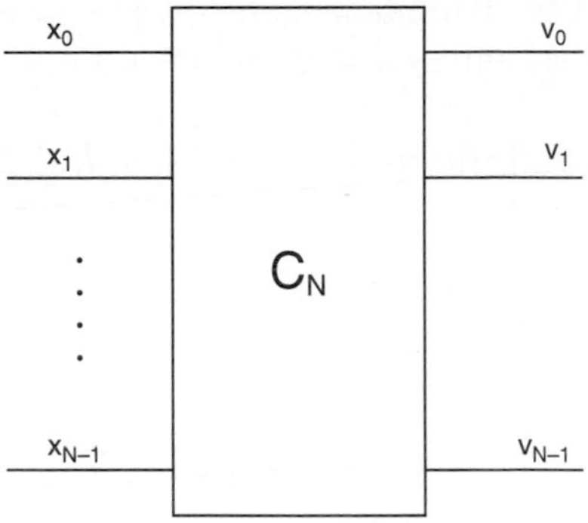

Fig. 7.33
Block representation of **DCT** computation

The implementation of a bank of bandpass filters employing the **DCT** as the principal computational element is shown in Fig. 7.31. N consecutive input samples are processed by the matrix multiplication depicted by $\mathbf{C}_N$ to provide one output sample for each of N channels. In most cases where the **DCT** is used the input is real-valued and, therefore, so is the output.

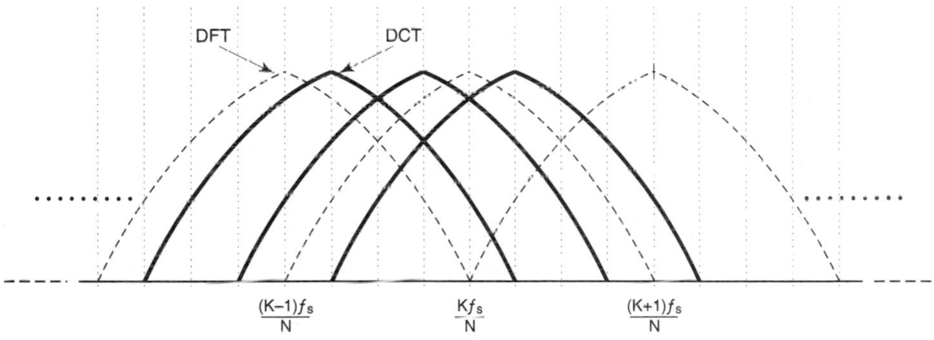

Fig. 7.34
Adjacent filters in the **(DFT)** and **(DCT) schemes** for covering $[0, f_s]$

Representing the filters in terms of ideal brickwall rectangular shapes conceals some of the information regarding frequency "coverage." A more "accurate" description is depicted in Fig. 7.34 for the two cases of N complex filters (**DFT**) and N real filters (**DCT**). Only the principal lobe of the response (see Fig. 7.25) is shown. The width of this lobe, between frequencies where it has its first zero crossings, is actually $2(f_s/N)$ in both cases (arising from the rectangular window $\{h_0(n)\}$. The coverage of the N real filters seems to be more *dense*, with greater overlap between adjacent filter responses. This is to be expected since we are fitting N such responses between $[0, f_s/2]$ with the **DCT** implementation while we are fitting N/2 such responses over the same frequency interval with the **DFT** implementation.

7.3.3 Formal Definitions of the DFT and DCT

Given N samples of a discrete-time signal, $\{x(n); n=0, 1, \ldots, (N-1)\}$ (we have dropped the notation T_s which specifies the sampling interval for notational convenience), the **DFT** is formally defined as the N-point sequence $\{X(k); k=0, 1, 2, \ldots, (N-1)\}$ where the forward and inverse relations are given by

$$X(k) = \sum_{n=0}^{N-1} x(n)\, e^{-j2\pi(\frac{nk}{N})} \tag{7.3.29a}$$

$$x(n) = \frac{1}{N} \sum_{k=0}^{N-1} X(k)\, e^{+j2\pi(\frac{nk}{N})} \tag{7.3.29b}$$

where the usual convention of using upper case for the "transform" has been applied. The definition of Eq. (7.3.29) satisfies one of the conditions we implicitly require of any transformation, namely it is unique and invertible.

The **DCT** is applied to real-valued signals only. The forward and inverse **DCT** relationships are given by

$$X(k) = \sum_{n=0}^{N-1} x(n) \cos\left(\frac{2k+1}{2N} \pi n\right) \tag{7.3.30a}$$

$$x(n) = \frac{1}{N} \sum_{k=0}^{N-1} X(k) \cos\left(\frac{2k+1}{2N} \pi n\right) \tag{7.3.30b}$$

Proofs that the transform definitions of Eq. (7.3.29) and (7.3.30) are consistent are simple and based on the identities:

$$\sum_{n=0}^{N-1} e^{-j2\pi(\frac{nk}{N})} = N\,\delta(k) \tag{7.3.31a}$$

$$\sum_{k=0}^{N-1} \cos\left(\frac{(2k+1)\pi(n+m)}{2N}\right) = 0 \quad (n, m > 0) \tag{7.3.31b}$$

$$\sum_{k=0}^{N-1} \cos\left(\frac{(2k+1)\pi(n-m)}{2N}\right) = N\,\delta(n-m) \qquad (7.3.31c)$$

7.3.4 Relationship between the DFT and the DTFT

Why do we consider the **DFT** sum as a "transform," especially when we have an alternative that we defined in Chapter 3 called the Discrete Time Fourier Transform (**DTFT**)? Recall from there that the **DTFT** is obtained as

$$X(f) = \sum_{n} x(n)\, e^{-j2\pi n f T_s} \qquad (7.3.32)$$

The **DFT** can be viewed as a special case of the **DTFT**. In fact there are two cases where the two are "equivalent" and each case yields considerable insight into the nature of the **DFT** vis-à-vis the **DTFT**.

Case A. Suppose that {x(n)} was a **finite-length sequence** such that

$$x(n) = \begin{cases} 0 \text{ for } n < 0 \\ 0 \text{ for } n \geq N \end{cases} \qquad (7.3.33)$$

That is, {x(n)} has at most N nonzero samples. Then the **DTFT** is the finite sum

$$X(f) = \sum_{n=0}^{N-1} x(n)\, e^{-j2\pi n f T_s} \qquad (7.3.34)$$

If X(f) is computed at N frequencies, equally spaced in the frequency domain, that is, $\{X(f_k);\ f_k = kf_s/N,\ k = 0, 1, \ldots, (N-1)\}$, and we denote $X(f_k)$ by X(k), then

$$X(f_k) = X(k) = \sum_{n=0}^{N-1} x(n)\, e^{-j2\pi\left(\frac{nk}{N}\right)} \qquad (7.3.35)$$

which is none other than the **DFT** sum. Thus, for a finite-length sequence, the **DFT** is indeed "equivalent" to the **DTFT** on the chosen frequency grid.

Case B. Suppose that we know {x(n); n=0, 1, ... , (N–1)}. For sample indices outside this window of N samples we need to make some assumption on the nature of {x(n)}. In Case 1 we assume these samples are zero. For Case 2 we assume that {x(n)} is **periodic with a period of N samples**. Then {x(n)} is indeed determined completely by the N known samples. Because of this periodic behavior, we know from the results derived in Chapter 3, that $\{x(n)\}$ can be written as the sum of N complex sinusoids as

$$x(n) = \sum_{n=0}^{N-1} \alpha_k \exp\left(j2\pi n T_s\left(\frac{k f_s}{N}\right)\right) = \sum_{n=0}^{N-1} \alpha_k e^{j2\pi\left(\frac{nk}{N}\right)} \qquad (7.3.36)$$

Comparing Eq. (7.3.36) with (7.3.30) we can make the association

$$X(k) = N\alpha_k \qquad (7.3.37)$$

That is, the **DFT** sum provides the Fourier series components of the signal $\{x(n)\}$ under the assumption of periodicity.

7.3.5 Convolution and the DFT

In dealing with transforms in a general sense, we see that they have the property of converting convolution in the time domain to multiplication in the transform domain. Does the **DFT** possess such a property? It turns out that the notion of convolution associated with the **DFT** does not represent the output of a regular linear time-invariant system except in certain special cases. The equivalent time domain operation associated with multiplication in the **DFT** transform domain is ***cyclic convolution***.

Given two N-point sequences $\{x(n), n=0, 1, \ldots, (N-1)\}$ and $\{h(n), n=0, 1, \ldots, (N-1)\}$, the cyclic convolution is defined as

$$y(n) = \sum_{k=0}^{N-1} x((n-k))h(k)\,;\, 0 \le n \le (N-1) \qquad (7.3.38)$$

where the notation ((n)) stands for "n modulo N."

The **DFT** and cyclic convolution are related in the following way. If we denote the **DFT** of the N-point sequences $\{x(n)\}$, $\{h(n)\}$, and $\{y(n)\}$ by $\{X(k)\}$, $\{H(k)\}$, and $\{Y(k)\}$, respectively, then

$$Y(k) = X(k)H(k)\,;\, 0 \le k \le (N-1) \qquad (7.3.39)$$

and indeed we can associate convolution in the time domain with multiplication in the transform domain with the caveat that the convolution is to be interpreted as cyclic. The **DMT** scheme described in Section 7.5 explicitly uses this property of the **DFT**.

7.3.6 Symmetry Properties of the DFT

In dealing with the Fourier transform (Section 2.4) we notice that there are certain symmetry properties that can be exploited. The **DFT** exhibits certain symmetries as well.

a. If $\{x(n); n=0, 1, \dots, (N-1)\}$ is real-valued then since

$$X(k) = \sum_{n=0}^{N-1} x(n)\, e^{-j2\pi\left(\frac{nk}{N}\right)} \tag{7.3.40}$$

we can see that $X(N-k)$ is given by

$$X(N-k) = \sum_{n=0}^{N-1} x(n)\, e^{-j\frac{2\pi n}{N}(N-k)} = \sum_{n=0}^{N-1} x(n)\, e^{+j2\pi\left(\frac{nk}{N}\right)} = X^*(k) \tag{7.3.41}$$

Since $x(n)$ is real-valued,

$$X(N-k) = X^*(k);\ \ (X(0) = \text{real-valued}) \tag{7.3.42}$$

b. If $\{x(n)\}$ is purely imaginary then we can show that

$$X(N-k) = -X^*(k);\ (X(0) = 0) \tag{7.3.43}$$

The similarity of the forward and inverse **DFT** can be used to verify that symmetry in the time domain implies a "real" or "imaginary" property in the transform domain.

c. If $\{x(n)\}$ is conjugate symmetric, then $X(k)$ is real. That is,

$$x(n) = x^*(N-n) \Rightarrow X(k) = \text{real-valued} \tag{7.3.44}$$

d. If $\{x(n)\}$ is conjugate assymmetric, then $X(k)$ is imaginary. That is,

$$x(n) = -x^*(N-n) \Rightarrow X(k) = \text{purely imaginary} \tag{7.3.45}$$

The principal use of such properties of symmetry is to aid us in computing the **DFT** in an efficient manner. For example, if we can compute the **DFT** of an N-point complex sequence (efficiently), we can compute two **DFT**s, of two real-valued sequences, simultaneously. To see this, suppose $\{a(n); n = 0, 1, \dots, (N-1)\}$ and $\{b(n); n = 0, 1, \dots, (N-1)\}$ are real-valued N-point sequences. We can construct a complex-valued sequence $\{c(n); n = 0, 1, \dots, (N-1)\}$ as

$$c(n) = a(n) + j\,b(n)\ ;\ n = 0, 1, \dots, (N-1) \tag{7.3.46}$$

Using uppercase notation to represent the **DFT**, we have

$$C(k) = A(k) + j\,B(k)\ ;\ k = 0, 1, \dots, (N-1) \tag{7.3.47}$$

The $\{C(k)\}$ are obtained by the said efficient procedure. Now

$$\begin{aligned} C(N-k) &= A(N-k) + j\,B(N-k) \\ &= A^*(N-k) + j\,B^*(N-k) \end{aligned} \tag{7.3.48}$$

because of property (**a**) above. Therefore, taking the complex conjugate of both sides,

$$C^*(N-k) = A(k) - j\,B(k) \tag{7.3.49}$$

from which we can see that

$$\begin{aligned} A(k) &= 0.5\,[\,C(k) + C^*(N-k)] \\ B(k) &= -\,0.5j\,[C(k) - C^*(N-k)] \end{aligned} \tag{7.3.50}$$

Consequently, by one application of an efficient procedure to compute the **DFT** of a complex-valued N-point sequence, we can, with some additional computation, obtain the **DFT**s of two real-valued sequences. One attraction of the **DFT** is that there are efficient algorithms available for calculating it. Fast Fourier Transform (**FFT**) is the name given to algorithms that are computationally efficient for evaluating the **DFT**. The computational complexity associated with the computation of the **DFT** increases with N, at a rate proportional to N-squared if the calculation is done in a "brute-force" manner, or as Nlog(N) if an **FFT** algorithm is employed. There are several books and articles that have been written on the subject of **FFT**s, a small sample of **FFT** literature is provided in the bibliography.

Given a single N-point real-valued sequence, $\{a(n);\ n = 0, 1, \ldots, (N-1)\}$, it is possible to compute its **DFT** using two applications of an **FFT** of size (N/2) along with some additional manipulations. This is achieved by first recognizing that since

$$\begin{aligned} A(2m) &= \sum_{n=0}^{N-1} a(n)\, e^{-j2\pi(\frac{n(2m)}{N})} \\ &= \sum_{n=0}^{N/2-1} [\,a(n) + a(n+\tfrac{N}{2})\,]\, e^{-j2\pi(\frac{nm}{N/2})} \end{aligned} \tag{7.3.51}$$

we can obtain the **DFT** of $\{a(n)\}$ for even indices $\{k{=}2m;\ m = 0, 1, \ldots, ((N/2)-1)\}$, by taking the (N/2)-point **DFT** of the (N/2)-point sequence

$$b(n) = a(n) + a(n+(N/2));\ n = 0, 1, \ldots, ((N/2)-1) \tag{7.3.52}$$

For odd indices $\{k{=}(2m+1),\ m = 0, 1, \ldots, ((N/2)-1)\}$, the **DFT** of $\{a(n)\}$ is given by

$$\begin{aligned} A(2m+1) &= \sum_{n=0}^{N-1} a(n)\, e^{-j2\pi(\frac{n(2m+1)}{N})} \\ &= \sum_{n=0}^{N/2-1} e^{-j\frac{2\pi n}{N}}\,[\,a(n) - a(n+\tfrac{N}{2})\,]\, e^{-j2\pi(\frac{nm}{N/2})} \end{aligned} \tag{7.3.53}$$

and thus we can obtain this by the (N/2)-point **DFT** of the (N/2)-point sequence

$$c(n) = e^{-j\frac{2\pi n}{N}} [\, a(n) - a(n + \frac{N}{2}) \,] \qquad (7.3.54)$$

Generally speaking, splitting an N-point computation into two (N/2)-point computations reduces the complexity.

7.3.7 The DFT and Bandpass Filter Banks

Earlier we associated the **DFT** computation as the principal computational element for implementing a bank of bandpass filters. The frequency response of the filters was not arbitrary but related to the response of a lowpass "prototype," namely the rectangular window **FIR** filter. In this subsection we extend the notion to more general frequency responses.

Consider the specification of a lowpass filter of nominal passband of f_s/N shown in Fig. 7.35. The passband width is comprised of $(f_s/2N)$ between $[0, f_s/2]$ for "positive" frequency and $(f_s/2N)$ between $[f_s/2, f_s]$ for "negative" frequency. Achieving this response would probably take an **FIR** filter of length much greater than N. Let us assume that these specifications can be met with an **FIR** filter of length M=RN, where R is an integer. We shall denote this "prototype" filter by $H_p(f)$. The impulse response sequence $\{h_p(n)\}$ is assumed to be real-valued.

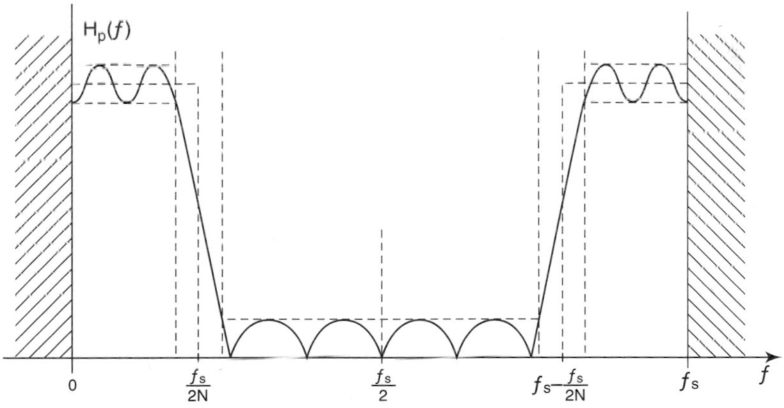

Fig. 7.35
Frequency response of prototype lowpass filter

Then with

$$H_k(f) = H_p(f - \frac{k f_s}{N}) \text{ for } 0 \le k \le (N-1) \qquad (7.3.55)$$

we achieve a covering of the frequency window $[0, f_s]$ as shown in Fig. 7.27.

The complex-valued impulse response of filter $H_k(f)$ is given by

$$h_k(n) = h_p(n)e^{+j\frac{2\pi nk}{N}} \quad \text{for } 0 \leq n \leq (M-1) \tag{7.3.56}$$

and the corresponding output given by (compare Eq. (7.3.8))

$$y_k(mT_s) = \sum_{n=0}^{M-1} x[(m-n)T_s]\, h_p(n)e^{+j\frac{2\pi nk}{N}} \tag{7.3.57}$$

Splitting the summation index, n, into an ordered pair (r,s) by

$$n = rN + s;\ r = 0, 1, \ldots, (R-1);\ s = 0, 1, \ldots, (N-1) \tag{7.3.58}$$

the convolution expression for $y_k(mT_s)$ is

$$y_k(mT_s) = \sum_{s=0}^{N-1} \left\{ \sum_{r=0}^{R-1} x[(m-rN-s)T_s]\, h_p(rN+s) \right\} e^{+j\frac{2\pi sk}{N}} \tag{7.3.59}$$

where we utilize the periodic property of the complex exponential. The term in parentheses is similar to an **FIR** filter of length R. Further, there are N such mini-**FIR** filters, indexed by s. These are "common" to all N outputs and the N outputs are comprised of the **DFT** computation applied to the outputs of these N mini-**FIR** filters. Since each leg of the **DFT** requires an R-point **FIR** filter, the scheme is said to need "R active taps." The impulse response of each of these mini-**FIR** filters is obtained by taking every Nth sample of the M-point **FIR** response $\{h_p(n); n = 0, 1, \ldots, (M-1)\}$:

$$\eta_s(r) = h_p(rN+s);\ 0 \leq r \leq (R-1);\ 0 \leq s \leq (N-1) \tag{7.3.60}$$

Of special interest is when the output $\{y_k(mT_s)\}$, considering it has a bandwidth of (nominally) f_s/N, is undersampled by a factor of N. That is, we are interested only in those indices m such that $m = \nu N$. Hence,

$$y_k(\nu NT_s) = \sum_{s=0}^{N-1} \left\{ \sum_{r=0}^{R-1} x[(\nu-r)NT_s - sT_s]\, \eta_s(r) \right\} e^{+j\frac{2\pi sk}{N}} \tag{7.3.61}$$

This computation can be represented by the polyphase schematic of Fig. 7.36. If the prototype lowpass filter is of length N (R = 1), then each of the mini-FIR filters is simply a multiplicative constant. The polyphase schematic in this case represents the action of "windowing" associated with the **DFT** computation in spectral analysis. If the undersampling ratio is changed from N to, say (N/2) or (N/4), then the output sampling rate of each leg of the **DFT** output is larger than that required for signals of bandwidth f_s/N, but may be appropriate considering the nonideal nature of the filters. Such scenarios are generally called "sliding-window **DFT**" spectral analyzers.

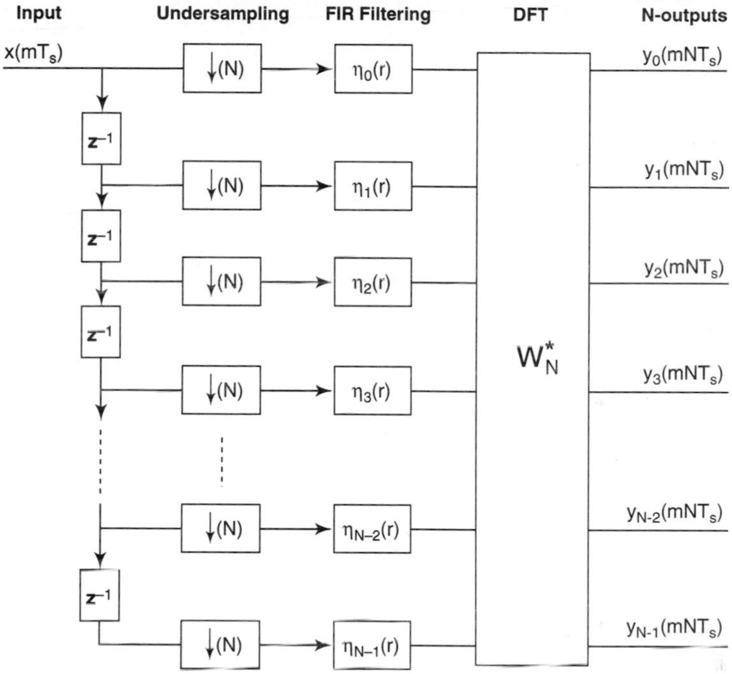

Fig. 7.36
Polyphase representation of bank of N bandpass filters using the **DFT**

7.3.8 The DCT and Bandpass Filter Banks

If we are dealing with real-valued signals and want real-valued outputs, then we can organize the covering as shown in Fig. 7.32. Our prototype lowpass filter then has a narrower bandwidth. With reference to Fig. 7.35, the cutoff frequency is reduced to $f_s/4N$ to allow for a (positive frequency) bandwidth of $f_s/2N$ for each of the N bandpass filters. Using the same notation as before for this case (see Eq. (7.3.23) and (7.3.24)) we get

$$g_k(n) = h_p(n) \cos\left(\frac{2k+1}{2N}\pi n\right) \qquad (7.3.62)$$

The bandpass filter outputs can be written as

$$v_k(mT_s) = \sum_{n=0}^{M-1} x[(m-n)T_s]\, h_p(n) \cos\left(\frac{2k+1}{2N}\pi n\right) \qquad (7.3.63)$$

Simplification of this expression by using a double index scheme is a little more complicated than for the **DFT**. We assume first that R is even (M is an even multiple of N) and thus

$$n = 2rN + s;\ r = 0, 1, \dots, (R/2 - 1); \quad s = 0, 1, \dots, (2N-1) \qquad (7.3.64)$$

from which we can write

$$v_k(mT_s) = \sum_{s=0}^{2N-1} \sum_{r=0}^{(R/2)-1} x[(m-2rN-s)T_s]\, h_p(2rN+s)(-1)^r \Big\} \cos\left(\frac{2k+1}{2N}\pi s\right) \qquad (7.3.65)$$

To simplify the notation, let

$$W(m, s) = \sum_{r=0}^{(R/2)-1} x[(m-2rN-s)T_s]\, h_p(2rN+s)(-1)^r; \quad 0 \le s \le (2N-1)$$

$$= \sum_{r=0}^{(R/2)-1} x[(m-2rN-s)T_s]\, \gamma_s(2rN)(-1)^r; \quad 0 \le s \le (2N-1) \qquad (7.3.66)$$

With this substitution, Eq. (7.3.65) becomes

$$v_k(mT_s) = \sum_{s=0}^{2N-1} W(m, s) \cos\left(\frac{2k+1}{2N}\pi s\right) \qquad (7.3.67)$$

which is similar, but not quite equal to the Discrete Cosine Transform (**DCT**). Utilizing the fact that

$$\cos\left(\frac{2k+1}{2N}\pi N\right) = 0 \qquad (7.3.68)$$

we can show that

$$v_k(mT_s) = \sum_{s=0}^{N-1} V(m, s) \cos\left(\frac{2k+1}{2N}\pi s\right) \qquad (7.3.69)$$

where

$$V(m, 0) = W(m, 0) \quad (s = 0)$$
$$V(m, s) = W(m, s) - W(m, (2N-s)); \quad s = 1, 2, \dots, (N-1) \qquad (7.3.70)$$

Eq. (7.3.69) does express the output of the bank in terms of the **DCT**.

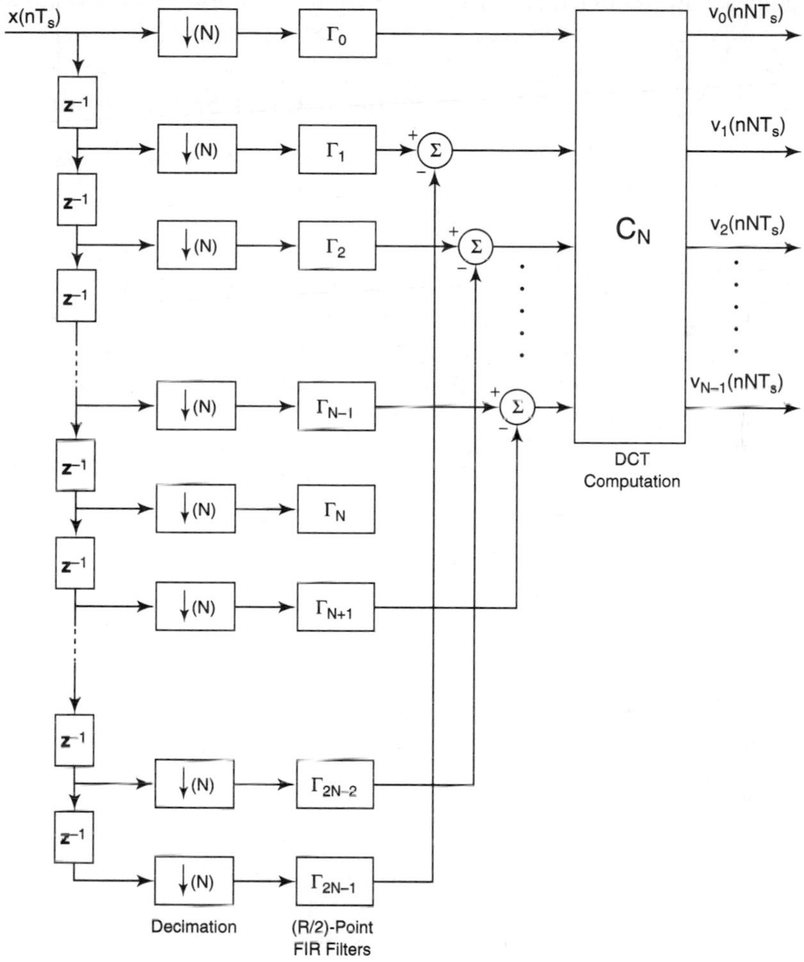

Fig. 7.37
Polyphase schematic of bank of bandpass filters incorporating decimation

If we include the undersampling of the outputs by a factor of N, we can represent the overall operation by the polyphase schematic of Fig. 7.37. The blocks labeled "Γ_k," $k = 0, 1, \ldots, (2N-1)$, in the figure correspond to mini-**FIR** filters that implement Eq. (7.3.66).

The principal reason for describing the bandpass filtering operation via polyphase schematics is that the reverse process can be obtained quite easily. These schematics can be "inverted" by reversing the directions of the arrows, which describe the signal flow, replacing summation points by branches, branching nodes by summing points, interpolation by decimation, and vice versa. The similarity of the forward and inverse transforms is key for simplifying the reversal process.

7.3.9 Concluding Remarks on the Discrete Fourier Transform

The acronym **FFT** is probably the most recognized acronym associated with digital signal processing. Unfortunately it is probably the most misunderstood as well. First, the Fast Fourier Transform is not a transform. Cooley and Tukey, in a pioneering paper on the **FFT**, refer to it as an "algorithm for the machine computation of the complex Fourier series." That is, the **FFT** is an algorithm for the computation of the Discrete Fourier Transform (**DFT**). Second, it is not a single entity. Any algorithm that significantly reduces the complexity of computation of the **DFT** can be called an **FFT**. When one views the **DFT** as a matrix multiplication, it is clear that a "brute-force" approach would require of the order of N^2, written as $O(N^2)$, arithmetic operations to compute the N terms of the **DFT** sequence. That is, as the size of the **DFT** increases, the number of computations increases as N^2. However, the matrix has considerable structure, as evidenced by the properties discussed in Section 7.3.6, and further, the matrix entries are all complex exponentials related to the Nth root of unity. By taking advantage of such structure, algorithms can be formulated that reduce the computational complexity to $O(N\log(N))$, a considerable reduction compared to a general matrix multiplication. For an algorithm to be classified as an **FFT** we require that the number of computations increases as $N\log(N)$ rather than N^2. The bibliography contains several references that describe the theory of **FFT**s, algorithms, and applications.

The principal usage of the **DFT** to date has been in spectral analysis as applied to such diverse endeavors as oil exploration, speech analysis, the search for extra-terrestrial intelligence, and analysis of radar signals to map the surface of planets. The traditional approach for applying the **DFT** sum for spectral analysis consists of segmenting the input signal (to be analyzed) in blocks or "windows" of N points. The **DFT** is computed over each such segment. The terminology applied to each frequency sample is "bin," indicating that the signal component within a narrow bandwidth around the frequency point collects in the "bin." These N-point windows are moved over the entire input record and the **DFT** recomputed each time. The numerical values for each bin are then averaged in some sense over the several calculations to provide an estimate of the (average) strength of the signal component at the bin frequency.

Fig. 7.31 is a representation of this process when the N-point (rectangular) window is moved over the input record by 1 sample at a time. Viewing the bandwidth of the corresponding "bin signal" as a factor of (1/N) of the sampling frequency, one can envision undersampling the bin signal by a factor of as much as N. For this "maximally decimated" case, the window slides over the input record by N samples between each **DFT** computation. This is equivalent to Fig. 7.36 with each mini-**FIR** filter corresponding to a one-tap **FIR** filter, namely just a constant gain term. The gains need not be the same. When the gains are such as to accentuate the "middle" and deweight the endpoints we get the notion of a "windowed spectral analysis" and some of the commonly used windows are the "Hamming window," the "Kaiser window," and so on. The scheme depicted in Fig. 7.36 is sufficiently general to accommodate window lengths greater than N, the size of the transform. In particular, each mini-FIR filter could be R long, corresponding to a weighting window of length RN samples. A variation of the scheme in Fig. 7.36, whereby the downsampling factor is (N/2) achieves a weighting window of length RN with the observation window (of effective length RN) sliding over (N/2) samples of the input record after each **DFT** calculation.

The **DFT** is a widely studied concept and several publications are available that discuss properties and applications. The bibliography lists but a few of several references where additional material on the **DFT** can be found.

7.4 THE TRANSMULTIPLEXER

Long-haul transmission, at least prior to about 1986, has been predominantly analog in nature. Using the techniques of Single Side Band (**SSB**), voice channels are assigned 4-kHz slots (8 kHz considering both positive and negative frequencies) in Frequency Division Multiplex (**FDM**) *assemblies*. The **FDM** "unit" in North America is the ***Group*** signal, comprising 12 channels stacked in a frequency band between 60 and 108 kHz. The **FDM** "unit" in European and rest-of-the-world networks is the ***Supergroup*** comprising 60 channels stacked in the frequency band 312 to 552 kHz. These units can be further stacked in frequency to create ***Mastergroups***, ***Jumbogroups***, and so on, in what is called "the Frequency Division Multiplex Hierarchy," to match the available bandwidth of the radio/cable transmission channel.

In the 1970s there was a rapid trend toward the deployment of digital switching systems. By their very nature, these switching systems are oriented toward digital trunking between switches. The digital trunks employ Time Division Multiplex (**TDM**) *assemblies*. At that time the trunking fabric was still predominantly analog, creating the need to provide bidirectional connectivity between digital trunks and analog trunks. This functionality is achieved using digital signal processing techniques in devices called *transmultiplexers* since they translate between the two multiplexing schemes. With the advent of fiber optic transmission capability, geared to the transmission of digital data (bit-streams), the nature of the long-haul transmission network is now trending toward being completely digital. The need for transmultiplexers has thus diminished to practically zero. Nevertheless, the transmultiplexer embodies several of the principal concepts underlying digital signal processing, which we point out here.

From the viewpoint of telecommunications, the transmultiplexer is no longer actively deployed for new trunks in North America. With the advent of fiber optic transmission of digital information, even long-haul trunk routes are being implemented using fiber. This obviates the need for transmultiplexers. In developing countries, where there still is a great deal of analog trunk capacity while the switching infrastructure is being converted to digital, the deployment of transmultiplexers continues.

From the viewpoint of signal processing, on the other hand, the transmultiplexer is an excellent example for several different concepts. These include, at the heart of the device, the bank of filters implementing the channel separation. It is interesting that the "same" filters can be used in both the **TDM-FDM** and **FDM-TDM** directions. Multirate filters are a natural requirement in the transmultiplexer. The "equivalence" of signals in the analog and digital domains is exemplified. The use of polyphase schematics, and in particular the property of "inversion" is used in the design. The conversion between analog and digital domains used the notion of passband sampling, a technique that is not well publicized in other areas where digital signal processing is employed. The requirement for filters with prescribed zeros of transmission occurs widely but rarely with the specificity as seen in a transmultiplexer and as a result the technique for designing such filters is not as widely studied as it should be. Finally, the transmultiplexer concept can be somewhat generalized to various applications including the transmission of digital data, as exemplified by the Discrete multitone scheme described in the next section (Section 7.5).

In the following discussion we shall ignore much of the fine details that specify the total functionality of a transmultiplexer as a piece of transmission equipment but shall concentrate on the underlying principles. Also, we shall use the North American **DS1**-to-**Group** application for specificity. Extension of the concepts to the **E1**-to-**Supergroup** conversion is straightforward.

7.4.1 Outline of a 24-Channel Transmultiplexer

In its most basic form the transmultiplexer achieves a bilateral conversion between analog and digital formats maintaining the informational content. It is assumed that the traffic consists of voice-band signals. This is depicted by the block diagram of Fig. 7.38. In the **TDM-FDM** direction the transmultiplexer extracts the 24 information channels from the incoming digital stream (**DS1**). These are split into two collections of 12 digital signals each. Each collection of 12 is multiplexed, in a frequency domain manner, into a "digital group" signal which is then converted into the desired analog form using a D/A converter. In the **FDM-TDM** direction, the incoming analog group signal is converted into digital format using an A/D converter. This "digital group" signal is then decomposed into a collection of 12 individual channels. Two such collections, for a total of 24 channels, are multiplexed in a time domain manner to construct the outgoing **DS1** signal.

The principal filtering done within the transmultiplexer is achieved by employing a bank of bandpass filters. In the **FDM-TDM** direction, the digital group signal is filtered by a bank that extracts each channel from its frequency slot. This is called the analysis bank or demodulation bank. The process of undersampling, and the attendant aliasing achieves the demodulation of the signal from its frequency slot down to baseband. In the **TDM-FDM** direction each channel is interpolated and the associated spectral replication positions one replica in the allocated frequency slot. The filter bank then attenuates the unwanted spectral replicates. The filter bank in this direction is called the synthesis bank or modulation bank. The terminology *analysis* and *synthesis* is actually quite recent and stems from the application of such filter banks in coding speech signals by splitting the signal into "bands" and encoding each band separately.

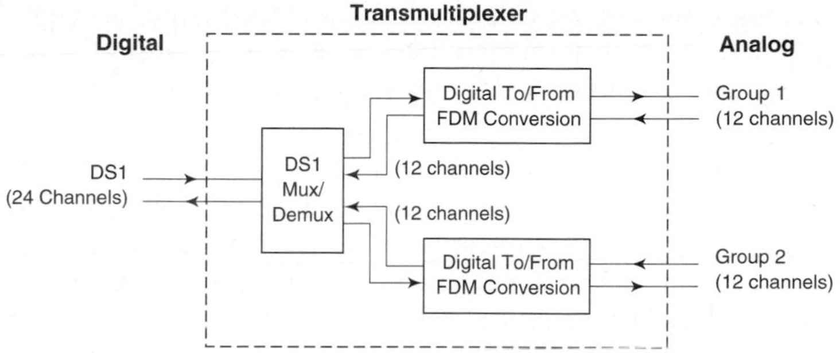

Fig. 7.38
Block diagram of a 24-channel transmultiplexer

Since the processing for the two groups is the same we shall talk about the first group, which takes channels 1 through 12 of the **DS1**. Operations for the second group, channels 13 through 24, are exactly the same.

7.4.2 Analog-to-Digital and Digital-to-Analog Conversion

Conventional wisdom in A/D conversion is to sample the signal at a frequency that is twice the highest frequency present in the signal. With this philosophy, the analog group signal must be sampled at greater than 216 kHz since the upper limit of the group band is 108 kHz. However, the Sampling Theorem specifies that the sampling frequency needs to be equal to the total bandwidth (considering both positive and negative frequencies) of the signal. This is the notion of bandpass sampling in contrast with conventional lowpass sampling. This is depicted in Fig. 7.39. The spectral content of the group signal covers the range from 60 to 108 kHz. Allowing for suitable "guard bands" on either side of this frequency range, we actually sample the signal at f_{s2} = 112 kHz. The anti-aliasing (analog bandpass) filter must achieve the characteristic shown in Fig. 7.39.

The group signal gets translated to baseband because of the inherent nature of sampling, which creates spectral replicates. The spectral composition of the digital group signal, considering a frequency window of $[0, f_{s2}]$ is shown in Fig. 7.40. In the **FDM-TDM** direction we then pick off each channel by the application of bandpass filters operating at a sampling rate of f_{s2} = 112 kHz.

The **TDM-FDM** direction (digital to analog) is similar. We use filter banks operating at a sampling rate of f_{s2} = 112 kHz to pick the appropriate spectral replicate of each channel to construct the digital group signal of the form shown in Fig. 7.40. The digital-to-analog conversion will generate this spectrum and a multitude of replicates. By positioning the passband of the post D/A replicate selection filter, in accordance with the bandpass filter characteristic shown in Fig. 7.39, we can generate the correct analog group signal.

Between [0, 56] kHz there is room for 14 4-kHz slots. Thus, in principle, we can use this scheme for the multiplexing of 14 channels. In practice two channels, numbered 0 and 13, are sacrificed as guard bands to allow for a region of rolloff for the analog filters. If we examine Fig. 7.39, taking into account these guard bands, it is seen that the transition region for the analog bandpass filter, for example around 112 kHz, can extend over a frequency range of 8 kHz, between 108 kHz and 116 kHz.

The wordlength of the converters needs to be at least 14 bits. This estimate is arrived at by the following reasoning. Since the sampling rate is 112 kHz, the A/D quantization noise is spread out over this band and only (1/14) of this noise will be in any channel. The signal power level at the group converter is greater than that of any individual channel by a factor of 12 (almost 14). Since the quantization noise is about 6**B** dB below the peak power level, assuming a quantizer of **B** bits, it follows that quantizing the Group signal to **B** bits is virtually the same as quantizing each channel separately to **B** bits. Since the m-law codec is "equivalent" to a 13-bit uniform codec, from the viewpoint of the smallest stepsize, it follows that we require that **B** 13. Since the transmultiplexer is in series with the transmission path, the additive noise inherent in the device should be *less* than the ambient noise in the channel. Consequently, we set the A/D wordsize to be 14 bits or greater. This reasoning is qualitative and does not take into account certain factors like traffic loading (not all the channels will be carrying live, loud, telephone conversations) but does provide a reasonable estimate. All reputable commercial transmultiplexers use this wordlength in the conversion process.

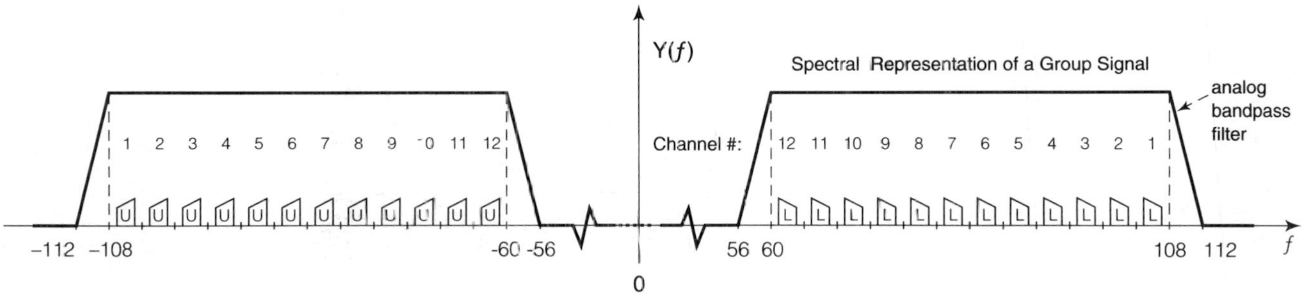

Fig. 7.39
Spectrum of a 12-channel **SSB FDM** group

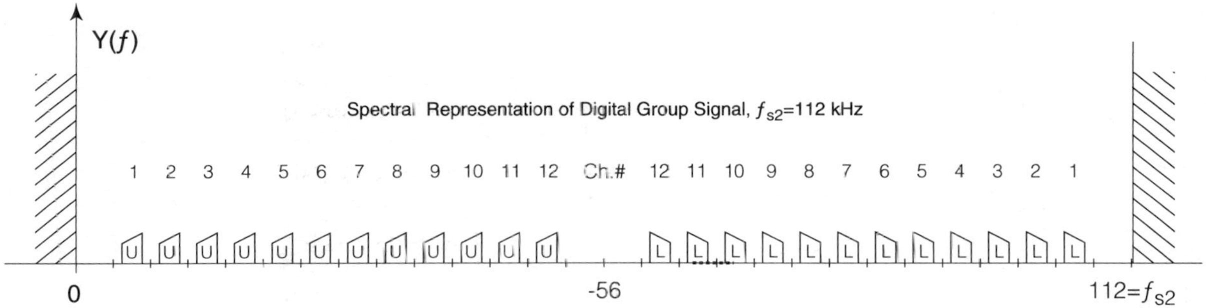

Fig. 7.40
Frequency domain representation of sampled group signal

7.4.3 Overview of the Signal Processing in the Transmultiplexer

The signal processing involved in the **TDM-FDM** direction is depicted in Fig. 7.41.

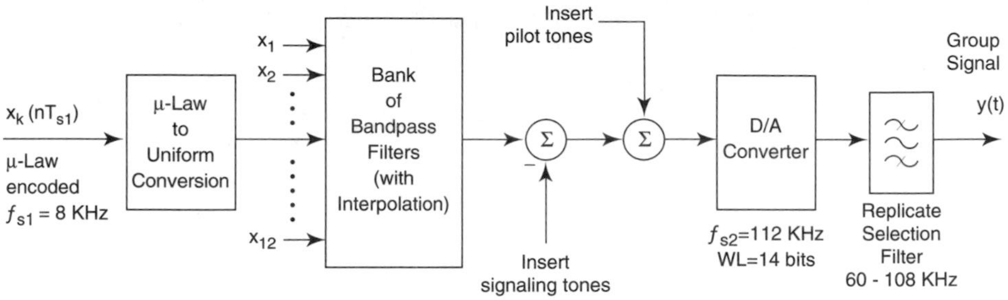

Fig. 7.41
Processing involved in converting **TDM** to **FDM**.

The channel signals are extracted from the **DS1** stream as 24 individual 64-kbps signals. For purposes of processing, the μ-law representation is not conducive to arithmetic operations and thus we convert the μ-law format into a uniform format in accordance with the wordlength used in the digital signal processing elements (greater than 13 bits). Signals from 12 channels are treated together using a bank of filters to create the digital group signal. Channels 1,3,5, ... ,11 have an additional operation that will be discussed later.

The diagram shows the addition of additional signals called pilots and signaling tones. Pilots are single-frequency signals that are used to convey certain information to the far end and also serve as a means for level control of the overall group signal. The frequencies employed are of the form **xx**.08 kHz, where **xx** is 84 or 104. Signaling tones are used to represent, for each channel, the supervisory state (on-hook, off-hook, etc.). These tones, if we view them as part of the channel signal, are placed at 3825 Hz. This technique, called channel associated signaling (**CAS**), because the supervisory information uses the same path as the channel itself, is not always used. An alternative, called Common Channel Interoffice Signaling, or **CCIS** for short, uses separate facilities to convey the signaling state of the channel and these signaling tones are not present in the group signal.

The signal processing in the **FDM-TDM** direction is shown in Fig. 7.42.

The functions are dual to those in the other direction. The digital group signal is bandpass filtered and undersampled to yield the 12 information channels. Channels 1, 3, 5, ... , 11 have an additional process that is not explicitly indicated. At the 8-kHz sampling rate the signal is "cleaned up" whereby the remnants of any power due to the pilot and/or the signaling tones is removed. The uniform code used in the arithmetic is converted into the μ-law format for subsequent assembly into the **DS1**.

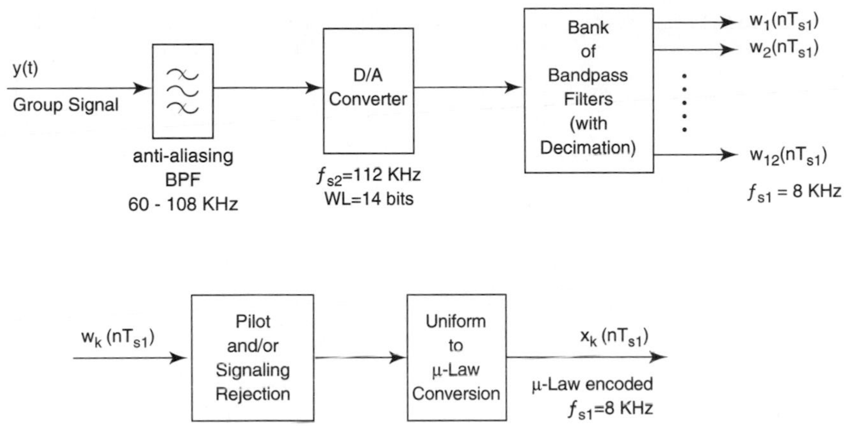

Fig. 7.42
Processing involved in converting **FDM** to **TDM**

7.4.4 The Frequency Response of the Bandpass Filter Bank

The bandpass filters implement the required channel separation. In the **FDM-TDM** direction the filtering extracts a channel from the assembly, whereas in the **TDM-FDM** direction the filter removes replicates prior to the summing of the channels to create the **FDM** assembly. The bandpass nature of the filter frequency characteristic is depicted in Fig. 7.43.

In either direction, the filters are a bank of filters with center frequencies given by

$$f_k = \frac{(2k+1)\,f_{s2}}{4N}\ ; \ N = 14;\ 0 \le k \le (N-1) \qquad (7.4.1)$$

of which the filters $G_0(f)$ and $G_{13}(f)$ are associated with the guard bands. The nominal bandwidth of each filter is 4 kHz (positive frequencies) and, being real-valued, exhibit the symmetry indicated (recall that the range $[-f_{s2}/2,0]$ maps into $[f_{s2}/2,\ f_{s2}]$).

An immediate observation is that while channels 2, 4, ... , 12 map correctly to and from the digital Group signal, channels 1, 3, ... , 11 do not because the incorrect sideband is chosen. This requires that for the odd-numbered channels we need to flip the sidebands around. Achieving this is surprisingly easy. Suppose we define

$$u_k(nT_{s1}) = (-1)^n x_k(nT_{s1}) \qquad (7.4.2)$$

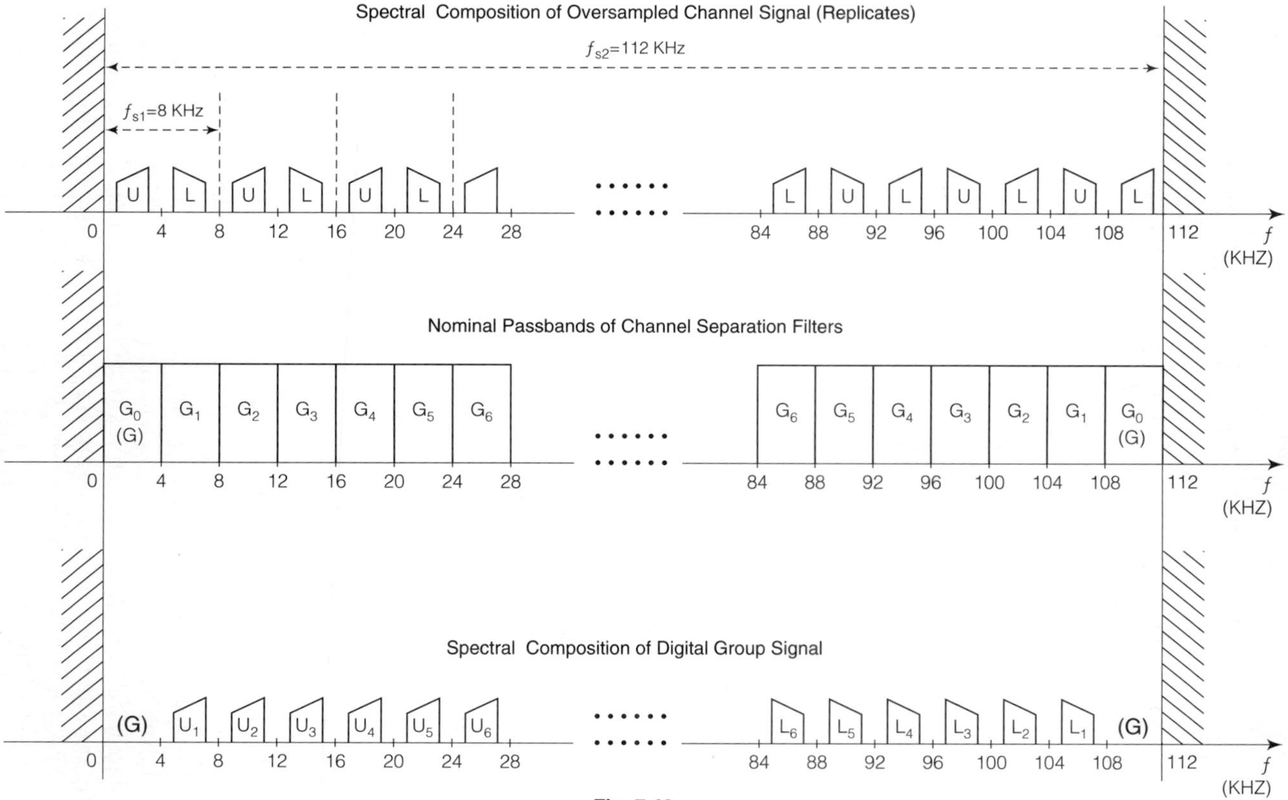

Fig. 7.43

Bandpass filter representation of transmultiplexer operation

then $\{u_k(nT_{s1})\}$ is the side-band flipped version of $\{x_k(nT_{s1})\}$. To prove this, we can write

$$u_k(nT_{s1}) = x_k(nT_{s1})\, e^{j2\pi n T_{s1} f_\alpha} \text{ where } f_\alpha = \frac{f_{s1}}{2} \qquad (7.4.3)$$

Since multiplication in the time domain is convolution in the frequency domain, the action of multiplication by the "4-kHz tone" (f_{s1} = 8 kHz), which is the same as the negation of alternate samples, the spectrum of $\{u_k(nT_{s1})\}$ is given by

$$U_k(f) = X_k(f - \frac{f_{s1}}{2}) \qquad (7.4.4)$$

When viewed in the frequency domain, as shown in Fig. 7.44, this frequency shift corresponds to flipping the sidebands.

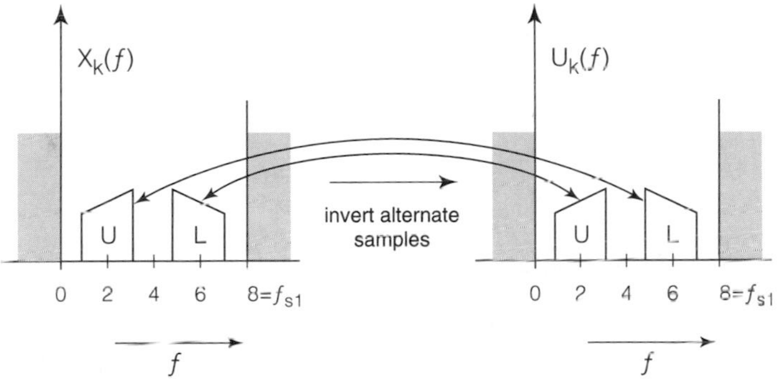

Fig. 7.44
Swapping upper and lower sideband by alternate sample inversion

From the viewpoint of the bandpass filters, the chain of **TDM-FDM-TDM** appears as shown in Fig. 7.45. In drawing shown in the figure we have ignored the fact that the group signal gets converted from digital to analog, is processed in the analog domain by the transmission plant, and reconverted back to digital. We have implicitly assumed that this processing is not supposed to alter the information content of the signal. The blocks labeled "**INV**" refer to the process of inverting alternate samples for the odd-numbered channels.

The two extra signal, $\{u_0(nT_{s1})\}$ and $\{u_{13}(nT_{s1})\}$ are "phantom" signals. In implementing the filter banks using the **DCT**, we use a 14-point **DCT** and thus will need to introduce these channels. In the synthesis filter bank these are set to zero. In the analysis bank these are discarded.

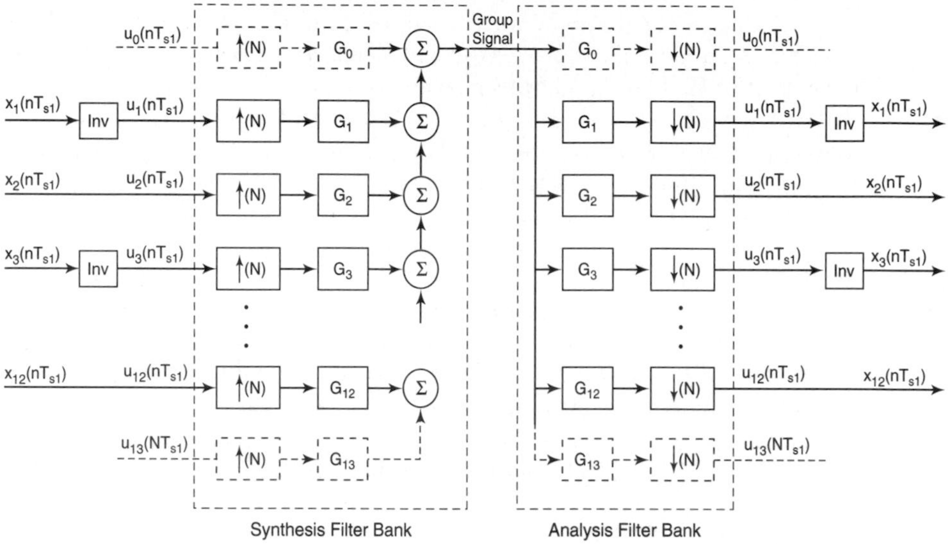

Fig. 7.45
The transmultiplexer as banks of bandpass filters

In order that the **DCT** be applicable for the implementation of the filter bank, it is necessary that all the filters be the "same," i.e., shifted versions of the same prototype lowpass filter. The manner in which the requirements for this prototype filter, $H_p(f)$, are obtained is depicted in Fig. 7.46.

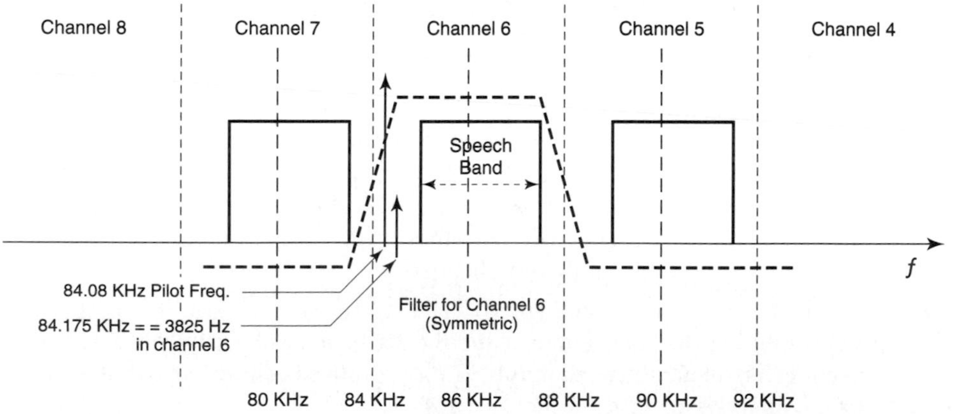

Fig. 7.46
Frequency response of channel separation bandpass filter

Consider channel 6, which is nominally placed between 84 and 88 kHz. The frequency band allocated to a channel, 300 Hz to 3400 Hz, appears in the figure in reverse order. This is an artifact since the band shown is in the range $[f_{s2}/2, f_{s2}]$, which corresponds to negative frequency. The filter must pass this band of frequencies but attenuate the same band from adjacent frequency slots (other channels). Recognize that the speech band is not symmetrically placed within the 84 through 88 kHz frequency slot. Since $H_p(f)$ is a real-valued filter, and since the center frequency of $\mathbf{G}_6$ is 86 kHz (midpoint of 84 and 88), this means that the passband extends between (86+1.7) and, from considerations of symmetry, (86–1.7) kHz. The transition band is defined by the edges, 300 Hz and 3400 Hz, of the adjacent slots. Consequently, the stopband edge on the left is at (84–0.3) kHz and, because of symmetry, the stopband edge is at (88+0.3) kHz. This translates into the requirements for the prototype filter shown in Fig. 7.47. The stopband attenuation is a measure of the *crosstalk* rejection and must be in excess of 70 dB. The passband ripple is a measure of how much frequency distortion is introduced into the channel itself and is usually kept to less than 0.1 dB. These requirements can be attained by a linear-phase **FIR** filter of length 560. We shall denote the length of this filter as M where M = RN. N is the number of channels of the filter bank, 14, and R > 1 indicates that the filter length is greater than the size of the **DCT** itself.

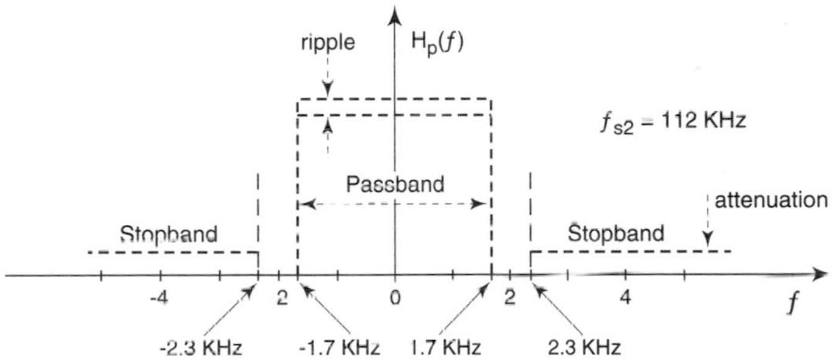

Fig. 7.47
Frequency response specification of lowpass prototype

7.4.5 Implementation of the Synthesis Filter bank

In Section 7.3 we saw that the analysis filter bank can be implemented using the **DCT** in an efficient manner. In Section 7.2 we stated that the synthesis bank could be obtained by manipulating the polyphase schematic of the analysis bank. Here we show that this is indeed true by manipulating the expressions for the output of the synthesis bank.

Denote by $\{y_k(mT_{s2})\}$ the output of the filter $\mathbf{G}_k$ in Fig. 7.45. The impulse response of the filter is given by

$$g_k(m) = h_p(m)\cos\left(\frac{2k+1}{2N}\pi m\right) \tag{7.4.5}$$

To simplify the notation we will drop the notation T_{s2} that indicates the sampling rate (112 kHz). We next denote the output of the interpolator which increases the sampling rate of $\{u_k(mT_{s1})\}$, as

$$w_k(m) = u_k(mT_{s1})\,\delta(m - nN) \tag{7.4.6}$$

Then $\{y_k(m)\}$ is given by

$$y_k(n) = \sum_{m=0}^{M-1} w_k(n-m)\,h_p(m)\cos\left(\frac{2k+1}{2N}\pi m\right) \tag{7.4.7}$$

The Group signal output is composed of the sum of the outputs of the individual filters of the bank,

$$y(n) = \sum_{k=0}^{M-1} y_k(n) \tag{7.4.8}$$

The time index, n, is written as a double index as

$$n = (aN + b);\ 0 \le b \le (N-1) \tag{7.4.9}$$

indicating that between every two sampling epochs of the lower rate, f_{s1}, there are (N–1) epochs at the high sampling rate $f_{s2} = Nf_{s1}$. With this substitution,

$$y_k(\alpha N+\beta) = \sum_{m=0}^{M-1} w_k(\alpha N+\beta-m)\,h_p(m)\cos\left(\frac{2k+1}{2N}\pi m\right) \tag{7.4.10}$$

We split up the summation index in a similar manner

$$m = (rN + s);\ s = 0, 1, \ldots, (2N-1);\ r = 0, 1, \ldots, (R-1) \tag{7.4.11}$$

With this notation, and recognizing that $w_k(n)$ will be zero if n is not a perfect multiple of N, we get

$$y_k(\alpha N+\beta) = \sum_{r=0}^{R-1} u_k[(\alpha-r)T_{s1}]\,h_p(rN+\beta)\cos\left(\frac{(2k+1)\pi(rN+\beta)}{2N}\right) \tag{7.4.12}$$

If we split the index r into odd and even values by

$$\begin{aligned} r &= 2\eta \quad (\text{even } r);\ \eta = 0, 1, \ldots, (R/2 - 1) \\ r &= 2\gamma+1 \quad (\text{odd } r);\ \gamma = 0, 1, \ldots, (R/2 - 1) \end{aligned} \tag{7.4.13}$$

then we can combine the expression for y(n) in terms of $y_k(n)$ from Eq. (7.4.8) with Eq. (7.4.12) and, after much algebra, get

$$\begin{aligned} y(\alpha N+\beta) = &\sum_{k=0}^{N-1}\sum_{\eta=0}^{R/2-1} u_k[(\alpha-2\eta)T_{s1}]h_p(2\eta N+\beta)\cos(\frac{(2k+1)\pi\beta}{2N})(-1)^{\eta} \\ &- \sum_{k=0}^{N-1}\sum_{\gamma=0}^{R/2-1} u_k[(\alpha-2\gamma-1)T_{s1}]h_p[(2\gamma+1)N+\beta]\cos(\frac{(2k+1)\pi(N-\beta)}{2N})(-1)^{\gamma} \end{aligned} \tag{7.4.14}$$

The manipulation to derive Eq. (7.4.14) uses the results

$$\cos[(2k+1)\eta\pi + \theta] = (-1)^{\eta}\cos(\theta) \tag{7.4.15a}$$

$$\cos(\frac{(2k+1)\pi(N-\beta)}{2N}) = -\cos(\frac{(2k+1)\pi(N+\beta)}{2N}) \tag{7.4.15b}$$

$$\cos(\frac{(2k+1)\pi(N+\beta)}{2N}) = 0 \text{ for } \beta = 0 \tag{7.4.15c}$$

We can recognize the **DCT** computation embedded in Eq. (7.4.14). In particular, we can write

$$U_{\beta}(\kappa) = \sum_{k=0}^{N-1} u_k(\kappa T_{s1})\cos(\frac{2k+1}{2N}\pi\beta);\ 0 \le \beta \le (N-1) \tag{7.4.16}$$

which, other than a constant, is the inverse **DCT** summation. Defining a quantity $U_N = 0$, we can substitute $U_{\beta}(\ )$ into Eq. (7.4.14) to get

$$y(\alpha N+\beta) = \sum_{\eta=0}^{R/2-1} U_{\beta}(\alpha-2\eta)\, p_{\beta}(2\eta) - \sum_{\eta=0}^{R/2-1} U_{N-\beta}(\alpha-2\eta-1)\, q_{\beta}(2\eta)$$

$$p_{\beta}(2\eta) = h_p(2\eta N+\beta)(-1)^{\eta};\ q_{\beta}(2\eta) = h_p[(2\eta+1)N+\beta](-1)^{\eta} \tag{7.4.17}$$

This result shows that the group digital signal can be obtained by first applying the (inverse) **DCT** summation to the N inputs $\{u_k(mT_{s1});\ k=0, 1, \ldots, 13\}$, and then applying mini-**FIR** filters to the N **DCT** outputs and combining the outputs of these filters. The polyphase schematic for the synthesis bank is shown in Fig. 7.48.

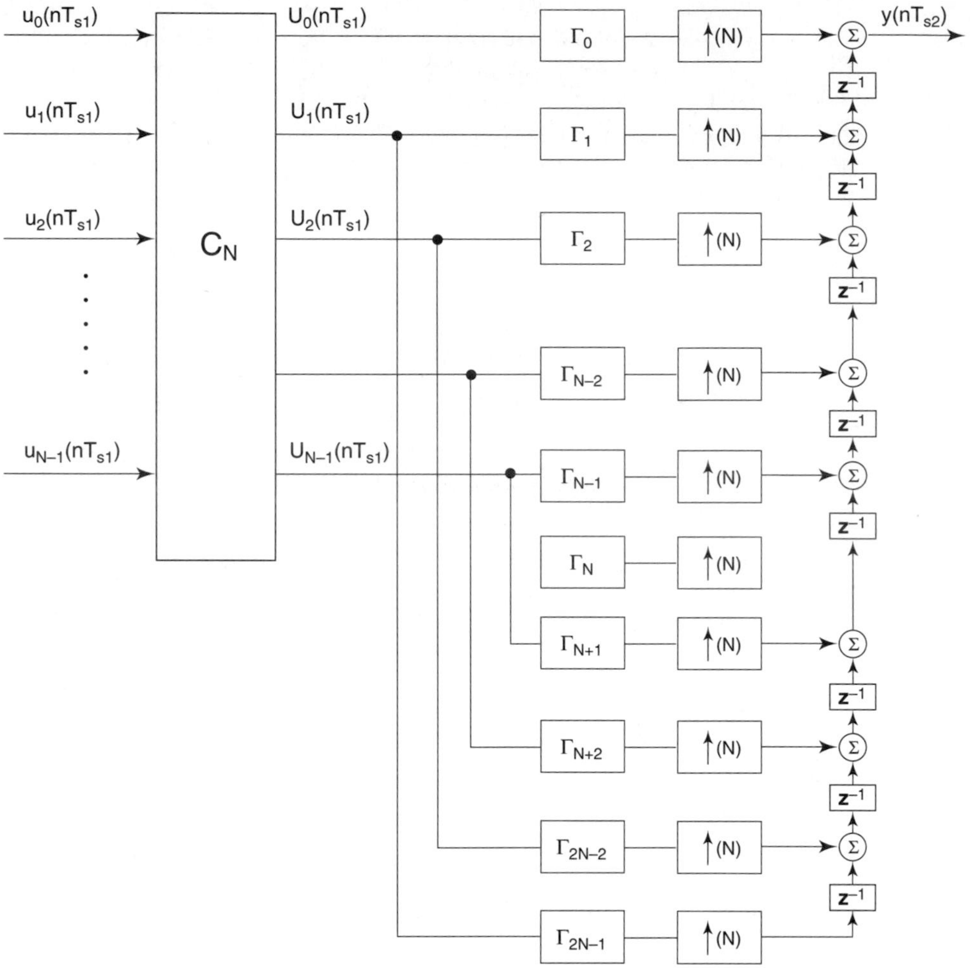

Fig. 7.48
Polyphase representation of bandpass filter bank in **TDM-FDM** direction

Examination of the polyphase schematic indicates that it is indeed the dual of the analysis bank described in Section 7.3 (Fig. 7.37). Further it can be discerned that the **FIR** filters are each of order (R/2) and operate on sample-delays equivalent to a sampling rate of $f_{s1}/2$, half the lower sampling frequency. This observation indicates that, effectively, for each leg of the **DCT** operation in either the analysis or synthesis banks, there are R multiply-accumulate operations, corresponding to the terminology *R active taps.*

If the analysis bank is implemented by "brute-force" techniques, each filter Γ_k requires an **FIR** filter of length M = RN. Some reduction in complexity may be obtained by recognizing that the filter is accompanied by decimation (interpolation in the case of the synthesis bank) and thus only those outputs that are actually required need be computed (in the synthesis bank the zero-valued samples do not require any associated computation). Nevertheless, it can be shown that the **DCT** approach is the most efficient, especially since there are **fast** algorithms for the **DCT** as there are for the **DFT**.

7.4.6 Per-Channel Filters in the Transmultiplexer

In the **FDM-TDM** direction, some additional filtering is often required to remove the remnants of the pilot tones and signaling tones. Examination of Fig. 7.46 shows that these tones appear in any particular channel at the frequencies 3825 Hz, 3920 Hz, 175 Hz, and 80 Hz. The 3825 Hz and 3920 Hz for the signaling and pilot tones appear in the channel of interest directly. The 175 Hz and 80 Hz tones appear because of aliasing from an adjacent channel. These frequencies lie within the transition band of the channel separation filter and are thus not completely attenuated.

A suitable frequency response requirement, to remove all four tones, is shown in Fig. 7.49. This is not required for all channels since there are a limited number of pilot tones applied. The zeros at 175 Hz and 3825 Hz that remove the remnants of the signaling tone are required if the signaling scheme is not channel associated signaling. We cover the design of such filters, with flat passbands and prespecified zeros of transmission in Chapter 9.

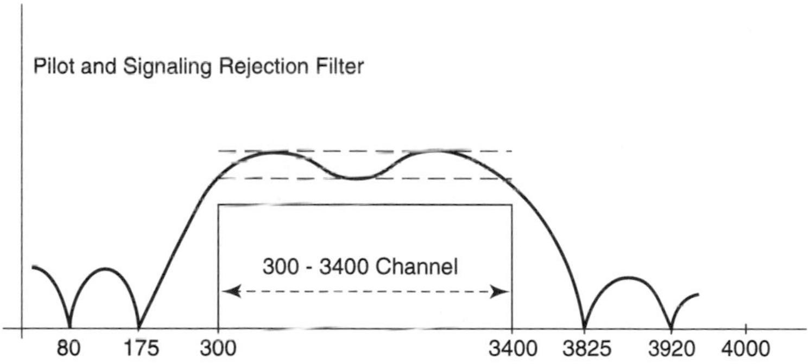

Fig. 7.49
Per-channel filter to remove pilots and signaling tones (f_{s1}=8 kHz)

7.4.7 Concluding Remarks on the Transmultiplexer

Though obsolete in practice, the transmultiplexer embodies several aspects of signal processing. These include A/D and D/A conversion, digital filtering, sampling rate conversions, polyphase representations, and so on. These concepts have found new applications and Vaidyanathan [1.5] describes some of these. Additional material can be found in the publications listed in Section 7.7.2.3 of the bibliography.

7.5 DISCRETE MULTITONE TRANSMISSION (DMT)

Consider the problem of transmitting data over a channel that has severe frequency distortion. That is, the frequency response of the channel is far from "flat," such as the frequency response $H_C(f)$ depicted in Fig. 7.50. The "bandwidth" of the channel is shown as f_C, where f_C is somewhat arbitrary. For frequencies less than f_C, the amplitude response of the channel has "satisfactory" attenuation; beyond f_C the attenuation is too great for substantive transmission of information.

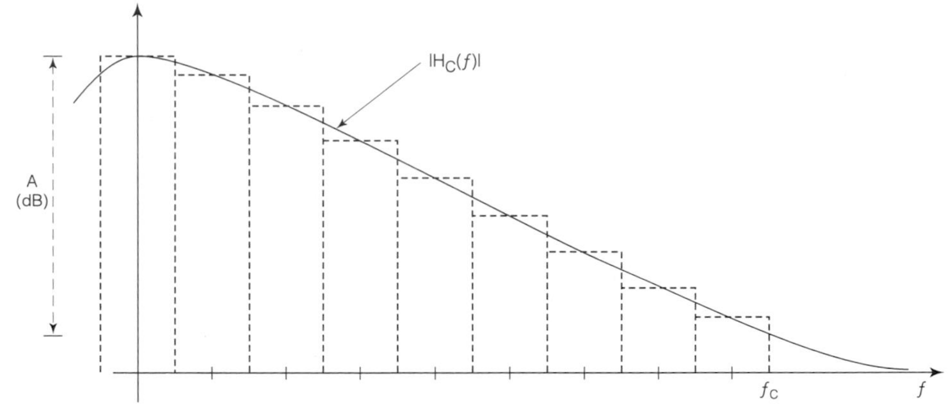

Fig. 7.50
Frequency response of a far-from-ideal channel

In order to transmit data reliably over this channel we need equalization, which is clearly of a highpass nature. The problem is that this equalization enhances any additive noise in the channel. Also, if the range A (dB) is large, the equalization filter may not be practically feasible. The ***Discrete Multitone Transmission*** (**DMT**) scheme addresses this problem by dividing the channel into several "mini-channels" by breaking up the channel bandwidth into several manageable chunks. Over each of these bandpass channels, the channel response is expected to be approximately "flat" and thus equalization is not required.

The **DMT** scheme has generated considerable interest since it is the most viable scheme for transmitting high speed data over subscriber loops. We shall consider just the principle of the scheme to show how the **DFT** is employed. Details of the scheme can be found in the literature, especially in the various contributions made by Cioffi et. al., to the **T1E1** subcommittee that is responsible for setting standards for the U.S. telecommunications industry.

Consider M individual signals at a sampling rate of $f_{s1} = 1/T_{s1}$, $\{a_k(nT_{s1});\ k = 0, 1, \ldots, (M-1)\}$. From this we construct $N = 2M$ signals in the following manner.

$$\alpha_k(nT_{s1}) = a_k(nT_{s1}) \text{ for } 0 \leq k \leq (M-1)$$
$$\alpha_k(nT_{s1}) = a_{N-k}(nT_{s1}) \text{ for } M \leq k \leq (N-1) \tag{7.5.1}$$

We shall assume that $\{a_0(nT_{s1})\}$ is real-valued. Examination of Eq. (7.5.1) indicates that there are only (N/2) different signals, the remaining (N/2) being created by symmetry. From $\{a_k(nT_{s1})\}$ we generate a new set of signals $\{x_i(nT_{s1})\}$ by taking the (inverse) **DFT** sum, using k as the index of summation. That is,

$$x_i(nT_{s1}) = \sum_{k=0}^{N-1} \alpha_k(nT_{s1}) e^{+j\frac{2\pi ik}{N}} \tag{7.5.2}$$

Because of the symmetry imposed on the signals $\{a_k(nT_{s1})\}$, we are sure that $\{x_i(nT_{s1})\}$ are all real-valued.

Now consider the process of interpolation and interleaving of the $\{x_i(nT_{s1})\}$ to generate a new, composite, signal, $\{x(mT_{s2})\}$, with the underlying sampling rate of $f_{s2} = N f_{s1}$. This is achieved by using the following index mapping. Writing m as

$$m = (nN + i); \; i = 0, 1, \ldots, (N-1) \tag{7.5.3}$$

we identify $\{x(mT_{s2})\}$ as

$$x((nN+i)T_{s2}) = x_i(nNT_{s2}) = x_i(nT_{s1}) \tag{7.5.4}$$

The operations involved in creating $\{x(mT_{s2})\}$ are depicted in Fig. 7.51.

The signal $\{x(mT_{s2})\}$ is converted into analog form using a D/A converter and this analog signal, $x_A(t)$, is launched over the channel. The received signal, $y_A(t)$, is converted into digital form using an A/D converter to provide $\{y(mT_{s2})\}$. The sampling rate involved, f_{s2}, is assumed to be $2f_C$. The modification of the signal is diagramed in Fig. 7.52. The block labeled **DAC** includes the D/A conversion and the replicate rejection filters; the block labeled **ADC** includes anti-aliasing filtering and the A/D conversion.

Denoting by $H(f)$ the composite frequency response, we get

$$Y(f) = H(f)\,X(f); \; H(f) = H_1(f)H_C(f)H_2(f) \tag{7.5.5}$$

and if the channel response is indeed flat, and the analog filters in the conversion process are "ideal," the $\{y(mT_{s2})\}$ are nominally the same as $\{x(mT_{s2})\}$. In this case, if we take N consecutive samples of the signal $\{y(mT_{s2})\}$ and apply the **DFT** summation, we regain the signals $\{x_k(nT_{s1})\}$. That is, if we define

$$m = (nN + i); \; y_i(nT_{s1}) = y(mT_{s2}) = y((nN+i)T_{s2}) \tag{7.5.6}$$

and compute

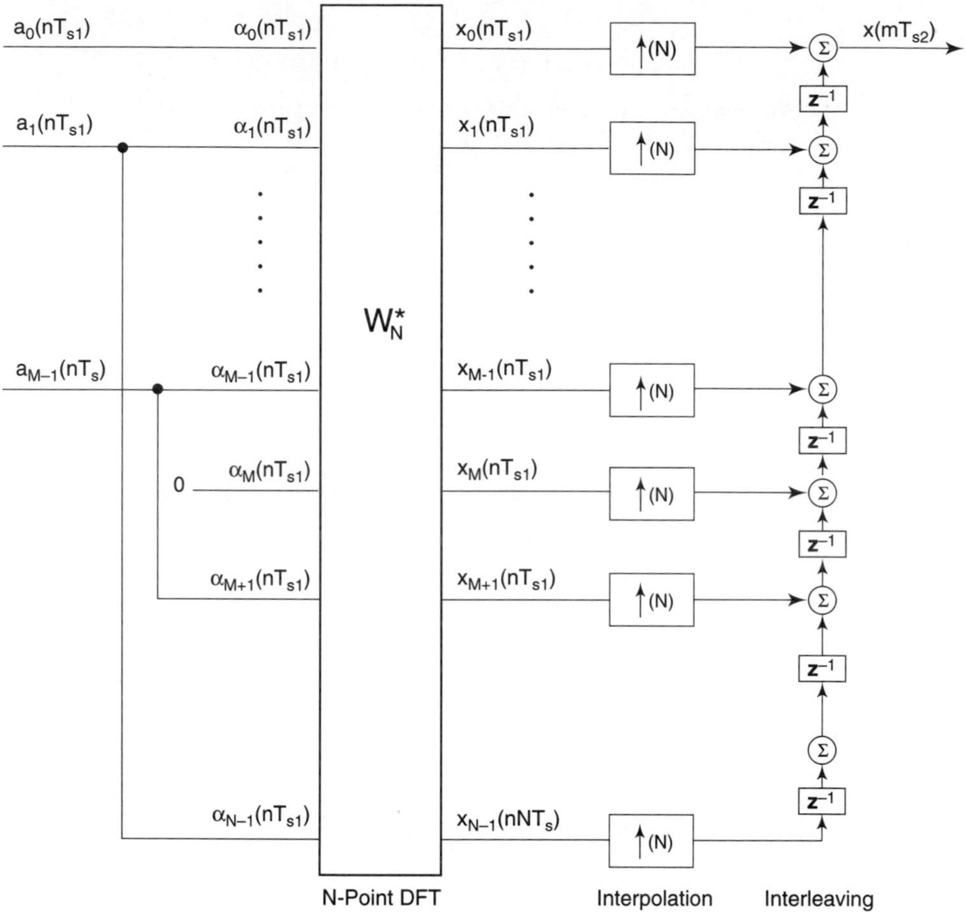

Fig. 7.51
Creation of N real-valued samples of **DMT** output from M complex input samples (M=N/2)
(a_0 is real-valued)

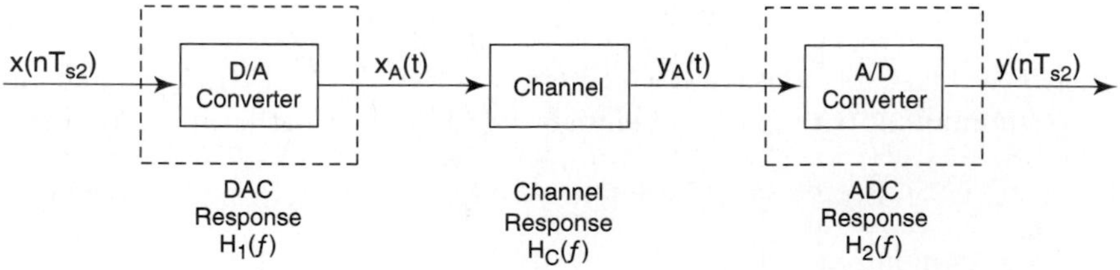

Fig. 7.52
Filtering encountered by the **DMT** signal between transmitter and receiver

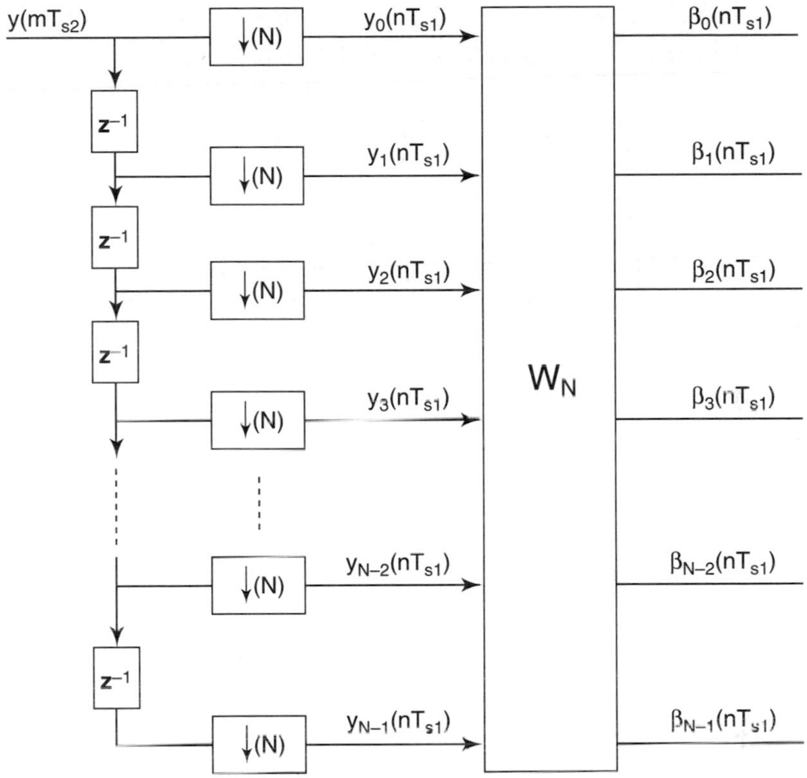

Fig. 7.53
Analysis of **DMT** receive signal into N=2M subsymbols

$$\beta_k(nT_{s1}) = \frac{1}{N}\sum_{i=0}^{N-1} y_i(nT_{s1})e^{-j\frac{2\pi ik}{N}} \tag{7.5.7}$$

then the $\{\beta_k(nT_{s1})\}$ are the nominally the same as the set $\{\alpha_k(nT_{s1})\}$. The creation of the signals $\{\beta_k(nT_{s1})\}$ from $\{y(mT_{s2})\}$ is depicted in Fig. 7.53.

Why would the set $\{\beta_k(nT_{s1})\}$ differ from $\{\alpha_k(nT_{s1})\}$? There are several reasons for this. To illustrate some of the causes for the difference, we introduce the model shown in Fig. 7.54.

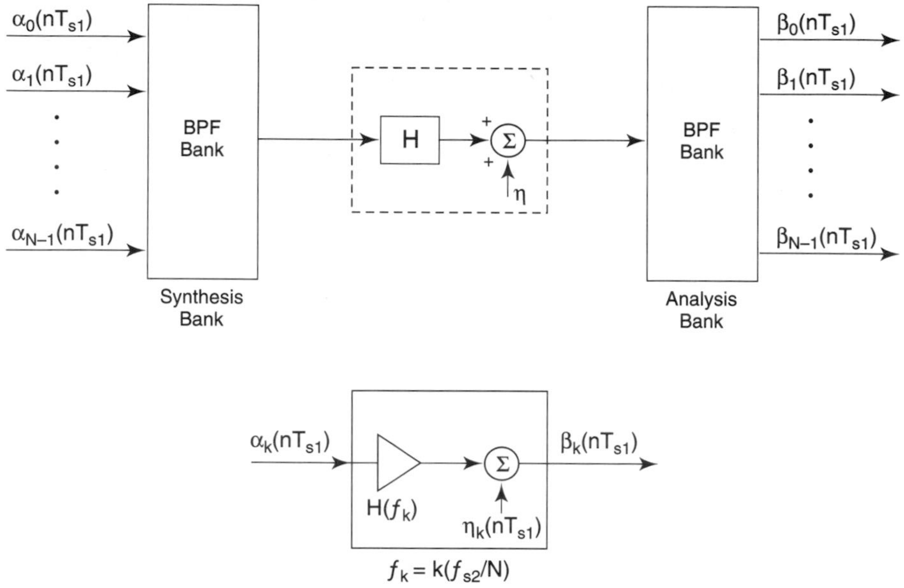

Fig. 7.54
Model of **DMT** as N separate channels

The effect of the analog filters is combined into $H(f)$ and the impact of A/D conversion noise, aliasing, and any additive noise in the channel is modeled as an additive component $\{\eta(nT_{s2})\}$. These are the principal reasons the $\{\beta_k\}$ would be different from the $\{\alpha_k\}$. The interpretation of the **DFT** as a bank of filters allows us to visualize the **DFT** computations at the "transmitter" and the "receiver" as a synthesis bank and an analysis bank, respectively. With this interpretation, we can visualize each $\{\beta_k(nT_{s1})\}$ being transmitted over a bandpass channel of bandwidth $f_{s2}/N = f_{s1}$, centered at $f_k = k(f_{s2}/N)$. To a first approximation, these bandpass filters are "flat" over their passband and modeled as a "gain" of $H(f_k)$. Each link between $\{\alpha_k(nT_{s1})\}$ and $\{\beta_k(nT_{s1})\}$ can be treated as a channel as depicted in Fig. 7.54.

This model is indeed considered to be accurate if the bandpass filters are "ideal" and can ensure that there is no "crosstalk" between subchannels. Treating the channel as "flat" over the passband is allowed if the bandwidth of each channel is small (large N). Since the bandpass filters implied by the **DFT** are shifted version of a prototype lowpass filter corresponding to an N-point rectangular window, we know that the bandpass filters are far from ideal. Consequently, the model depicted in Fig. 7.54 is not entirely accurate.

Assume that the noise component is ignored. Then the composite response $H(f)$ can be expressed as a filter with (infinite) impulse response $\{h(nT_{s2})\}$ and thus

$$y(mT_{s2}) = \sum_{k=0}^{\infty} x[(m-k)T_{s2}]\, h(kT_{s2}) \qquad (7.5.8)$$

Converting the single indices into double indices by

$$m = (nN + r)\ ;\ k = (iN + s)\ ;\ s,r = 0, 1, 2, \ldots, (N-1) \qquad (7.5.9)$$

Eq. (7.5.8) can be manipulated to yield

$$y_r(nT_{s1}) = \sum_{i=0}^{\infty} \sum_{s=0}^{N-1} x[(n-i)T_{s1} + (r-s)T_{s2}]\, h_s(iT_{s1}) \qquad (7.5.10)$$

where we define N subfilters $\{h_s(iT_{s1})\}$ by

$$h_s(iT_{s1}) = h[iT_{s1} + sT_{s2}] \qquad (7.5.11)$$

Taking into account the manner in which $\{x(mT_{s2})\}$ is built up of $\{x_r(nT_{s1})\}$ by interpolation and interleaving of samples, we can write

$$y_r(nT_{s1}) = \sum_{i=0}^{\infty} x_r[(n-i)T_{s1}]\, h_r(iT_{s1}) \qquad (7.5.12)$$

and, with some manipulation, we can express $\{\beta_k(nT_{s1})\}$ as

$$\beta_k(nT_{s1}) = \frac{1}{N} \sum_{s=0}^{N-1} \sum_{i=0}^{\infty} \alpha_s[(n-i)T_{s1}] \left\{ \sum_{r=0}^{N-1} h_r(iT_{s1}) e^{j\frac{2\pi r(s-k)}{N}} \right\} \qquad (7.5.13)$$

Note that $\beta_k(nT_{s1}) \neq \alpha_k(nT_{s1})$ because of two effects. The notion of interchannel crosstalk is demonstrated by $\beta_k(nT_{s1})$ having some components related to the $\alpha_r(nT_{s1})$ with $r \neq k$. Second, we have the notion of intrachannel crosstalk, or inter symbol interference, by noting that $\beta_k(nT_{s1})$ has components from $\alpha_k((n-i)T_{s1})$ for i=1, 2, ... The underlying precept of the **DMT** scheme is that we can remove these crosstalk contributions.

The **DMT** scheme is applicable if the channel has a finite impulse response. Suppose the **ADC** and **DAC** operated at a sampling rate of $f_s = ((N+S)/N)f_{s2}$, with $T_s = 1/f_s$. We suppose now that the impulse response length of the channel (plus all the analog filters) was less than (or equal to) S samples at this sampling frequency. That is,

$$H(f) \rightarrow \{h(mT_s)\};\ 0 \le m \le (S-1);\ T_s = \frac{N}{N+S}T_{s2} = \frac{1}{f_s} \qquad (7.5.14)$$

where the length of the impulse response, S, is less than N. The signal applied to the **DAC**, $\{x(mT_s)\}$ is created from $\{x_k(nT_{s1})\}$ by interleaving, with the following additional twist:

$$m = n(N+S)+k;\ k = 0, 1, \ldots, (N+S-1);\ n = \ldots, -1, 0, 1, 2, \ldots$$

$$x(mT_s) = \begin{cases} x_k(nT_{s1});\ k = 0, 1, \ldots, (N-1) \\ x_{N-k}(nT_{s1});\ k = N, (N+1), \ldots, (N+S-1) \end{cases} \tag{7.5.15}$$

We can see that the intent is to create a *periodic extension* of the N-point sequence $\{x_k(nT_{s1});\ k = 0, 1, \ldots, (N-1)\}$, but only over (N+S) samples. The reason for doing so is that if we choose the correct N samples of $\{y(mT_s)\}$ to form $\{y(mT_{s2})\}$, then $\{y(mT_{s2})\}$ will *appear to be a cyclic convolution.* In fact, since

$$y[(n(N+S)+r)T_s] = \sum_{k=0}^{S-1} x[(n(N+S)+r-k)T_s]\, h(kT_s) \tag{7.5.16}$$
$$r = S, (S+1), \ldots, (N+S-1)$$

we can, for r = S, (S+1), ... , (N+S–1), associate, with q = (S–r),

$$y_0(nT_{s1}) = y[(n(N+S)+r)T_s] = \sum_{k=0}^{S-1} x_{S-k}(nT_{s1})\, h(kT_s) \tag{7.5.17a}$$

$$y_q(nT_{s1}) = y[(n(N+S)+S+q)T_s] = \sum_{k=0}^{S-1} x_{S+q-k}(nT_{s1})\, h(kT_s) \tag{7.5.17b}$$

$$y_{N-1}(nT_{s1}) = y[(n(N+S)+S+N-1)T_s]$$
$$= \sum_{k=0}^{S-1} x_{S+N-1-k}(nT_{s1})\, h(kT_s) \tag{7.5.17c}$$

Extending the length of the impulse response to N samples by appending zeros,

$$h(kT_s) = 0\ ;\ k = S, (S+1), \ldots, (N-1) \tag{7.5.18}$$

we can write

$$y_i = \sum_{k=0}^{N-1} x_{((S+i-k))_N} h_k = \sum_{k=0}^{N-1} h_{((S+i-k))_N} x_k \tag{7.5.19}$$

where the explicit dependence on time (T_{s1}, T_s) has been suppressed solely for notational convenience and to emphasize the cyclic convolution nature of the $\{y_i\}$.

With this expression for the $\{y_i(nT_{s1})\}$, the **DFT** computation to get the $\{\beta_r(nT_{s1})\}$ becomes

$$\beta_r = \frac{1}{N} \sum_{i=0}^{N-1} y_i e^{-j\frac{2\pi ri}{N}}\ ;\ r = 0, 1, \ldots, (N-1) \tag{7.5.20}$$

Substituting Eq. (7.5.19), (7.5.15), and (7.5.2) into Eq. (7.5.20) and simplifying, we get

$$\beta_r = H_r \alpha_r e^{j\frac{2\pi rs}{N}} \tag{7.5.21}$$

where various properties of the **DFT** have been invoked. The term H_r in Eq. (7.5.21) is the **DFT** sum of the impulse response coefficients $\{h(kT_s)\}$,

$$H_r = \sum_{k=0}^{N-1} h(kT_s e^{-j\frac{2\pi rk}{N}} \quad ; \; r = 0, 1, \ldots, (N-1) \tag{7.5.22}$$

Assuming N is large, much greater than S, then we can approximate N (N+S) (or T_s T_{s2}) and thus associate H_r with the frequency response of the channel as

$$H_r \approx \sum_{k=0}^{N-1} h(kT_{s2} e^{-j\frac{2\pi rk}{N}} = H(r\frac{f_{s2}}{N}) = H(f_r) \tag{7.5.23}$$

where f_r is the "center frequency of channel r." Thus analyzing Eq.(7.5.21) and Eq.(7.5.23) we have achieved the form shown in Fig. 7.54, at least from the viewpoint of the frequency response effect of the channel. The term exp(j2πrS/N) corresponds to a fixed phase shift and is not considered an impairment since it can be corrected for.

The operation of the **DMT** method of transmission can be encapsulated by the following observations.

1) The transmission channel is divided into M units, where each unit can be considered a passband channel with bandwidth f_{s2}/N (positive frequencies, f_{s2}/M considering both positive and negative frequencies). The sampling rates of the subchannels, f_{s1} is related to f_{s2} by $f_{s2} = Nf_{s1}$.
2) If the channel response is not flat over $[-f_{s2}/2, f_{s2}/2]$, but can be modeled as having a finite impulse response of length S samples, then the interchannel and intrachannel crosstalk can be eliminated by using a sampling frequency of $f_s = (N+S)f_{s1}$ for conversion to and from analog for the purposes of channel transmission. Thus we would be transmitting (N+S) symbols (samples) instead of N. The additional S samples are created by periodically extending the N samples to fit a window of (N+S) time slots.

There are several other facets to **DMT** that need to be considered but are beyond the scope of this book. These include methods for creating the symbols (samples) $\{\alpha_k(nT_{s1})\}$ in terms of the data to be transmitted; amplitude equalization to account for the $H(f_k)$, training sequences to determine what an appropriate value for S should be; some "pre-equalization" at the f_s sampling rate to "force" the effective response of the channel to be **FIR**; and so on. The interested reader is referred to the various contributions from Cioffi et. al to the **T1E1** Subcommittee examining different methods for transmitting high speed data on subscriber loops and to the *IEEE Transactions on Selected Areas in Communications*, Special Issue on High-Speed Digital Subscriber Lines (August 1991, Vol. 9, No. 6).

7.6 EXERCISES

7.1. The input to an N-tap **FIR** filter, H(**z**), is the signal {x(n)}, which is periodic with period M samples. Denote the output by {y(n)}.

a) Show that {y(n)} is periodic.

b) Show that {y(n)} can be written as the cyclic convolution of {x(n)} and a derivative of {h(n)}, where {h(n)} is the impulse response of the filter. Consider cases where N = M, N = RM, and M = SN (where R and S are positive integers).

c) Show the **DFT** can be used to compute the output.

7.2. The input and output of a filter, {x(n)} and {y(n)}, respectively, are observed (a reasonably large number, N, of samples of each are collected). Show how the **DFT** can be used to compute the frequency response (magnitude). If we are allowed to choose {x(n)}, what are the properties we will look for {x(n)} to have? Consider cases where the filter is **FIR** and **IIR**.

7.3. Suppose the size of a **DFT** satisfies N = RS, where R and S are integers. Suppose the "time" index is split into a double index as n = aS + b, and the "frequency" index split into a double index as $k = \alpha R + \beta$. What are the limits for the double indices? Show that the N-point **DFT** can be split into "stages," where the first "stage" comprises R S-point **DFT**s and where the N outputs of the first stage are multiplied by some constants (the second "stage") and the third "stage" comprises S R-point **DFT**s.

7.4. Suppose we have at our disposal a routine that computes the **DFT** of an N point sequence. Now suppose that we have to compute the frequency response of a digital filter at N equally spaced points in frequency spanning $[0, f_s]$, where f_s is the implied sampling frequency. How can this be accomplished ? Consider the cases of an **FIR** filter of M taps with M < N, M = N, and M > N. Also consider the case of an **IIR** filter with an Mth-order numerator and a Dth-order denominator.

7.5. Consider the following technique for estimating the signal-to-noise ratio when the signal is a tone:

a) N consecutive samples (the "window") are processed by an N-point **DFT** sum. Show that if $\{x_n\}$ is a pure tone with $x_n = A\cos(2\pi f_0 n)$, where the frequency $f_0 = (k_0/N)$ (f_s is the sampling frequency that is normalized to unity) then only two **DFT** "samples" of $\{X_k\}$ will be nonzero. Which samples? Relate the value of these nonzero samples to the amplitude of the tone.

b) Show that if $\{x_n\}$ is white noise of power σ_x^2 then $\Sigma|X_k|^2 = c\ \sigma_x^2$. Establish the constant c.

c) From parts **a)** and **b)** indicate how the signal-to-noise ratio can be computed from the **DFT** sum over N samples.

d) What happens if f_0 is not an integer multiple of (1/N)? What modification to step **c)** is recommended in this case?

e) If we have RN samples of the signal $\{x(n)\}$, and our computational element is an N-point **DFT**, then how can we go about estimating the **SNR** (the answer is not unique) and what pitfalls may we encounter?

f) Suppose we weight the **DFT** by applying an N-point "window," such as the Hamming window, to the N-samples prior to computing the N-point **DFT**. What are the advantages and disadvantages of using such a window relative to the unweighted case ("rectangular window")?

g) Draw polyphase schematics to represent the processing in cases **e)** and **f)**.

7.7 REFERENCES AND BIBLIOGRAPHY

7.7.1 References

[1.1] Cioffi, J. M., "A multicarrier primer," *ANSI T1E1.4 Committee Contribution*, Boca Raton, Nov. 1991.

[1.2] Crochiere, R. E., and Rabiner, L. R., *Multirate Digital Signal Processing*, Prentice Hall, Inc., Englewood Cliffs, NJ, 1983.

[1.3] Narasimha, M. J., and Peterson, A. M., "Design of a 24-channel transmultiplexer," *IEEE Trans. Acoust. Speech and Signal Proc.*, vol. 27, Dec. 1979.

[1.4] Vaidyanathan, P. P., "Multirate digital filters, filter banks, polyphase networks, and applications: a tutorial, " *Proc. IEEE*, Vol. 78, Jan. 1990.

[1.5] Vaidyanathan, P. P., *Multirate Systems and Filter Banks,* Prentice Hall, Inc., Englewood Cliffs, NJ, 1993.

7.7.2 Bibliography

7.7.2.1 General

[2.1] Bellanger, M., *Digital Signal Processing, Theory and Practice*, John Wiley and Sons, New York, 1984.

[2.2] Childers, D. G., Ed., *Modern Spectrum Analysis*, IEEE Press, New York, 1978.

[2.3] Digital Signal Processing Committee, IEEE Acoustics, Speech, and Signal Processing Society, Editors, *Digital Signal Processing, II*, IEEE Press, New York, 1976.

[2.3.1] McClellan, J. H., Parks, T. W., and Rabiner, L. R., "A computer program for designing optimum FIR linear phase digital filters," *IEEE Trans. on Audio and Electroacoustics*, Dec. 1973.

[2.3.2] Rabiner, L. R., McClellan, J. H., and Parks, T. W., "FIR digital filter design using weighted Chebyshev approximation," *Proc. IEEE*, April 1975.

[2.3.3] Kaiser, J. H., "Nonrecursive digital filter design using the I_0-sinh window function," *Proc. 1974 IEEE International Symposium on Circuits and Systems*, April 1974.

[2.3.4] Schafer, R.W., and Rabiner, L. R., "A digital signal processing approach to interpolation," *Proc. IEEE*, June 1973.

[2.4] Digital Signal Processing Committee of the IEEE Acoustics, Speech, and Signal Processing Society, Editors, *Programs for Digital Signal Processing*, IEEE Press, New York, 1979.

[2.5] Kesler, S. B., Ed., *Modern Spectrum Analysis, II*, IEEE Press, New York, 1986.

[2.6] Oppenheim, A.V., and Schafer, R.W., *Discrete-Time Signal Processing*, Prentice Hall, Inc., Englewood Cliffs, NJ, 1989.

[2.7] Oppenheim, A.V., *Applications of Digital Signal Processing*, Prentice Hall, Inc., Englewood Cliffs, NJ, 1978.

[2.8] *PC-MATLAB*: Software package from The MathWorks, Inc., Also, the *Signal Processing Toolbox,* which can be used with *MATLAB*.

[2.9] Rabiner, L.R., and Gold, B., *Theory and Application of Digital Signal Processing*, Prentice Hall, Inc., Englewood Cliffs, NJ, 1975.

7.7.2.2 DFT, DCT, and Fast Algorithms

[2.10] Ahmed, N., Natarajan, T., and Rao, K. R., "Discrete cosine transform," *IEEE Trans. Computers*, vol. C-23, Jan. 1974.

[2.11] Brigham, E. O., *The Fast Fourier Transform and its Applications*, Prentice Hall, Inc., Englewood Cliffs, NJ, 1988.

[2.12] Blinn, J. F., "What's that deal with the DCT?" *IEEE Computer Graphics and Appl.*, Vol. 13, No. 4, July 1993.

[2.13] Deller, J. R., Jr., "Tom, Dick, and Mary discover the DFT," *IEEE Signal Processing Mag.*, Vol. 11, No. 2, Apr. 1994.

[2.14] Kraniauskas, P., "A plain man's guide to the FFT," *IEEE Signal Processing Mag.*, Vol. 11, No. 2, Apr. 1994.

[2.15] Makhoul, J., "A fast cosine transform in one and two dimensions," *IEEE Trans. Acoustics, Speech, and Signal Proc.*, Vol. ASSP-28, Feb. 1980.

[2.16] Narasimha, M. J., and Peterson, A. M., "On the computation of the Discrete Cosine Transform," *IEEE Trans. Communications*, Vol. COM-26, June 1978.

[2.17] Rabiner, L.R., and Rader, C.M., Editors, *Digital Signal Processing*, IEEE Press, New York, 1972.

[2.17.1] Cooley, J. W., and Tukey, J. W., "An algorithm for the machine calculation of complex Fourier series," *Mathematics of Computation*, 1965.

[2.17.2] Bergland, G. D., "A guided tour of the fast fourier transform," *IEEE Spectrum*, July 1969.

[2.17.3] IEEE Group on Audio and Electroacoustics Subcommittee on Measurement Concepts, "What is the FFT?," *IEEE Trans. Audio and Electroacoustics*, June 1967.

[2.18] Vetterli, M., and Nussbaumer, H., "Simple FFT and DCT algorithms with reduced number of computations," *Signal Processing*, Vol. 6, Aug. 1984.

[2.19] Yang, P. P. N., and Narasimha, M. J., "Prime factor decomposition of the Discrete Cosine Transform and its hardware realization," *Proceedings of the IEEE International Conference on Acoustics, Speech, and Signal Processing*, ICASSP '85.

7.7.2.3 Filter Banks and Transmultiplexers

[2.20] Bellanger, M., Bonnerot, G., and Coudreuse, M., "Digital filtering by polyphase network: application to sample rate alteration and filter banks," *IEEE Trans. Acoust. Speech and Signal Proc.*, Vol. 24, Apr. 1976.

[2.21] Bingham, J. A. C., "Multicarrier Modulation for Data Transmission: an idea whose time has come," *IEEE Communications Magazine*, May 1990.

[2.22] Chow, J. S., Tu, J. C., and Cioffi, J. M., "A computationally efficient adaptive transceiver for high-speed digital subscriber lines," *Proceedings 1990 International Conference on Communications*, Boston, MA, June 1990.

[2.23] Chu, P. L., "Quadrature mirror filter design for an arbitrary number of equal bandwidth channels," *IEEE Trans, Acoustics, Speech, and Signal Processing*, Vol. ASSP-33, No. 1, Jan. 1985.

[2.24] Darlington, S., "On digital single-sideband modulators," *IEEE Trans. Circuit Theory*, Vol. CT-17, Aug. 1970.

[2.25] Koilpillai, R. D., Nguyen, T. Q., and Vaidyanathan, P. P., "Some results in the theory of crosstalk-free transmultiplexers," *IEEE Trans. on Signal Processing*, Vol. 39, No. 10, Oct. 1991.

[2.26] Koilpillai, and Vaidyanathan, P. P., "Cosine-modulated FIR filter banks satisfying perfect reconstruction," *IEEE Trans. on Signal Processing*, Vol. 40, No. 4, Apr. 1992.

[2.27] Nayebi, K., Barnwell, T. P., III, and Smith, M. J. T., "Time domain filter bank analysis: A new design theory," *IEEE Trans. on Signal Processing*, Vol. 40, No. 6, June. 1992.

[2.28] Ruiz, A., Cioffi, J. M., and Kasturia, S., "Discrete multiple tone modulation with coset coding for the spectrally shaped channel," *IEEE Trans. Communications*, Vol. 40, June 1992.

[2.29] Smith, M. J. T., and Barnwell, T. P., III, "A New Filter Bank Theory for Time-Frequency Representation," *IEEE Trans, Acoustics, Speech, and Signal Processing*, Vol. ASSP-35, No. 3, Mar. 1987.

[2.30] Vaidyanathan, P. P., "Theory and design of M-channel maximally decimated quadrature mirror filters with arbitrary M, having the perfect-reconstruction property," *IEEE Trans, Acoustics, Speech, and Signal Processing*, Vol. ASSP-35, No. 4, Apr. 1987.

[2.31] Vetterli, M., and Le Gall, D., "Perfect reconstruction FIR filter banks: some properties and factorizations," *IEEE Trans, Acoustics, Speech, and Signal Processing*, Vol. 37, No. 7, Jul. 1989.

[2.32] Vetterli, M., "A theory of multirate filter banks," *IEEE Trans, Acoustics, Speech, and Signal Processing*, Vol. ASSP-35, No. 3, Mar. 1987.

DELTA SIGMA MODULATION

8

8.1 INTRODUCTION

The use of Delta Sigma Modulation (**ΔΣM**) techniques in A/D and D/A conversion is gaining in popularity primarily because of its simplicity. With the continual decrease in the cost of digitally integrated circuits, and the phenomenal advances in packing of logic circuits into microchips, it is natural that the burden of signal processing be shifted from the analog domain to the digital domain. In doing so one gains the advantages provided by digital signal processing, namely, repeatability, predictability, stability with time, linear phase filters, and so on.

Delta Sigma Modulators are a class of converters termed "oversampled." The sampling rate at which the conversion is performed is much higher than the (minimum) rate required. That is, the bandwidth of the underlying (analog) signal is much less than the rate at which the **ΔΣM** operates.

In Section 8.2 we discuss some of the advantages of oversampling. If f_{s1} is the required sampling rate, based on the signal bandwidth, the actual A/D conversion is performed at the rate $f_{s2} = Nf_{s1}$. Digital filters operating at this higher rate then condition the signal to allow a sampling rate reduction by a factor of N. The same principles apply in the reverse direction. The sampling rate is increased by a factor of N, from f_{s1} to $f_{s2} = Nf_{s1}$, and digital filters perform the necessary replicate rejection. The conversion to analog is performed at this higher rate.

Inherent to the applicability of oversampled converters is the notion of noise shaping. The higher rate permits the shaping of the quantization noise into spectral regions outside the band occupied by the signal of interest. The principles of noise shaping are introduced in Section 8.3, which first examines a technique called Delta Modulation (**ΔM**) which is somewhat easier to visualize than Delta Sigma Modulation. The latter can be "derived" from the former by manipulating the constituent functional blocks.

In Section 8.4 we address some implementations of **ΔΣM**s. What will become obvious is that **ΔΣM** structures are very simple. Implementing such circuits is straightforward and requires just the normal care associated with any high speed electronic design. Extended mathematical analyses are not warranted in such cases since the behavior of such devices must be established by actual breadboarding and experimentation.

In Section 8.5 a suitable simple model is developed for a **ΔΣM**. We start by modeling a first-order structure for simplicity and extend the analysis to higher-order **ΔΣM**s. Of special importance are the concepts of the Signal Transfer Function (**STF**) and the Noise Transfer Function (**NTF**). The **STF** describes the impact of the **ΔΣM** on the signal being converted by modeling the frequency response experienced by the desired signal between the input and the output of the modulator. Ideally, the **STF** should be a constant (unity). The **NTF** provides a means for estimating the power of the quantization noise in the **ΔΣM** output. The **NTF** also provides an indication of the manner in which the quantization noise is spread out in the frequency domain. When viewed as a frequency response, the ideal behavior of the **NTF** is such that the response is small (zero) over the bandwidth of the desired signal. Since the preponderance of **ΔΣM**s used are of the single-bit variety, such a device is discussed in detail. The viewpoint taken is that of designing an A/D converter for telephony applications.

Often overlooked in the study of **ΔΣM**s is the need for a post-A/D digital filter to remove the (shaped) quantization noise. This aspect is covered in Section 8.6. Since the sampling rate is usually very high, typically of the order of MHz, the filters must be simple to implement. Two candidates, the "rectangular window" and the "triangular window" are given special attention. Section 8.7 addresses these filters from the viewpoint of implementation. A high-level design for such filters is provided, one that is quite suitable for implementation in an **ASIC**.

8.2 PRINCIPLES OF OVERSAMPLING

To see how we can transfer complexity from the analog to digital domain, consider the requirements of an A/D converter of the type used in telecommunications. The A/D converter in commercial **PCM** applications is capable of quantization to an accuracy of 13 bits and will be preceded by an anti-aliasing filter. The sampling rate of the converter is 8 kHz and the nominal cutoff frequency of the analog filter is 4 kHz. In practice, the encoder will be of the μ-law or A-law variety and the filter will have a transition band from 3.4 kHz to 4.6 kHz. This is depicted in Figs. 8.1 and 8.2. The requirements of the anti-aliasing filter are discussed in Chapter 1 and typically a fifth- or sixth-order elliptic filter is used. The complexity of the encoder can be expressed as the need for 256 (assuming a μ-law encoder) comparators capable of resolving a step-size proportional to 2^{-13}, which is about 100 μV on a scale of 1 volt; further, *all* 256 comparators need to have this accuracy.

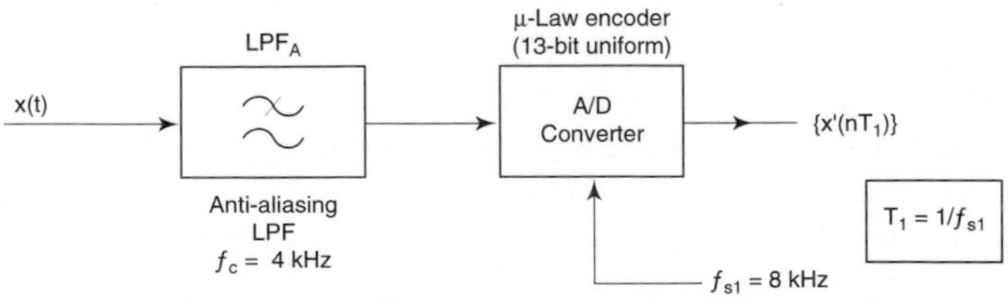

Fig. 8.1
Conventional A/D converter used in telephony

The oversampling approach to reducing the complexity of the analog circuitry required is shown in Fig. 8.3. The initial sampling is performed at a higher rate, $f_{s2} = Nf_{s1}$, and is followed by a digital filter to remove signal components above $0.5f_{s1}$.

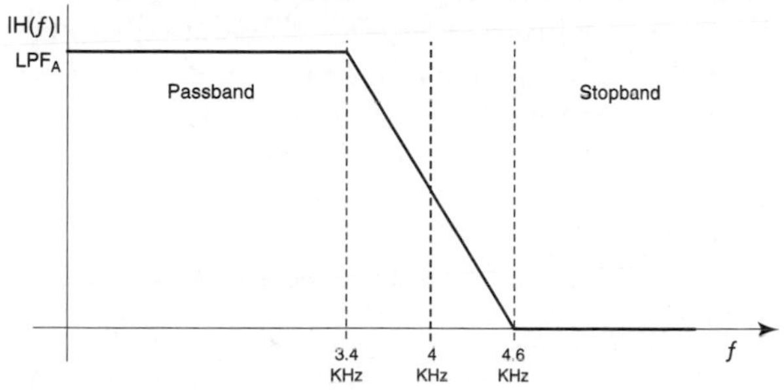

Fig. 8.2
Nominal frequency response of anti-aliasing lowpass filter in telephony applications

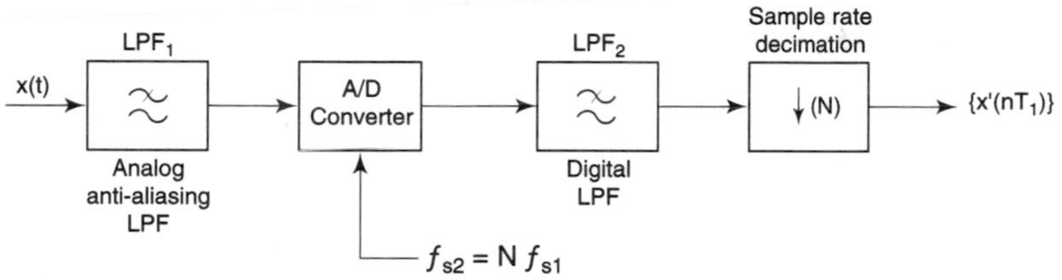

Fig. 8.3
Scheme for reducing analog filter complexity by increasing the sampling rate

An anti-aliasing analog lowpass filter is required to remove frequency components above $(f_{s2}/2)$, where the sampling frequency f_{s2} of the A/D converter is much greater than the eventual sampling rate ($f_{s2} = Nf_{s1}$). The frequency response requirements of this anti-aliasing filter, labeled LPF_1 in Fig. 8.3, are depicted in Fig. 8.4.

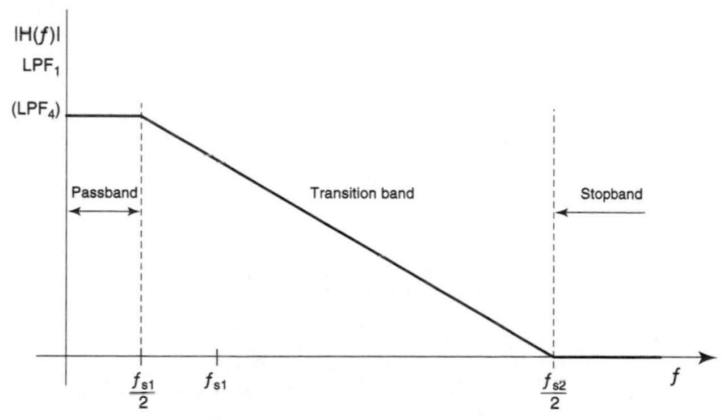

Fig. 8.4
Anti-aliasing frequency response requirement—increasing sampling rate allows a wider transition band.

Clearly, the requirements on the analog filter are reduced considerably, as evidenced by the increased width of the transition band allowed. This can be achieved with a much simpler filter than that required by Fig. 8.2.

One expects that the A/D converter is simpler as well. This is indeed the case. It will be shown that sampling rate of operation can be traded for wordlength. Since a suitable rule of thumb is that the complexity of an A/D converter increases "linearly with sampling rate but exponentially with wordlength," if the A/D converter in Fig. 8.3 requires $\log_2(N)$ fewer bits than that in Fig. 8.1 then there is indeed a saving. In the following sections it is explained how, by using appropriate feedback schemes one can reduce the wordlength by 1.5 $\log_2(N)$ or even 2.5 $\log_2(N)$ (approximately) bits.

Needless to say, the overall frequency response of the combination of LPF_1 (analog) and LPF_2 (digital) must still satisfy the requirements of an anti-aliasing filter, requiring that LPF_2 have a frequency response of the form shown in Fig. 8.5. We can see that the complexity in terms of frequency response is constant—just moved from the analog to the digital domain. The analog filter provides the attenuation above 0.5 f_{s2}, the digital filter provides attenuation above 0.5 f_{s1}.

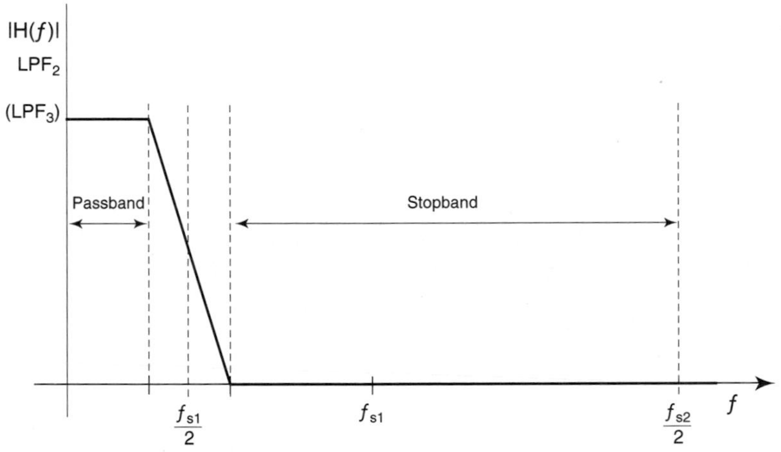

Fig. 8.5
Frequency response requirements of digital lowpass filter of Fig. 8.3

A similar situation presents itself in the D/A direction. Consider a digital signal $\{x(nT_1)\}$ that must be converted into analog form and the analog filter used to attenuate the ever-present spectral replicates. Again, assuming the case of a μ-law converter, the scheme is shown in Fig. 8.6 and Fig. 8.7, illustrates the requirement on the lowpass filter. The **DSP** approach, namely oversampling, is shown in Fig. 8.8.

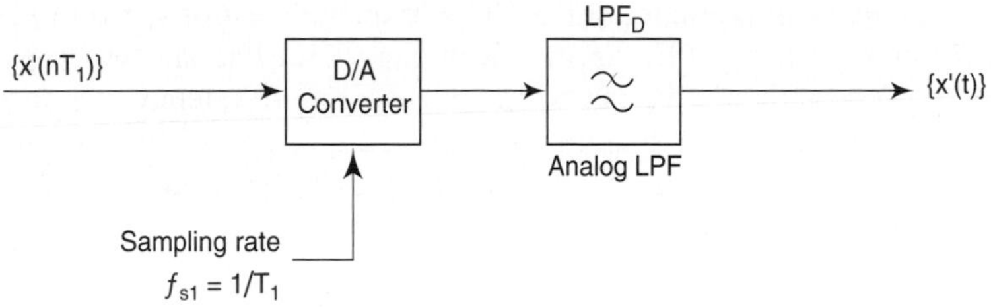

Fig. 8.6
Nominal conversion from digital to analog at the sampling rate f_{s1}

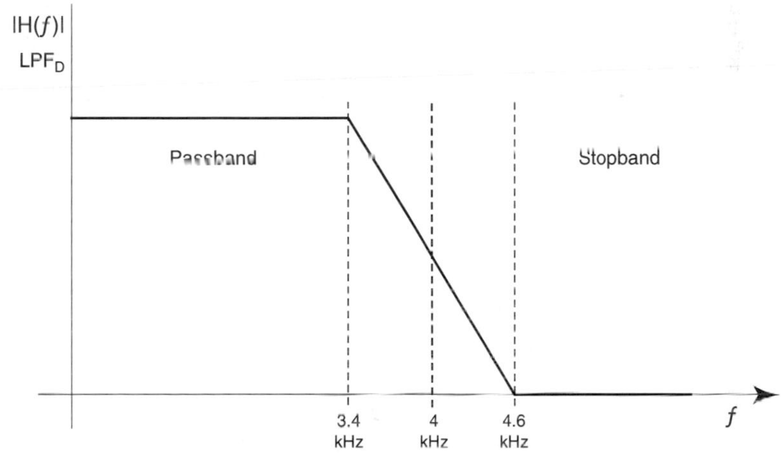

Fig. 8.7
Frequency response requirement of replicate rejection filter LPF$_D$ shown in Fig. 8.6

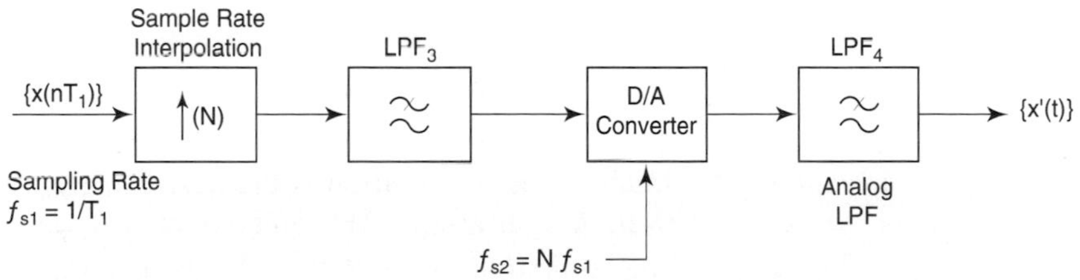

Fig. 8.8
Scheme for reducing analog filter complexity by increasing sampling rate

The digital signal $\{x(nT_1)\}$ is upconverted in sampling rate to $f_{s2} = Nf_{s1}$ by the insertion of zero samples. LPF_3 is a digital filter that attenuates the spectral replicates inherent in the interpolated signal. The frequency response required of LPF_3 is similar to that required of LPF_2, as shown in Fig. 8.5. The spectral replicate suppression by the analog filter LPF_4 can be achieved with a frequency response depicted in Fig. 8.4. Again, the overall response of LPF_3 (digital) and LPF_4 (analog) must still meet the specifications of Fig. 8.7, but the complexity has been shared between the analog and digital domains.

The approach depicted in Fig. 8.8 is used in the audio world of **CD** (compact disc) players. The terms "2* oversampling" and "4* oversampling" used in describing **CD** players is indicative of the sampling rate increase used. The use of digital filters has an added benefit that is particularly significant in hi-fi audio applications. The sharp cutoff analog filter required, for example, LPF_D in Fig. 8.7, tends to introduce significant phase distortion, especially close to the edge of the passband. Digital filters, however, can be of the **FIR** variety and introduce no phase distortion. Since the analog filter, LPF_4 in Fig. 8.8, can have a wide transition band, it can be designed to have a minimal amount of phase distortion in the passband.

8.3 DELTA MODULATION (ΔM) AND A SIMPLIFIED APPROACH TO ΔΣM

A fundamental requirement for **ΔΣM** to be efficient, or even applicable, as a means of analog-to-digital conversion is that the sampling rate at which the **ΔΣM** operates must be much higher than the bandwidth of the analog signal being digitized. That is, if x(t) is the analog signal being converted to digital form, then

$$|X(f)| \quad 0; \quad |f| > f_c \tag{8.3.1a}$$

or

$$S_{xx}(f) \quad 0; \quad |f| > f_c \tag{8.3.1b}$$

where the two forms of Eq. (8.3.1) indicate that there is essentially no component beyond a frequency of f_c that is required to fully describe the signal x(t). Eq. (8.3.1a) is appropriate for deterministic signals while Eq.(8.3.1b) is appropriate when **x**(t) is a random signal. In **ΔΣMs** the conversion is done at a high sampling rate, high enough to satisfy

$$f_s >> f_c \tag{8.3.2}$$

At this high sampling rate, the change in value of x(t) from sampling instant to sampling instant will be small. This observation is the basis for **Delta Modulation** (**ΔM**). The operation of **Delta Modulator** (**ΔM**) can be explained using Fig. 8.9 for reference. The signal x(t) is compared to a signal x'(t), which can be viewed as the estimate of x(t–T), where $T = 1/f_s$ is the sampling interval. The difference signal is quantized to 1 bit and stored in a flip-flop (i.e., a one-bit register that can be viewed as unit delay) clocked at the sampling rate f_s. The combination of comparator and flip-flop comprises a 1-bit A/D converter that follows the rule

$$b(nT) = \begin{cases} 1; & w(nT) > 0 \\ 0; & w(nT) < 0 \end{cases} \qquad (8.3.3a)$$

or, equivalently,

$$b(nT) = \begin{cases} 1; & x(nT) > x'(nT) \\ 0; & x(nT) < x'(nT) \end{cases} \qquad (8.3.3b)$$

Considering that x'(t) is an approximation of x(t–T), the difference signal [x(t) – x'(t)] [x(t) – x(t-T)] is an approximation to the scaled version of the derivative of x(t) with respect to time. Delta modulation can hence be viewed as an entity that quantizes the slope of the signal rather than the signal itself. The reconstruction of the signal is achieved using an integrator. Recalling the discussion in Chapter 5 on predictive encoders, the ΔM is obtained by taking the notion of quantization of levels to the limit, namely 1 bit. By using a very high sampling rate, the correlation between adjacent samples will be high; ΔM uses a = 1. Further, the ΔM uses analog methods to accomplish the reconstruction rather than the "all digital" representation considered in Chapter 5.

This is achieved in Fig. 8.9. The Level Generator block is simply a 1-bit D/A converter that outputs a voltage of ±B (volts) according as the digital signal is 1 or 0 (HIGH/LOW, LOGIC–1/LOGIC–0, etc.). That is, a precise analog voltage is produced in response to a logic level. Representative waveforms of the signals w(t) and the sequence {b(nT)} are shown in Fig. 8.10. Recognize that the sampling rate is much higher than the bandwidth of the signal x(t). Thus, over a few sampling intervals x(t) can be modeled as a constant with a small superimposed component comprising the higher frequencies. The reconstructed signal x'(t) will have the same "constant" component along with some high frequency content since w'(t) oscillates between two levels ±B at the sampling rate. This high frequency component is attenuated by G. The ΔM quantizes the difference, w(t), to 1 bit. This 1-bit quantization, representing a signal just by its sign, is apparent from Fig. 8.10.

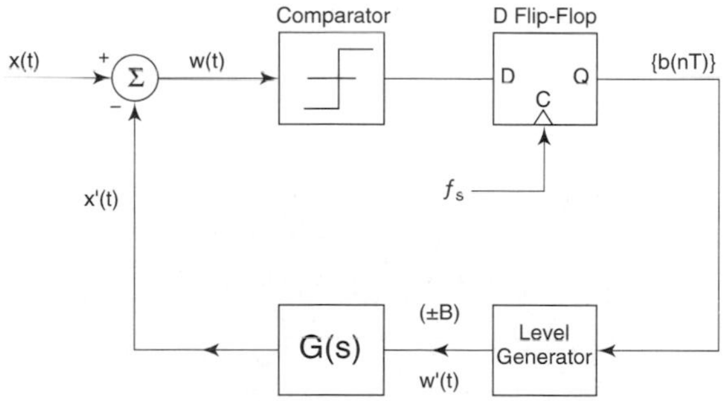

Fig. 8.9
Principal components of a Delta Modulator

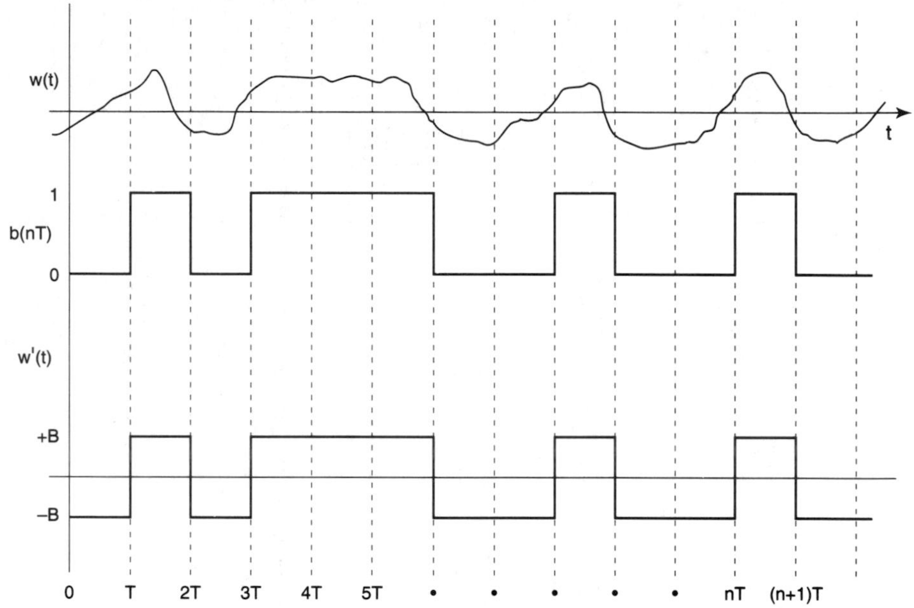

Fig. 8.10
Waveforms of signals in the Delta Modulator of Fig. 8.9

The reconstruction process comprises an integrator. That is,

$$x'(t) = G \int_{-\infty}^{t} w'(\tau)\, d\tau \tag{8.3.4}$$

At the sampling instants, $t = nT$, $x'(t)$ can be expressed as

$$x'(nT) = x'[(n-1)T] + G \int_{(n-1)T}^{nT} w'(\tau)\, d\tau \tag{8.3.5}$$

Recognizing that $w'(t)$ has a special form, namely, it is constant over the interval $[(n-1)T, nT]$, Eq. (8.3.5) yields

$$x'(nT) = x'[(n-1)T] + GT\, w'[(n-1)T] \tag{8.3.6a}$$

or

$$x'(nT) = x'[(n-1)T] + \begin{cases} +\,GTB; & w[(n-1)T] > 0 \\ -\,GTB; & w[(n-1)T] < 0 \end{cases} \tag{8.3.6b}$$

Thus when viewed at the sampling instants, x'(t) changes by an amount ±GTB or ±Δ, where

$$\Delta = GTB \tag{8.3.7}$$

is the ***stepsize*** associated with the Delta Modulator.

When the Delta Modulator is "tracking" the signal, the sampled values { x'(nT)} will tend to bracket the analog waveform of x(t). This is indicated in Fig. 8.11. The **ΔM** is considered to be tracking if the difference between x'(nT), the reconstructed value, and x(nT), the "exact" sampled value, is less than 2Δ.

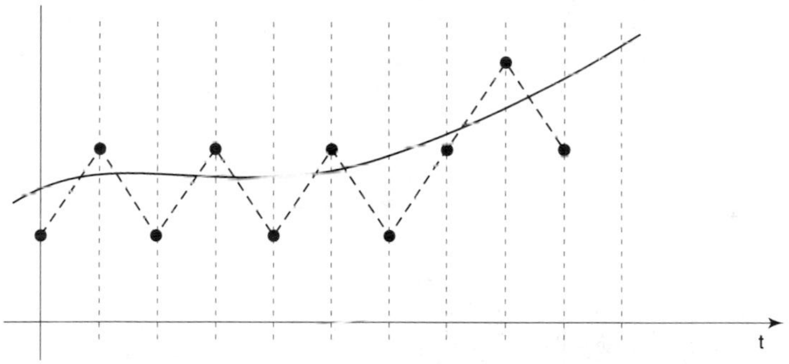

Fig. 8.11
Manner in which the Delta Modulator tracks the signal x(t)

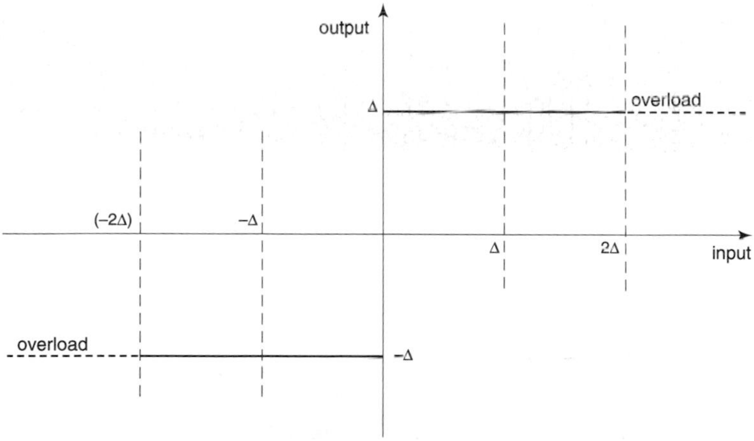

Fig. 8.12
Quantizer characteristic of A/D converter in a ΔM

In Fig. 8.12 overload is construed as the inability of the quantizer to maintain a difference between its input and output of magnitude less than Δ, the equivalent stepsize. This translates to the inability of the **ΔM** to track the signal and, in overload, the difference between x(t) and x'(t) exceeds 2Δ. There is another form of overload though. Practical implementations of the integrator will have an implicit limitation on the maximum voltage of the output signal. Ignoring this latter limitation, the *crashpoint* phenomenon in a ΔM is associated with the inability to represent the derivative of x(t) and is hence referred to as ***slope overload***. This dependence on slope is less preferable than overload based solely on amplitude. For example, if the ΔM goes into overload when the input signal is a sinusoid of frequency 100 Hz and amplitude A, then the maximum amplitude of a 1-kHz sinusoid is (A/10) to keep out of the overload state.

To illustrate how the dependence on slope can be transformed into a dependence solely on amplitude, consider the chain of ΔM encoder and reconstruction depicted in Fig. 8.13. The reconstruction is achieved using the equivalent level generator (denoted by LG) and the integrator that is used in the modulator itself. The final lowpass filter smooths out the saw-tooth nature of x'(t) and thus suppresses the replicates that are inherent in a sampled-data system.

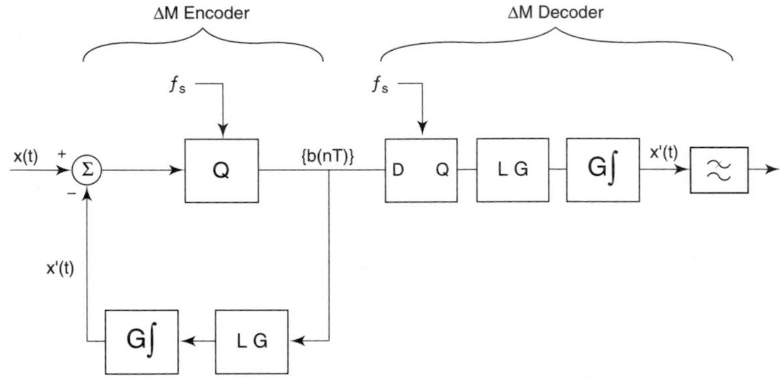

Fig. 8.13
Block diagram representation of ΔM encoder and decoder

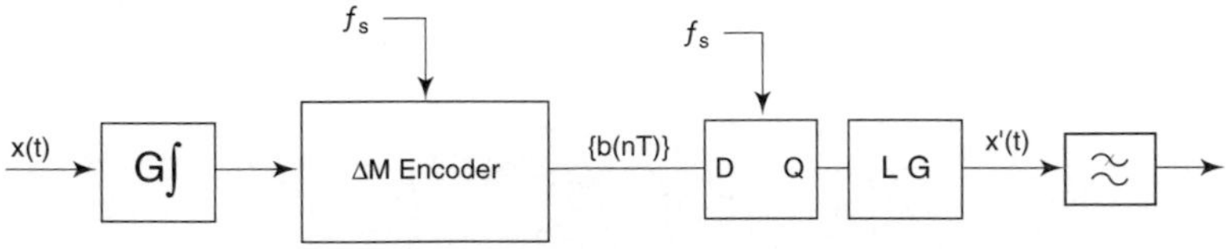

Fig. 8.14
Block manipulation of Fig. 8.13 to obtain a Sigma Delta Modulator

If all the blocks in Fig. 8.13 correspond to linear time-invariant operations, then they can be rearranged as shown in Fig. 8.14. Since the quantizing operation is inherently nonlinear, the block rearrangement shown in Fig. 8.14 is not, strictly speaking, equivalent to Fig. 8.13. Nonetheless, it displays one useful attribute. The integration function prior to the ΔM provides a *de-emphasis* function whereby the *slope overload* of the ΔM encoder, with respect to y(t), is equivalent to a *level overload* with respect to x(t). The reconstruction also seems simpler in Fig. 8.14 than in Fig. 8.13 since the complexity of the integrator has been shifted from the reconstruction to part of the encoder. The combination of integrator and ΔM is called a Sigma Delta Modulator, or ΣΔM. By further block manipulation, whereby the pre-integrator and the loop-integrator are combined, we get Fig. 8.15. The term **D**elta **S**igma **M**odulator (ΔΣM) for the combination of integrator and ΔM is preferred since, as seen in Fig. 8.15, the processing sequence is indeed difference followed by integration.

One viewpoint of Delta Sigma Modulation is in terms of a variation of Delta Modulation. A different viewpoint of a ΔΣM can be obtained by considering how the waveform y(t) in Fig. 8.15 can be viewed as a representation of x(t). The waveform y(t) can take on only two distinct voltage values, ±B, which correspond to the output of the level generator. x(t), on the other hand, is an analog signal that can take on a continuum of voltage values. How then can y(t) be considered a representation, or approximation, of x(t)? The key is that since x(t) has negligible high frequency content ($f_c << f_s$) it can be considered "constant" over (short) intervals of time. Now over a time period of N samples, the *average* value of y(t) can be written as

$$\langle y(t) \rangle = \frac{B}{N}(N_p - N_m) = B\frac{(N_p - N_m)}{(N_p + N_m)} \tag{8.3.8}$$

where, over the N sampling periods, y(t) takes on the value +B N_p times and the value –B N_m times. This average value can take on values between +B and –B and these intermediate values would have a granularity that depends on the interval of observation, N (samples). Thus y(t) can be viewed as a stream of pulses where the density of +B 's (positive pulses) is representative of the amplitude of a DC or slowly varying signal.

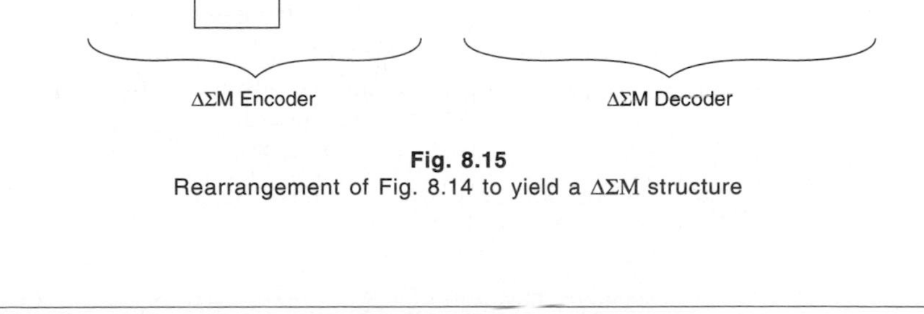

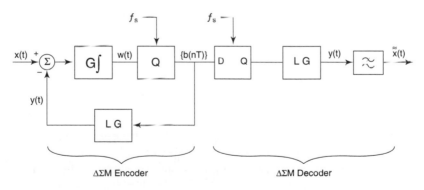

Fig. 8.15
Rearrangement of Fig. 8.14 to yield a ΔΣM structure

The $\Delta\Sigma M$ (encoder) achieves this pulse density by generating at sample (N+1) the equivalent of a +B pulse if the average difference between y(t) and x(t) were less than zero and a –B if the said difference were positive. The output of the integrator is representative of the difference between x(t) and y(t) averaged over time. Thus the average difference, including sample (N+1) would be closer to zero than the average difference up to and including sample N.

Strictly speaking, the pulse density explanation for the working of a $\Delta\Sigma M$ supposes that x(t) is (almost) a constant. This is a reasonable assumption if the sampling rate, f_s, is much greater than the maximum frequency, f_c, at which the signal, x(t), has appreciable power. If x(t) is a sinusoid of frequency f_c, then the interval (N samples) over which x(t) can be considered "constant" is proportional to the ratio (f_s/f_c). The greater is this ratio, the more the signal "looks like a DC signal." Since the granularity achieved is better for larger N, the $\Delta\Sigma M$ provides a better representation of the amplitude of lower frequency signals. Since good granularity is synonymous with low quantizing noise, the $\Delta\Sigma M$ can be viewed as a device that introduces quantization noise whose spectral content is essentially null at DC and increases with increasing frequency.

A third viewpoint of a **$\Delta\Sigma$M**, one that seems more apt in terminology, is to view a **$\Delta\Sigma$M** as a form of modulator. With no input, i.e., x(t) = 0, the waveform y(t) will oscillate between ±B. If at any sampling instant the input to the comparator is positive, then the value of y(t) in the subsequent interval would be +B. The negative nature of the feedback ensures that the input to the comparator would then be driven (eventually) negative, maybe after a few sampling intervals. This would make the output y(t) go negative to –B and initiate the cycle in reverse. Thus y(t), with no input to the **$\Delta\Sigma$M**, would appear to be a sequence of the following type

$$y(nT) = \ldots, -B, +B, -B, +B, -B, +B, \ldots \qquad (8.3.9a)$$

or

$$y(nT) = \ldots, -B, -B, +B, +B, -B, -B, +B, +B, \ldots \qquad (8.3.9b)$$

or

$$y(nT) = \ldots, -B, -B, +B, -B, +B, +B, \ldots \qquad (8.3.9c)$$

corresponding to fundamental frequencies of $(f_s/2)$, $(f_s/4)$, and $(f_s/6)$, respectively.

The effect of the input signal x(t) (nonzero) is to disturb this free-running pattern and thus the terminology "modulation." A positive x(t) would bias the input to the comparator so as to produce a greater number of positive pulses, causing an asymmetry in the number of positive and negative pulses which in turn can be related to the amplitude of the signal x(t). Thus the modulating signal introduces a "DC component" in {y(nT)} that is related to the value of the signal x(t), allowing the demodulation, or recovery of x(t), to be achieved by a lowpass filter.

Earlier it was stated that the quantization noise has a spectral content that is null at DC and increases with frequency. A heuristic explanation for this behavior can be developed by viewing the ΔΣM as a control loop. From this viewpoint the action of the feedback is to drive the output of the integrator, w(t), the input to the quantizer, toward zero. This is interpreted as the minimization of the power of w(t) by removing the "deterministic" components and thus decorrelating w(t) from sample to sample, making w(t) look like white noise. [The notions of minimization of power and whitening of the spectrum are closely linked and often equivalent.] Denoting the quantization error signal by e(t),

$$e(t) = x(t) - y(t) \tag{8.3.10}$$

and the action of the feedback loop is to make w(t) white, where

$$w(t) = G\int_{-\infty}^{t} e(\tau)\,d\tau \approx \text{white noise} \tag{8.3.11}$$

appear as white noise. Consequently, at least at lower frequencies, we can express the power spectral density of the quantization noise as

$$S_{ee}(f) = C\,(2\pi f)^2 \quad (C = \text{constant}) \tag{8.3.12}$$

The reason for the caveat "at lower frequencies" is that Eq. (8.3.12) is clearly not valid at frequencies beyond f_s since it does not express the inherent periodicity in the frequency domain associated with sampled-data systems. It is shown later that, with the appropriate (and reasonable) assumptions, Eq. (8.3.12) is indeed representative of the error spectra at low frequencies and that certain observations derived from Eq. (8.3.12) are indeed generalizable.

The power spectrum of the quantization noise exhibits the behavior indicated in Fig. 8.16 (only positive frequencies have been shown) along with the nominal response of the lowpass filter used for demodulation. Note that

i) the error spectrum has a null at DC and increases with a slope of 6 dB/octave;

ii) since the lowpass filter must remove the high frequency quantization noise, its cutoff slope of the lowpass filter must be greater than 6 dB per octave, implying that the filter must be of second or higher order;

iii) if the lowpass filter is ideal, with a cutoff frequency of f_c, then the quantization noise power after filtering, as indicated by the shaded area under the $S_{ee}(f)$ curve in Fig. 8.16, is proportional to the cube of f_c. Thus the (in-band) quantizing noise power exhibits a 9 dB per octave behavior with respect to f_c.

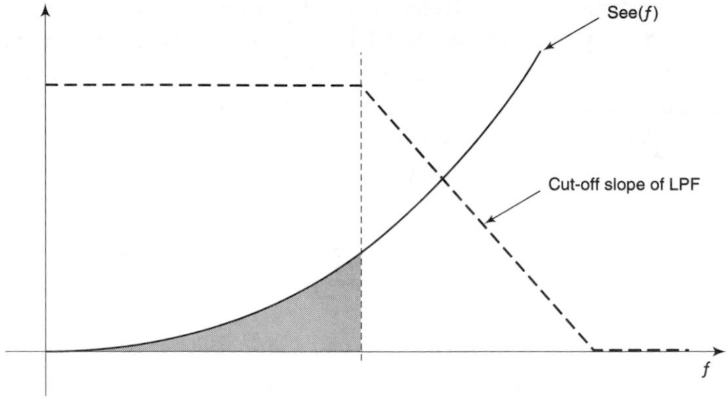

Fig. 8.16
Spectrum of quantizing noise depicting in-band noise power

It is shown later that the last observation, relating the quantization noise to filter bandwidth, is better expressed as

$$\sigma_e^2 = k\left(\frac{f_c}{f_s}\right)^3 \tag{8.3.13}$$

indicating that the noise power is proportional to the cube of the ratio of f_c to the sampling rate f_s. The filter bandwidth, that is, f_c, is determined by the nature of the signal x(t) and the information content therein. The sampling frequency, f_s, is a "design parameter" and can be chosen to meet certain performance criteria. Eq. (8.3.13) indicates that the noise power is reduced by 9 dB for every doubling of the sampling frequency. Equivalently, the **SNR** demonstrates a 9 dB per octave relationship with the sampling frequency.

This 9 dB per octave figure is characteristic of a ***first-order*** ΔΣM. The notion of *order* of a ΔΣM is related to the order of the filter (integrator) used in the feedback loop. In a second-order ΔΣM we expect the feedback loop to minimize (or whiten) the integral of the integral of the quantization noise. This yields an error power spectrum of the form

$$S_{ee}(f) = C\,(2\pi f)^4 \quad (C = \text{constant}) \tag{8.3.14}$$

which implies that the quantization noise power in the frequency band up to f_c is given by

$$\sigma_e^2 = k\left(\frac{f_c}{f_s}\right)^5 \tag{8.3.15}$$

Thus, for a second-order ΔΣM, doubling the sampling frequency provides a 15-dB increase in **SNR**. Note however, that the requirement placed on the demodulating lowpass filter is more stringent than that following a first-order **ΔΣM**.

It appears that by increasing the order of the ΔΣM we get a better **SNR** for a given sampling frequency. This is true, up to a point. Increasing the order of a **ΔΣM** has some significant stability implications and a second-order **ΔΣM** is probably the highest order we implement using feedback techniques.

There are not that many ways one can draw a block diagram of a first-order **ΔΣM**. The actual technology of implementation may vary (analog, switched-capacitor, etc.), but the basic block diagram of a first-order **ΔΣM** is shown in Fig. 8.17.

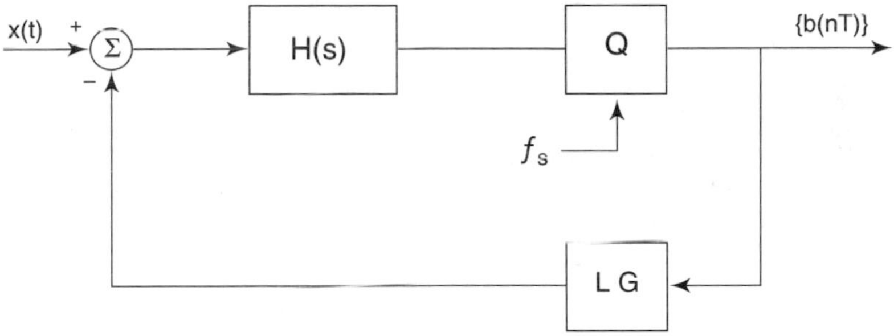

Fig. 8.17
Principal components of a ΔΣM are a filter, H(s), a quantizer, Q, which introduces a delay, and a level generator which performs a D/A function

A first-order **ΔΣM** embeds the quantizer (and associated level generator) in a feedback loop. The filtering is described by a first-order transfer function.

$$H(s) = \frac{G}{s} \tag{8.3.16}$$

The same block diagram is applicable to second-order **ΔΣM**, wherein the loop filter is described by a second-order transfer function of the form

$$H(s) = \frac{G}{s}\frac{s+b}{s+a} \tag{8.3.17}$$

In Eq. 8.3.17, the pole ***a*** is nominally close to zero and the zero ***b*** is large (high frequency). However, it is not advisable to assume H(s) is of the form G/s^2, a perfect double integrator. This is because such a transfer function is practically impossible to achieve and second the loop is "unstable" from a control theoretic viewpoint.

There are other variations possible in the case of second-order ΔΣMs. For example, Fig. 8.18 depicts a ***double loop*** **ΔΣM** in which each of the filtering elements $H_1(s)$ and $H_2(s)$ are each of first-order.

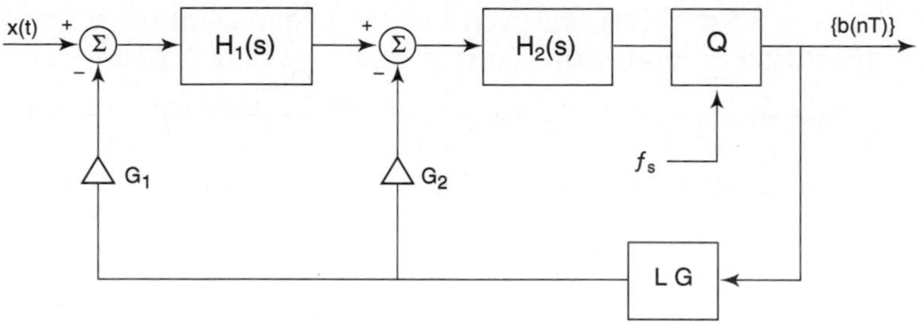

Fig. 8.18
Splitting H(s) to obtain a double-loop ΔΣM

8.4 ΔΣM STRUCTURES

8.4.1 A Typical ΔΣM Circuit

One of the nice features of a ΔΣM is its simplicity of implementation. A first-order ΔΣM is shown in Fig. 8.19.

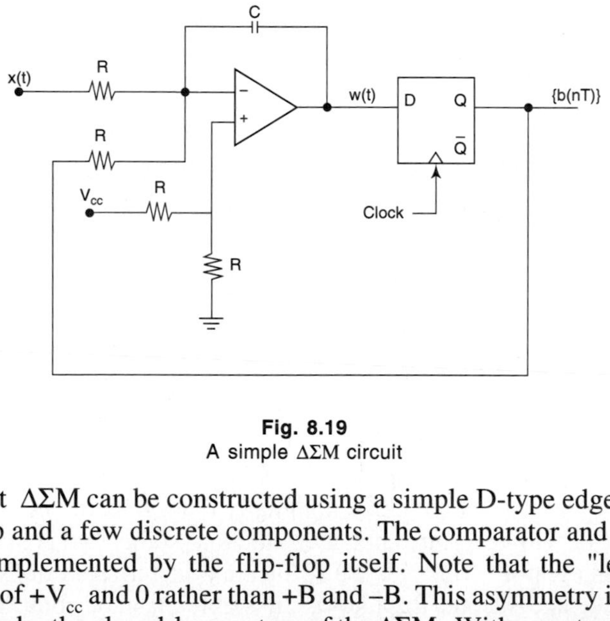

Fig. 8.19
A simple ΔΣM circuit

A single-bit ΔΣM can be constructed using a simple D-type edge-triggered flip-flop, an op-amp and a few discrete components. The comparator and level generator functions are implemented by the flip-flop itself. Note that the "level generator" provides levels of $+V_{cc}$ and 0 rather than +B and –B. This asymmetry is automatically compensated for by the closed-loop nature of the ΔΣM. With an actual flip-flop, such as a conventional 74HC74 device, the high and low voltages will not be quite +5 V and 0 V but the ΔΣM will self-bias itself such that the "zero-input" voltage at the op-amp output will be (nominally) 2.5 V, the decision threshold of the logic device. The input voltage range will be, nominally, [0, 5] volts. Because of the variability from device to device, the circuit in Fig. 8.19 is not used as is if the performance specifications and tolerance limits are stringent, but the circuit is very simple to construct and, apart from being a good demonstration of the ΔΣM concept, is surprisingly very good. For observing the behavior of the conversion process and examining the spectra of the noise, etc., a simple demodulator (lowpass filter) is shown in Fig. 8.20.

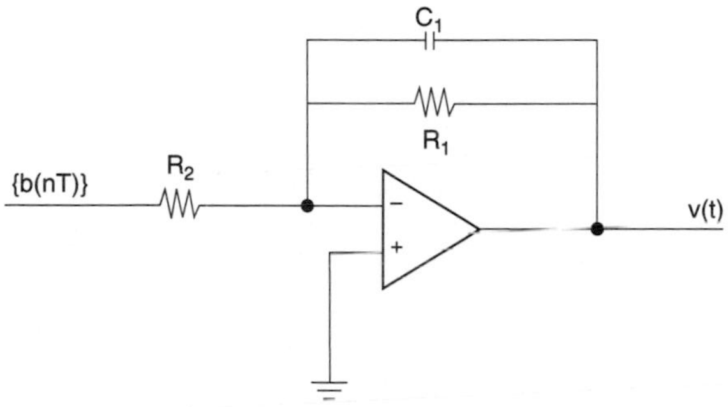

Fig. 8.20
A lowpass filter circuit to reconstruct the ΔΣM output in analog form

The output v(t) in of the circuit in Fig. 8.20 will contain a DC component because the input signal {b(nT)} is a logic level signal between $+V_{cc}$ and 0 V, but at other frequencies v(t) is representative of the analog equivalent of the digital representation, {b(nT)}, of the signal x(t). It is especially useful to observe v(t) in the frequency domain using a spectrum analyzer. Assuming that x(t) is a sinewave, the simplest form of test signal available in a laboratory environment, the spectrum of v(t) viewed on a spectral analyzer takes the form shown in Fig. 8.21.

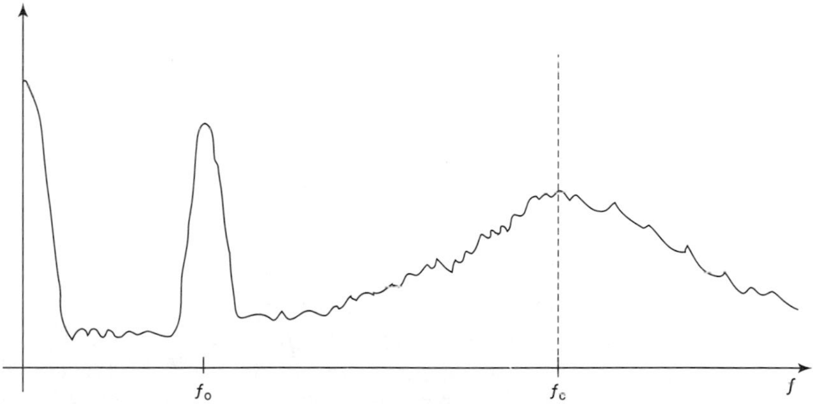

Fig. 8.21
Typical spectrum of a sinusoidal signal at frequency f_0 using ΔΣM depicted in Fig. 8.19

There is a peak at DC, corresponding to the DC offset present in v(t); a peak at frequency f_0, the frequency of the test tone; and a general background noise spectrum, the quantization noise spectrum, which appears as an increasing function of frequency up to a frequency f_c related to the cutoff frequency of the demodulating lowpass filter in Fig. 8.20. If the spectrum analyzer is applied directly to the "square-wave" waveform corresponding to {b(nT)}, it is seen that the quantizing noise spectrum increases all the way up to ($f_s/2$), where f_s is the sampling frequency, i.e., the frequency of the signal used to clock the D flip-flop.

What are appropriate values for R and C in Fig. 8.19? A simple model of a ΔΣM indicates that a suitable choice for the time constant, RC, is given by

$$RC = T = 1/f_s \tag{8.4.1}$$

For experimental purposes, the clock can be made as high as 1 MHz before the "non-ideal" behavior of the 74HC74 becomes significant. Depending on the choice of op-amp, the maximum rate may be somewhat lower. So that the demodulating LPF does not obscure the nature of the noise spectrum, a suitable cutoff (3 dB down) frequency is about ($f_s/10$).

8.4.2 Generalized Delta Sigma Modulation

The basic principle of a ΔΣM is to trade wordlength or, equivalently simplicity in quantizer design, for sampling rate. This is achieved by embedding the quantizer, or noise source, in a feedback loop. The basic configuration of a ΔΣM is shown in Fig. 8.22.

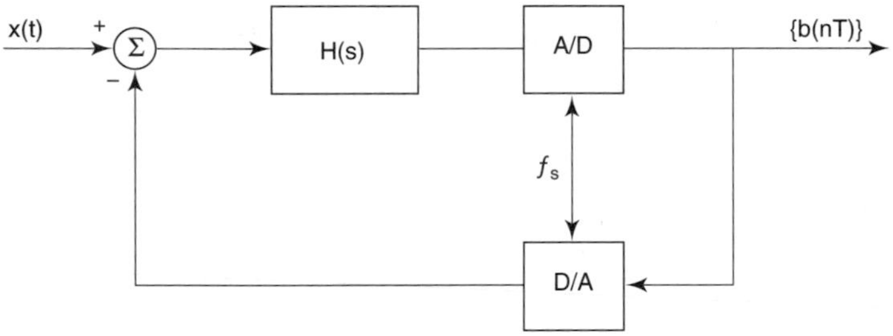

Fig. 8.22
Structure of a multibit ΔΣM

In conventional, that is 1-bit ΔΣMσ, the A/D and D/A converters are quite simple. Considering that ΔΣMs normally operate at sampling rates in excess of 1 MHz, this restriction to 1-bit used to be a limitation imposed by available technology. Advances in solid-state devices, switched-capacitor technology, thin film circuits, hybrid circuits, and a variety of other electronic technologies have removed this restriction—it is often economical to use A/D and D/A converters of 8 to 12 bits operating at sampling rates in excess of 10 MHz. For telephony application the single-bit converter suffices.

The ***order*** of a ΔΣM is determined by the order of H(s); for a first-order ΔΣM

$$H(s) = \frac{g}{s} \tag{8.4.2}$$

and, for a second-order ΔΣM, H(s) can take the form

$$H(s) = \frac{g}{s}\frac{(s+b)}{(s+a)} \tag{8.4.3}$$

It is not necessary that H(s) be a true "analog" filter. For instance, if a ΔΣM is implemented using switched-capacitor technology, then it is appropriate to use a discrete-time transfer function H(**z**) to represent the loop filter.

8.4.3 Switched-Capacitor Implementation of ΔΣM

A suitable representation for a single-bit ΔΣM implemented using switch capacitor techniques is shown in Fig. 8.23. The configuration is ***second-order*** in the sense that there are two integrators, and ***double loop*** in the sense that the quantized value is fed back in two places.

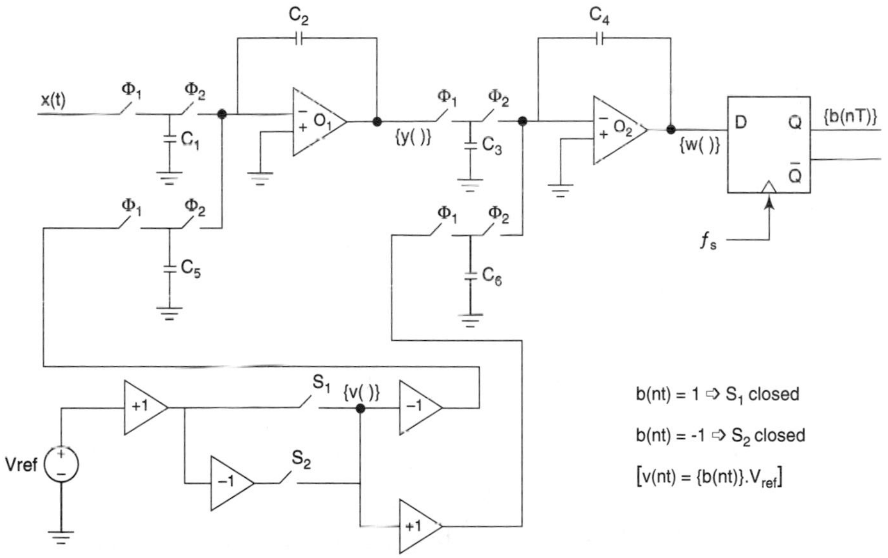

Fig. 8.23
Switched-capacitor implementation of a double-loop one-bit ΔΣM

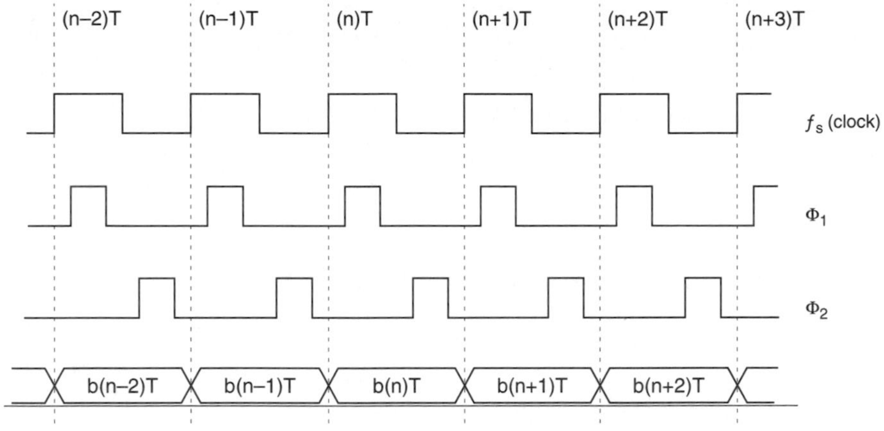

Fig. 8.24
Timing waveforms associated with the switched-capacitor ΔΣM of Fig. 8.23

The operation of the switched-capacitor ΔΣM will be explained using the timing diagram depicted in Fig. 8.24. The rising edges of the clock (f_s) load the (quantized) version of {w(nT)} into the flip-flop. The flip-flop output, {b(nT)}, controls the complementary switches S_1 and S_2. The voltage {v(nT)} is thus $\pm V_{ref}$, where V_{ref} is a precise reference voltage, which we shall assume is 1, i.e., we shall normalize all voltages with respect to V_{ref}. The phases Φ_1 and Φ_2 of the clock control the switches as indicated in the figure. Thus, during the interval of Φ_1, the voltage on the capacitor C_1 charges to the value of x(nT), assuming that x(t) is constant at x(nT) over the interval [nT,(n+1)T]. During the interval of Φ_2, assuming the operational amplifier O_1 is ideal, all the charge resident on C_1 is transferred to C_2, causing an increment in the output voltage of O_1. With ideal switches and amplifiers, the output voltage of amplifier O_1 at time t = (n+1)T can be written as

$$y[(n+1)T] = y(nT) - \left(\frac{C_1}{C_2}\right) x(nT) - \left(\frac{C_5}{C_2}\right) v(nT) \tag{8.4.4}$$

Similarly, the output voltage of amplifier O_2 at time t = (n+1)T can be written as

$$w[(n+1)T] = w(nT) - \left(\frac{C_3}{C_4}\right) y(nT) - \left(\frac{C_6}{C_4}\right) v(nT) \tag{8.4.5}$$

Thus in a switched-capacitor arrangement, the ΔΣM can be represented by the block diagram shown in Fig. 8.25. Determination of H(**z**) requires some manipulation of Eqs. (8.4.4) and (8.4.5).

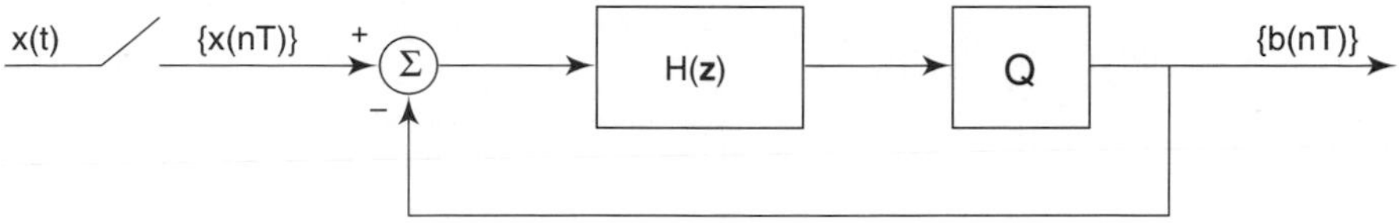

Fig. 8.25
Discrete-time equivalent of switched-capacitor ΔΣM

8.4.4 A Structure for a Digital ΔΣM

The principle of Delta Sigma Modulation, namely trading bandwidth for wordlength can also be used in the D/A process. If {x(nT)} is a digital signal, with a wordlength of B bits, and the sampling rate, $f_s = (1/T)$, is much greater than the minimum, $2f_c$, called out by the sampling theorem, the wordlength can be reduced. By embedding this wordlength reduction in a feedback loop, the noise-shaping capability of a **ΔΣM** can be used to minimize the wordlength required in the D/A converter, while maintaining an acceptable in-band quantization noise power. The block diagram of a digital **ΔΣM** is shown in Fig. 8.26.

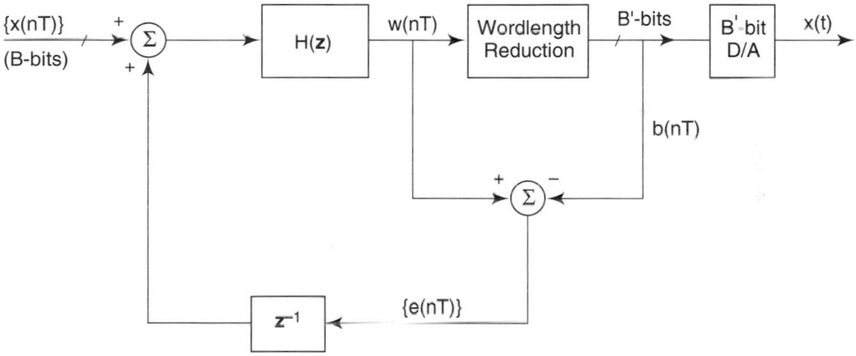

Fig. 8.26
Digital ΔΣM providing noise-shaped wordlength reduction

In a digital ΔΣM, the error introduced in the wordlength reduction process is readily available and is fed back to modify the next sample. The wordlength at the input, B bits, maybe quite large compared with the wordlength actually applied to the D/A converter, B' bits. The wordlength reduction, and computation of the error {e(nT)} is especially easy when the 2–s complement representation is used for representing the digital signals. If w(nT) is an N-bit word then b(nT) is composed of the B' most-significant bits and e(nT) is comprised of the (N–B') remaining bits. The wordlength reduction process is shown in Fig. 8.27.

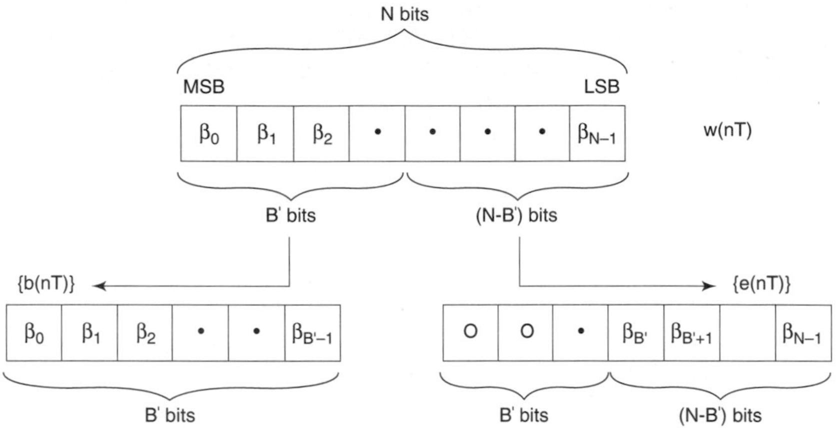

Fig. 8.27
Wordlength reduction achieved by choosing B' most-significant bits

In the simplest case, the equivalent H(**z**) in Fig. 8.26 is H(**z**) = 1, corresponding to a first-order ΔΣM. This technique was used to shape the roundoff noise in digital filters as explained in Chapter 4.

8.5 MODELING THE BEHAVIOR OF A DELTA SIGMA MODULATOR

In the previous section we considered several different structures for implementing **ΔΣM**s. In this section we analyze the behavior of **ΔΣM**s from the viewpoint of additive quantization noise. The output of the **ΔΣM** is modeled as the sum of the desired signal and quantization noise. We introduce the concept of the Signal Transfer Function (**STF**) to characterize the impact of the device on the desired signal. Similarly the Noise Transfer Function (**NTF**) provides the means for characterizing the quantization noise in the output of the **ΔΣM** by modeling the effect of the feedback on the quantization noise introduced by the quantizer itself as it passes on through to the output. This is the "linearized" model for the **ΔΣM**.

For simplicity these notions are introduced using a first-order **ΔΣM** as an example in Section 8.5.1. The analysis is then extended to higher-order **ΔΣM**s in Section 8.5.2. In the latter case a general analysis is more complex since there are a variety of structures that are used, whereas for the first-order case there is but one configuration. Based on the linearized model we define a figure of merit which provides a single number for assessing the efficacy of the feedback. The trade-off between sampling rate and wordlength is studied in the context of the linearized model and the figure of merit in Section 8.5.3.

As the front end for a telephony codec, a single-bit **ΔΣM** provides the best compromise between cost of implementation and performance. It is the simplest implementation that meets the signal-to-noise ratio requirements. In Section 8.5.4 we discuss the considerations involved in designing a single bit **ΔΣM** for telephony applications. From a signal-to-noise ratio, considering only in-band noise, a single-bit **ΔΣM** operating at 1-MHz sampling rate provides an **SNR** comparable to a +13-bit converter operating at 8 kHz and can therefore be used as the front-end converter for either a μ-law or A-law codec with some **SNR** margin.

A feature of the analysis is that it is done as a discrete-time system. Consequently, the results are applicable whether we are converting an analog signal into digital format or we are reducing the wordlength of a digital signal from many bits to a few (as low as 1) bits. In the D/A direction digital filters are used to interpolate the signal, increasing the sampling rate to that at which the actual digital-to-analog conversion is performed. A digital **ΔΣM**, such as depicted in Fig. 8.26, does the wordlength reduction prior to the conversion to analog.

8.5.1 Analysis of a First-Order ΔΣM.

A discrete-time model for a ΔΣM is developed in the same manner as for the ΔM in Section 8.3. For example, consider the ΔΣM shown in Fig. 8.19. The defining equations for the behavior of the circuit are:

$$b(nT) = \begin{cases} 1; & w(nT) > 0 \\ 0; & w(nT) \leq 0 \end{cases} \tag{8.5.1}$$

$$w[(n+1)T] = w(nT) - \left(\frac{T}{RC}\right)[\, x(nT) + b'(nT) \,] \tag{8.5.2}$$

where b'(nT) denotes the equivalent values attributed to the digital signal {b(nT)}. In the case of ΔM we took these values as +B and −B. In the case of the ΔΣM it is convenient to consider the voltages normalized with respect to the maximum expected (2.5 V with a 5 V swing for the logic levels used in Fig. 8.19) and thus b(nT) is +1 or −1 accordingly as b(nT) is 1 or 0 (logic levels). Thus in Eq. 8.4.2 it is assumed that $|x(nT)|$ 1.

Simulation of a ΔΣM based on Eqs. (8.5.1) and (8.5.2) is clearly very straightforward. However, it should be remembered that the results derived from such an exercise can be no better than the assumptions made in deriving these equations.

The relationship between b(nT) and w(nT), i.e., the I/O characteristic of the quantizer, is shown in Fig. 8.28. The notion of overload of the quantizer is when w(nT) exceeds 2 in magnitude. The nonlinear behavior of the quantizer makes an exact analysis of the ΔΣM quite intractable. A linearized model of the quantizer, on the other hand, makes the analysis quite simple. For the linearized model of the quantizer we can write

$$b'(nT) = w(nT) + \mathbf{e}(nT) \tag{8.5.3}$$

where **e**(nT) is the quantization error. This noise signal is clearly related to w(nT) in a deterministic although nonlinear fashion. Assuming that **e**(nT) is uncorrelated with w(nT) is clearly not true but does linearize the model. If **e**(nT) and w(nT) can be considered uncorrelated, then the quantizer can be modeled as a gain and additive noise as shown in Fig. 8.29. A suitable value for the gain, K, is unity, corresponding to line of slope = 1 in Fig. 8.28.

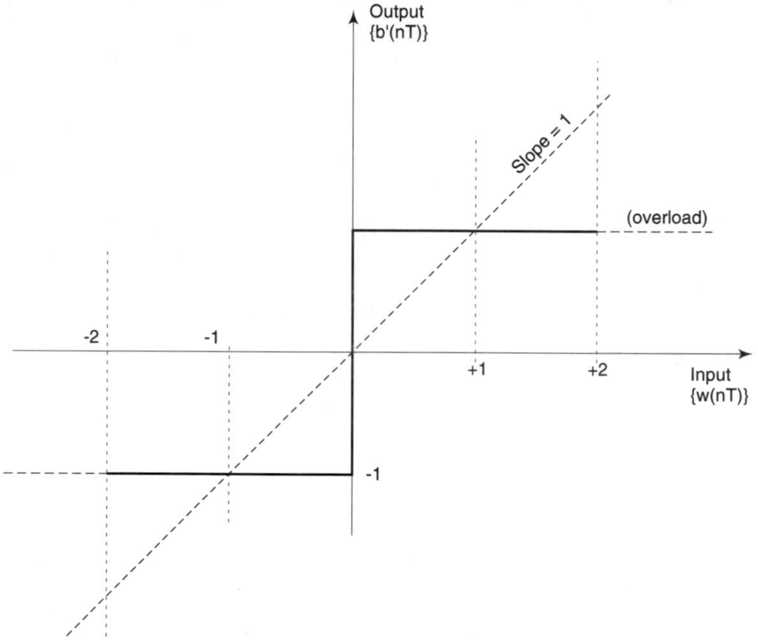

Fig. 8.28
Quantizer characteristic for a one-bit $\Delta\Sigma$**M**

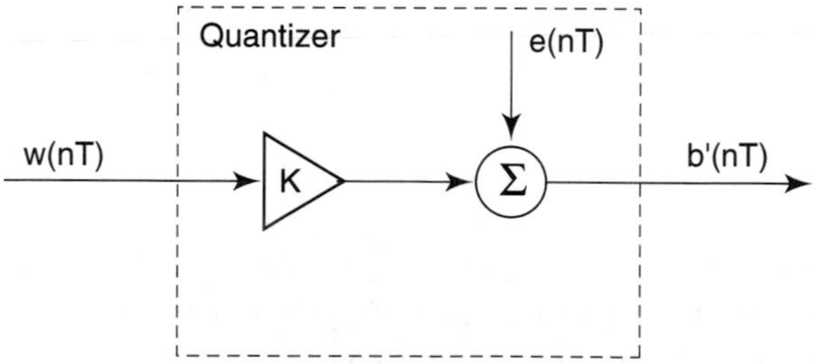

Fig. 8.29
Linearized model for the quantizer

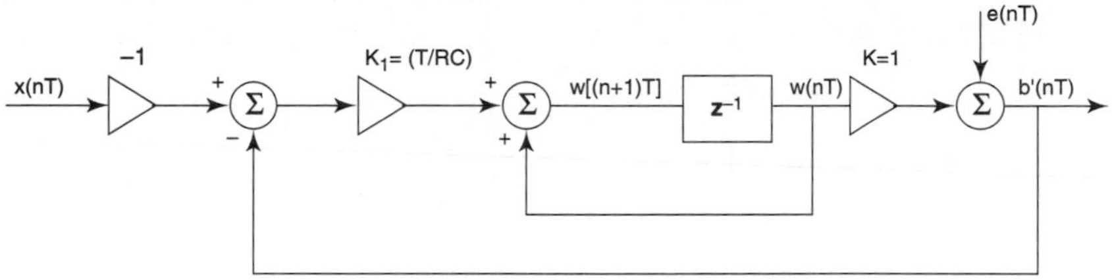

Fig. 8.30
Discrete-time linear system model for a first-order ΔΣM

With the quantizer modeled as shown, a discrete-time equivalent of the ΔΣM is developed as shown in Fig. 8.30. Clearly, the input gain term of –1 applied to x(nT) is an artifice of the op-amp implementation and can be ignored. Based on this block diagram we define two transfer functions. First, the *Signal Transfer Function*, or **STF**, relates how the output b(nT) depends on the input x(nT). The second, the *Noise Transfer Function,* or **NTF**, describes the manner in which the additive noise e(nT) contributes to the output b(nT). Simple manipulation yields

$$\textbf{STF}:\ S(\mathbf{z}) = \frac{B'(\mathbf{z})}{X(\mathbf{z})} = \frac{KK_1\mathbf{z}^{-1}}{1 + (KK_1 - 1)\mathbf{z}^{-1}} \tag{8.5.4}$$

$$\textbf{NTF}:\ N(\mathbf{z}) = \frac{B'(\mathbf{z})}{E(\mathbf{z})} = \frac{1 - \mathbf{z}^{-1}}{1 + (KK_1 - 1)\mathbf{z}^{-1}} \tag{8.5.5}$$

The $\mathbf{z}^{-1}$ in the numerator of Eq. (8.5.4) is just indicative of a delay between the input and output and is of little consequence. The equations do depend on K, the equivalent linearized gain of the quantizer, and K_1, which is related to the sampling rate and the time constant of the integrator. Assuming both are unity, the **STF** and **NTF** become

$$\textbf{STF}:\ S(\mathbf{z}) = \frac{B'(\mathbf{z})}{X(\mathbf{z})} = \mathbf{z}^{-1} \tag{8.5.6}$$

$$\textbf{NTF}:\ N(\mathbf{z}) = \frac{B'(\mathbf{z})}{E(\mathbf{z})} = 1 - \mathbf{z}^{-1} \tag{8.5.7}$$

We make the usual set of assumptions made when dealing with quantizers. That is, the additive noise, **e**(nT), is white, uncorrelated with the input (in this case w(nT) and, indirectly, x(nT)), and has a uniform **pdf** provided overload conditions do not exist. While these assumptions may be easier to justify when the quantizer has several more quantization steps than just two, we make them anyway. Assuming no overload, it can be seen from the quantizer characteristics of Fig. 8.28 that

$$|e(nT)| \quad 1 \tag{8.5.8}$$

and thus the uniform **pdf** assumption implies that the variance of the noise signal, σ_e^2, is given by

$$\sigma_e^2 = \frac{2^2}{12} = \frac{1}{3} \tag{8.5.9}$$

The assumption that **e**(nT) is white means that the autocorrelation sequence, $\{R_{ee}(n)\}$, of the noise signal is given by

$$R_{ee}(n) = \frac{1}{3}\delta(n) \tag{8.5.10}$$

and its power spectral density, $S_{ee}(f)$, can be expressed as

$$S_{ee}(f) = \frac{1}{3} \tag{8.5.11}$$

Let us denote by $\eta(nT)$ the quantization noise component as it propagates to the output. It is related to **e**(nT) via the **NTF**. Using Eq. (8.5.7),

$$\eta(nT) = \mathbf{e}(nT) - \mathbf{e}[(n-1)T] \tag{8.5.12}$$

and thus

$$\sigma_\eta^2 = \frac{2}{3} \tag{8.5.13}$$

$$R_{\eta\eta}(n) = \frac{2}{3}\delta(n) + \frac{1}{3}\delta(n-1) + \frac{1}{3}\delta(n+1) \tag{8.5.14}$$

and, after taking the discrete-time Fourier transform and some algebraic manipulation,

$$S_{\eta\eta}(f) = \frac{4}{3}\left[\sin\left(\frac{\pi f}{f_s}\right)\right]^2 \tag{8.5.15}$$

The fraction of the noise power that resides at low frequencies can be computed as

$$\mu = \frac{\int_0^{f_c} S_{\eta\eta}(f)\,df}{\int_0^{(\frac{1}{2})f_s} S_{\eta\eta}(f)\,df} \tag{8.5.16}$$

which, after substituting $S_{\eta\eta}(f)$ from Eq. (8.5.15) yields

$$\mu = \frac{1}{\pi}\left[\left(\frac{2\pi f_c}{f_s}\right) - \sin\left(\frac{2\pi f_c}{f_s}\right)\right] \qquad (8.5.17)$$

For $f_c << f_s$, the fraction of noise power at low frequencies can be written as

$$\mu = \frac{1}{\pi\, 3!}\left[\left(\frac{2\pi f_c}{f_s}\right)\right]^3 \qquad (8.5.18)$$

It is seen that the total quantization noise power, σ_η^2, can be greater than the quantization noise introduced by the quantizer, σ_e^2. However, the efficacy of the **ΔΣM** stems from the fact that only a (small) fraction of this noise power is actually within the frequency band of interest and that out-of-band noise can be removed using a lowpass filter. Of course, the extent to which the out-of-band noise can be removed is related to how good the LPF is; achieving Eq. (8.5.18) requires an ideal LPF.

8.5.2 The General Model for a ΔΣM

The noise performance of a **ΔΣM** is analyzed from the noise transfer function. Assuming that the quantizer introduces white noise, $\{\mathbf{e}(nT)\}$, of power σ_e^2, then the net quantization noise in the output is modeled as the output of a discrete-time **LTI** system with transfer function $N(\mathbf{z})$ and excited by the white noise sequence $\{\mathbf{e}(nT)\}$. This is depicted in Fig. 8.31.

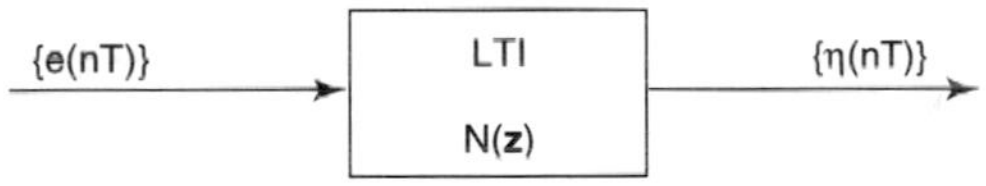

Fig. 8.31
Noise-shaping modeled as a transfer function N(**z**)

The noise power gain or G_N, is evaluated from the **NTF** as:

$$G_N = \frac{1}{2\pi j}\oint_{|\mathbf{z}|=1} N(\mathbf{z})N(\mathbf{z}^{-1})\mathbf{z}^{-1}\,d\mathbf{z} \qquad (8.5.19)$$

The fraction of the noise power in low frequencies, up to f_c, is computed as

$$\mu = \frac{\int_0^{f_c} |N(f)|^2\,df}{\int_0^{(\frac{1}{2})f_s} |N(f)|^2\,df} \qquad (8.5.20)$$

The quantizing noise power in the band of interest is, therefore, $\mu\, G_N\, \sigma_e^2$.

In order to ascertain the **STF** and **NTF**, we need to derive a discrete-time equivalent for the **ΔΣM**. In the case of a switched-capacitor configuration or a digital ΔΣM, this configuration is implicit and will be of the form shown in Fig. 8.32.

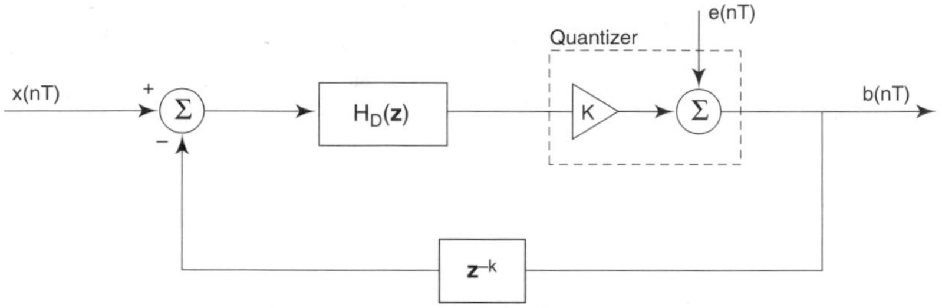

Fig. 8.32
Model for establishing the **STF** and **NTF** for a ΔΣM

The feedback path includes a delay of k samples. It is crucial to understand that such a delay is inevitable. In the case of a digital **ΔΣM** it is possible to ensure that k=1. In the case of the analog **ΔΣM**, determining a suitable value for k is often not obvious though it is clear that there will be some delay associated with the conversion from analog to digital and back to analog. The reason for this is that A/D and D/A converters often employ pipelined architectures internal to the devices. The implication of pipelining in an A/D converter is described with reference to the timing diagrams depicted in Fig. 8.33.

An A/D converter is visualized as having three internal functions—sampling, conversion, and output formatting. The sampling action captures the value of the analog signal, the conversion action determines where in the quantizer characteristic the sample value lies, and the output-formatting function assigns the appropriate code word to the quantized level. As shown in Fig. 8.34, the sample of the analog signal taken at t=nT may not be available at the output until after t=(n+1)T. This value is then clocked into the D/A and the conversion to analog cannot occur prior to (n+2)T, assuming that all transfers are done on the rising edge of the master clock. Thus the delay, k samples, comprising the effect of A/D and D/A conversion, can be as much as two and even greater, if the D/A has internal pipelining as well.

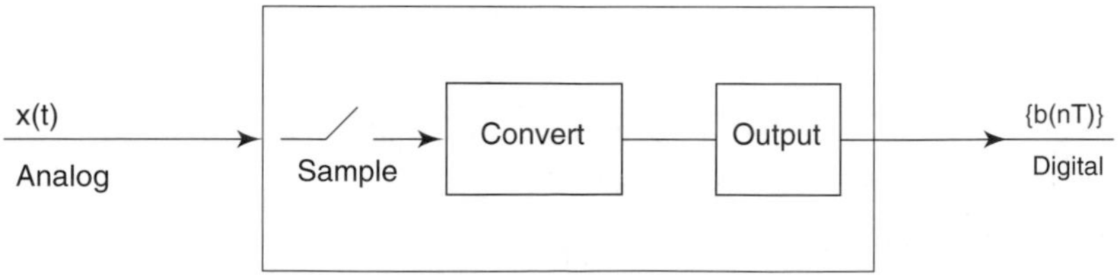

Fig. 8.33
Internal functions of an A/D converter

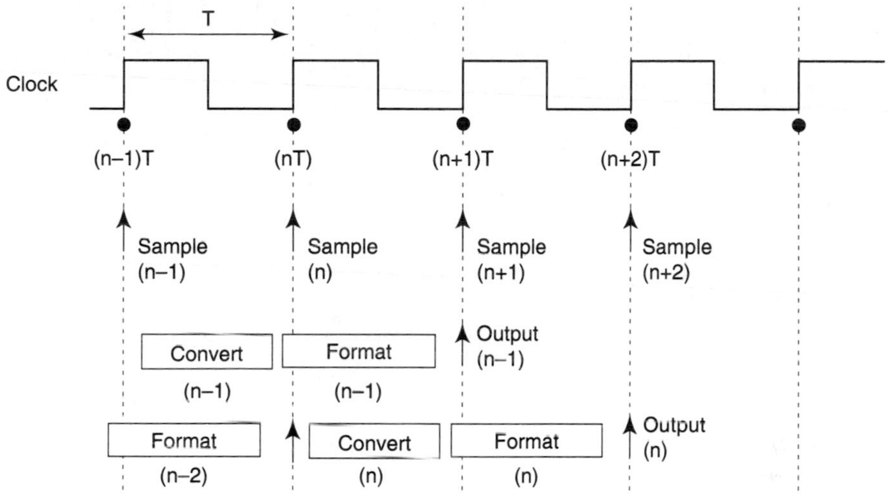

Fig. 8.34
Timing schematic illustrating delay introduced by an A/D converter

The discrete time equivalent, $H_D(\mathbf{z})$, of the analog loop filter, H(s), depends, to some extent, on the assumptions made regarding the behavior of the analog signals between sampling instants. A simple, though realistic, assumption is that the input analog signal and the output of the D/A remain constant over each sampling interval, T. With this assumption, consider the simple analog transfer functions $H_1(s)$ and $H_2(s)$ given by

$$H_1(s) = \frac{g}{s} \tag{8.5.21}$$

and

$$H_2(s) = \frac{g}{s+a} \tag{8.5.22}$$

We wish to obtain the discrete time equivalents $H_{1D}(\mathbf{z})$ and $H_{2D}(\mathbf{z})$, respectively. The equivalent differential equation corresponding to Eq. (8.5.21) is

$$\frac{dy(t)}{dt} = g\, x(t) \tag{8.5.23}$$

which is solved as, for t nT,

$$y(t) = y(nT) + g \int_{nT}^{t} x(\tau)\, d\tau \tag{8.5.24}$$

which yields

$$y[(n+1)T] = y(nT) + gT\,x(nT) \qquad (8.5.25)$$

In deriving the above equations we have assumed that x(t) is constant over the interval [nT, (n+1)T]. The equivalent discrete-time transfer function is, therefore,

$$H_{1D}(\mathbf{z}) = \frac{Y(\mathbf{z})}{X(\mathbf{z})} = \frac{gT\,\mathbf{z}^{-1}}{1 - \mathbf{z}^{-1}} \qquad (8.5.26)$$

Note, however, that we cannot assume that y(t) is constant over the interval [nT, (n+1)T].

The equivalent differential equation corresponding to Eq. (8.5.22) is

$$\frac{dy(t)}{dt} + a\,y(t) = g\,x(t) \qquad (8.5.27)$$

which can be solved as, for t nT,

$$y(t) = e^{-a(t-nT)}y(nT) + g\int_{nT}^{t} e^{-a(t-\tau)}x(\tau)\,d\tau \qquad (8.5.28)$$

which yields

$$y[(n+1)T] = e^{-aT}y(nT) + \frac{g}{a}(1-e^{-aT})\,x(nT) \qquad (8.5.29)$$

where it has been assumed again that x(t) is constant over the interval [nT, (n+1)T]. The equivalent discrete-time transfer function is therefore,

$$H_{2D}(\mathbf{z}) = \frac{\frac{g}{a}(1-e^{-aT})\,\mathbf{z}^{-1}}{1 - e^{-aT}\mathbf{z}^{-1}} \qquad (8.5.30)$$

Eqs. (8.5.26) and (8.5.30) provide suitable discrete-time replacements for first-order all-pole transfer functions given by Eqs. (8.5.21) and (8.5.22).

If $H_3(s)$ is a second-order transfer function of the form

$$H_3(s) = \frac{g\,(s+b)}{s\,(s+a)} \qquad (8.5.31)$$

then it can be split into a partial fraction expansion as

$$H_3(s) = \frac{g}{a}\left[\frac{b}{s} - \frac{b-a}{s+a}\right] \tag{8.5.32}$$

and the equivalent discrete-time transfer function is obtained as

$$H_{3D}(\mathbf{z}) = \frac{g}{a}\frac{b\mathbf{z}^{-1}}{1-\mathbf{z}^{-1}} - \frac{g}{a}\frac{(b-a)\mathbf{z}^{-1}}{1-e^{-aT}\mathbf{z}^{-1}}\frac{(1-e^{-aT})}{a} \tag{8.5.33}$$

The pole, a, will be close to zero. Writing $\alpha = e^{-aT}$, α is close to unity. With this substitution, Eq. (8.5.33) can be rearranged into the form

$$H_{3D}(\mathbf{z}) = \frac{K_1(\mathbf{z}-\eta)}{(\mathbf{z}-1)(\mathbf{z}-\alpha)} \tag{8.5.34}$$

where

$$\gamma = \frac{(1-e^{-aT})}{aT} \tag{8.5.35a}$$

$$\eta = \frac{\alpha b - \gamma(b-a)}{b - \gamma(b-a)} \tag{8.5.35b}$$

$$K_1 = \frac{gT}{a}[b - \gamma(b-a)] \tag{8.5.35c}$$

The terms α, η, and γ are all close to, but *not* equal to, unity. If we assume that, as before, gT is approximately 1 then even K_1, will be close to 1. With reference to Fig. 8.32, the signal transfer function and noise transfer function can be written as

$$\textbf{STF}:\ S(\mathbf{z}) = \frac{K\,H_D(\mathbf{z})}{1+\mathbf{z}^{-1}K\,H_D(\mathbf{z})} \tag{8.5.36}$$

$$\textbf{NTF}:\ N(\mathbf{z}) = \frac{1}{1+\mathbf{z}^{-1}K\,H_D(\mathbf{z})} \tag{8.5.37}$$

The gain term K represents the linearized gain of the quantizer. When the A/D and D/A converters have several bits of accuracy, that is the quantizer has several levels, assuming K=1 is quite accurate. As mentioned before, in the discussion of the single-bit case, we make this assumption regardless of the number of bits, since otherwise the analysis becomes too complicated to be worthwhile.

Substituting Eq. (8.5.34) into Eq. (8.5.36), the **STF** for the second-order **ΔΣM** with analog transfer function $H_3(s)$ can be written as

$$\textbf{STF}: S(\mathbf{z}) = \frac{K_1\mathbf{z}^{-1}(1-\eta\,\mathbf{z}^{-1})}{1 - (\alpha+1-K_1)\mathbf{z}^{-1} + (\alpha-\eta K_1\mathbf{z}^{-2})} ; (k=0)$$

(8.5.38)

$$\textbf{STF}: S(\mathbf{z}) = \frac{K_1\mathbf{z}^{-1}(1-\eta\,\mathbf{z}^{-1})}{1-(\alpha+1)\mathbf{z}^{-1}+(\alpha+K_1\mathbf{z}^{-2}) - \eta K_1\mathbf{z}^{-3}} ; (k=1)$$

Three points of interest can be gleaned from Eq. (8.5.38). First, incorporation of the transmission zero, ***b***, in the analog transfer function has a stabilizing effect on the loop as is evidenced from the poles of the **STF**, which determine the stability of the loop. For stability, the poles must be **inside** the unit circle. The presence of the zero forces the product of the poles to be $(\alpha - \eta K_1)$, nominally a small number, improving the chances that each pole is less than unity in magnitude. Second, the delay introduced by the A/D-D/A combination, k samples, has a distinct destabilizing effect by increasing the order of the loop. The product of the poles, for a delay of k=1 samples, is $K_1\eta$, which being close to unity, indicates that there is a good chance for one or more of the three poles to be outside the unit circle. Third, there is a spectral shaping applied to the desired signal. The signal component of the **ΔΣM** output will be a filtered version of the input.

The noise transfer function N(z) is given by

$$\textbf{NTF}: N(\mathbf{z}) = \frac{(1-\mathbf{z}^{-1})(1-\alpha\,\mathbf{z}^{-1})}{1 - (\alpha+1-K_1)\mathbf{z}^{-1} + (\alpha-\eta K_1\mathbf{z}^{-2})} ; (k=0)$$

(8.5.39)

$$\textbf{NTF}: N(\mathbf{z}) = \frac{(1-\mathbf{z}^{-1})(1-\alpha\,\mathbf{z}^{-1})}{1-(\alpha+1)\mathbf{z}^{-1}+(\alpha+K_1\mathbf{z}^{-2}) - \eta K_1\mathbf{z}^{-3}} ; (k=1)$$

Second-order **ΔΣM**s are usually associated with a noise transfer function

$$\textbf{NTF} : N(\mathbf{z}) = (1 - \mathbf{z}^{-1})^2 \qquad (8.5.40)$$

and if the pole, ***a***, is close enough to DC, the numerator of Eq. (8.5.39) does tend toward the N(**z**) of Eq. (8.5.40). Assuming for the moment that the ideal second-order noise transfer function is achievable, then the following holds true.

$$G_N = \underset{n}{} [h_N(n)]^2 = 6 \tag{8.5.41}$$

Assuming that the quantizer adds white noise, the power spectrum of the quantizing noise in the **ΔΣM** output is given by

$$S_{\eta\eta}(f) = \sigma_e^2 \left[6 - 8\cos\left(\frac{2\pi f}{f_s}\right) + 2\cos\left(\frac{4\pi f}{f_s}\right) \right] \tag{8.5.42}$$

where some common trigonometric identities have been used to obtain the form of the power spectral density $S_{\eta\eta}(f)$. The fraction of the noise power at low frequencies, less than f_c, is given by

$$\mu = \frac{1}{6\pi} \left[\left(\frac{12\pi f_c}{f_s}\right) - 8\sin\left(\frac{2\pi f_c}{f_s}\right) + \cos\left(\frac{4\pi f_c}{f_s}\right) \right] \tag{8.5.43}$$

For $f_c << f_s$, the fraction of noise power at low frequencies is given by

$$\mu \approx \frac{16}{5} \pi^4 \left(\frac{f_c}{f_s}\right)^5 \tag{8.5.44}$$

Thus, with the ideal second-order noise transfer function, the quantization noise present at low frequencies is estimated as

$$\sigma_N^2 = G_N \mu \sigma_e^2 \approx 6 \left\{ \frac{16}{5} \pi^4 \left(\frac{f_c}{f_s}\right)^5 \right\} \sigma_e^2 \tag{8.5.45}$$

In the case of switched-capacitor and digital ΔΣMs, the inherent discrete-time nature permits us to develop the **STF** and **NTF** models quite easily. For instance, Fig. 8.35 depicts a double-loop single-bit ΔΣM, which could be either a switched-capacitor version or a digital implementation. The signal levels are normalized such that the quantized signal takes on the numerical values ±1. The choice of the gain term of (1/2) in the feedback is deliberate, and the reason for this choice is explained shortly.

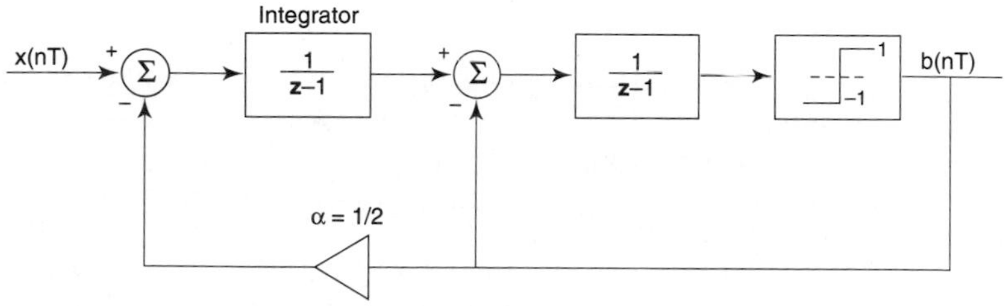

Fig. 8.35
Block representation of a double-loop second-order ΔΣM (digital or switched-capacitor)

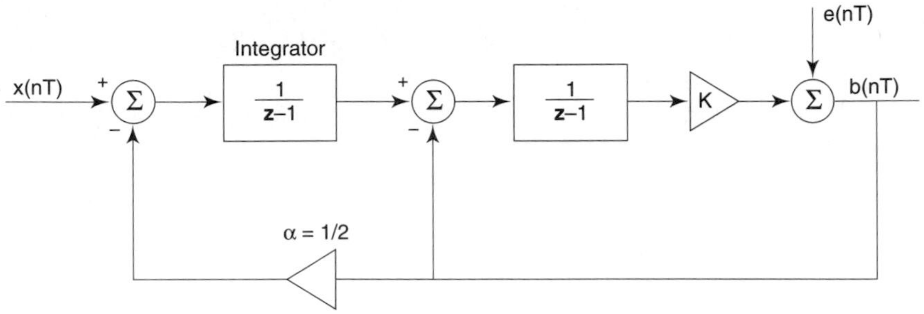

Fig. 8.36
Linearized model for analyzing ΔΣM of Fig. 8.35

Replacing the quantizer single bit in the case of Fig. 8.35 by a gain and additive noise yields the linearized equivalent shown in Fig. 8.36. Assuming K=1, the **STF** and **NTF** are derived as

$$S(\mathbf{z}) = \frac{1}{\mathbf{z}^2 - \mathbf{z} + \alpha} \tag{8.5.46}$$

$$N(\mathbf{z}) = \frac{(\mathbf{z} - 1)^2}{\mathbf{z}^2 - \mathbf{z} + \alpha} \tag{8.5.47}$$

Note that the numerator of N(**z**) does provide the ideal second-order noise shaping characteristic. Also, the signal transfer function does not exhibit the zero that the analog version, as shown in Eq. (8.5.38), has.

It is clear that the **NTF** of a practical second-order **ΔΣM** does not have the ideal behavior expressed in Eq. (8.5.40). First, obtaining the ideal double zero at **z** = +1 is not possible with an analog implementation. Second, there is the impact of the poles. Typical noise spectra are depicted in Fig. 8.37. The presence of poles has the effect of "early peaking" of the noise spectrum. This increases the burden on the subsequent lowpass filter whose job is to attenuate the noise components above f_c. If the poles of the **NTF** are $re^{j\theta}$, then $S_{\eta\eta}(f)$ will peak close to $f = (\theta/2\pi)f_s$ and the peak will be sharp if r is close to 1. If r >1 the structure is unstable. Also, the idle pattern will have a fundamental frequency close to $f = (\theta/2\pi)f_s$, especially if r is close to unity. From the viewpoint of lowpass filter requirements and idle channel behavior, it is advisable to keep r as small as possible and θ as close to π (radians) (180 degrees) as possible. That is, the poles should lie well within the unit circle and away from the voice-band.

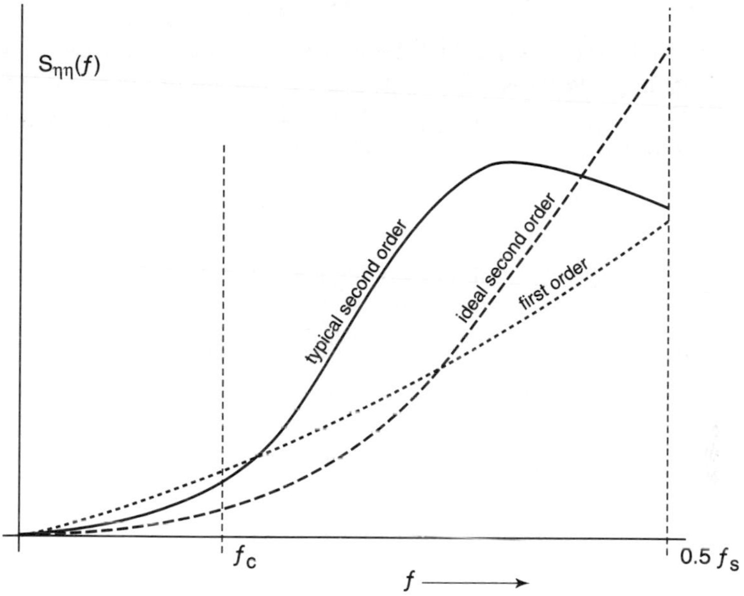

Fig. 8.37
Typical noise spectra for first- and second-order ΔΣMs

8.5.3 Figure of Merit and Bandwidth for Wordlength Trade-Off

A useful guide as to how well the **ΔΣM** is providing noise shaping of the quantization error is the Figure of Merit (**FOM**) defined as

$$\mathbf{FOM} = -\log_{10}[\, G_N \mu \,] \text{ dB} \tag{8.5.48}$$

The **FOM** is an indication of the signal-to-noise ratio improvement provided by the feedback. If the quantizer provides a signal-to-noise-ratio of $\mathbf{SNR_Q}$, the overall **SNR** is given by

$$\mathrm{SNR_{TOT}} = \mathrm{SNR_Q} + \mathbf{FOM} \quad \text{dB} \tag{8.5.49}$$

The usual assumption is that the quantizing noise introduced by an A/D converter is uncorrelated with the signal being quantized and exhibits properties associated with random white noise that is amplitude limited. That is, if an A/D converter is in its linear region, not overloaded, the additive noise $\{\mathbf{e}(n)\}$ is described statistically, by the following measures:

$$|\mathbf{e}(n)| \le \Delta \tag{8.5.50a}$$

$$S_{\mathbf{ee}}(f) = \sigma_{\mathbf{e}}^2 \tag{8.5.50b}$$

$$\sigma_e^2 = K\Delta^2 \tag{8.5.50c}$$

If we further assume that the **pdf** of the noise is uniform, the constant K in Eq. (8.5.50) is given by K=1/3. This noise is spread out in the frequency domain over $[-f_s/2, f_s/2]$ and of this noise, only a fraction lies in the range $[-f_c, f_c]$, which is the frequency band of interest. Thus the effective noise power is

$$\sigma_\eta^2 = 2\left(\frac{f_c}{f_s}\right)\sigma_e^2 \tag{8.5.51}$$

Increasing the sampling frequency by a factor of two halves the effective quantization noise. In other words, we get a 3-dB improvement in signal–to-noise ratio for every octave increase in sampling frequency.

For a first-order **ΔΣM** the in-band quantization noise can be quantified as, from Eqs. (8.5.13) and (8.5.18),

$$\sigma_\eta^2 = \frac{2}{3!\,\pi}\left(\frac{f_c}{f_s}\right)^3 \sigma_e^2 \tag{8.5.52}$$

Increasing the sampling frequency by a factor of two, reduces the noise power by a factor of 2^3. In other words, using a first-order feedback loop provides a 9-dB improvement in **SNR** for every octave increase in sampling frequency. For a second-order **ΔΣM**, the quantizing noise power in-band can be expressed as (from Eq. (8.5.45))

$$\sigma_\eta^2 = (\text{constant})\left(\frac{f_c}{f_s}\right)^5 \sigma_e^2 \tag{8.5.53}$$

indicating that we can get a 15-dB improvement in **SNR** for every octave increase in sampling frequency. Since we can reduce the in-band quantization noise power either by increasing the sampling rate or by increasing the resolution of the A/D converter, the rules of thumb shown in Table 8.1 can be applied.

Table 8.1
The sampling rate for wordlength trade-off

Configuration	Equivalence		
	Wordlength Increase		Sampling Rate Increase
No Feedback	1 Bit (6 dB/Bit)	≈	2 Octaves (3 dB/Octave)
First-Order ΔΣM	3 Bits (6 dB/Bit)		2 Octaves (9 dB/Octave)
Second-Order ΔΣM	5 Bits (6 dB/Bit)		2 Octaves (15 dB/Octave)

8.5.4 Considerations in the Design of Single-Bit ΔΣMs

The most commonly encountered **ΔΣM**s are the ones that quantize the signal to 1 bit, primarily because of the simplicity of implementation. Since **ΔΣM**s usually operate at high sampling rates, multibit A/D converters tend to be complex and expensive. Furthermore, such high speed converters tend to have pipelined implementations—increasing the delay in the loop with the ensuing destabilizing effect. The design methodology applicable to single-bit **ΔΣM**s will be described with reference to a specific example—design of an A/D converter for speech signals in telephony applications.

In telephony, the A/D conversion must meet the specification depicted in Fig. 8.38. For signal levels between 0 and –30 dBm0 the **SNR** must exceed 35 dB; the "worst-case" specification is an **SNR** of 25 dB at –45 dBm0, which when linearly extrapolated, corresponds to an **SNR** of 73 dB at a signal level of +3 dBm0.

It is convenient to divide the signal range into three distinct regions: high input or *overload*, medium input or *normal*, and low input or *idle channel*. The plot in Fig. 8.38, which displays signal-to-noise ratio versus input signal power, delineates these three regions. The normal region is characterized by a linear relation between **SNR** and input power. The peak input signal is arbitrarily defined as 3 dBr. That is, a sinewave of peak allowable amplitude V_{ref} has a power of 3 dBr. This peak amplitude is normalized to unity. Note that in Fig. 8.38, the abscissa has two different scales, dBr and dBm0. The dBm0 representation of signal level is a system specification whereas dBr is a component specification relating specifically to the A/D converter. They are related via 0 dBm0 = – **x** dBr where **x** (0) is also a design parameter. The designer of the **ΔΣM** (i.e., the A/D converter) has the freedom of choosing **x**, in effect shifting the specification template to lie entirely under the **SNR** curve of the **ΔΣM**, preferably with some **margin** to allow for nonideal behavior of circuit components. **x** is therefor, the margin.

The system requirements can be translated into two quantities: the peak **SNR**, 73 dB in the above example, and an overload point, nominally 3 dBm0. How does the overload point influence the design parameters μ, G_N, and f_s? Since the output of the ΔΣM is bilevel, ±1, the output power is fixed at unity, independent of input power. The output power must be shared between "signal" and "noise." If the share of the noise is large, the dynamic range or overload point of the ΔΣM will be small.

For elaboration, let us assume, as a first approximation, that the additive noise {e(n)} has a uniform probability density function over (–1, 1), corresponding to a variance σ_e^2, of 1/3. The output noise power is then $G_N/3$. Clearly, $G_N < 3$ is a necessary restriction on the noise power gain. This line of reasoning suggests that the ΔΣM is in an "overload" condition when the input signal power exceeds $(1 - G_N/3)$. Actually, the ΔΣM goes into overload when the comparator input exceeds 2 (normalized units), implying an instantaneous quantization error greater than 1. Simulation and experimental results indicate that the input signal power is roughly $(1/10)(1 - G_N/3)$ when the ΔΣM just enters overload and, furthermore, that this onset of overload condition is not catastrophic.

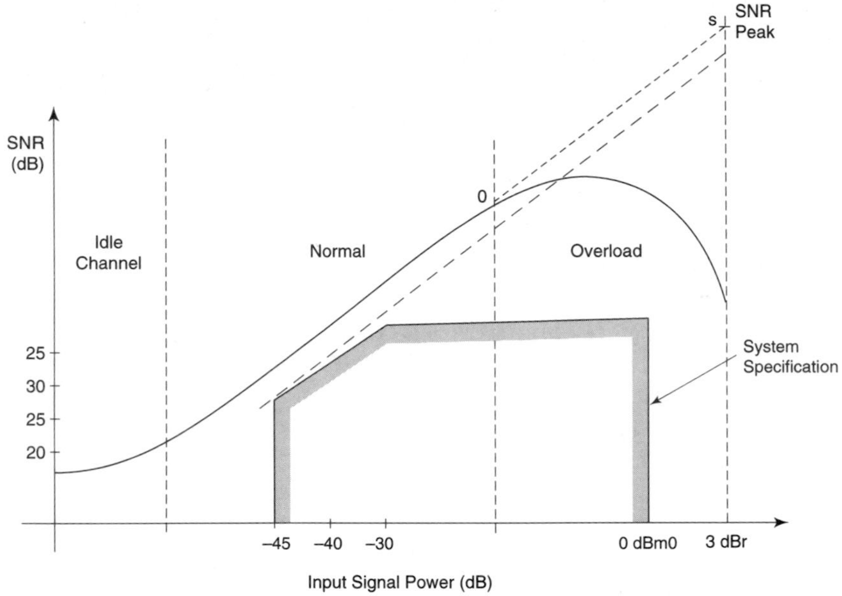

Fig. 8.38
Performance of a converter expressed in terms of **SNR** as a function of signal power

The determination of μ, G_N, and f_s is an iterative procedure and, since these parameters are interrelated, a consistency check is required. Furthermore, it is preferable to obtain a range, rather than a fixed number, for these parameters to account for any constraints imposed by circuit implementation. A suitable first step is to choose the sampling frequency. For the example considered, which is derived from a telephony application, it is known that a 13-bit uniform coder at 8 kHz will meet the requirements. Assuming a second-order **ΔΣM** and using the rules of thumb of *15 dB/octave* and *6 dB/bit*, the sampling frequency should be about 256 kHz (actually greater). In comparison, a first-order **ΔΣM** needs to operate at a sampling frequency in excess of 1 MHz. We shall assume that the sampling frequency is set at 1 MHz and a second-order structure is employed. This is to allow for adequate margin.

The next step is to ascertain a suitable ballpark figure for G_N. Assuming that we provide a 6-dB margin (**x**=6), then as a first approximation,

$$10 \log_{10}\left(1 - \frac{G_N}{3}\right) \approx -6 \quad \text{(dB scale)} \tag{8.5.54a}$$

and thus

$$G_N \approx 2.25 \tag{8.5.54b}$$

The peak signal power we can sustain, assuming that the desired signal is a sinusoid, is

$$S_{peak} = \left(\frac{1}{2}\right) \qquad (8.5.55)$$

since the voltages are normalized such that the peak is unity. Based on our discussion of the **FOM**, the in-band noise power is

$$\sigma_\eta^2 = G_N \mu \sigma_e^2 = G_N \mu \left(\frac{1}{3}\right) \qquad (8.5.56)$$

The peak signal-to-noise ratio is thus

$$SNR_p = -10 \log_{10}\left(\frac{2}{3}\right) - 10 \log_{10}(G_N \mu) \qquad (8.5.57)$$

and with the value of noise power gain given in Eq. (8.5.54), the approximate value for the fraction, μ, of in-band noise power, can be obtained by setting $\mathbf{SNR}_p = 79$ dB in Eq. (8.5.57) (73 dB + 6-dB margin), which indicates that the value of μ is, approximately

$$\mu \approx 8.4 \times 10^{-9} \qquad (8.5.58)$$

For any choice of sampling frequency, it must be verified that pole-zero locations, or circuit element values, can be determined to meet the requirement of peak **SNR**. The rules of thumb used earlier indicate that a sampling frequency in excess of 256 kHz is required. We shall assume that a value of 1 MHz (or 1.024 MHz) is chosen.

Apart from the peak **SNR**, there are some other considerations that govern the choice of design parameters. These are idle-channel noise and the requirements on the subsequent lowpass filter that attenuates the high frequency noise components prior to undersampling. To see how these requirements interact, consider the configuration of Fig. 8.22 with a single-bit quantizer and H(s) given by Eq. (8.4.3). The zeros of the **NTF** should be kept as close to $\mathbf{z} = +1$ as possible since the noise shaping parameter, μ, is minimized when all the zeros of the **NTF** are at $\mathbf{z} = +1$, which corresponds to a (multiple) transmission zero at DC. If the **NTF** has zeros close to $\mathbf{z} = +1$, the mean power gain, G_N, is reduced by moving the poles of the **NTF** closer to $\mathbf{z} = 1$ (at the expense of μ). Unfortunately this causes an "early" peaking of the noise transfer function, requiring that the subsequent lowpass filter have a sharp rolloff. If the poles of the **NTF** are $re^{j\theta}$, then $S_{\eta\eta}(f)$ will peak close to $f = (\theta/2\pi)f_s$ and the peak will be sharp if r is close to 1. If r >1 the structure is unstable. Also, the idle pattern will have a fundamental frequency close to $f = (\theta/2\pi)f_s$, especially if r is close to unity. From the viewpoint of lowpass filter requirements and idle-channel behavior, it is advisable to keep r as small as possible and θ as close to π (radians) (180 degrees) as possible. That is, the poles should lie well within the unit circle and away from the voice-band (at the expense of G_N and, consequently, the dynamic range).

If $\boldsymbol{a}$ (the pole of the second-order transfer function in Eq. (8.4.3) is lowered, μ decreases but G_N increases. That is, by lowering $\boldsymbol{a}$, we shall reduce the overload point and degrade the stability. In contrast, if $\boldsymbol{b}$ is decreased, μ increases and G_N goes down, which corresponds to better stability. While choosing the circuit parameters, it is useful to remember the above trade-offs.

A suitable strategy for choosing circuit parameters is necessarily dependent on the application. In the telephony situation we are considering we shall assume that the **ΔΣM** operates at a 1-MHz clock rate and is followed by a *simple* finite impulse response (**FIR**) lowpass filter whose output is resampled at 32 kHz. After subsequent filtering, the signal is again resampled at 8 kHz. The chain is required to have the noise performance of a 13-bit uniform coder (at 8 kHz). One possible approach for selecting appropriate values for the design parameters $\boldsymbol{a}$, $\boldsymbol{b}$, and $\boldsymbol{g}$ is as follows.

First we choose the sampling rate. As mentioned before, 1 MHz is appropriate. Then, for any given ($\boldsymbol{g}$, $\boldsymbol{a}$, $\boldsymbol{b}$) the following are computed:

1) the poles $re^{j\theta}$.
2) mean power gain G_N.
3) the **FOM** from Eq. (8.5.48).

The given triplet ($\boldsymbol{g}$, $\boldsymbol{a}$, $\boldsymbol{b}$) is then accepted or rejected based on the following procedure:

1) reject if the angle of the pole is too small; a cutoff angle of 22.5 degrees (64 kHz) is chosen.
2) reject if $r > 0.75$.
3) reject if **FOM** < 80.
4) reject if $G_N > 3$.

The guidelines chosen above are based on the following heuristic arguments. Since the ideal **SNR** performance of the **ΔΣM** should exceed that of a 13-bit coder operating at 8 kHz, the peak **SNR** must exceed 78 dB. The acceptable pole locations are based on stability and the "knee" of the noise spectrum. Since $r < 1$ is necessary for stability, r 0.75 provides some margin. The angle of the pole determines where the noise spectrum peaks, which to ease requirements on the lowpass (decimating) filter, should be at as high a frequency as possible. Upon resampling the lowpass filter outputs at 32 kHz, the noise energy centered around multiples of 32 kHz folds back, or aliases, into the baseband. If the angle of the pole is greater than 22.5 degrees (64 kHz), the *early peaking* of the noise spectrum will be only a second-order effect.

Assuming $\boldsymbol{g} = 1/T$, acceptable ranges for the pole and zero locations, $\boldsymbol{a}$ and $\boldsymbol{b}$ are 2 $\boldsymbol{a}$ 4 (kHz) and 75 $\boldsymbol{b}$ 100 (kHz), as shown in Table 8.2. Clearly, the **ΔΣM** is relatively insensitive to the parameter values $\boldsymbol{a}$ and $\boldsymbol{b}$, permitting deviations of 15 percent from nominal. In Table 8.2, rather than show the peak **SNR**, the figure of merit, or **FOM**, is used since the **FOM** is more representative of the **SNR** improvement obtained when the quantizer is embedded in a feedback loop. The **FOM** required in this case is **FOM** 80 dB (approximately), which provides a margin of approximately 3 dB over the peak **SNR** of 78 dB required. To reduce the dimensionality of the search, Table 8.2 is generated assuming $\boldsymbol{g}$ T=1. The program to do so is quite simple and is based on Eqs. (8.5.35).

Table 8.2
Circuit parameters for a second-order analog $\Delta\Sigma$**M**, indicating acceptable choices of pole and zero

Ratio (b/a)	10	20	30	40	50	Comments
Pole, a (KHz)						
0.5	unstable	unstable	unstable	unstable	unstable	All real poles
1.0	unstable	unstable	unstable	unstable	unstable	All real poles
1.5	unstable (real poles)	unstable (real poles)	unstable (real poles)	θ=16 (reject)	Accept	θ=angle of pole (deg)
2.0	unstable (real poles)	unstable (real poles)	θ=13.5 (reject)	Accept	Accept	-
2.5	unstable (real poles)	unstable (real poles)	Accept	Accept	G=3.2 (reject)	G=Noise Power Gain
3.0	unstable (real poles)	θ=13.5 (reject)	Accept	G=3.1 (reject)	G=3.6 (reject)	-
3.5	unstable (real poles)	Accept	Accept	G=3.4 (reject)	G=4.2 (reject)	Values of G greater than 4 usually
4.0	unstable (real poles)	Accept	G=3.1 (reject)	G=3.8 (reject)	G=4.2 (reject)	have a pole whose
4.5	unstable (real poles)	Accept	G=3.3 (reject)	G=4.3 (reject)	G=6.0 (reject)	magnitude is greater than 0.75
5.0	unstable (real poles)	Accept	G=3.6 (reject)	G=4.9 (reject)	G=7.6 (reject)	
5.5	unstable (real poles)	Accept (marginal)	G=3.9 (reject)	G=5.7 (reject)	G=10.5 (reject)	
6.0	FOM=77 (reject)	G=3.05 (reject)	G=4.2 (reject)	G=6.7 (reject)	G=16.9 (reject)	reject all
6.5	FOM=77 (reject)	G=3.2 (reject)	G=4.6 (reject)	G=8.3 (reject)	G=41.8 (reject)	reject all
7.0	FOM=77 (reject)	G=3.3	G=5.1	G=10.2	unstable	reject all
7.5	FOM=77 (reject)	G=3.5	G=5.7	G=15.6	unstable	
8.0	FOM=77 (reject)	G=3.7	G=6.5	G=27.6	unstable	
9.0	FOM=77 (reject)	G=4.1	G=8.9	unstable	unstable	
10.0	FOM=77 (reject)	G=4.6	G=14.0	unstable	unstable	

Those entries in Table 8.2 that are labeled *Accept* correspond to choices of the pole and zero which yield a stable noise transfer function, N(**z**); have a noise power gain of less than 3; and have complex poles with magnitude less than 0.75, angle greater than 22.5 degrees, and an **FOM** greater than 80 dB. For comparison, the **NTF** of Eq. (8.5.47), which describes the behavior of the switched-capacitor or digital ΔΣM modeled by Fig. 8.35, provides an **FOM** greater than 85 dB, a noise power gain of about 2.4, and has complex poles at a radius of about 0.7 and angle of 45 degrees. The structure, operating at 1 MHz, will provide the necessary performance for the application being considered.

We have considered overload mainly from the viewpoint of the quantizer. In practice, the implementation of the H(s) would require active devices. Ideally, these should never exhibit saturation. Experimental and simulation results have shown that a supply voltage of 4 V_{ref} is appropriate for the cases labeled "Accept." Provided G 3, the instances where the amplifiers exhibit clipping are sufficiently infrequent.

8.6 ANALYZING THE IMPACT OF DECIMATION IN A ΔΣM BASED A/D CONVERTER

When a **ΔΣM** is used in an A/D converter application, the eventual sampling rate is usually much lower than the sampling rate employed in the **ΔΣM** itself. This decimation in frequency must be accompanied with an appropriate lowpass filtering operation that removes the high frequency components, principally quantization noise, which otherwise aliases into the lower frequency band of interest. The equivalent signal processing chain is shown in the block diagram in Fig. 8.39 which assumes that the lower, decimated, sampling frequency is $f_{s1} = (f_s/N)$ ($T_1 = NT$ for the sampling interval). The underlying sampling rate at which the lowpass filter operates is f_s. At this high sampling rate it is difficult to build complex, in terms of order, digital filters in an economical fashion. In other words, the attenuation of the high frequency noise cannot be assumed as infinite and must be accounted for and the noise power will, in general, be greater than $\mu G_N \sigma_{\mathbf{e}}^2$.

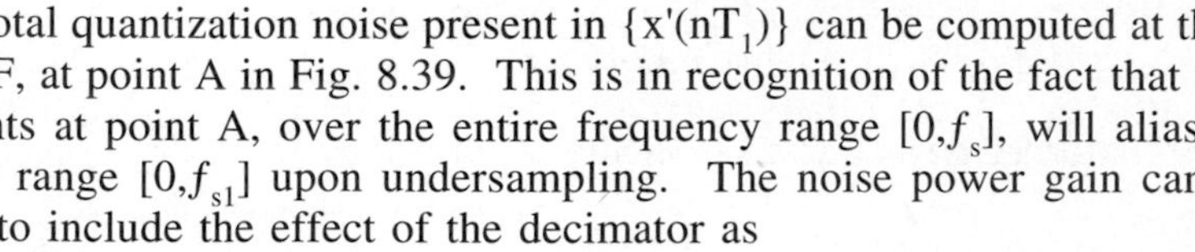

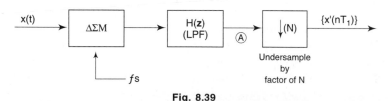

Fig. 8.39
Noise power at LPF output is equal to the total quantization noise power after undersampling

The total quantization noise present in $\{x'(nT_1)\}$ can be computed at the output of the LPF, at point A in Fig. 8.39. This is in recognition of the fact that all signal components at point A, over the entire frequency range $[0, f_s]$, will alias into the frequency range $[0, f_{s1}]$ upon undersampling. The noise power gain can thus be modified to include the effect of the decimator as

$$G_N = \frac{1}{f_s}\int_0^{f_s} |N(f)|^2 |H(f)|^2 \, df \qquad (8.6.1)$$

and the quantizing noise, power is evaluated as

$$\sigma_\eta^2 = G_N \sigma_e^2 \qquad (8.6.2)$$

The reduced sampling rate $f_{s1} = (f_s/N)$ may yet be higher than the eventual sampling rate (which could be as low as $2f_c$). Thus an additional stage, or stages, of decimation may be employed. However, since the sampling rate is now low (or lower, anyway) the later stages of lowpass filtering can be more complex and provide significant stopband attenuation.

The fraction of the noise power in the band up to f_c, including the aliased components can be written as

$$\mu = \frac{\sum_{k=0}^{N-1} \int_0^{f_c} \left| N\left(f - \frac{k}{N} f_{s1}\right) \right|^2 \left| H\left(f - \frac{k}{N} f_{s1}\right) \right|^2 df}{\sum_{k=0}^{N-1} \int_0^{\frac{f_{s1}}{2}} \left| N\left(f - \frac{k}{N} f_{s1}\right) \right|^2 \left| H\left(f - \frac{k}{N} f_{s1}\right) \right|^2 df} \qquad (8.6.3)$$

where $H(f)$ represents the combined frequency response of all lowpass filtering. Assuming that any subsequent stages of decimation use filters that provide substantial attenuation beyond f_c, the total in-band noise power is evaluated as

$$\sigma_\varepsilon^2 = \mu G_N \sigma_e^2 \qquad (8.6.4)$$

Since the lowpass filter shown as the first stage of decimation operates at the high sampling rate f_s, it is likely that it will be simple and, specifically, either a rectangular window or triangular window or a combination thereof. As will be shown in the next section, such lowpass filters are exceptionally simple to implement. The effective noise power gain given in Eq. (8.6.1) can be evaluated quite easily in these circumstances. In the following, we consider the cases of first- and second-order ΔΣMs in conjunction with rectangular and triangular windows.

8.6.1 First-Order ΔΣM

8.6.1.1 Rectangular Window LPF

A first-order **ΔΣM** has a noise transfer function

$$N(\mathbf{z}) = 1 - \mathbf{z}^{-1} \qquad (8.6.5)$$

When the lowpass filter $H(\mathbf{z})$ is a rectangular window of M taps, its transfer function can be written as

$$H_R(\mathbf{z}) = \frac{1}{M}(1 + \mathbf{z}^{-1} + \ldots + \mathbf{z}^{-(M-1)}) = \frac{1}{M}\frac{1 - \mathbf{z}^{-M}}{1 - \mathbf{z}^{-1}} \tag{8.6.6}$$

The equivalent noise power gain G_{N1} is calculated as

$$G_{N1} = \sum_{n=0}^{\infty} [h(n)]^2 \tag{8.6.7}$$

where $\{h(n)\}$ is the equivalent time sequence with **Z**-transform equal to $N(\mathbf{z})H_R(\mathbf{z})$. Since

$$N(\mathbf{z})H_R(\mathbf{z}) = \frac{1}{M}(1 - \mathbf{z}^{-M}) \tag{8.6.8}$$

it follows that

$$G_{N1} = \frac{2}{M^2} \tag{8.6.9}$$

It may appear that by choosing the rectangular window length, M, appropriately, we can reduce the total quantization noise to any desired level. This is only partially true. Since the decimation filter also affects the desired signal, increasing M, or in effect reducing the passband width of the filter, may significantly attenuate desired signal components as well as the noise. For example, if the desired signal band extends to $f_c = 4$ kHz and the sampling rate is $f_s = 1.024$ MHz, Table 8.3 shows the attenuation of the 4-kHz component for different values of M (the number of taps). If the desired signal was of extremely low frequency, practically DC, then by making M very large we can get almost arbitrary precision.

Table 8.3

Attenuation at 4 kHz as a function of number of taps at a sampling frequency of 1.024 MHz

M = (# of TAPS)	16	32	64	128
Attenuation (dB) at 4 kHz	0.06	0.22	0.91	3.92

Digital filters operating at the lower sampling rate, f_{s1}, can probably compensate for some of the passband shaping introduced by the decimating filter. However, from Table 8.3 it can be deduced that, given the assumption of f_s and f_c, a window length of greater than 64 is not advisable.

8.6.1.2 Triangular Window LPF

The transfer function of a triangular window with 2(M – 1) taps can be written as

$$H_\Delta(\mathbf{z}) = \left[\frac{1}{M}\left(1 + \mathbf{z}^{-1} + \ldots + \mathbf{z}^{-(M-1)}\right)\right]^2 \tag{8.6.10}$$

and thus, with N(**z**) from Eq. (8.6.5)

$$N(\mathbf{z})\,H_\Delta(\mathbf{z}) = \frac{1}{M^2}\left(1 + \mathbf{z}^{-1} + \ldots + \mathbf{z}^{-(M-1)} - \mathbf{z}^{-M} - \ldots - \mathbf{z}^{-(2M-1)}\right) \tag{8.6.11}$$

Computing the effective power gain using Eq. (8.6.7) yields

$$G_{N2} = \frac{2}{M^3} \tag{8.6.12}$$

It will be shown in the next section that the implementation of a triangular window is simple if the decimation in sampling frequency is by a factor of M. Thus, when using a triangular window, the length is constrained by two factors. One is the decimation factor and the second relates to any untoward passband shaping of the desired signal. Table 8.4 depicts these considerations assuming f_s = 1.024 MHz and f_c = 4 kHz.

Table 8.4
Attenuation at 4 kHz and reduced sampling rate assuming an initial sampling rate of 1.024 MHz and a triangular window of 2(M–1) taps

M = (# of TAPS)	16	32	64	128
Attenuation (dB) at 4 kHz	0.12	0.44	1.8	7.8
f_{s1} (kHz)	64	32	16	8

8.6.1.3 Combination of Rectangular and Triangular Windows

The implementation of a rectangular window filter is quite simple even if there is no decimation in sampling rate. Consequently, cascading the two types filters is possible. The effective filter transfer function is

$$H(\mathbf{z}) = \frac{1}{M_R}\frac{1 - \mathbf{z}^{-M_R}}{1 - \mathbf{z}^{-1}}\left[\frac{1}{M_\Delta}\frac{1 - \mathbf{z}^{-M_\Delta}}{1 - \mathbf{z}^{-1}}\right]^2 \tag{8.6.13}$$

Just for the convenience of analysis, we shall assume that $M_R = 2\,M_\Delta$. The product N(**z**)H(**z**) is then given by

$$N(\mathbf{z})H(\mathbf{z}) = \frac{\mathbf{z}}{2M_R^2}\left\{ \sum_{k=0}^{M_\Delta} k\mathbf{z}^{-1} - \sum_{k=0}^{M_\Delta} k\mathbf{z}^{-(k+2M_\Delta)} \right.$$

$$\left. + \sum_{k=M_\Delta+1}^{2M_\Delta} (2M_\Delta - k)\mathbf{z}^{-1} - \sum_{k=M_\Delta+1}^{2M_\Delta} (2M_\Delta - k)\mathbf{z}^{-(k+2M_\Delta)} \right\} \quad (8.6.14)$$

The assumption $M_R = 2M_\Delta$ is not necessary but does simplify the algebra involved in deriving Eq. (8.6.14) and is thus used as an example since the result is available in closed form. The noise power gain, following Eq. (8.6.7), is given by

$$G_{N3} = \frac{2M_\Delta^3 - M_\Delta^2 + M_\Delta}{4M_\Delta^6} \approx \frac{1}{2M_\Delta^3} \quad (8.6.15)$$

It is observed that when the **ΔΣM** is of first-order, cascading a rectangular window and triangular window is only marginally better than using the triangular window by itself, from the viewpoint of noise, and is worse from the viewpoint of passband shaping. The attenuation at 4 kHz, assuming an initial sampling rate of 1.024 MHz, is obtained by summing the attenuation (in dB) of the rectangular and triangular windows using Table 8.3 and Table 8.4.

8.6.2 Second-Order ΔΣM

8.6.2.1 Rectangular Window LPF

We will approximate the noise transfer function of a second-order **ΔΣM** by

$$\mathbf{NTF}\text{: } N(\mathbf{z}) = (1 - \mathbf{z}^{-1})^2 \quad (8.6.16)$$

even though it was pointed out in the previous sections that such an **NTF** is not entirely appropriate. The conclusions that we will draw, considering the ideal **NTF**, will however, for all practical purposes, be the same even if the poles of the **NTF** are included. Using N(**z**) of the Eq.(8.6.16) does simplify the analysis and can provide closed form results.

With a rectangular window of M taps, the product N(**z**)H(**z**) is given by

$$N(\mathbf{z})H(\mathbf{z}) = \frac{1}{M}(1 - \mathbf{z}^{-1} - \mathbf{z}^{-M} + \mathbf{z}^{-(M+1)}) \quad (8.6.17)$$

Consequently,

$$G_{N4} = \frac{4}{M^2} \quad (8.6.18)$$

8.6.2.2 Triangular Window LPF

With a triangular window of length 2(M–1), the product N(**z**)H(**z**) is

$$N(\mathbf{z})H(\mathbf{z}) = \frac{1}{M^2}(1 - 2\mathbf{z}^{-M} + \mathbf{z}^{-2M}) \qquad (8.6.19)$$

which yields

$$G_{N5} = \frac{6}{M^4} \qquad (8.6.20)$$

8.6.2.3 Rectangular and Triangular Window

Again assuming $M_R = 2M_\Delta$, the product N(**z**)H(**z**) becomes

$$N(\mathbf{z})H(\mathbf{z}) = \frac{1}{2M^2}(1 - \mathbf{z}^{-M_\Delta})(1 - \mathbf{z}^{-2M_\Delta})(1 + \mathbf{z}^{-1} + \ldots + \mathbf{z}^{-(M_\Delta - 1)})$$

$$= \frac{1}{2M_\Delta^3} \sum_{k=0}^{4M_\Delta - 1} a_k \mathbf{z}^{-k} \qquad (8.6.21)$$

where a_k is either 1 or –1. Applying Eq. (8.6.7) yields

$$G_{N6} = \frac{1}{M_\Delta^5} \qquad (8.6.22)$$

8.6.3 FOM Calculations

The window-type lowpass filters usually comprise a first stage of decimation, accompanying the reduction of sampling rate from f_s to f_{s1}. A subsequent stage of filtering accompanies the decimation from f_{s1} to $f_{s2} = 2 f_c$. Since this second stage can be a more complicated filter, we can assume that it is "ideal." This filter provides an **SNR** improvement of 10 $\log_{10} (f_{s1}/f_{s2})$ provided that the noise is white. To a first approximation the noise introduced by the **ΔΣM**, which is biased toward high frequency, will alias into the band $[0, f_{s1}]$ in a manner weighted by the first decimation lowpass filter, to provide an approximately flat spectrum. This approximation is conservative because the **ΔΣM** noise transfer function has a zero at DC and the window LPF can be engineered to have transmission zeros at all multiples of f_{s1}. Thus the noise power around multiples of f_{s1}, which will alias into a band around DC, will be small.

The overall **FOM**, combining the effect of the window (LPF) and the subsequent filter can be evaluated as

$$\mathbf{FOM} = -10\log_{10}(G_{Nx}) + 10\log_{10}\left(\frac{f_{s1}}{f_{s2}}\right) \qquad (8.6.23)$$

The noise performance is computed assuming $f_s = 1.024$ MHz and $f_{s2} = 2f_c = 8$ kHz. Table 8.5 shows the results for a first and second-order **ΔΣM**s assuming that the intermediate sampling rate is $f_{s1} = f_s/M$.

Table 8.5
Figure of merit calculations combining the **ΔΣM** and subsequent lowpass filter

	f_{s1} (KHz)	Rectangular Window			Triangular Window			Combination $M_R = 2M_\Delta$, $M_\Delta = M$		
			Order of ΔΣM 1st	2nd		Order of ΔΣM 1st	2nd		Order of ΔΣM 1st	2nd
		Atten. (dB) @4KHz	FOM (dB)	FOM (dB)	Atten. (dB) @4KHz	FOM (dB)	FOM (dB)	Atten. (dB) @4KHz	FOM (dB)	FOM (dB)
M = 16	64	0.06	30	27	0.12	42	49	0.34	48	69
M = 32	32	0.22	33	30	0.442	48	58	1.35	54	81
M = 64	16	0.91	36	33	1.8	54	67	5.7	66	93
M = 128	8	3.92	39	36	7.8	60	76	∞	–	–

Colloquially, a rectangular window is referred to as a *first-order-hold* and a triangular window as a *second-order-hold*. Thus we ascribe an "order" to the lowpass filter doing the decimation. With this association, we can see from Table 8.5 that the order of the decimating filter should be equal to or greater than the order of the **ΔΣM**. In fact, a combination of first-order filter and second-order **ΔΣM** is about 3-dB *worse* than if the **ΔΣM** was first-order. This interaction of noise shaping and decimation filtering is the reason we impose the requirement that the peak of the error spectrum occur at as high a frequency as possible.

8.7 RECTANGULAR AND TRIANGULAR WINDOWS USED IN INTERPOLATION AND DECIMATION

The processes of decimation, i.e., reduction of sampling frequency, and interpolation, i.e., increase of sampling frequency, both involve the use of lowpass filters. The former requires the lowpass characteristic to reduce the impact of aliasing while the latter requires a lowpass filter to attenuate the frequency domain replicates inherent in the oversampling process. In situations where the sampling frequency is high (of the order of 100 kHz or greater) and since it is necessary to keep the cost of the implementation low, it helps to have simple filters from the viewpoint of efficient implementation. Rectangular and triangular windows are especially useful in such circumstances. For example, when **ΔΣM** techniques are used in a codec for telecommunications, the sampling rates are of the order of 1 MHz. Economic reasons dictate that the implementation be suitable for inclusion in an Application-Specific Integrated Circuit (ASIC). It is shown here that the triangular and rectangular windows are especially well suited for such an application.

8.7.1 Impulse and Frequency Response

The rectangular window filter is an **FIR** filter defined by

$$h_R(n) = \begin{cases} 1; & 0 \le n \le N-1 \\ 0; & \text{otherwise} \end{cases} \tag{8.7.1}$$

A plot of the impulse response is shown in Fig. 8.40 and the shape discerned therein is what gives the filter its name.

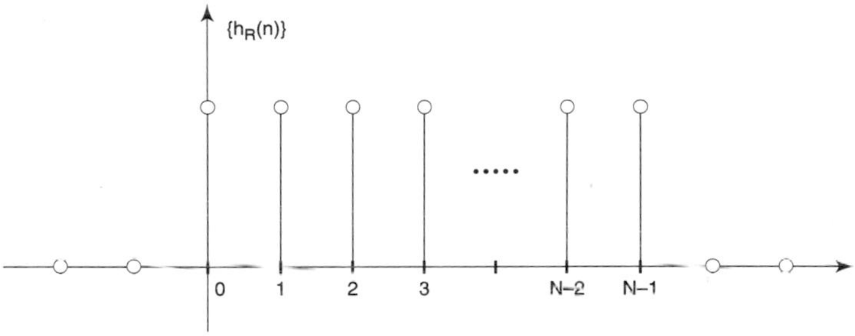

Fig. 8.40
Impulse response of rectangular window

The transfer function $H_R(\mathbf{z})$ can be written as

$$H_R(\mathbf{z}) = 1 + \mathbf{z}^{-1} + \ldots + \mathbf{z}^{-(N-1)} = \frac{1 - \mathbf{z}^{-N}}{1 - \mathbf{z}^{-1}} \tag{8.7.2}$$

and hence $H_R(f)$ can be seen to be

$$H_R(f) = e^{-j\pi(N-1)f'} \left(\frac{\sin(N\pi f')}{\sin(\pi f')}\right) \text{ where } f' = \frac{f}{f_s} \tag{8.7.3}$$

The rectangular window can be viewed as a filter with a (modified) linear phase response corresponding to a delay of (N – 1)/2 samples and with a magnitude response that follows a sin(Nx)/sin(x) pattern. The response has a maximum at $f = 0$ (DC), of value N, and transmission zeros at integer multiples of (f_s/N). That is,

$$H_R(k\frac{f_s}{N}) = \begin{cases} N; & k = 0 \\ 0; & k \neq 0 \end{cases} \tag{8.7.4}$$

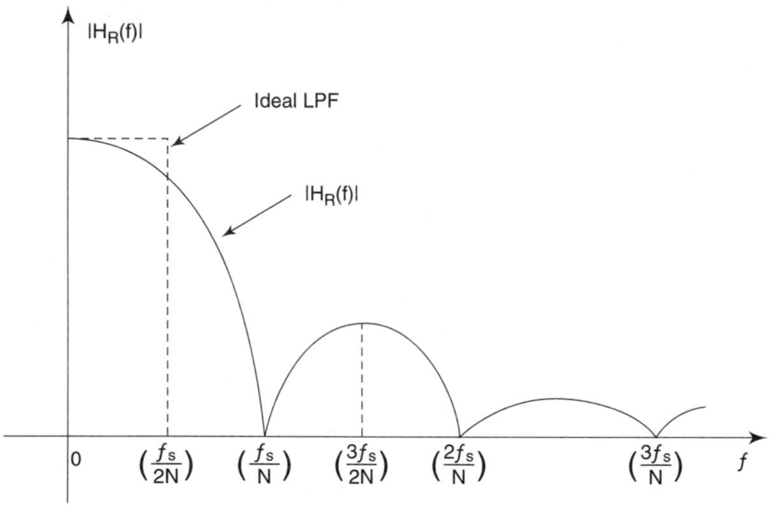

Fig. 8.41
Frequency response of a rectangular window

The magnitude response of a rectangular window filter is shown in Fig. 8.41 (not to scale). Also depicted is the frequency response of an ideal lowpass filter with cutoff frequency ($f_s/2N$). The frequency response of the rectangular window is nominally that of a lowpass filter, though not a very good lowpass filter. For example, consider the first lobe, which occurs at ($3f_s/2N$). The response at this frequency, relative to the response at f=0 is given by, for large N,

$$\left(\frac{1}{N}\right)\left|H_R\left(\frac{3f_s}{2N}\right)\right| = \left(\frac{1}{N}\right)\frac{\left|\sin\left(\frac{3\pi}{2}\right)\right|}{\left|\sin\left(\frac{3\pi}{2N}\right)\right|} \approx \frac{2}{3\pi} \tag{8.7.5}$$

Thus the first lobe of the rectangular window is only about 13.5 dB down from its peak (at f=0) value.

The triangular window can be derived from the rectangular window. Denoting the **Z**-transforms of the impulse responses by $H_R(\mathbf{z})$ and $H_\Delta(\mathbf{z})$,

$$H_\Delta(\mathbf{z}) = \left(H_R(\mathbf{z})\right)^2 \tag{8.7.6}$$

A plot of the impulse response of the triangular window is shown in Fig. 8.42.

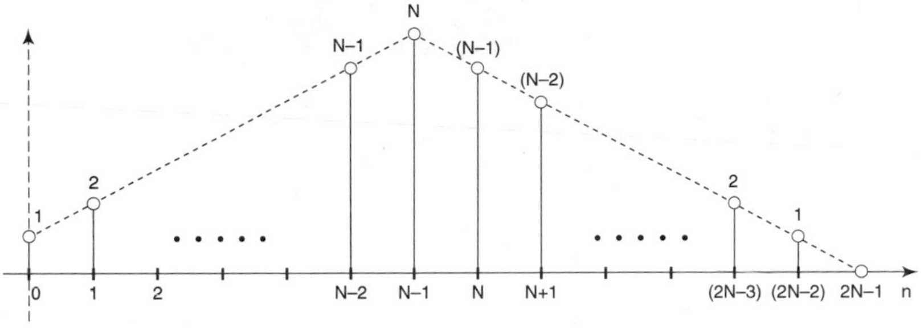

Fig. 8.42
Impulse response of a triangular window

The impulse response of the rectangular window has N nonzero samples; for the triangular window there are (2N–1) nonzero samples. The frequency response, $H_\Delta(f)$, of the triangular window is given by

$$H_\Delta(f) \;=\; e^{-j\pi(N-1)f}\left(\frac{\sin(N\pi f')}{\sin(\pi f')}\right)^2 \text{ where } f' = \frac{f}{f_s} \qquad (8.7.7)$$

A plot of the frequency response, ($H_\Delta(f)$) is similar to that of a rectangular window except that, considering the square, we double the response (in dB) relative to the response at DC. Hence the triangular window provides greater attenuation in the "stop-band." Both windows have the same transmission zeros. The first lobe of the triangular window is about 27 dB down from its peak (i.e., DC) value, a significant improvement over the rectangular window.

If we redefine the triangular window by incorporating a delay of one sample, i.e.,

$$H_\Delta(\mathbf{z}) \;=\; \mathbf{z}^{-1}\left(H_R(\mathbf{z})\right)^2 \qquad (8.7.8)$$

then the frequency selectivity is unchanged but the impulse response takes the shape in Fig. 8.43.

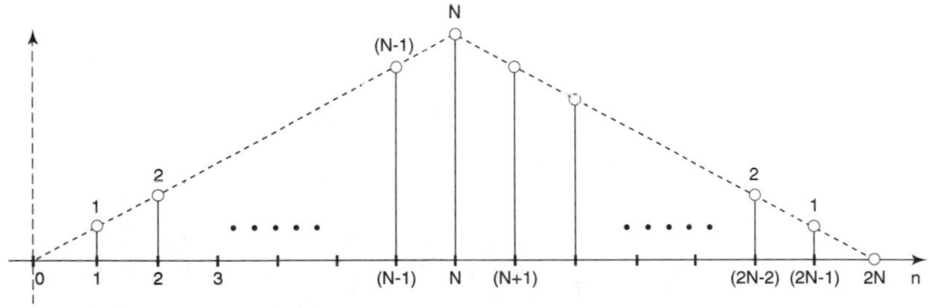

Fig. 8.43
Shifted triangular window has a linear-phase delay of an integer number of samples

The nomenclature of "triangular" window arises from the graph of the impulse response for obvious reasons.

8.7.2 Implementing a Decimating Rectangular Window

Based on Eq. (8.7.1)., one implementation of a rectangular window is depicted in Fig. 8.44.

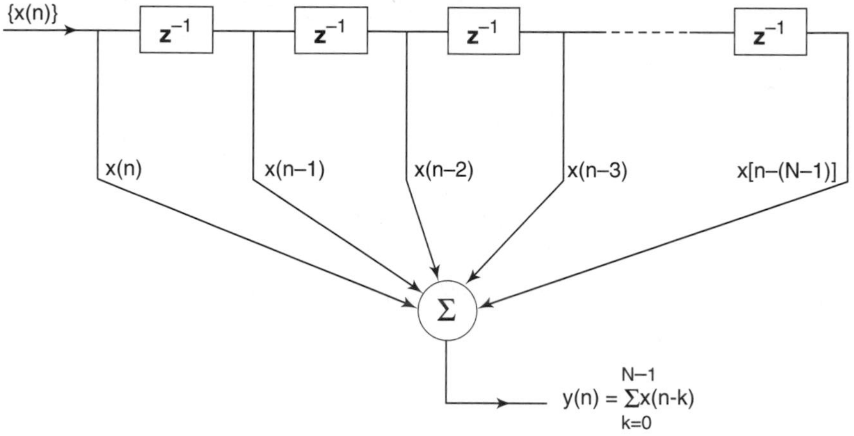

Fig. 8.44
Tapped delay line representation of a rectangular window

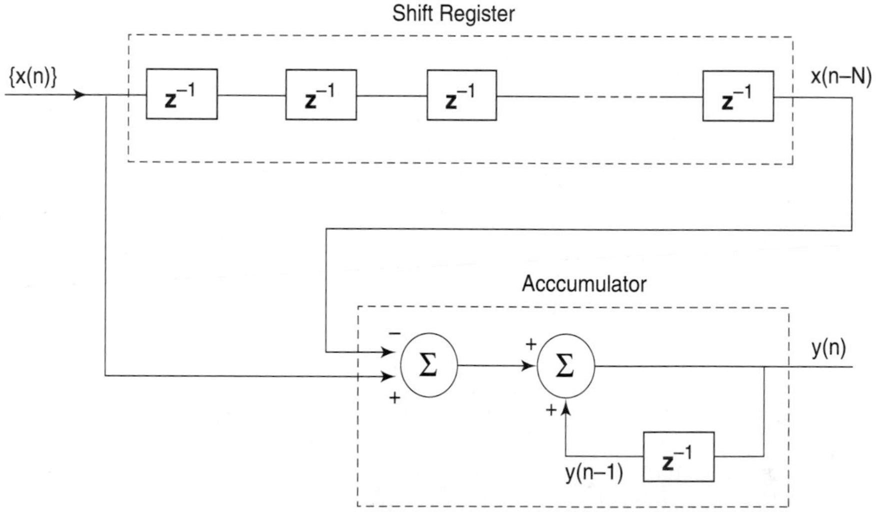

Fig. 8.45
Simplified implementation of a rectangular window requiring 2 adds per input sample.

The filter output is the sum of the N most recent input samples, x(n), x(n–1), ..., x(n–(N–1)). When implemented as shown, computing each output sample requires

N additions. This is quite inefficient, especially when one recognizes the fact that (N–1) of the N samples summed up in the generation of y(n) were also summed up in the generation of y(n–1). In fact, the difference between y(n) and y(n–1) is just the new sample x(n) and an old sample x(n–N). Thus

$$y(n) \;=\; y(n-1) \;+\; x(n) \;-\; x(n-N) \qquad (8.7.9)$$

which is equivalent to writing the transfer function as in Eq. (8.7.2). Such an implementation appears to have a pole at **z** = 1, on the unit circle, and is thus potentially unstable. This "instability" is equivalent to having infinite memory. If, at any time, y(n) is not equal to the sum of the N most recent input samples, this error propagates in time. By appropriate initialization procedures this error can be eliminated. Based on Eq. (8.7.9), a simple implementation of a rectangular window is shown in Fig. 8.45.

The chain of delays is implemented by a shift register of length N and width commensurate with the wordlength used. The accumulator implements the arithmetic

$$y(n) \;=\; (\, y(n-1) \;-\; x(n-N)\,) \;+\; x(n) \qquad (8.7.10)$$

and requires only two addition operations per output sample.

The initialization operation required to prevent error propagation is that the contents of the accumulator, at "power-on" be equal to the sum of the contents of the shift register at "power-on." This is easily accomplished, assuming 2–s complement arithmetic, by setting all shift register contents and accumulation value to zero in the power-on initialization procedure.

It should be noted that this implementation provides y(n) at the *same* sampling rate as x(n). The complexity of computation is two additions per input sample, regardless of whether the output is resampled at a lower sampling rate. A second observation is that any arithmetic overflow in the accumulation process can introduce an error that will propagate for all time, a characteristic of the recursive implementation of the conditionally stable (unstable) pole at **z** = 1. This drawback can be circumvented by using an accumulator wordlength of $[B + \log_2(N)]$ bits where B is the wordlength associated with x(n). Third, the complexity in terms of shift-register storage is NB bits, which is not usually a big consideration when the input is the small wordlength from a ΔΣ**M**.

When a rectangular window accepts the output of a 1-bit ΔΣ**M**, the complexity of a rectangular window is quite low. Since both x(n) and x(n–N) are 1-bit words, the accumulator can be replaced by an up-down counter.

8.7.3 Implementing a Decimating Triangular Window

Based on Eq. (8.7.6), a triangular window can be implemented as the cascade of two rectangular windows of the type shown in Fig. 8.45. Such an implementation has an overall complexity given by

arithmetic	:	four additions per input sample
storage	:	$NB + N(B + \log_2(N))$ bits
Accumulator (1)	:	$[B + \log_2(N)]$ bits wide
Accumulator (2)	:	$[B + 2\log_2(N)]$ bits wide

The complexity in terms of storage requirements, can be reduced if the decimation is by a factor of N and only the required samples, at the lower rate, are computed.

The output of the triangular window can be written as

$$y(n) = \sum_{m=0}^{2N-1} x(n-m)\,h(m) \tag{8.7.11}$$

Now if the output is decimated, only every N-th sample is required. That is

$$y(nN) = \sum_{m=0}^{2N-1} x(nN-m)\,h(m) \tag{8.7.12}$$

Using the impulse response coefficients of a triangular window depicted in Fig. 8.43., y((n+1)N) can be written as

$$y((n+1)N) = \sum_{k=1}^{N} k\,x(nN+(N-k)) + \sum_{k=1}^{N-1} k\,x(nN-(N-k)) \tag{8.7.13}$$

Consider the configuration shown in Fig. 8.46 comprising two accumulators whose outputs are labeled w_1 and w_2. The accumulators are reset, i.e., cleared, every N samples. Consequently, the contents of the accumulators can be expressed as

$$w_1(nN) = 0 + x(nN) \text{ and } w_2(nN) = 0 + w_1(nN) \tag{8.7.14a}$$

$$\begin{aligned} w_1(nN+1) &= w_1(nN) + x(nN+1) \\ w_2(nN+1) &= w_2(nN) + w_1(nN+1) \end{aligned} \tag{8.7.14b}$$

. . . .

$$\begin{aligned} w_1(nN+(N-1)) &= \sum_{k=0}^{N-1} x(nN+k) \\ w_2(nN+(N-1)) &= \sum_{k=0}^{N-1} k\,x(nN+(N-k)) \end{aligned} \tag{8.7.14c}$$

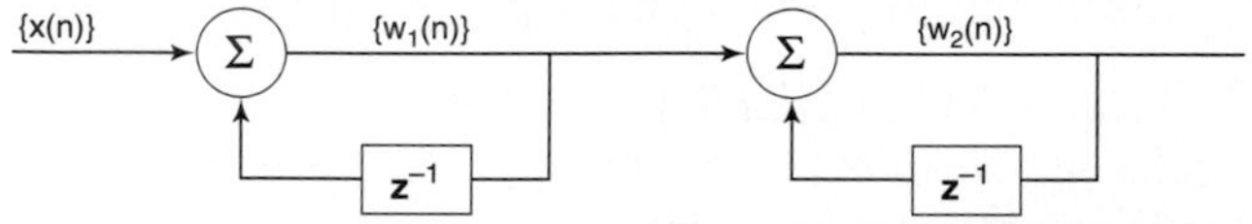

Fig. 8.46
Double accumulation used in an efficient implementation of a triangular window

Define $s_1(nN)$, $s_2(nN)$, and $s_3(nN)$ by

$$s_1(nN) = w_1(nN + (N-1)) \tag{8.7.15a}$$

$$s_2(nN) = w_2(nN + (N-1)) \tag{8.7.15b}$$

$$s_3(nN) = s_2(nN) - Ns_1(nN) \tag{8.7.15a}$$

That is, the values $s_1(nN)$ and $s_2(nN)$ correspond to the contents of the accumulators just prior to the time the accumulators are cleared. $s_1(nN)$ is equal to the sum of the N input samples between time epochs nN and $[nN + (N-1)]$; $s_2(nN)$ can be recognized as the linearly weighted sum of the same N input samples corresponding to the first term on the right-hand side of Eq. (8.7.13). Now $s_3(nN)$ can be written as

$$s_3(nN) = -\sum_{k=1}^{N-1} k\,x(nN+k) \tag{8.7.16}$$

Therefore, the previous value computed for s_3, i.e., $s_3((n-1)N)$ is equal to

$$s_3((n-1)N) = -\sum_{k=1}^{N-1} k\,x(nN-(N-k)) \tag{8.7.17}$$

which is seen to correspond to the (negative) of the second term on the right-hand side of Eq. (8.7.13). Therefore, the output of the triangular window, incorporating the decimation in sampling rate is provided by

$$y((n+1)N) = s_2(nN) - s_3((n-1)N) \tag{8.7.18}$$

In terms of operations per output sample, this implementation has approximately the same arithmetic complexity as the cascade of two rectangular windows but requires considerably less storage. Other than temporary storage for s_2 and s_1, only $s_3(nN)$ needs to be stored in preparation for the computation of the next output sample.

8.7.4 Implementing an Interpolating Rectangular Window

Interpolation is a two step process. The first step is to increase the sampling rate by inserting (N–1) zero samples between every two input samples. The second step is to apply a filter that will remove the unwanted spectral replicates. Thus if the input signal is $\{x(nT_1)\}$, where T_1 corresponds to the sampling interval at the lower rate, the interpolated signal $\{y'(nT)\}$, where T ($= T_1/N$) corresponds to the higher rate, can be described as

$$y'(nT) = \begin{cases} x(mT_1); & n = mN \\ 0; & \text{otherwise} \end{cases} \tag{8.7.19}$$

Samples of y', at time indices that are an exact multiple of N, the ratio of sampling rates, are equal to the corresponding input sample while the others are zero. This is depicted in Fig. 8.47(a) and (b) for N=4.

If the length of the rectangular window is N, the ratio of sampling frequencies, the implementation of the filter is "trivial," involving the repetition of samples. In particular, if we call the output of the filter {y(nT)} then the second step is described by

$$y(nT) = \sum_{k=0}^{N-1} y((n-k)T) \tag{8.7.20}$$

and y(nT) is the sum of the current y' and (N–1) most recent y'. The nature of the zero sample insertion ensures that only one nonzero sample of y occurs in the sum. This is equivalent to repeating the y'(mNT) sample (N–1) times as depicted in Fig. 8.47 (for N=4). The nomenclature of "rectangular hold" used for such a filtering process is clearly appropriate.

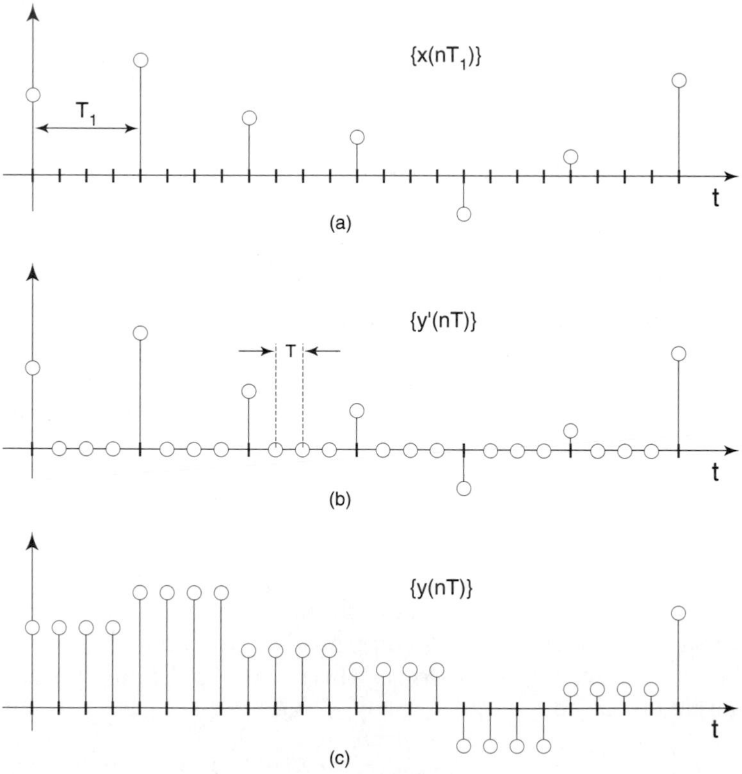

Fig. 8.47
Illustration of an interpolating rectangular window implemented as a *hold*

When the filter length is a multiple of N the implementation is still quite simple, if not trivial. Consider the case when the filter length is AN, i.e.,

$$y(nT) = \sum_{k=0}^{AN-1} y\,((n-k)T) \tag{8.7.21}$$

Replacing the indices n and k with index pairs that reflect the change in sampling rate as

$$n = rN+s;\; k = aN+b;\; 0 \le s, b \le (N-1)\,;\; 0 \le a, r \le (A-1) \tag{8.7.22}$$

the filter output can be expressed as

$$y((rN+s)T) = \sum_{a=0}^{A-1}\sum_{b=0}^{N-1} y\,((r-a)NT + (s-b)T) \tag{8.7.23}$$

The zero sample insertion process forces y' to be zero for (s–b) $\neq$ 0 (modulo N). Therefore,

$$y((rN+s)T) = \sum_{a=0}^{A-1} y\,((r-a)NT)\,;\; 0 \le s \le (N-1) \tag{8.7.24}$$

Eq. (8.7.24) indicates that a rectangular window of length AN can be implemented as an **FIR** filter of length A at the lower rate followed by a rectangular hold. If we denote a delay at the high sampling rate by $\mathbf{z}^{-1}$, then a unit delay at the low rate is $\mathbf{z}^{-N}$. Fig. 8.48 depicts a suitable implementation.

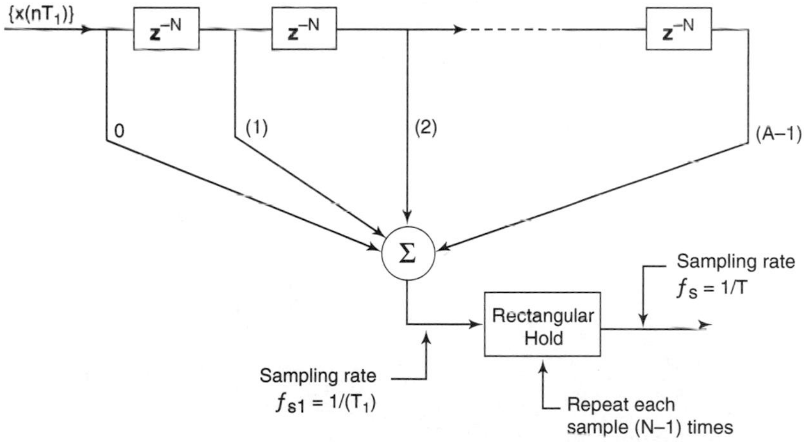

Fig. 8.48
Interpolating rectangular window implemented as an **FIR** filter at the lower sampling rate followed by a *hold*

8.7.5 Implementing an Interpolating Triangular Window

Once again, the first step is the generation of {y(nT)} as given in Eq. 8.7.19. We shall assume that the triangular window parameter N_Δ is equal to the ratio of sampling rates N. In this case the triangular window has a simple implementation.

From Fig. 8.43, which shows the impulse response of a triangular window, we can make the following observation. Each input sample x(nNT), i.e., y(nT) for n=mN, contributes fully to the output sample at time (m+1)NT and its contribution to output samples at the interpolated time instants in between drops off in a linear manner. Therefore,

$$y((m+1)NT) = N\,x(mNT) \text{ and } y((m+2)NT) = N\,x((m+1)NT) \quad (8.7.25)$$

For sample instants between [(m+1)NT] and [(m+2)NT] the output has a contribution from x(mNT) weighted linearly (falling) and x((m+1)NT) weighted linearly (rising). Hence the output can be written as

$$\begin{aligned} y((m+1)NT) &= N\,x(mNT) \\ y((m+1)NT+T) &= (N-1)\,x(mNT) + x((m+1)NT) \\ y((m+1)NT+2T) &= (N-2)\,x(mNT) + 2x((m+1)NT) \\ &\cdots \\ y((m+1)NT+kT) &= (N-k)\,x(mNT) + k\,x((m+1)NT) \quad (8.7.26) \\ &\cdots \\ y((m+2)NT) &= N\,x((m+1)NT) \end{aligned}$$

The implementation can take the form

$$\begin{aligned} y((m+1)NT) &= N\,x(mNT) \\ y((m+1)NT+T) &= y((m+1)NT) + \Delta(m) \\ y((m+1)NT+2T) &= y((m+1)NT+T) + \Delta(m) \\ &\cdots \\ y((m+1)NT+kT) &= y((m+1)NT+(k-1)T) + \Delta(m) \end{aligned} \quad (8.7.27)$$

where

$$\Delta(m) = x((m+1)NT) - x(mNT) \quad (8.7.28)$$

This particular implementation involves a "gain" of N as is evidenced by the first entry of Eq. (8.7.27), which may require an increase in wordlength. This gain can be made unity by scaling all coefficients by N. This yields the implementation

$$
\begin{aligned}
y((m+1)NT) &= x(mNT) \\
&\cdots \\
y((m+1)NT+kT) &= y((m+1)NT+(k-1)T) + \Delta_1(m) \\
&\text{for } (k = 1, 2, \ldots, (N-1)) \\
y((m+2)NT) &= N\,x((m+1)NT)
\end{aligned}
\tag{8.7.29}
$$

where

$$
\Delta_1(m) = \frac{1}{N}\left[x((m+1)NT) - x(mNT)\right] \tag{8.7.30}
$$

This description, while having "unity" gain has a drawback. Computation of $\Delta_1(m)$ will inevitably be inexact because of finite wordlength whereas $\Delta(m)$ in Eq. (8.7.28) is not affected so. Any error in $\Delta_1(m)$ propagates through (N–1) samples, as is clear from Eq. (8.7.29). The outputs at multiple of NT are, however, free from error.

This error involved with calculating the increment can be reduced, or "smeared" by computing Δ as

$$
\Delta_2(m) = \frac{1}{N}\left[x((m+1)NT) - y((m+1)NT)\right] \tag{8.7.31}
$$

and evaluating the output samples as

$$
\begin{aligned}
y((m+1)NT+kT) &= y((m+1)NT+(k-1)T) + \Delta_2(m) \\
&\text{for } (k = 1, 2, \ldots, N)
\end{aligned}
\tag{8.7.32}
$$

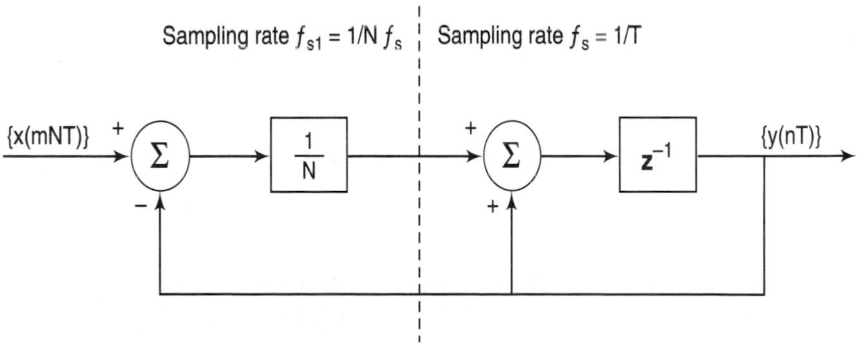

Fig. 8.49
Interpolating triangular window implementation using closed-loop linear interpolation

The implementation described by Eqs. (8.7.31) and (8.7.32) is similar to that described by Eqs. (8.7.29) and (8.7.30) with the principal differences being that first there are no samples that are "free from error" and second that any error in computing Δ at instant mNT is "fed back" to the computation of Δ at time instant (m+1)NT. This feedback of error provides an error shaping of the form $(1-\mathbf{z}^{-1})$ and thus moves the error arising from the finite wordlength effect *away* from low frequencies where the desired signal has its principal components.

A block diagram that indicates the implementation of Eqs. (8.7.31) and (8.7.32) is shown in Fig. 8.49.

8.8 CONCLUDING REMARKS ON ΔΣM

Delta Sigma Modulators were first employed as a 1-bit encoders for converting analog signals to digital format, for transmission purposes, principally for telemetry. Conversion back to analog was performed simply by lowpass filtering. Using such devices as "front ends" for multibit converters became popular in the late 1970s and principally for the implementation of telephony codecs. The use of this technique in general-purpose converters is now common, with commercially available devices available from semiconductor houses such as Analog Devices, Inc., and Crystal Semiconductor, Inc.

Implementing a **ΔΣM** is, at least in principle, quite straightforward. Analyzing the behavior mathematically is quite another story. The presence of a nonlinear element, the quantizer, embedded in a feedback loop, makes the analysis complicated, to say the least. It has been the author's experience that the simpler the analysis, the more is gained in terms of intuition. Such intuition, colloquially "gut-feel", is necessary for explaining why a circuit behaves the way it does. However comprehensive, a mathematical model and hence an analysis or simulation, will not be able to account for all the nuances of even such a "simple" device or take into account all the peculiarities of the circuit elements, without becoming too cumbersome to interpret the results. Simulation is helpful, but not definitive. Assumptions made in order to keep the computer time reasonable will certainly mask some of the circuit behavior.

Again, the author's experience has been that the best way to "analyze" a **ΔΣM** is to construct the circuit and actually observe the performance using such tools as oscilloscopes, spectrum analyzers, and so on. The circuits are simple enough that such laboratory experimentation is often less time consuming than computer simulation using circuit simulators that are commercially available. Simple computer simulations, based on the difference equations are, to a large extent, quite sufficient to obtain a picture of the expected voltage levels that we might observe on the lab bench.

8.9 EXERCISES

8.1. Write a program to simulate the behavior of the switched-capacitor $\Delta\Sigma$M shown in Fig. 8.23. Use a pure tone as an input signal. Devise an output format that shows the density of +1s and –1s and convince yourself that the density of +1s and –1s reflects the amplitude of the input signal.

8.2. The simulation program of Exercise 8.1 can be enhanced in the following ways:

a) Follow the $\Delta\Sigma$**M** with a decimator such as the rectangular or triangular window. By doing so the sampling rate can be reduced (which saves compute time).

b) Apply a **SNR** measurement to the output of the decimator using either the notch filter technique or the method based on the **DFT** (based on exercises in Chapter 3, 4, and 7).

c) Generate **SNR** estimates for different signal powers to develop a **SNR** versus signal-power curve. Extend the linear part of the curve to establish a value for $\mathbf{SNR}_{peak}$.

8.3. The simulation program can be used to investigate the impact of amplifier behavior, especially saturation.

a) Keep track of the maximum values of the amplifier outputs. Consider saturation of these outputs at some value, A (a suitable value for A is about 4, assuming that all voltages are normalized such that the reference voltage is unity).

b) Other, "more accurate" models for amplifiers can be considered. In particular, when the gain of the amplifier is not infinite, the charge transfer between capacitors is not complete; one effect of saturation is to effectively reduce the gain further and the charge transfer is much less complete; and so on. Devise a suitable model for the overall behavior that does not require an exorbitant ammount of compute time.

c) Convince yourself that the "accuracy" of the model for the amplifier is a second-order effect and that even by using the notion of an ideal amplifier with clipping at A=4, the value of $\mathbf{SNR}_{peak}$ is within 3 to 6 dB of the value obtained by applying more complex models for amplifier behavior.

8.4. Use the simulation program to evaluate the impact of comparator hysteresis. Hysteresis can be modeled by changing the decision threshold of the comparator, nominally zero, to $\Delta_{n+1} = -\varepsilon\, b_n$. That is, if the current decision is +1, then the threshold for the next sample is $-\varepsilon$. Show that the impact of hysteresis is a degradation in $\mathbf{SNR}_{peak}$ and try to draw a relationship between ε and the deterioration in $\mathbf{SNR}_{peak}$.

8.5. Consider the case where there is an imbalance in the levels used for +1 and –1. Demonstrate that the principal impact is the presence of a DC offset in the output (the **DFT** method is useful in this case).

8.6. For each variation in model parameters, consider the output when the input is identically zero. Note that every deviation from "ideal" for any parameter has the effect of making the idle pattern have a longer repetition interval. This in turn implies that the noise spectrum peaks earlier and thus more quantization noise appears at the output of the decimating filter (which is not ideal). Try to draw a correlation between the idle-pattern period and $\mathbf{SNR}_{peak}$.

8.10 BIBLIOGRAPHY

[1] Candy, J. C., and Temes, G. C., Ed., *Oversampling Delta-Sigma Data Converters. Theory, Design, and Simulation*, IEEE Press, New York, 1992.

[2] Candy, J. C., "A use of double integration in sigma-delta modulation," *IEEE Trans. on Communications*, Vol. 33, Mar. 1985.

[3] Engineering Staff of Analog Devices, Inc., Sheinhold, D.H., Ed., *Analog-Digital Conversion Handbook*, Prentice Hall, Inc., Englewood Cliffs, NJ, 1986.

[4] Goodman, D. J., and Carey, M. J., "Nine digital filters for decimation and interpolation," *IEEE Trans. on Acoustics, Speech, and Signal Processing*, Vol. ASSP—25, Apr. 1977.

[5] Jayant, N. S., and Noll, P., *Digital Coding of Waveforms*, Prentice Hall, Inc., Englewood Cliffs, NJ, 1984.

[6] Jayant, N. S., Ed., *Waveform Quantization and Coding*, IEEE Press, New York, 1976.

[7] Inose, H., Yasuda, Y., and Marakami, J., "A telemetering system by code modulation, delta-sigma modulation," *IRE Transactions on Space, Electronics and Telemetry*, SET-8, Sept. 1962.

[8] Inose, H., and Yasuda, Y., "A unity bit coding method by negative feedback," *Proceedings of the IEEE*, Vol 51, Nov. 1963.

[9] Park, S., "Principles of sigma-delta modulation for analog-to-digital converters," *MOTOROLA Inc. Application Note APR8/D*, 1990.

[10] Shenoi, K., and Agrawal, B. P., "Selection of a PCM coder for digital switching," *IEEE Trans. on Acoustics, Speech, and Signal Processing*, vol. ASSP-28, Oct. 1980.

[11] Shenoi, K., and Agrawal, B. P., "Design methodology for sigma-delta modulators," *IEEE Trans. on Communications*, vol. COM-31, March 1983.

[12] Tewksbury, S. K., and Hallock, R. W., "Oversampled, linear predictive, and noise-shaping coders of order *N*>1," *IEEE Trans. on Circuits and Systems*, Vol. CAS-25, July 1978.

DESIGN OF RECURSIVE (IIR) DIGITAL FILTERS

9

9.1 INTRODUCTION

We design a digital filter to obtain the coefficients in order to achieve a specified objective response. That is, given a function of frequency, G(*f*) or G(ω), we need to obtain the coefficients of a digital filter H(**z**) such that

$$H(e^{j\omega}) \approx G(\omega) \tag{9.1.1}$$

If we define an ***error function***, E(ω) by

$$E(\omega) = \left| H(e^{j\omega}) - G(\omega) \right| \tag{9.2.2}$$

then the implication of *approximation* is that we intend to choose the coefficients so as to minimize some functional of E(ω). Typical error measures are the ***Chebyshev*** measure

$$\mathbf{E}_\infty = \max_{\omega} \left| E(\omega) W(\omega) \right| \tag{9.3.3}$$

and the ***squared-error*** criterion

$$\mathbf{E}_2 = \frac{1}{2\pi} \int_{-\pi}^{+\pi} \left| E(\omega) W(\omega) \right|^2 d\omega \tag{9.1.4}$$

The function W(ω) is a nonnegative function that allows us to weight the approximation, indicating that certain frequency regions may be more important than others. Minimizing the measure **E** is called ***min-max*** approximation; minimizing $\mathbf{E}_2$ is called ***least-squares*** approximation.

Sometimes we do not explicitly minimize an error measure. Rather, we obtain the coefficients by some means that, intuitively, provides a digital filter H(**z**) that approximates the desired response. Following this step we can calculate the frequency response $H(e^{j\omega})$ and compare it with $G(e^{j\omega})$ and see how good the approximation actually is. This method of "trial and error" is quite effective since software packages are available that permit the rapid computation of frequency response and provide adequate graphics support so that we can, literally, *see* how good the approximation is.

The target response may not be provided in terms of a function of frequency. For example, a lowpass filter response may be specified as a "template," which shows the allowable region for the frequency response $H(e^{j\omega})$. This method of describing the acceptable frequency response is shown in Fig. 9.1 using a lowpass filter as an example. The passband is specified in terms of a ***ripple***, ε, and the stopband in terms of an attenuation, **D**. The passband extends from 0 through ω_p, in normalized radian frequency, and the stopband from ω_s to π.

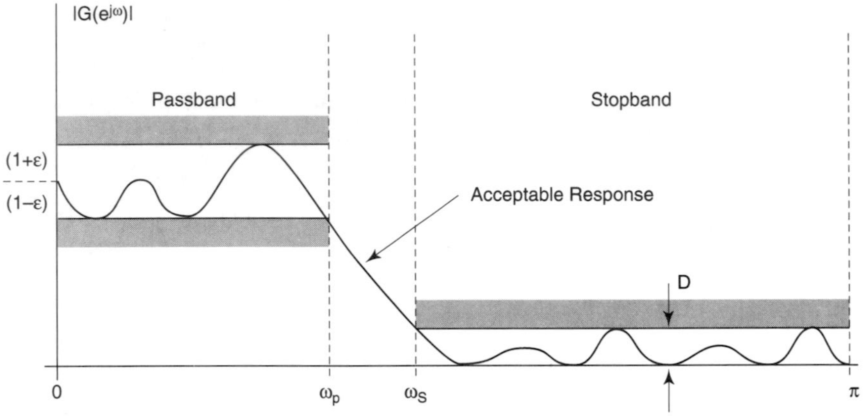

Fig. 9.1
Specifying the target frequency response of a lowpass filter

At other times the desired response may be specified in the time domain by a sequence $\{g(n);\ n = 0,\ 1, \ldots\}$ and the filter impulse response, $\{h(n)\}$, is supposed to approximate this sequence. Again, we can define error measures in a variety of ways. Three common measures in the time domain are

$$E_1 = \sum_{n=0}^{\infty} \left| h(n) - g(n) \right| \tag{9.1.5a}$$

$$E_2 = \sum_{n=0}^{\infty} \left| h(n) - g(n) \right|^2 \tag{9.1.5b}$$

$$E_\infty = \max_n \left| h(n) - g(n) \right| \tag{9.1.5c}$$

For a discipline that is hardly two decades old, the science of digital filter design is quite well developed. Several computer programs are available for solving the approximation problem in the context of digital signal processing. The algorithms and programs in [1.8] are available readily; commercial packages such as the Signal Processing Toolbox available with MATLAB [1.6] have automated the process of filter design to a significant degree. If the approximation problem is ***well posed***, that is the approximation error measure can be calculated for a given choice of filter coefficients in a unique manner, canned routines for approximation using gradient and smart search techniques can be employed. For example, Press [1.7] provides a suite of routines that can be used for the approximation problem.

The form of the transfer function introduces a constraint in the approximation procedure. For H(**z**), a function of **z**, to be a **transfer function**, it **must** be of the form

$$H(\mathbf{z}) = \frac{\sum_{k=0}^{N} a_k \mathbf{z}^{-k}}{1 + \sum_{k=1}^{D} b_k \mathbf{z}^{-k}} \tag{9.1.6}$$

for an **IIR** filter and of the form

$$H(\mathbf{z}) = \sum_{k=0}^{N-1} a_k \mathbf{z}^{-k} \tag{9.1.7}$$

for an **FIR** filter. The design of a filter is the process or procedure we use to obtain these coefficients. Notice that, in order to serve as a transfer function, H(**z**) must be a ratio of polynomials in $\mathbf{z}^{-1}$. Transcendental functions of **z**, or forms that are not polynomials, are **not valid** transfer functions. Thus g(**z**) of the form

$$g(\mathbf{z}) = e^{\mathbf{z}} \text{ or } g(\mathbf{z}) = \sqrt{\mathbf{z}} \tag{9.1.8}$$

for example, is NOT a transfer function.

In the sections that follow we will not treat the design of filters in its most general form. Further, we restrict our discussion to recursive, i.e., **IIR** filters. We shall assume that the target is specified in magnitude response form. We then pose the design problem as one of obtaining the **squared-magnitude function**, which approximates the desired (squared-) magnitude response. The filter coefficients, or, equivalently the constituent second-order sections for a cascade implementation, are obtained from the squared-magnitude function by extracting the roots of the polynomials describing the numerator and denominator as described in Chapter 3. The focus of this chapter is on lowpass filters. Extensions to other forms of pass/stop filters is straightforward. Also, we do not specifically minimize an error criterion. The design is achieved in a manner more intuitive than derived.

In Section 9.2 we consider certain classical techniques for designing filters that approximate a piece-wise constant target frequency response. These methods are based on the premise that a design (or transfer function) is known *a priori* but needs to be modified to, for example, change the passband edge frequency.

In Section 9.3 we provide a formulation of the design problem for lowpass (and bandpass, etc.) filters and present it as one of establishing pairs of polynomials whose ratio is either zero (i.e., very small) or infinite (very large). We use this formulation to design "polynomial" filters, those that achieve their frequency selectivity based on the fact that polynomials "blow up" as x^N for values of x far removed from the (real) roots of the polynomial. The digital counterparts of traditional analog filters, such as Butterworth and Chebyshev, are derived in the context of polynomial filters.

In Section 9.4 we discuss the primary contribution of this chapter, the method for designing filters that have an equi-ripple passband and have transmission zeros that are prescribed in advance. Such a situation occurs quite often when tones, used for signaling or as pilots (see Chapter 7), need to be removed while keeping the signal frequency band "untouched." The design procedure is based on Darlington's definition of ***rational Chebyshev polynomials*** [1.2]. An example of a filter designed with this technique is described in Section 9.5 and is used to illustrate the notion of pole-zero pairing that is necessary for designing a cascade implementation of second-order sections introduced in Chapter 3.

9.2 DESIGN OF DIGITAL FILTERS BY TRANSFORMATIONS

The idea of using a "transformation" is based on the assumption that we already have a filter design that "almost" meets our requirements. Transformations are typically used in the design of **IIR** filters that attempt to meet the piece-wise constant target responses associated with lowpass, highpass, bandpass, and bandstop filters. The "known" filter is either a digital lowpass "prototype" filter or, more commonly, an analog filter. The design of analog filters is a very mature field and several designs already exist, usually in tabular form, that provide the design parameters of a wide variety of filters.

We provide a brief description of such transformations for two reasons. The first is that they are arguably the most widely used methods for designing **IIR** filters. Second is that a transformation, or mapping, of the independent variable (frequency) is a very useful tool and is used frequently in subsequent sections.

9.2.1 Digital-to-Digital Transformations

Suppose we have a digital filter, $G(\eta)$, that is a lowpass prototype with adequate passband ripple and stopband attenuation and a nominal cutoff frequency θ_p in normalized radian frequency. We have used the variable η and θ instead of the traditional **z** and ω in order to allow us to maintain the distinction between the prototype filter, which is an "abstraction," and the filter we are designing, $H(\mathbf{z})$. The frequency response of the prototype filter, $G(e^{j\theta})$, is obtained by setting $\eta = e^{j\theta}$ in $G(\eta)$. Assuming $G(\eta)$ is a lowpass filter, the magnitude response of G, as a function of the normalized radian frequency θ, could appear as shown in Fig. 9.2. The filter is shown as having a passband edge, θ_p and a stopband edge θ_s.

The ***all-pass*** mapping

$$\eta^{-1} = \frac{\mathbf{z}^{-1} + \mathbf{a}}{1 + \mathbf{a}\mathbf{z}^{-1}} \tag{9.2.1}$$

provides a one-to-one mapping between the unitcircle in the η-plane and the unit circle in the **z**-plane and a one-to-one mapping between the radian frequencies θ and ω, both in the range $[0, 2\pi]$. Consequently,

$$H(\mathbf{z}) = G(\eta) = G\left(\frac{\mathbf{z} + \mathbf{a}}{1 + \mathbf{a}\,\mathbf{z}}\right) \tag{9.2.2}$$

defines a transfer function H(**z**) that is also lowpass in nature. This is also shown in Fig. 9.2. The mapping between the frequency variables can be written as

$$\omega = \theta - \rho(\theta) = \theta - 2\arctan\left(\frac{\mathbf{a}\sin(\theta)}{1 - \mathbf{a}\cos(\theta)}\right) \tag{9.2.3}$$

Since we have one parameter, **a**, to choose, we can do so such that

$$\omega_p = \theta_p - \rho(\theta_p) \Rightarrow \mathbf{a} = \frac{\sin\left(\frac{1}{2}(\omega_p - \theta_p)\right)}{\sin\left(\frac{1}{2}(\omega_p + \theta_p)\right)} \tag{9.2.4}$$

and the passband edge of H(**z**) is then the desired value ω_p. Since we have only one disposable parameter, we cannot choose where the stopband edge, ω_s, occurs. The stopband edge is given by

$$\omega_s = \theta_s - \rho(\theta_s) \tag{9.2.5}$$

Clearly, we can choose **a** such that the stopband edge is the design parameter and in which case the passband edge cannot be controlled. Since we control only one parameter, this mapping is suitable for filters such as the Butterworth, or Chebyshev Type 1, which define a "passband" but not a distinct "stopband," or Chebyshev Type 2, which defines a "stopband" but not a distinct "passband." Daniels [1.1] and Vlach [1.13] provide easy-to-read discussions of these types of filters.

The digital-to-digital transformation method is used not just for a lowpass-lowpass mapping but for lowpass-highpass, lowpass-bandpass, and lowpass-bandstop as well. Constantinides [1.5.3] provides a complete set of such mappings.

The lowpass-highpass mapping is similar to Eq. (9.2.1). Obtaining a bandpass filter H(**z**) from the prototype lowpass filter G(η) requires a second-order all-pass transformation of the form

$$\eta^{-1} = \frac{\mathbf{z}^{-2} - \alpha\mathbf{z}^{-1} + \beta}{1 - \alpha\mathbf{z}^{-1} + \beta\mathbf{z}^{-2}} \tag{9.2.6}$$

Of interest is that the order of H(**z**) is twice that of G(η). A lowpass-bandstop mapping will also be of the form of Eq. (9.2.6). Thus the lowpass-lowpass and lowpass-highpass preserve the order of the prototype while the lowpass-bandpass and lowpass-bandstop double the order.

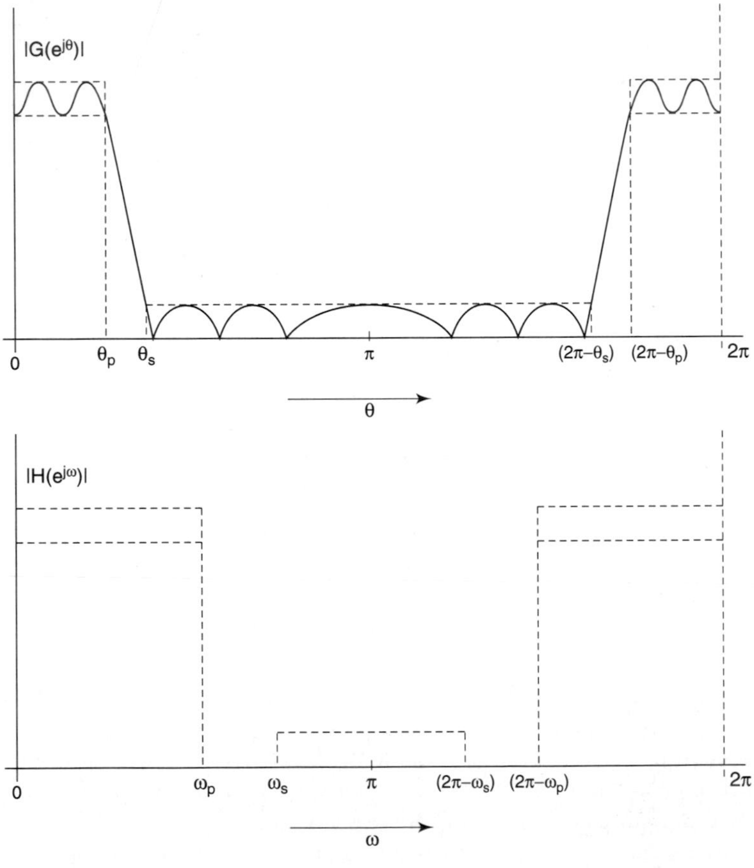

Fig. 9.2
Frequency mapping of θ to ω from Eq. (9.2.3) (lowpass-to-lowpass)

9.2.2 An Analog-Digital Transformation

The ***bilinear z transformation***, given by

$$\mathbf{s} = \frac{\mathbf{z} - 1}{\mathbf{z} + 1} \quad \text{or} \quad \mathbf{z} = \frac{1 + \mathbf{s}}{1 - \mathbf{s}} \tag{9.2.7}$$

provides a mapping between the **z**-plane and the **s**-plane. The mapping is one-to-one and the unit circle maps onto the $\mathbf{s} = j\theta$ axis with $\mathbf{z} = 1$ mapping into $\mathbf{s} = j$ or $\mathbf{s} = -j$ according to whether $\mathbf{z} = 1$ is approached from above or below. The implied sampling frequency of the digital filter is $f_s = 1/T$. The mapping is depicted in Fig. 9.3.

Now suppose we had an analog filter prototype with transfer function G(**s**), where **s** is the complex frequency in the Laplace transform domain. The frequency response of this analog filter is obtained by setting $\mathbf{s} = j\theta$, where θ is the analog frequency in radians/sec (or equivalent). The substitution according to Eq. (9.2.7) provides us a digital transfer function H(**z**) where

$$H(\mathbf{z}) = G\left(\frac{\mathbf{z} - 1}{\mathbf{z} + 1}\right) \tag{9.2.8}$$

The equivalent mapping of frequency response is depicted in Fig. 9.4, which assumes that G(**s**) is a lowpass filter with passband edge θ_p and stopband edge θ_s. Clearly, H(**z**) is also lowpass in nature with passband and stopband edges given by ω_p and ω_s, respectively. The nature of the mapping ensures that the passband ripple and stopband attenuation for the digital filter and the analog prototype match. These frequencies are given by

$$\theta = \tan\left(\frac{\omega}{2}\right) = \tan\left(\frac{\pi f}{f_s}\right) \tag{9.2.9}$$

which are obtained by setting $\mathbf{s} = j\theta$ and $\mathbf{z} = e^{j\omega}$ in Eq. (9.2.7). Note that ω is the *normalized radian frequency*. Eq. (9.2.9) also indicates the relationship between the analog domain and the digital domain with the frequency in the latter expressed in the usual units (f Hz), normalized by the sampling frequency f_s (Hz).

The relationship expressed in Eq. (9.2.9) is called "frequency warping." In order to design a digital filter, we first have to "pre-warp" the frequencies to obtain the correct values for the θ domain. The filter design is accomplished in the θ domain and then transformed back to the ω domain. For example, if we wished to obtain a digital filter with passband edge ω_p and stopband edge ω_s, we design an analog filter with the frequency edges given by θ_p and θ_s where

$$\theta_p = \tan\left(\frac{\omega_p}{2}\right) \quad \text{and} \quad \theta_s = \tan\left(\frac{\omega_s}{2}\right) \tag{9.2.10}$$

to account for this nonlinear frequency mapping. Analog transfer functions are always ratios of polynomials in **s**, with the order of the numerator no greater than that of the denominator. That is, G(**s**) is of the form

$$G(\mathbf{s}) = \frac{\sum_{k=0}^{N} c_k \mathbf{s}^k}{\sum_{k=0}^{D} d_k \mathbf{s}^k} \quad (N \leq D) \tag{9.2.11}$$

Substitution of Eq. (9.2.8) will yield H(**z**) given by

$$H(\mathbf{z}) = \frac{\sum_{k=0}^{N} c_k \left(\frac{\mathbf{z} - 1}{\mathbf{z} + 1}\right)^k}{\sum_{k=0}^{D} d_k \left(\frac{\mathbf{z} - 1}{\mathbf{z} + 1}\right)^k} = \frac{\sum_{k=0}^{D} a_k \mathbf{z}^{-k}}{1 + \sum_{k=1}^{D} b_k \mathbf{z}^{-k}} \tag{9.2.12}$$

Since G(**z**) is a ratio of polynomials, we are guaranteed that H(**z**) will likewise be a ratio of polynomials of order D. Knowledge of G(**s**) is tantamount to knowledge of the coefficients $\{c_i\}$ and $\{d_i\}$ and Eq. (9.2.12), indicating that the digital filter coefficients can be obtained from these, albeit with some computation. Often the analog filter is expressed in terms of its poles and zeros. Suppose that "p" is a pole of G(**s**). Then the bilinear transformation maps this pole into H(**z**) as

$$\frac{1}{\mathbf{s} - p} \rightarrow \frac{1}{1 - p} \frac{1 + \mathbf{z}^{-1}}{1 - \xi \mathbf{z}^{-1}} \quad \text{where } \xi = \frac{1 + p}{1 - p} \tag{9.2.13}$$

If "p" is a zero of G(**s**) then this maps into H(**z**) as

$$(\mathbf{s} - p) \rightarrow (1 - p) \frac{1 - \xi \mathbf{z}^{-1}}{1 + \mathbf{z}^{-1}} \quad \text{where } \xi = \frac{1 + p}{1 - p} \tag{9.2.14}$$

Further, since for analog filters N D, the digital filter will have a (DN)th -order zero at **z** = -1 (half the sampling frequency).

The availability of design methods for analog filters has made the bilinear transformation quite popular. For example, Gray and Markel [1.4] provide a program for designing digital elliptic filters by transforming the design of an analog elliptic filter. Such a program is also available in MATLAB [1.6]. However, the transform is just as useful to design analog transfer functions from digital filter prototypes.

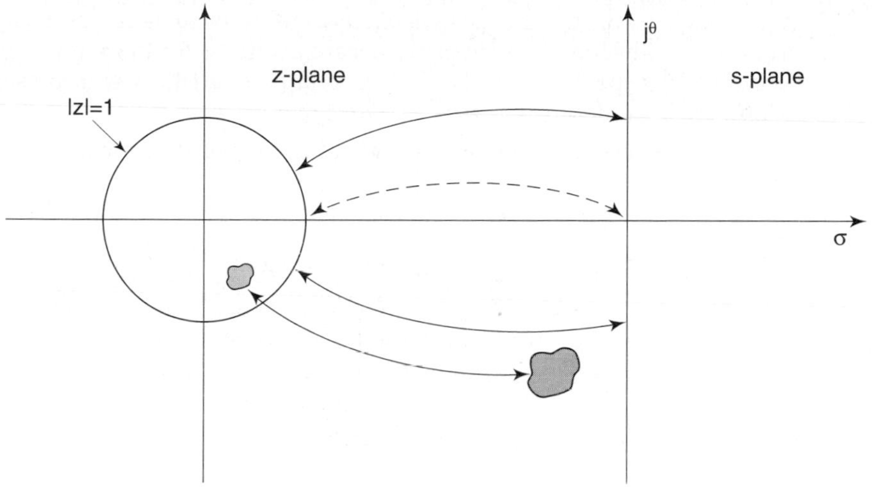

Fig. 9.3
z-plane-to-s-plane mapping of Eq. (9.2.7), the bilinear z-transformation

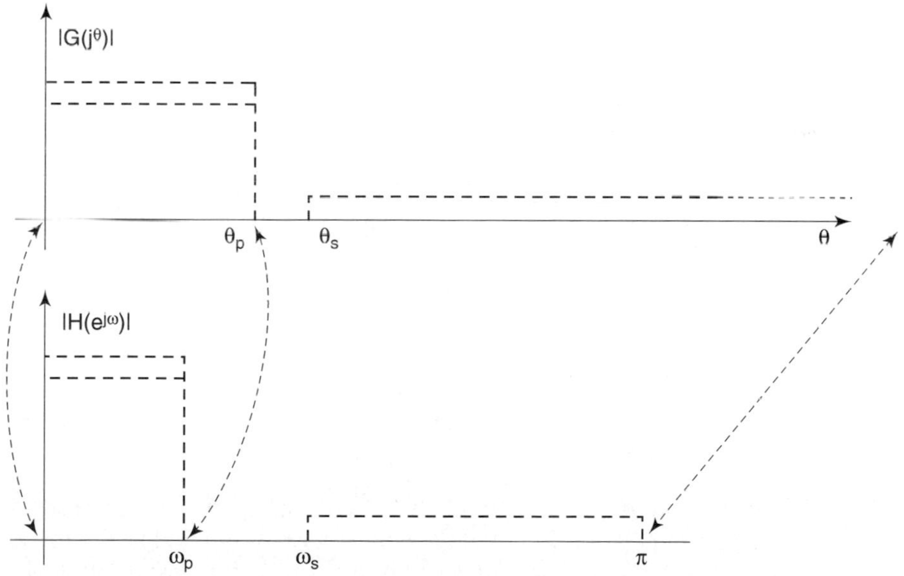

Fig. 9.4
Frequency warping associated with the bilinear z-transformation and Eq. (5.1.19)

9.3 DESIGN OF IIR FILTERS

The approach taken for the design of **IIR** filters is to pose the design problem appropriately. Several software packages are available for the actual approximation procedure. In a special case, the design of *passband* filters with prescribed transmission zeros, we describe a useful design method.

We shall concentrate on the design of **IIR** filters that approximate a given magnitude response. Unlike **FIR** filters where, by using symmetric coefficients, we could guarantee a linear-phase response, **IIR** filters are constrained by the notions of stability and causality to have a relationship between the magnitude and phase responses. By designing to a magnitude specification the phase response cannot be controlled. Following the design procedure the phase response can be evaluated to see if it is satisfactory. If necessary an allpass filter is used to "correct" the phase response.

9.3.1 The Squared-Magnitude Function

The transfer function $H(\mathbf{z})$ and the squared-magnitude function $H(\mathbf{z})H(\mathbf{z}^{-1})$ can be written as

$$H(\mathbf{z}) = \frac{\sum_{k=0}^{N} a_k \mathbf{z}^{-k}}{1 + \sum_{k=1}^{D} b_k \mathbf{z}^{-k}} \tag{9.3.1a}$$

$$S(\mathbf{z}) = H(\mathbf{z})H(\mathbf{z}^{-1}) = \frac{\sum_{k=0}^{N} \alpha_k (\mathbf{z}^{-k} + \mathbf{z}^{+k})}{\sum_{k=0}^{D} \beta_k (\mathbf{z}^{-k} + \mathbf{z}^{+k})} \tag{9.3.1b}$$

where we will assume that N D.

Given the coefficients $\{a_i\}$ and $\{b_i\}$ we can compute the coefficients $\{\alpha_i\}$ and $\{\beta_i\}$ in a straightforward manner by

$$\alpha_i = \sum_{k=0}^{N-i} a_k a_{k+i} \quad \text{for } 0 \le i \le N \tag{9.3.2a}$$

$$\beta_i = \sum_{k=0}^{D-i} b_k b_{k+i} \quad \text{for } 0 \le i \le D \tag{9.3.2b}$$

The reverse, that is, obtaining the $\{a_i\}$ and $\{b_i\}$ from $\{\alpha_i\}$ and $\{\beta_i\}$, is not that straightforward.

The roots of say, the denominator of $S(\mathbf{z})$, are related to those of $H(\mathbf{z})$ in a special way. Each pole of $H(\mathbf{z})$, say $\mathbf{p}$, generates two poles for $S(\mathbf{z})$, at $\mathbf{p}$ and $1/\mathbf{p}$. Since we assume that the coefficients will be real, this implies that each complex-conjugate pair of poles of $H(\mathbf{z})$ will generate a foursome of poles for $S(\mathbf{z})$. Conversely, if we extract the poles of $S(\mathbf{z})$, we expect to see this foursome grouping. Of this foursome one complex-conjugate pair is inside the unit circle and is chosen for $H(\mathbf{z})$. Therefore, going from $S(\mathbf{z})$ to $H(\mathbf{z})$ requires the extraction of the roots of denominator of $S(\mathbf{z})$. A similar situation arises for the numerator. The difference is that for the numerator we are not constrained to choose the zeros inside the unit circle.

Substituting $\mathbf{z} = e^{j\omega}$ to get the frequency response as a function of the normalized radian frequency ω, we get

$$S(e^{j\omega}) = \frac{\sum_{k=0}^{N} \alpha_k \cos(k\omega)}{\sum_{k=0}^{D} \beta_k \cos(k\omega)} \tag{9.3.3}$$

which shows that the squared-magnitude function $S(e^{j\omega})$ can be written as the ratio of trigonometric polynomials. Since this is a symmetric function of ω, we only need to consider frequency values in the range $[0, \pi]$.

Now consider the substitution

$$\mathbf{x} = \frac{1}{2}(\mathbf{z}^{-1} + \mathbf{z}^{+1}) \tag{9.3.4}$$

For values of frequency in the range $[0,\pi]$, setting $\mathbf{z} = e^{j\omega}$ yields

$$\mathbf{x} = \cos(\omega);\ -1 \le \mathbf{x} \le +1\ ;\ 0 \le \omega \le \pi \tag{9.3.5}$$

The mapping of the frequency variable ω into the variable $\mathbf{x}$ is depicted in Fig. 9.5. Low frequencies, for ω close to zero, map into the region close to $\mathbf{x} = 1$; high frequencies, with ω close to π, map into the region close to $\mathbf{x} = -1$. With this substitution we can express the squared-magnitude function as a function of $\mathbf{x}$ and this takes the form

$$S(\mathbf{x}) = \frac{\sum_{k=0}^{N} \alpha_k T_k(\mathbf{x})}{\sum_{k=0}^{D} \beta_k T_k(\mathbf{x})} = \frac{N(\mathbf{x})}{D(\mathbf{x})} \tag{9.3.6}$$

Note that S($\mathbf{x}$) is a ratio of polynomials in $\mathbf{x}$. $T_k(\mathbf{x})$ in the above equation is the kth Chebyshev polynomial of the first kind.

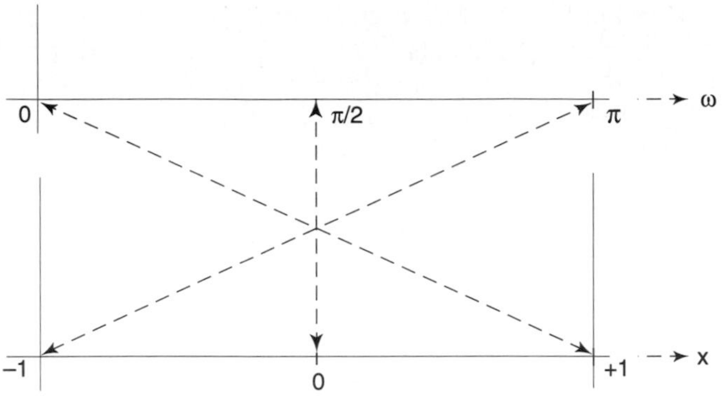

Fig 9.5
Mapping of independent variable by Eq. (9.3.5).

9.3.2 Statement of the Design Problem

Suppose that the target frequency response is specified by a function F(ω), which via the substitution $\mathbf{x} = \cos(\omega)$, is transformed into a function of $\mathbf{x}$ that we shall, with some abuse of notation, call F($\mathbf{x}$). Since F(ω) is to be a squared-magnitude function, it is constrained to be nonnegative. The approximation problem can be written as:

Find N($\mathbf{x}$)   0 and D($\mathbf{x}$) > 0 such that

$$\frac{N(\mathbf{x})}{D(\mathbf{x})} \approx F(\mathbf{x}) \quad \text{for} \quad \mathbf{x} \in [-1, +1] \tag{9.3.7}$$

where the approximation error, or "goodness of fit," is suitably defined.

Now suppose that the approximation procedure provides us with the polynomials N($\mathbf{x}$) and D($\mathbf{x}$). Obtaining the equivalent digital filter from the "$\mathbf{x}$-domain" description can be implemented in two ways. We shall consider the denominator since it is the more crucial choice and has implications of stability. The first, which is more illustrative than expedient, is to expand D($\mathbf{x}$) in terms of Chebyshev polynomials:

$$D(\mathbf{x}) = \sum_{i=0}^{D} \eta_i T_i(\mathbf{x}) \tag{9.3.8}$$

Then the poles of H($\mathbf{z}$) are the roots of $D_1(\mathbf{z})$ that lie inside the unit circle where $D_1(\mathbf{z})$ is given by

$$D_1(\mathbf{z}) = \mathbf{z}^{+D}\{2\eta_0 + \sum_{k=1}^{D} \eta_k(\mathbf{z}^{-k} + \mathbf{z}^{+k})\} \tag{9.3.9}$$

Observe that $D_1(\mathbf{z})$ is a polynomial that exhibits symmetry in its coefficients. Such a polynomial is called a ***mirror image polynomial*** and its roots are guaranteed to occur in reciprocal pairs. Since the coefficients are real, these roots will occur in foursomes, allowing the choice of complex-conjugate pairs that lie inside the unit circle. This method requires the extraction of the roots of a polynomial of degree 2D. Some computational efficiency may be gained from the knowledge of the symmetry of the coefficients.

However, the second, and simpler, method is to extract the roots of the polynomial $D(\mathbf{x})$ directly. This root extraction is done on a polynomial of degree D. If $\mathbf{x}_i$, i= 1, 2, ... , D, are the roots of $D(\mathbf{x})$, then the poles of $H(\mathbf{z})$ are given by

$$\mathbf{z}_i = \mathbf{x}_i \pm \sqrt{(\mathbf{x}_i^2 - 1)} \tag{9.3.10}$$

where the sign of the square root is chosen to make $|\mathbf{z}_i| < 1$. The derivation of Eq. (9.3.10) is based on Eq. (9.3.4), which relates the "**x**-domain" to the "**z**-domain." It can be shown that if any of the roots $\mathbf{x}_i$ is real and lies between -1 and +1, then the corresponding $\mathbf{z}_i$ will lie on the unit circle. Since it is not advisable to have poles on the unit circle we have to constrain our approximation procedure to ensure that $D(\mathbf{x})$ does not have such roots. This is the reason for the constraint $D(\mathbf{x}) > 0$ (strictly greater than 0) associated with Eq. (9.3.7).

9.3.3 IIR Lowpass Filters

We will be talking, for the most part, about lowpass (or "passband") filters that have the notion of a "stopband," where the objective is to provide infinite attenuation. This suggests that the filter $H(\mathbf{z})$ have all its transmission zeros on the unit circle, and in the stopband. A transmission zero at the frequency ω_0, $0 < \omega_0 < \pi$, gives rise to a second-order term in the numerator of $H(\mathbf{z})$ of the form

$$\text{zero at } \omega_0\text{: } 1 - 2\cos(\omega_0)\mathbf{z}^{-1} + \mathbf{z}^{-2} \tag{9.3.11}$$

This implies that $N(\mathbf{x})$, which is related to $H(\mathbf{z})H(\mathbf{z}^{-1})$, has a double root at $\mathbf{x}_0 = \cos(\omega_0)$. That is,

$$N(\mathbf{x}) = (\mathbf{x} - \mathbf{x}_0)^2 N_1(\mathbf{x}) \tag{9.3.12}$$

If $H(\mathbf{z})$ has a zero at DC or half the sampling frequency, then the induced root(s) of $N(\mathbf{x})$ are

$$\text{zero at } \omega = 0\text{: } (1 - \mathbf{x}) \text{ and zero at } \omega = \pi\text{: } (1 + \mathbf{x}) \tag{9.3.13}$$

Thus for a lowpass filter (which will not have a zero at DC), and which has distinct (nonrepeated zeros), the form of N(**x**) is

$$N(\mathbf{x}) = Q(\mathbf{x})^2 \quad \text{for even order} \tag{9.3.14a}$$

$$N(\mathbf{x}) = (1 + \mathbf{x})\,Q(\mathbf{x})^2 \quad \text{for odd order} \tag{9.3.14b}$$

where Q(**x**) is a polynomial of N/2 (N even) or (N-1)/2 (N odd). For a highpass filter the first-order term, for odd N, is (1-**x**) rather than (1+**x**).

Since S(**x**) is a ratio of polynomials, we can write S(**x**) in the form

$$S(\mathbf{x}) = \frac{N(\mathbf{x})}{D(\mathbf{x})} = \frac{1}{1 + [\frac{P(\mathbf{x})}{N(\mathbf{x})}]} \tag{9.3.15}$$

and we can visualize the design of a lowpass (or passband) filter as the procedure to obtain P(**x**) and N(**x**) such that

$$\textit{passband}: \frac{P(\mathbf{x})}{N(\mathbf{x})} \approx 0 \tag{9.3.16a}$$

$$\textit{stopband}: \frac{P(\mathbf{x})}{N(\mathbf{x})} \approx \infty \tag{9.3.16b}$$

9.3.4 Design of Polynomial Lowpass Filters

Methods similar to those used in classical (analog) filter design (see Daniels [1.1] or Vlach [1.13]) can be used to provide recursive digital filter transfer functions. Consider the design of such a filter with nominal cutoff frequency ω_c and let $\mathbf{x}_c = \cos(\omega_c)$.

If an all-pole filter is desired, then N(**x**) = 1 (constant). If we define

$$P(\mathbf{x}) = \mathbf{g}\,(1 - \mathbf{x})^D \tag{9.3.17}$$

then S(**x**) takes the form

$$S(\mathbf{x}) = \frac{1}{1 + P(\mathbf{x})} = \frac{1}{1 + g\,(1 - \mathbf{x})^D} \tag{9.3.18}$$

This choice of P(**x**) corresponds to a ***Butterworth*** approximation. P(**x**) "approximates" zero in the band [$\mathbf{x}_c$, 1] by making the approximation error and up to D derivatives zero at **x** = 1. This behavior is called ***maximally flat*** for obvious reasons. Since all the zeros of P(**x**) are in the band [$\mathbf{x}_c$, 1] (actually all are at **x** = 1), the function P(**x**), for values of **x** far removed from [$\mathbf{x}_c$, 1] will be large. This is how the intent of Eq. (9.3.16) is achieved. The constant **g** is chosen to yield the desired response at $\mathbf{x} = \mathbf{x}_c$ or, equivalently, at $\omega = \omega_c$. Clearly, the behavior in both the passband and the stopband is monotonic.

In general, a polynomial filter is obtained by setting

$$P(\mathbf{x}) = g \prod_{k=1}^{D} (\mathbf{x} - \mathbf{x}_k) \tag{9.3.19}$$

where the $\mathbf{x}_i$ are all contained in the interval $[\mathbf{x}_c, 1]$, which corresponds to the passband. With this form, $P(\mathbf{x})$ will be small (being close to its zeros) in the passband and, for values of $\mathbf{x}$ distant from $[\mathbf{x}_c,1]$, will increase monotonically in magnitude. This will achieve the desired behavior expressed by Eq. (9.3.16). The constant g is chosen to make the frequency response equal to a prescribed value at a prescribed frequency. Since we require that $S(\mathbf{x})$ be nonnegative over [-1, 1], the sign of g is chosen to make $P(\mathbf{x})$ positive over $[-1, \mathbf{x}_c]$. Since all its zeros are in $[\mathbf{x}_c, 1]$, $P(\mathbf{x})$ will not change sign in $[-1, \mathbf{x}_c]$.

Chebyshev filters can be obtained by setting $P(\mathbf{x})$ to a scaled and shifted version of the Dth Chebyshev polynomial. The Chebyshev polynomials are defined in the following manner. Over the interval [-1, +1], $T_D(\mathbf{y})$ is given by

$$T_D(\mathbf{y}) = \cos(D \cos^{-1}(\mathbf{y})) \tag{9.3.20}$$

which using common trigonometric identities can be modified to the conventional representation of a polynomial, as a linear combination of powers of $\mathbf{y}$, which is valid for all $\mathbf{y}$. The first few Chebyshev polynomials are shown in Fig. 9.6 and have been normalized to 1 at $\mathbf{y} = +1$. The terminology "equi-ripple" or "equi-ripple approximation to zero," or "min-max approximation to zero," etc., are quite apt, considering that the functional value is constrained to be +1 at $\mathbf{y} = +1$. Chebyshev polynomials are used in the design of filters that exhibit either an equi-ripple passband, called Chebyshev Type 1, or equiripple stopband, called Chebyshev Type 2.

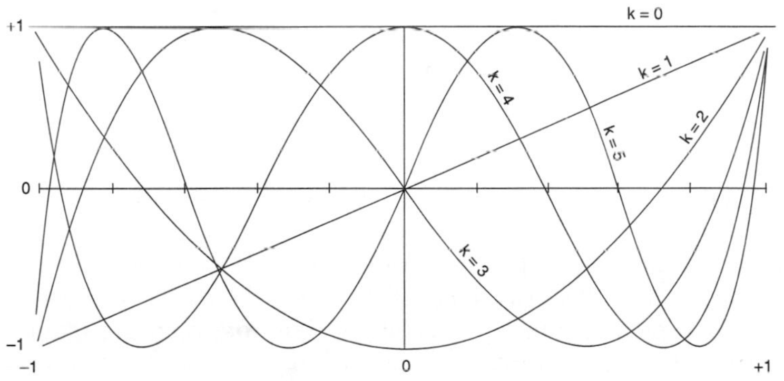

Fig. 9.6
Chebyshev polynomials $T_k(\mathbf{y})$ over [−1, +1] for k = 0, 1, 2, 3, 4, 5

The mapping $\mathbf{y} = f(\mathbf{x})$ given by

$$\mathbf{y} = f(\mathbf{x}) = \frac{2\mathbf{x}}{1 - \mathbf{x}_c} - \frac{1 + \mathbf{x}_c}{1 - \mathbf{x}_c} \qquad (9.3.21)$$

"stretches" $[\mathbf{x}_c, 1]$ into [-1, 1]. This is depicted in Fig. 9.7.

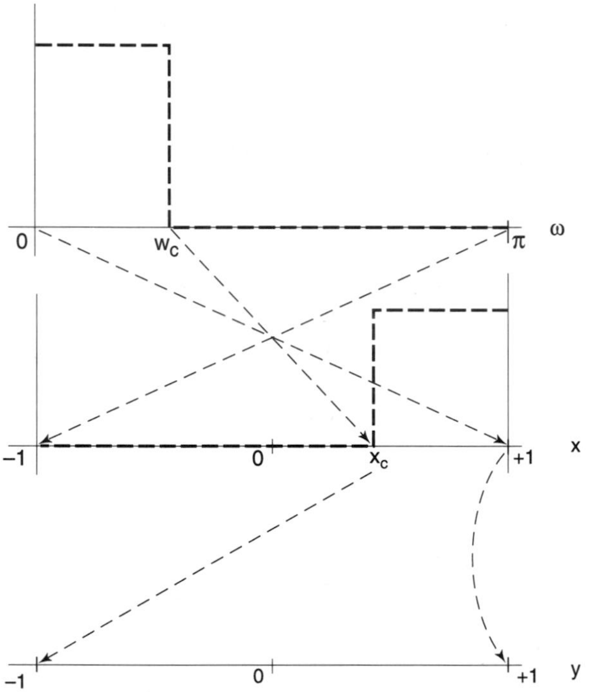

Fig. 9.7
Two-stage mapping of passband into [−1, 1]

We can then obtain a Chebyshev filter, analogous to the analog Type 1 Chebyshev filter, by setting

$$P(\mathbf{x}) = \varepsilon\, T_D(f(\mathbf{x})) \qquad (9.3.22)$$

Since the Chebyshev polynomials exhibit an equi-ripple approximation to zero over the range [-1, 1], it follows that $P(\mathbf{x})$ is an equi-ripple approximation to zero in the band $[\mathbf{x}_c, 1]$. The quantity ε defines the passband ripple. The constraint that $D(\mathbf{x}) > 0$ requires that we choose the sign of ε to ensure that $\varepsilon P(\mathbf{x}_c) > 0$. With this choice, $S(\mathbf{x})$ is given by

$$S(\mathbf{x}) = \frac{1}{D(\mathbf{x})} = \frac{1}{1 + \varepsilon\, T_D(f(\mathbf{x}))} \qquad (9.3.23)$$

The polynomial filters described in Eq. (9.3.18), (9.3.19), and (9.3.25) are "all-pole" and provide monotonic behavior in the stopband. An improvement in the magnitude characteristic can be achieved by introducing transmission zeros in the stopband. For an **IIR** filter, since the numerator degree is usually less than, or equal to, the degree of the denominator, N(**x**) will take the form of Eq. (9.3.14), where we have assumed the selection of L complex zeros and the associated L complex conjugates to write Q(**x**) as

$$Q(\mathbf{x}) = \prod_{k=0}^{L-1} (\mathbf{x} - \mathbf{x}_k); \; 2L \leq D \tag{9.3.24}$$

A filter with these zeros and maximally flat behavior in the passband can be readily obtained by setting S(**x**) as

$$S(\mathbf{x}) = \frac{1}{1 + \gamma \dfrac{(1 - \mathbf{x})^D}{N(\mathbf{x})}} \tag{9.3.25}$$

where N(**x**) is given by Eq. (9.3.14). Again, the constant γ is chosen to give the desired attenuation at a chosen frequency, usually at $\mathbf{x} = \mathbf{x}_c$ ($\omega = \omega_c$). As a special case, if we set all the zeros of the filter to be at half the sampling rate, then $N(\mathbf{x}) = (1+\mathbf{x})^N$ and

$$S(\mathbf{x}) = \frac{1}{1 + \gamma \dfrac{(1 - \mathbf{x})^D}{(1 + \mathbf{x})^N}} \tag{9.3.26}$$

and as a special case of this, if N = D, then

$$S(\mathbf{x}) = \frac{1}{1 + \gamma \left[\dfrac{1 - \mathbf{x}}{1 + \mathbf{x}}\right]^D} \tag{9.3.27}$$

It can be shown that the bilinear transformation of an analog Butterworth filter will yield a filter equivalent to one special case of Eq. (9.3.27). The advantage of the approach described here is that we can directly choose the attenuation at any prescribed frequency.

If we are allowed to have transmission zeros, then we can modify the mapping of Eq. (9.3.21) to **y** = f(**x**) where

$$\mathbf{y} = f(\mathbf{x}) = \frac{r\mathbf{x} + s}{1 + \mathbf{x}} \tag{9.3.28}$$

where r and s are chosen such that the interval $[\mathbf{x}_c, 1]$ in x maps into $[-1, 1]$ in $\mathbf{y}$ and the point $\mathbf{x} = -1$ maps into - in $\mathbf{y}$. By setting

$$P(\mathbf{x}) = \varepsilon\, T_D(f(\mathbf{x})) \tag{9.3.29}$$

we get a function $P(\mathbf{x})$ that is not a polynomial but is of the form

$$P(\mathbf{x}) = \varepsilon \frac{P_1(\mathbf{x})}{(1 + \mathbf{x})^D} \tag{9.3.30}$$

which is a ratio of polynomials. That is, $N(\mathbf{x}) = (1+\mathbf{x})^D$. The resulting squared-magnitude function, $S(\mathbf{x})$ is

$$S(\mathbf{x}) = \frac{(1 + \mathbf{x})^D}{(1 + \mathbf{x})^D + \varepsilon P_1(\mathbf{x})} \tag{9.3.31}$$

The parameter ε defines the passband ripple. It can be shown that the bilinear transformation of an analog Chebyshev Type 1 filter yields a digital filter equivalent to that obtained from Eq. (9.3.31).

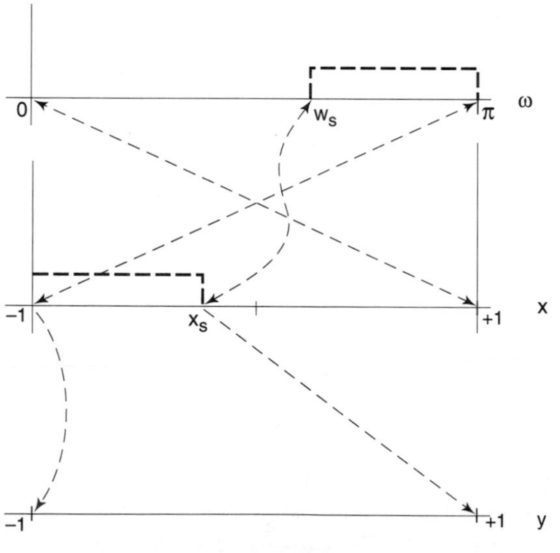

Fig. 9.8
Two-stage mapping of stopband into [–1, 1]

Filters such as the Chebyshev Type 2 exhibit an equi-ripple behavior in the stopband. This can be accommodated in the following manner. First we ascertain the stopband edge ω_s. This maps into the $\mathbf{x}$-domain as $\mathbf{x}_s = \cos(\omega_s)$. We next determine the mapping $\mathbf{y} = f(\mathbf{x})$ given by

$$\mathbf{y} = f(\mathbf{x}) = \frac{r_1 \mathbf{x} + s_1}{1 - \mathbf{x}} \tag{9.3.32}$$

In Eq. (9.3.32), r_1 and s_1 are chosen such that f(**x**) maps the stopband [-1, $\mathbf{x}_s$] into the range [-1, 1] for the variable **y** and the point **x** = 1 maps to . The mapping sequence from ω to **x** to **y** is shown in Fig. 9.8.

We then obtain Chebyshev Type 2 behavior by setting

$$P(\mathbf{x}) = \frac{\gamma}{1 + T_D(f(\mathbf{x}))} \tag{9.3.33}$$

Again, P(**x**) is not a polynomial but is of the form

$$P(\mathbf{x}) = \frac{\gamma(1 - \mathbf{x})^D}{Q(\mathbf{x})^2} \quad \text{for even D} \tag{9.3.34a}$$

$$P(\mathbf{x}) = \frac{\gamma(1 - \mathbf{x})^D}{(1 + \mathbf{x})Q(\mathbf{x})^2} \quad \text{for odd D} \tag{9.3.34b}$$

where Q(**x**) is a polynomial of degree n = D/2 if D is even, or n = (D-1)/2 if D is odd. From this we obtain S(**x**) as

$$S(\mathbf{x}) = \frac{Q(\mathbf{x})^2}{Q(\mathbf{x})^2 + \gamma(1 - \mathbf{x})^D} \quad \text{for even D} \tag{9.3.35a}$$

$$S(\mathbf{x}) = \frac{(1 + \mathbf{x})Q(\mathbf{x})^2}{(1 + \mathbf{x})Q(\mathbf{x})^2 + \gamma(1 - \mathbf{x})^D} \quad \text{for odd D} \tag{9.3.35b}$$

which is exactly the form we need. It can be verified that the bilinear transformation applied to an analog Chebyshev Type 2 filter yields a digital filter equivalent to the squared-magnitude function in Eq. (9.3.35).

To illustrate the type of magnitude response we obtain by these methods, three examples are shown below.

Example 1. The response shown in Fig. 9.9 is of a sixth-order filter (D = 6) obtained using Eq. (9.3.27) with $\omega_p = 0.25\pi$. The value of γ is chosen such that $P(\mathbf{x}_p) = 0.1$. Note that both the pass and stopbands are monotonic.

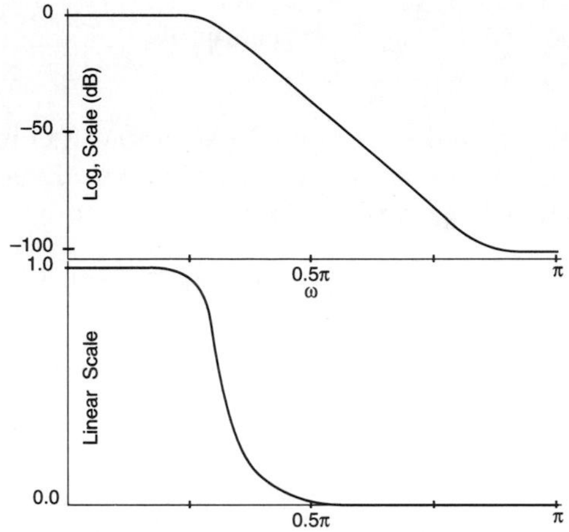

Fig. 9.9
Response of maximally flat lowpass filter of Example 1

Example 2. Fig. 9.10 shows the response of a sixth-order Chebyshev Type 1 filter obtained using Eq. (9.3.31). The ripple parameter chosen is $\varepsilon = 0.1$. Observe the equi-ripple passband and the monotonic stopband.

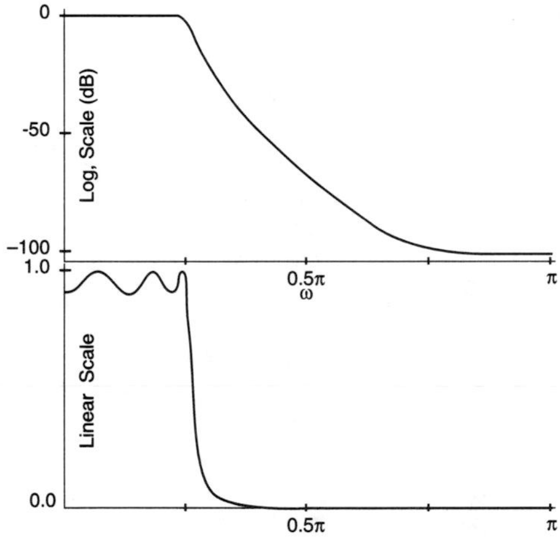

Fig. 9.10
Response of Chebyshev Type 1 filter of Example 2

Example 3. The response of a sixth-order Chebyshev Type 2 filter with $\omega_s = 0.75\pi$ and $P(\mathbf{x}_s) = 0.01$ using equations Eq. (9.3.33) and (9.3.35) is depicted in Fig. 9.11.

The mathematical operations involved in designing digital filters based on the properties of polynomials are simple to automate. These involve root extraction, substitution of a polynomial for the independent variable of another polynomial, addition of polynomials, multiplication of polynomials, expressing a polynomial as a series of Chebyshev polynomials, and so on.

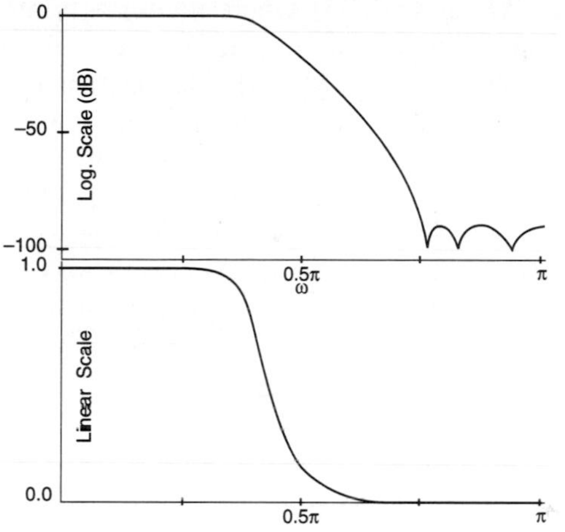

Fig. 9.11
Response of Chebyshev Type 2 filter of Example 3

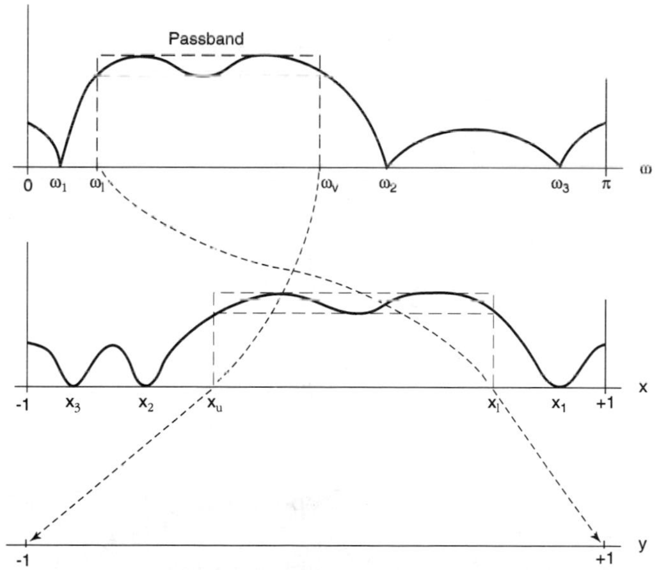

Fig. 9.12
Target response with prescribed zeros at ω_1, ω_2 and ω_3

9.4 EQUI-RIPPLE FILTERS WITH PRESCRIBED TRANSMISSION ZEROS

In certain cases we apply filters to get rid of signals, usually tones, regarding which we have some information. For example, in telephony applications, we know that the frequency band for the speech signal is prescribed as [0.3, 3.4] KHz. The digital signal is available as $\{x(nT_s)\}$ where the underlying sampling rate is 8 KHz. This signal may contain significant quantities of power-line hum, signals at 60 Hz or 50 Hz, which must be removed. Long-haul transmission of this channel signal in analog form using Single Side Band (**SSB**) techniques could introduce artifacts at 80 Hz and or 3920 Hz. In Chapter 7 we discuss transmultiplexers, where we explain the notion of ***pilot*** tones that accompany the **SSB** signal. These pilot tones are usually at a frequency of xxx.08 KHz and appear at 3920 Hz in one channel and at 80 Hz in the adjacent channel of the frequency division multiplex assembly. For signaling purposes tones, nominally at 3825 Hz, are inserted in each channel. These will appear at 3825 Hz in the principal channel and at 175 Hz in the adjacent channel. In short, we have to design a filter such that the passband is [0.3, 3.4] KHz with transmission zeros at prescribed frequencies.

The form of the filter design will follow Eq. (9.3.15). Since the transmission zeros are prescribed, the polynomial $N(\mathbf{x})$ is known. $P(\mathbf{x})$ has to be determined such that $P(\mathbf{x})/N(\mathbf{x})$ is an equi-ripple approximation to zero in the pass band $[x_u, x_l]$. The situation is depicted in Fig. 9.12, which shows three (positive frequency) specified zeros. The symmetry of the frequency response implies that there are three zeros at negative frequencies at $-\omega_1$, $-\omega_2$, and $-\omega_3$. The equi-ripple requirement means that the **ratio** $P(\mathbf{x})/N(\mathbf{x})$ must behave like a Chebyshev polynomial in the interval $[x_u, x_l]$.

$P(\mathbf{x})$ can be obtained using the concept of a "rational Chebyshev polynomial" which is described next in Section 9.4.1. In Section 9.4.2, we use the notion of rational Chebyshev polynomials to outline a design procedure for obtaining equi-ripple passband filters with predetermined transmission zeros.

9.4.1 Rational Chebyshev Polynomials

Darlington [1.2] devised analytical techniques for determining rational function approximations to zero that yield error extrema in the interval [–1, 1] of equal amplitude. These are termed ***Rational Chebyshev polynomials***.

The theory of rational Chebyshev polynomials uses many of the same principles applied in general discrete-time signal processing theory and the **Z**-transform. To avoid confusion we shall use the variable η instead of the traditional **z**. Let

$$R_N(\mathbf{y}) = \frac{U(\mathbf{y})}{V(\mathbf{y})} \tag{9.4.1}$$

where $U(\mathbf{y})$ is an Nth degree polynomial and $V(\mathbf{y})$ is a polynomial of degree M, where M N. $V(\mathbf{y})$ is prescribed in advance and does not have any real zeros in the interval [–1, 1]. We wish to determine $U(\mathbf{y})$ such that $R_N(\mathbf{y})$ behaves like a Chebyshev polynomial, namely, has (N–1) extrema within (–1, 1) and extrema at the end-points, ± 1. These extrema must be of equal magnitude and, as we traverse the interval from –1 to 1, these extrema must have alternating signs.

To facilitate the determination of U(**y**), we consider the mapping

$$\mathbf{y} = 0.5\,(\eta^{-1} + \eta^{+1}) \tag{9.4.2}$$

which maps the interval [–1, 1] into the upper (or lower) half of the unit circle, $|\eta| = 1$, in the η-plane. When η is on the unit circle we can set $\eta = e^{j\theta}$, which transforms Eq. (9.4.2) into

$$\mathbf{y} = \cos(\theta) \tag{9.4.3}$$

The warping of the **y**-scale via Eq. (9.4.3) transforms the equal extrema of the Nth-order Chebyshev polynomial, $T_N(\mathbf{y})$, into a function whose extrema are equal in magnitude, alternating in sign, and equally spaced. Specifically

$$T_N(\mathbf{y}) = \cos(N\theta) \tag{9.4.4}$$

Equivalently, $T_N(\mathbf{y})$ can be expressed as

$$T_N(\mathbf{y}) = 0.5\,(\eta^{-N} + \eta^{+N}) = 0.5\,\mathrm{Re}\{G(\eta)\} \tag{9.4.5}$$

where $G(\eta) = \eta^N$. On the unit circle the magnitude of $G(\eta)$ is unity and its argument (i.e., phase angle) varies from 0 through $2N\pi$ as θ goes from 0 through 2π.

The mapping of Eq. (9.4.2) permits us to express U(**y**) as a sum

$$U(\mathbf{y}) = u(\eta) + u(\eta^{-1}) \tag{9.4.6}$$

where the coefficients of $u(\eta)$ uniquely determine the coefficients of U(**y**). The mapping of Eq. (9.4.2) also implies that the roots of V(**y**), when expressed as a function of η, will form reciprocal pairs. Therefore we can express V(**y**) as the product of two polynomials:

$$V(\mathbf{y}) = v(\eta)\,v(\eta^{-1}) \tag{9.4.7}$$

Here we assume that the roots of $v(\eta)$ lie outside the unit circle in the η-plane. The roots of $v(\eta^{-1})$ will therefore lie inside the unit circle.

Substituting Eqs (9.4.6) and (9.4.7) into (9.4.1) yields

$$R_N(\mathbf{y}) = \frac{u(\eta)}{v(\eta)\,v(\eta^{-1})} + \frac{u(\eta^{-1})}{v(\eta)\,v(\eta^{-1})} = T(\eta) + T(\eta^{-1}) \tag{9.4.8}$$

where $T(\eta)$ is a ratio of polynomials involving η and η^{-1}. Now if $T(\eta)$ is an *all-pass* filter, and it mimics, in some sense, the term η^N, then comparison with Eq. (9.4.5) indicates that $R_N(\mathbf{y})$ has the desired equal-extrema (equi-ripple) behavior. If $T(\eta)$ is to be all-pass, then its poles and zeros must have a reciprocal relationship. The following choice of $T(\eta)$,

$$T(\eta) \;=\; c\,\eta^{+N}\frac{v(\eta^{-1})}{v(\eta)} \tag{9.4.9}$$

achieves this condition. The polynomial $u(\eta)$ is then determined by setting

$$u(\eta) + u(\eta^{-1}) \;=\; c\left[\eta^{+N} v^2(\eta^{-1}) + \eta^{-N} v^2(\eta)\right] \tag{9.4.10}$$

and equating the coefficients of like powers of η and η^{-1}. The constant **c** is chosen so that $R_N(\mathbf{y})$ is unity (or any other desired value) at $\mathbf{y} = 1$. The form of Eq. (9.4.10) indicates that $V(\mathbf{y})$ can be of the form $V(\mathbf{y}) = (V_1(\mathbf{y}))^{0.5}$, where $V_1(\mathbf{y})$ is a polynomial of degree M 2N. In particular, relating $V_1(\mathbf{y})$ to $v_1(\eta)$ and $v_1(\eta^{-1})$ as in Eq. (9.4.7), we have

$$u(\eta) + u(\eta^{-1}) \;=\; c\left[\eta^{+N} v_1(\eta^{-1}) + \eta^{-N} v_1(\eta)\right] \tag{9.4.11}$$

9.4.2 Obtaining Filters with Prescribed Zeros

Choosing a and b appropriately, the mapping

$$\mathbf{y} \;=\; a\,\mathbf{x} + b \tag{9.4.12}$$

maps the passband $[\mathbf{x}_u, \mathbf{x}_l]$ into [–1,1]. Similarly,

$$\mathbf{x} \;=\; \alpha\,\mathbf{y} + \beta \tag{9.4.13}$$

allows the reverse mapping of variables. Since we have prescribed transmission zeros, $N(\mathbf{x})$ is known, which provides

$$V(\mathbf{y}) \;=\; N(a\mathbf{y}+b) \tag{9.4.14}$$

We have the choice of mapping the polynomial $N(\mathbf{x})$ via Eq. (9.4.14) to get the coefficients of $V(\mathbf{y})$ or we could map the roots of $N(\mathbf{x})$, which are known from the specification of the transmission zeros, into the roots of $V(\mathbf{y})$. Since the transmission zeros are definitely not in the passband, we are guaranteed that the zeros of $V(\mathbf{y})$ will not be in the interval [–1, 1]. Application of Darlington's method, described in the previous section allows us to generate

$$R_N(\mathbf{y}) \;=\; \frac{U(\mathbf{y})}{V(\mathbf{y})} \approx 0 \;\text{ for }\; \mathbf{y} \in [-1, +1] \tag{9.4.15}$$

By transforming the variable back to **x**, we can write

$$P(\mathbf{x}) = \varepsilon\, U(a\,\mathbf{x} + b) \tag{9.4.16}$$

and the squared magnitude of the desired filter is given by S(**x**), where

$$S(\mathbf{x}) = \frac{N(\mathbf{x})}{N(\mathbf{x}) + P(\mathbf{x})} \tag{9.4.17}$$

The parameter ε defines the passband ripple. The poles of H(**z**) can be obtained by extracting the roots of D(**x**) = N(**x**) + P(**x**) and using Eq. (9.3.10).

While appearing quite complicated, with several mappings and so on, the method is quite straightforward to automate. The key routines for the program are the following.

a) **Change of Variable.** Given a polynomial P(**x**) of degree N, specified by the coefficients $\{p_i\}$,

$$P(\mathbf{x}) = \sum_{i=0}^{N} p_i \mathbf{x}^i \tag{9.4.18}$$

obtain the polynomial Q(**y**), i.e., the coefficients $\{q_i\}$, where

$$Q(\mathbf{y}) = P(\alpha\,\mathbf{x} + \beta) = \sum_{i=0}^{N} p_i(\alpha\,\mathbf{x} + \beta)^i = \sum_{i=0}^{N} q_i \mathbf{y}^i \tag{9.4.19}$$

b) **Express a polynomial as a Chebyshev series.** Given a polynomial of degree N as in Eq. (9.4.18), obtain the coefficients $\{t_i\}$ such that

$$P(\mathbf{x}) = \sum_{i=0}^{N} p_i \mathbf{x}^i = \sum_{i=0}^{N} t_i T_i(\mathbf{x}) \tag{9.4.20}$$

c) **Root Extraction.** Given the polynomial P(**x**) as in Eq. (9.4.18), find the N roots of P(**x**).

d) **Polynomial evaluation.** Given P(**x**) as in Eq. (9.4.20), and given $\mathbf{x} = x_e$, evaluate

$$P(x_e) = \sum_{i=0}^{N} p_i x_e^i = \sum_{i=0}^{N} t_i T_i(x_e) \tag{9.4.21}$$

Such routines are readily available in, for example, Press et al. [1.7].

9.5 A DESIGN EXAMPLE

Consider the following requirements on a digital bandpass filter:

i) Sampling rate = 8 KHz.

ii) Passband: 300 Hz to 3.4 KHz.

iii) Passband ripple: less than ±0.15 dB.

iv) Transmission zeros at 80 Hz, 3825 Hz, and 3920 Hz.

This is a "typical" requirement for a bandpass filter in telephony (see Chapter 7 and the discussion on transmultiplexers).

The design method described in Section 9.4 is used for obtaining the transfer function. Choosing $\varepsilon = 0.025$ (in Eq. 9.4.16) achieves a passband ripple of ±0.11 dB, providing some leeway relative to the ±0.15 dB requirement. The transmission zeros are chosen "dead-on" but in practice are only as close as the quantization of coefficients will allow. The sixth-order filter is described by the following poles and zeros (numbers shown to five decimal places):

zeros: $\mathbf{1 + 2\cos(\theta)\, z^{-1} + z^{-2}}$ (9.5.1)

A.	angle = ±3.07876 (rad)	(3)
B.	angle = ±0.06283 (rad)	(2)
C.	angle = ±3.00415 (rad)	(1)

poles: $\mathbf{1 + 2r\cos(\theta)\, z^{-1} + r^2 z^{-2}}$ (9.5.2)

A.	mag. (sq) = 0.85379;	angle = ±2.71196 (rad)	(1)
B.	mag. (sq) = 0.37275;	angle = ±2.65958 (rad)	(3)
C.	mag. (sq) = 0.83738;	angle = ±0.13674 (rad)	(2)

The frequency response is depicted in Fig. 9.13, which shows the frequency range from zero to 4 KHz, half the sampling frequency. The passband behavior is shown in Fig, 9.14, where the frequency range is 300 Hz to 3.4 KHz.

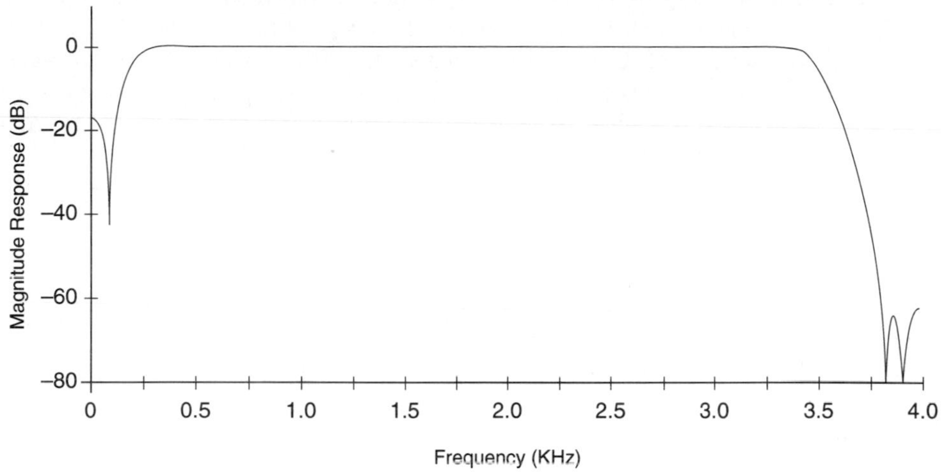

Fig. 9.13
Frequency response of the sixth-order bandpass filter with prescribed transmission zeros

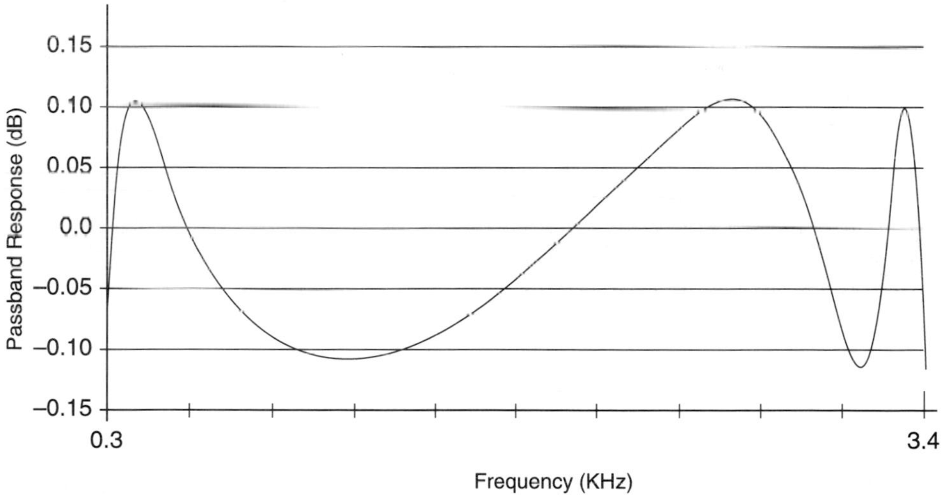

Fig. 9.14
Frequency response of filter showing the equi-ripple passband behavior

For implementation, we shall choose a cascade of second-order sections in the manner described in Chapter 4. The second-order sections are of the noncanonical form shown in Fig. 4.33. As per the guidelines outlined there, the poles and zeros are associated in the following manner. First, the poles are ordered according to the closeness to the unit circle. The order is A, C, B and is indicated in parentheses. The zero closest to pole-A is zero-C and thus zero-C is tagged with (1). Similarly, the zero closest to pole-C is zero-B with pole-B and zero-A paired by default.

The three (unnormalized) second-order sections are:

$$H_1(\mathbf{z}) = \frac{1 + 1.981138\,\mathbf{z}^{-1} + \mathbf{z}^{-2}}{1 + 1.680065\,\mathbf{z}^{-1} + 0.853792\,\mathbf{z}^{-2}} \tag{9.5.3a}$$

$$H_2(\mathbf{z}) = \frac{1 - 1.996053\,\mathbf{z}^{-1} + \mathbf{z}^{-2}}{1 - 1.813082\,\mathbf{z}^{-1} + 0.837377\,\mathbf{z}^{-2}} \tag{9.5.3b}$$

$$H_2(\mathbf{z}) = \frac{1 + 1.906053\,\mathbf{z}^{-1} + \mathbf{z}^{-2}}{1 + 1.0819335\,\mathbf{z}^{-1} + 0.372745\,\mathbf{z}^{-2}} \tag{9.5.3c}$$

The normalized frequency responses of the three second-order sections is shown in Fig. 9.15. The association of poles and closest zero tends to make the second-order section frequency responses as "flat" as possible, keeping the spectral peak as small as possible.

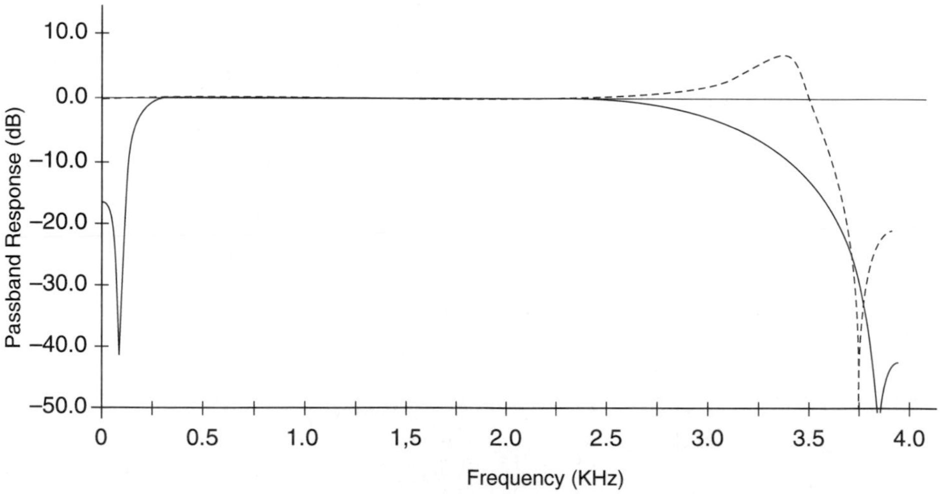

Fig. 9.15
Frequency responses of the three second-order sections of Eq. (9.5.3)

To decide the order in which the sections are implemented, we compute G_i for the three sections, i = 1, 2, 3. Recall that G is a measure of the "growth" of a single-frequency component and is equal to the peak of the frequency response

$$G_{\infty i} = \max_{\omega} \left| H_i(\omega) \right| \tag{9.5.4}$$

The section with the smallest G is made the first section (to the left in Fig. 4.33). Computing $|H_i(\omega)|$ over a reasonably dense grid in frequency (usually between 512 and 1024 points suffices) yields

$$G_1 = 2.80 \text{ at } f = 3.42 \text{ KHz} \tag{9.5.5a}$$

$$G_2 = 1.14 \text{ at } f = 3.91 \text{ KHz} \tag{9.5.5b}$$

$$G_3 = 1.63 \text{ at } f = 0.0 \text{ KHz} \tag{9.5.5c}$$

The first section is, therefore, $H_2(\mathbf{z})$ and, with the scaling factor incorporated, is given by

$$H_2^S(\mathbf{z}) = \frac{1}{1.14} \frac{1 - 1.996053\,\mathbf{z}^{-1} + \mathbf{z}^{-2}}{1 - 1.813082\,\mathbf{z}^{-1} + 0.837377\,\mathbf{z}^{-2}} \tag{9.5.6}$$

To decide which is the second section we compute G for the cases $H_2(\mathbf{z})H_1(\mathbf{z})$ and $H_2(\mathbf{z})H_3(\mathbf{z})$ to decide which is smaller:

$$G_{21} = 3.06 \text{ at } f = 3.42 \text{ KHz} \tag{9.5.7a}$$

$$G_{23} = 1.85 \text{ at } f = 0.39 \text{ KHz} \tag{9.5.7b}$$

This suggests that the second section must be $H_3(\mathbf{z})$. Incorporating the scale factors, the second section is given by

$$H_3^S(\mathbf{z}) = \frac{1.14}{1.85} \frac{1 + 1.906053\,\mathbf{z}^{-1} + \mathbf{z}^{-2}}{1 + 1.0819335\,\mathbf{z}^{-1} + 0.372745\,\mathbf{z}^{-2}} \tag{9.5.8}$$

Taking the three sections together, computing G yields

$$G_{231} = 5.11 \text{ at } f = 3.4 \text{ KHz} \tag{9.5.9}$$

and thus the scaled version of the third section is

$$H_1^S(\mathbf{z}) = \frac{1.85}{5.11} \frac{1 + 1.981138\,\mathbf{z}^{-1} + \mathbf{z}^{-2}}{1 + 1.680065\,\mathbf{z}^{-1} + 0.853792\,\mathbf{z}^{-2}} \tag{9.5.10}$$

Normally the gain of the cascade is chosen as 0 dB at a reference frequency, typically 1 KHz in telephony applications. The gain of the (unnormalized) cascade is +6.4 dB at 1 KHz. An equivalent loss of 6.4 dB can be spread out over the three sections. However, it is advantageous to apply this loss separately, after the output of the third section.

9.6 CONCLUDING REMARKS

The prime focus of this chapter is a design method for obtaining **IIR** digital filters with an equi-ripple passband behavior and with prescribed transmission zeros. In addition, we consider methods for obtaining lowpass filters that provide frequency selectivity based on the behavior of polynomials. The key to these methods is their simplicity and computer programs to automate the procedures are quite simple to develop.

The translation of normalized frequency to a variable x by x = cos(w) is used in the design of **FIR** filters as well. In particular, McClellan et al. [1.3.1] have shown how this substitution permits the use of "conventional" approximation methodology, in particular, the Remez Exchange algorithm, for designing **FIR** filters. Shenoi et. al. [1.9, 1.10, 1.11] show how other approximation methods can be applied in the case of **IIR** filters, where the Remez Exchange is difficult to apply. Most of these methods tend to be iterative in nature and can be accelerated by having a good "initial guess." Such guesses can be obtained using simple "least-square" methods, which involve just the solution of linear simultaneous equations. Some effective least-square methods are described in Shenoi et al. [1.12].

9.7 EXERCISES

9.1. Consider the frequency response of an N-point **FIR** filter expressed in terms of its coefficients $\{a_i;\ i=0, 1, \ldots, (N-1)\}$ and assume that the coefficients exhibit symmetry (there are four cases depending on whether N is even or odd and whether the symmetry is odd or even). Show that the transformation $\mathbf{x} = \cos(\omega)$ converts the frequency response expression into a polynomial in **x** (even symmetry) or a weighted polynomial in **x** (odd symmetry). Pose the approximation problem as one of a weighted approximation of a function by a polynomial.

9.2. A polynomial, P(**x**), of Nth degree has (N+1) coefficients. Convince yourself that, given an arbitrary function F(**x**), we can find the coefficients of P(**x**) such that the "error," (P(**x**) - F(**x**)) can be made zero at (N+1) arbitrary points, $\{\mathbf{x}_i;\ i = 1, 2, \ldots, (N+1)\}$. Show that the coefficients of the polynomial can be found by solving a system of linear simultaneous equations.

9.3. Consider the (squared-magnitude) function H(**x**) given by

$$N(\mathbf{x}) = \left(1 + \varepsilon \frac{(1 - \mathbf{x})^N}{(1 - \mathbf{x})^N}\right)^{-1}$$

What type of filter does H(**x**) represent, given the relation **x** = cos(ω)? What is the significance of ε? Can we choose ε in such a way as to force the frequency response to be a prescribed value at a given frequency? How? Now consider the function H(**x**) given by

$$N(\mathbf{x}) = \left(1 + \varepsilon \frac{(x_0 - \mathbf{x})^N}{(1 - \mathbf{x})^N}\right)^{-1}$$

What type of filter does H(**x**) represent, given the relation **x** = cos(ω)? What is the significance of ε? Can we choose ε in such a way as to force the frequency response to be a prescribed value at a given frequency? How?

9.4. Write a program to design a bandpass filter with prescribed transmission zeros that is maximally flat at a prescribed frequency f_0. The selectable parameters are the center frequency and attenuation (A dB) at a selected frequency f_1.

9.5. Write programs to automate the design of Butterworth, Chebyshev Type 1, and Chebyshev Type 2 filters. The selectable parameters are the passband edge frequency (stopband for the case of Chebyshev Type 2), and attenuation (A dB) at a selected frequency f_1.

9.6. Write a routine to generate P(**x**) of degree N, such that P(**x**)/Q(**x**) is a Darlington rational Chebyshev polynomial over [x_p, 1], given Q(**x**), a polynomial of degree M < N, which has no real zeros in [x_p, 1].

9.7. If $T_N(\mathbf{x})$ is the Nth degree Chebyshev polynomial normalized such that $T_N(1) = 1$, show that $(T_N(x)+1)$ is of the form $(Q_K(\mathbf{x}))^2$, or $(1+\mathbf{x})(Q_K(\mathbf{x}))^2$, where $Q_K(\mathbf{x})$ is a polynomial of degree (N/2) or ((N–1)/2), according as N is odd or even, respectively.

9.8. Write a routine to generate Q(**x**) of degree N, such that Q(**x**)/ P(**x**) is a Darlington rational Chebyshev polynomial over [–1, x_s], given P(**x**), a polynomial of degree M < 2N, which has no real zeros in [–1, x_s].

9.9. Develop a method for generating an **IIR** filter that is equi-ripple in the stopband as well as equi-ripple in the passband using an iterative scheme as follows. First choose any set of transmission zeros. This determines N(**x**). Then obtain P(**x**) such that the filter will be equi-ripple in the passband. With this choice of P(**x**), find N(**x**) such that the filter will exhibit equi-ripple behavior in the stopband (use Exercises 9.7 and 9.8). Next, use this new N(**x**) to find a "better" P(**x**) and so on. The procedure is terminated when P(**x**) (or N(**x**)) does not change significantly from one iteration to the next.

9.10. Use the Gray and Markel program [1.4] or MATLAB [1.6] to generate an elliptic filter given a passband edge, ω_c, and a stopband edge, ω_s, and a given passband ripple (say 0.1 dB, approximately). Use Exercise 9.9, with $\mathbf{x}_s = \cos(\omega_s)$ and $\mathbf{x}_c = \cos(\omega_c)$ to obtain a filter that is equi-ripple in both the passband and stopband. Compare the two filters.

9.11. For an N-point **FIR** filter, $H(\mathbf{z}) = \Sigma a_i \mathbf{z}^{-i}$, with coefficients $\{a_i, i=0, 1, 2, \ldots, (N-1)\}$, write the frequency response as a (complex) function of ω. Given a target (complex) frequency response, $G(\omega)$, develop an expression for the "error" at frequency ω_k as $E(\omega_k) = |G(\omega_k) - H(\omega_k)|W(\omega_k)$, where $W(\omega_k)$ is a suitable non-negative weighting function of frequency. Define a "total error" as the sum of the (squared) magnitude of the "error" over a frequency grid $\{\omega_k, k = 1, 2, \ldots, (M-1)\}$. Show that the "optimal" choice of the filter coefficients $\{a_i, i = 0, 1, 2, \ldots, (N-1)\}$ can be obtained by solving a set of linear simultaneous equations.

9.12. Suppose H(**z**) is an Nth-order **IIR** filter defined by the numerator coefficients $\{a_i, i = 0, 1, 2, \ldots, (N-1)\}$ and denominator coefficients $\{b_i, i = 1, 2, \ldots, (N-1)\}$. Develop a "total error" as in Exercise 9.11. Show that obtaining an "optimal" choice of filter coefficients requires the solution of a set of nonlinear simultaneous equations.

9.13. Consider the following modification of Exercise 9.12. Let $A(\omega_k)$ and $B(\omega_k)$ represent the frequency functions of the numerator and denominator, respectively. Define $E(\omega_k) = |B(\omega_k)G(\omega_k) - A(\omega_k)|W(\omega_k)$ where $W(\omega_k)$ is a suitable nonnegative weighting function of frequency. Defining the "total error" as before, show that the optimal choice of filter coefficients can be obtained by solving a set of linear simultaneous equations.

9.14. Extend Exercise 9.13 in the following manner. At the start define $W(\omega_k) = 1$. Obtain the appropriate set of coefficients $\{a_i\}$ and $\{b_i\}$. For the next iteration use $W(\omega_k) = 1/B(\omega_k)$ using the most recent set of coefficients $\{b_i\}$. This sets up an iterative procedure. The procedure terminates when coefficients do not change from iteration to iteration. Establish that when (and if) this iterative procedure terminates, the minimized error is related to the minimum "total error" defined in Exercise 9.13.

9.15. Suppose the target response was described in the time domain by the desired impulse response $\{g(n)\}$, not necessarily finite in extent. Now suppose that we are designing a filter H(**z**) with associated impulse response $\{h(n)\}$. The "instantaneous error" is given by e(n) = g(n)–h(n) and the "total error" given by the sum of $[e(n)]^2$ over all time indices n. Formally the error sequence is describeD by E(**z**) = G(**z**) – H(**z**).

a) For an N-point **FIR** filter, $H(\mathbf{z}) = \Sigma a_i \mathbf{z}^{-i}$, with coefficients $\{a_i, i = 0, 1, 2, \dots, (N-1)\}$, show that the optimal "least-squares" solution is obtained by setting $a_n = g(n)$, n = 0, 1, ... , (N–1).

b) Suppose H(**z**) was an all-pole filter defined by $H(\mathbf{z}) = a_0/B(\mathbf{z})$. Show that the optimal coefficients, those that minimize the "total error," are the solution of a set of nonlinear simultaneous equations.

c) Consider the following modification to part **b)**. Suppose we redefine our error sequence to read $E(\mathbf{z}) = G(\mathbf{z})B(\mathbf{z}) - a_0$. Establish that the coefficients $\{b_i\}$ that minimize the "total error" (the energy of the error sequence) in this case are the solution of a set of linear simultaneous equations. Further, the equations are identical in form to the normal equation developed in Chapter 5. Consequently we can solve for the coefficients in a manner that includes a built-in test for stability.

d) For a general **IIR** filter with both poles and zeros, consider the following extension of part **c)**. Define the error sequence as E(**z**) = G(**z**)B(**z**) – A(**z**). Establish that the coefficients $\{a_i\}$ and $\{b_i\}$ are the solution of a set of linear simultaneous equations. The equations are somewhat different from the Normal Equations in the all-pole case and thus we do not have a built-in stability check. One alternative is to obtain the coefficients of an all-pole filter first and then choose the **FIR** section by making the error sequence equal to zero for the first N terms.

9.8 REFERENCES AND BIBLIOGRAPHY

9.8.1 References

[1.1] Daniels, R. W., *Approximation Methods for Electronic Filter Design*, Bell Telephone Labs, McGraw-Hill Publishing Co., New York, 1974.

[1.2] Darlington, S., "Analytical approximation to approximations in the Chebyshev sense," *Bell System Tech. Journal*, Vol. 49, Jan. 1970.

[1.3] Digital Signal Processing Committee, IEEE Acoustics, Speech, and Signal Processing Society, Editors, *Digital Signal Processing, II*, IEEE Press, New York, 1976.

[1.3.1] McClellan, J. H., Parks, T. W., and Rabiner, L. R., "A computer program for designing optimum FIR linear phase digital filters," *IEEE Trans. on Audio and Electroacoustics*, Dec. 1973.

[1.3.2] Rabiner, L. R., McClellan, J. H., and Parks, T. W., "FIR digital filter design using weighted Chebyshev approximation," *Proc. IEEE*, Apr. 1975.

[1.3.3] Kaiser, J. F., "Nonrecursive digital filter design using the I_0-sinh window function," *Proc. 1974 IEEE International Symposium on Circuits and Systems*, April 1974.

[1.3.4] Deczky, A. G., "Synthesis of recursive digital filters using the minimum p-error criterion," *IEEE Trans. Audio and Electroacoustics*, Oct. 1972.

[1.4] Gray, A. H., and Markel, J. D., "A computer program for designing digital elliptic filters," *IEEE Trans. Acoustics, Speech, and Signal Processing*, Vol. ASSP-24, Dec. 1976.

[1.5] Rabiner, L.R., and Rader, C.M., Editors, *Digital Signal Processing*, IEEE Press, New York, 1972.

[1.5.1] Kaiser, J. F., "Design methods for sampled data filters," *Proc. 1st Annual Allerton Conf. on Circuits and System Theory*, 1963.

[1.5.2] Rader, C. M., and Gold, B., "Digital filter design techniques in the frequency domain," *Proc. IEEE*, Feb. 1967.

[1.5.3] Constantinides, A. G., "Spectral transformations for digital filters," *Proc. IEEE*, Aug. 1970.

[1.5.4] Rabiner, L. R., "Techniques for designing finite duration impulse response digital filters," *IEEE Trans. Comm. Tech.*, Apr. 1971.

[1.6] *PC-MATLAB*: Software package from The MathWorks, Inc., Also, the *Signal Processing Toolbox*, which can be used with *MATLAB*.

[1.7] Press, W. H., Flannery, B. P., Teukolsky, S. A., and Vetterling, W. T., *Numerical Recipes*, Cambridge Univ. Press, 1989. (Available in FORTRAN and C with source code on floppy disks for personal computers).

[1.8] *Programs for Digital Signal Processing*, edited by the Digital Signal Processing Committee of the IEEE Acoustics, Speech, and Signal Processing Society, IEEE Press, New York, 1979.

[1.9] Shenoi, K., and Agrawal, B. P., "On the design of recursive lowpass digital filters," *IEEE Trans. on Acoustics, Speech, and Signal Processing*, Vol. ASSP-28, Feb. 1980.

[1.10] Shenoi, K., and Agrawal, B. P., "A design algorithm for constrained digital filters," *IEEE Trans. on Acoustics, Speech, and Signal Processing*, Vol. ASSP-30, Apr. 1982.

[1.11] Shenoi, K., Narasimha, M. J., and Peterson, A. M., "On the design of recursive digital filters," *IEEE Trans. Circuits and Systems*, Vol. CAS-23, No. 8, Aug. 1976.

[1.12] Shenoi, K., Narasimha, M. J., and Peterson, A. M., "On the least-squares design of recursive digital filters," *IEEE Trans, Acoustics, Speech, and Signal Processing*, Vol. ASSP-24, No. 4, Aug. 1976.

[1.13] Vlach, J., *Computerized Approximation and Synthesis of Linear Networks*, John Wiley and Sons, New York, 1969.

9.8.2 Bibliography

[2.1] Antoniou, A., *Digital Filters: Analysis and Design*, McGraw-Hill Publishing Co., New York, 1979.

[2.2] Applications Engineering Staff of Analog Devices, DSP Division, Mar, A., Ed., *Digital Signal Processing Applications using the ADSP-2100 Family*, Prentice Hall, Inc., Englewood Cliffs, NJ, 1990.

[2.3] Hamming, R.W., *Digital Filters*, Prentice Hall, Inc., Englewood Cliffs, NJ, 1977.

[2.4] Lane, J., and Hillman, G., "Implementing IIR/FIR Filters with Motorola's DSP56000/DSP56001," *Motorola Application Note* APR7/D, 1990. [This is but one of several excellent Application Notes generated by the Digital Signal Processing Group at Motorola.]

[2.5] Williams, A. B., *Electronic Filter Design Handbook*, McGraw-Hill Publishing Co., New York, 1981.

APPENDIX

A list of commonly used acronyms is provided in this appendix.

ADC: Analog to Digital Converter. Also written as A/D.

ADSL: Asymmetric Digital Subscriber Loop. One version of a **DSL** where the data rates in the two directions are different. From the central office to the subscriber the data rate is high, of the order of 6 Mbps and in the other direction the data rate is lower, of the order of 500 kbps.

AGC: Automatic Gain Control

ANSI: American National Standards Institute

AR: Auto-Regressive. (see also **MA** and **ARMA**).

ARMA: Auto-Regressive, Moving Average. A general digital filter fas an **IIR** part (the Auto-Regressive part) and an **FIR** part (the Moving Average part). The term **ARMA** is also applied to stochastic processes. When a process is modeled as the output of a filter excited by white noise, the process is **AR** if the filter is **IIR**, is **MA** if the filter is **FIR**, and **ARMA** when the filter has both **FIR** and **IIR** parts.

ASIC: Application Specific Integrated Circuit

AWG: American Wire Gauge. A means of identifying the diameter of cable. A larger number corresponds to a smaller diameter.

BER: Bit Error Rate. A measure of probability of error in data transmission. Expresses the performance in terms of number of errors in a string of several transmitted bits. Typically written as 1E7 to indicate one error in 10^7 transmitted bits.

BPF: Bandpass Filter

BRI: Basic Rate Interface. One version of a **DSL** that specifies a full-duplex transport of 160 kbps. The 160 kbps are comprised of 2 "B" channels, each 64 kbps, one "D" channel of 16 kbps and 16 kbps of overhead for various purposes. Often referred to as "2B+D." The B channel is associated with speech or data (the "Bearer" of information) and the D channel associated with signaling information and, possibly, packet-switched data.

CAS: Channel Associated Signaling. A trunk where the information channel and the associated signaling (control) are transported together is called **CAS**.

CCITT: Consultative Committee for International Telephone and Telegraph (it is actually an acronym in French). The standards setting body of the International Telecommunications Union (**ITU**), an organization of the United Nations. It has been since replaced by the organization **ITU-TS**, where the latter TS stands for Telecommunications Sector.

CCS: Common Channel Signaling. A network feature whereby the signaling associated with a trunk, or trunk group, is carried separately from the information.

CDF or **cdf**: Cumulative Distribution Function

CDM: Code Division Multiplexing. Scheme whereby several individual digital (sub)channels can share a common channel, with each (sub)channel identified by its "code" or carrier waveform.

CDMA: Code Division Multiple Access. Wherein several (usually geographically separated) transmitters can communicate with a common receiver with each individual transmitter identified by its code as in **CDM**. **CDMA** is one of the standards for cellular telephony. All the individual transmitters can use the same frequency band. **CDMA** is thus applicable for point-to-multipoint communication.

CELP: Code Excited Linear Predictive (coding) or Code Excited Linear Prediction. A variation of Linear Predictive Coding whereby the prediction residual is quantized using a code-book approach.

CO: Central Office. Strictly speaking, the **CO** is the building that houses the equipment such as Switches, **DACS** machines, multiplexers, the **MDF**, other distribution frames, etc. Sometimes, the term is used to refer to equipment in the building, the nature of equipment obvious from context.

DAC: Digital to Analog Converter. Also written as D/A.

DACS: Digital Access and Cross-Connect System. Equipment in the **CO** that has interfaces at the **DS1** (and sometimes **DS3**) level and can perform a rearrangement of **DS0**s between **DS1**s.

DSL: Digital Subscriber Loop. A generic term used for schemes that use the conventional twisted-pair copper cable between the subscriber and the central office (or equivalent) for transporting digital data rather than analog (voice).

DSX: **D**igital **S**ignal Cross-Connect (X). A distribution frame where all cabling to equipment makes an appearance. Interconnecting equipment is done at the this distribution frame using regular wire-wrap or "patch cords." The type of signal is either **DS1** or **DS3** and the distribution frame is called the **DSX**-1 or **DSX**-3, accordingly.

DFE: Decision Feedback Equalizer

DFT: Discrete Fourier Transform

ΔM: Delta Modulation (or Modulator)

DMT: Discrete Multitone Transmission

DS0: Digital Signal Level 0 ; nominally a 64 kbps stream.

DS1: Digital Signal Level 1 ; nominally a 1544 kbps stream comprised of 24 DS0s plus 8 kbps of overhead for framing purposes. **E1** is the European equivalent, nominally a 2048-kbps stream comprised of 30 DS0s plus 128 kbps for overhead purposes.

DS2: Digital Signal Level 2 ; nominally a 6.312-kbps stream comprised of 4 DS1s plus 136 kbps of overhead for framing and "bit-stuffing" purposes.

DS3: Digital Signal Level 3 ; nominally a 44.736-kbps stream comprised of 7 DS2s plus 552-kbps overhead for framing and "bit-stuffing" purposes.

DSB: Double Side-Band (modulation)

DSBSC: Double Side-Band Supressed Carrier (modulation)

DSI: Digital Speech Interpolation

ΔΣM: Delta Sigma Modulation (or Modulator)

DSP: Digital Signal Processing, or Digital Signal Processor, or Discrete-time Signal Processing.

DTFT: Discrete-Time Fourier Transform

E1: see **DS1** above.

ERL: Echo Return Loss

ERLE: Echo Return Loss Enhancement

FDM: Frequency Division Multiplexing. Scheme whereby several individual (sub)channels can share a common (wideband) channel, with each (sub)channel allocated its own frequency slot. The bandwidth of the channels is (slightly greater than) the sum of the bandwidths of each channel.

FDMA: Frequency Division Multiple Access. Wherein several (usually geographically separated) transmitters can communicate with a common receiver with each individual transmitter identified by its frequency slot as in **FDM**. **FDMA** is thus applicable for point-to-multipoint communication.

FFT: Fast Fourier Transform. Any algorithm that reduces the computational complexity of the **DFT** significantly is called an **FFT**.

FIR: Finite Impulse Response (filter)

FOM: Figure of Merit

FSK: Frequency Shift Keying. A method for modulating a bit-stream wherein two carriers of different frequencies are used, one applied when the data bit is a "1" and the other when the data bit is a "0."

HDSL: High speed Digital Subscriber Loop. A scheme for transporting the equivalent of a **DS1**, 1.544 Mbps, over two pairs of subscriber cable. Each loop carries 772 kbps.

HPF: Highpass Filter

IEEE: Institute of Electrical and Electronics Engineers

IIR: Infinite Impulse Response (filter)

ISI: Inter-Symbol Interference

LCM: Least Common Multiple

LMS: Least-Mean Square (algorithm for adaptive filters)

LPC: Linear Predictive Coding

LPF: Lowpass Filter

LTI: Linear Time-Invariant (system) or Linear Time Invariance (the property).

MA: Moving Average. See **AR** and **ARMA**.

MDF: Main Distribution Frame. All external cabling for subscriber loops is brought to a "patch panel," where connections to different line circuits can be made. The **MDF** is invaluable from an administrative viewpoint for keeping track of the identity of the loop and the identity of the line circuit to which it is connected. Rearrangements are simple since the rewiring is done in a simple, sytemmatic manner. (See also **DSX**).

MSC: Mobile Switching Center

MTSO: Mobile Telephone Switching Office

MUX: Short form for multiplexer or multiplexing.

PAM: Pulse Amplitude Modulation. The representation of an (analog) signal by narrow pulses, ideally delta functions, whose amplitude (strictly speaking, the area under the pulse) is proportional to the amplitude of the signal at that instant.

PCM: Pulse Code Modulation. A digital version of **PAM**. The amplitude of the pulse is quantized and represented as a binary code. In practice, PCM (almost always) corresponds to the case where the sampling frequency is 8 KHz and the binary code is 8-bits, either A-law or μ-law, corresponding to a net bit rate of 64 kbps, i.e. a **DS0**.

pdf: probability density function

PRI: Primary Rate Interface. A specification for organizing the 24 channels of a DS1 as 23B+D, where each B (or Bearer) channel is 64 kbps as is the D channel which carries signaling information and, possibly, packet-switched data.

PSD: Power Spectral Density

PSK: Phase Shift Keying. A method of modulation of a carrier sinusoid by a bit stream wherein the "1"s and "0"s are encoded by the phase of the carrier. Variations include **QPSK** (Q for Quad) where bits, in pairs, are encoded into four possible phase values.

PSTN: Public Switched Telephone Network

QAM: Quadrature Amplitude Modulation. A modulation technique, principally for data transmission, where two phases of the carrier, offset by 90 degrees, are individually modulated using amplitude modulation.

RELP: Residual Excited Linear Predictive (coding) or Residual Excited Linear Prediction. A variation of Linear Predictive Coding whereby the prediction residual is quantized.

SDR: Signal-to-Distortion Ratio

SNR: Signal-to-Noise Ratio

SRL: Singing Return Loss

SSB: Single Side-Band (modulation)

SSS: Strict Sense Stationary. Stochastic processes whose joint **pdf**s, of all orders, and for all time instants, are not time dependent are **SSS**.

SS7: Signaling System 7. The specification of the protocols, messages, formats, and actions for implementing a **CCS** scheme. Since switches are essentially computer controlled, **SS7** can be viewed as a packet-oriented messaging protocol between these computers.

TDM: Time Division Multiplexing. A scheme wherein several digital (sub)channels are combined to form a composite bit-stream whose data rate is (slightly greater than) the sum of the data rates of the individual channels. The individual channels are identified by the temporal position relative to a framing epoch. Essentially each channel "bursts" its data during predetermined "time-slots."

TDMA: Time Division Multiple Access. A variant where several, usually geographically separated, transmitters can communicate with a central common point by identifying themselves by the time slot in which they transmit. In such a situation the central point is the "master," providing the timing synchronization required so that the individual transmitters do not overlap. **TDMA** is one of the standards for digital cellular telephony. All the individual transmitters use the same frequency band. **TDMA** is thus applicable for point-to-multipoint communication.

TLP: Transmission Level Power

VSB: Vestigial Sideband Modulation. A form of amplitude modulation that is between single sideband and double sideband. One of the sidebands is retained intact but the other is partially removed (or partially retained, depending on your point of view).

VSELP: Vector-Sum Excited Linear Prediction. A method for compressing speech signals.

WSS: Wide Sense Stationary. Stochastic processes whose mean and autocorrelation are not time-dependent are **WSS**.

INDEX

E

F

G, H

I

L

M

N, O

P

Q

R

S

T

U, Z